The Window of Life
A Theory of the Earth Based on Asteroid Impact

Part 1 Explaining the Trees of Axel Heiberg
Part 2 Surviving the Ice Age

By Ben Tripp M. A. Sc., P. Eng.
Illustrations and cover by Carolyn Tripp B.A.

Other books by the same author:

1 **Elements of Providence during the Genesis Flood**
    How did the Ark survive?
2. Fairytales for Adults
    Theories of the Earth in Disarray
3. Concerning the Birth of Christ
    A Discussion of the Timing of Christ's Birth
4. The Asteroid Theory of the Flood and the Ice Age
    The Necessity and Sufficiency of an Asteroid Shower to cause an Ice Age
5. The non-Myths of the Bible
    The co-relation between Scriptural Chronology and Nature
6. The Impossibility of Extraterrestrial Life
    The possible locations too few for life to exist elsewhere in the universe.
7. Time is Running Out
    Several of the circumstances necessary for our existence are petering out.
8. Climate Change and Holy Writ
    Both the Bible and nature indicate that our life on Earth will come to an end
9. Too Much CO2
    The carbon already in the atmosphere is the problem

(info at http://benatripp.wix.com/window-of-life)

About the author:

Ben earned Bachelor's and Master's of Applied Science degrees in Engineering from the University of Waterloo, Ontario, and has worked as a consulting engineer on such projects as controls for large telescopes and test equipment for the CanadArm. He holds patents for innovations involving the recycling of used tires into fence boards and a novel ground coil arrangement for geothermal heat pumps.

Ben's interest in the current topic, and his related background reading and research span several decades and have culminated in what he purports to be a credible and cohesive theory of the Earth which incorporates the evidence as it is currently known. It is his hope that these observations and opinions will be helpful to many in their own investigations.

Tripp, Ben, 2010
The Window of Life
(Alternately, The Window of Habitability of the Earth)
A Theory of the Earth based on Asteroid Impact
Part 1   Explaining the Trees of Axel Heiberg
Part 2   Surviving the Ice Age

1st printing 2011
2nd printing 2014
3RD printing 2014
Second Edition Jan. 2015
Third Edition Mar 2016

References have been included following the appendix. Whenever a reference is identified brackets have been used. For example (27) in chapter 6 will be the 27th reference for chapter 6. Illustrations have also been included. The illustration number is always the same as the paragraph number. Also an index of key words included at the end of the book enables particular discussions to be found quickly. If the reader wishes to skip to a point-form outline of the theory, he/she is referred to page 296 and 297.

Acknowledgements;

Fred Dyke, text review
Ed Mesesnel, text review
Wendy Speziali, formatting
Doug Garland, proofreading

ISBN 978-0-9936349-0-1   paperback
ISBN 978-0-9936349-1-8   electronic book

The Flow of Thought

1. Several currently popular notions of how nature operates are refuted. In particular the idea that an ice age would happen if the Earth became chilled is shown to be contrary to basic physics.
2. Life's enabling criteria are numerous, complex and simultaneously necessary. Several of these criteria ("Windows of Life") are discussed.
3. Trouble did come! The Earth was pummeled by an asteroid shower which totally disrupted most life-enabling factors.
4. The Great Ice Age developed. The absolutely necessary conditions for an ice age are spelled out and it is pointed out that the asteroid impact event would have generated these conditions.
5. How did anything survive? The literature is void of any survival discussion with only two exceptions. Now a third one has been added.

Some claims made and substantiated in: The Window of Life

1. There has only been one ice age
2. The Astronomical Theory of the Ice Age is false.
3. There will never be another ice age.
4. More than 2000 feet of ocean had to evaporate to have the Great Ice Age.
5. Once the Earth was both warm and illuminated from pole to pole.
6. The next impact of an asteroid with the Earth will terminate life on Earth.
7. There is no life elsewhere in the universe.
8. No land-based animals survived the ice age.
9. The Earth was once enveloped in a layer of water vapor.
10. Life on Earth will not continue indefinitely.
11. There certainly was a world-wide flood.
12. All of the mountain ranges of the Earth formed within a one-year time period.

The principle claim made and substantiated is:

Identification of the causal link between the asteroid impacts, the world-wide flood, the Great Ice Age and the absolute necessity to have both heat and cold both to have an ice age as well as retain the Greenhouse Effect.

This work is dedicated to my family

To my dear wife:
Judith Anne
The love of my life

To my children:
Bryan, Rebecca, Daniel and Carolyn
The great blessing of my life

To my dear grandchildren:
Evelyn, Ayla, Zoe, Izzy and Ben

And to my bonus children:
Andrea

May this humble epistle assist them in their search for truth.

TABLE OF CONTENTS

Part 1 - Explaining the Trees of Axel Heiberg

Foreword

If one is to have a theory of the Earth whether it involves ocean currents, volcanic activity, crustal plate movement, mountain formation, the great Ice Age, the evolution of animal life, or any other aspect of the physical world, it is absolutely necessary that the Earth-impacting asteroids be recognized. Immanuel Velikovsky had a theory of the Earth which was well supported by an extensive list of observations which are included in his book entitled "Earth in Upheaval". (published by Dell in 1955) While he did know about the Chubb Crater of northern Labrador and 'several other large meteoric craters '... in Australia, Arabia, and Mexico (and of) tens of thousands of oval formations on the Atlantic coast of the United States.' the only large impact crater that he was aware of was the Manicouagan Crater of Quebec. I believe that had he known that numerous other very large impact sites (including Sudbury, Chicxulub, Chesapeake Bay, Vredefort, Popigai, Czech Republic and Ishim), also existed and that the world-wide disruption which would have resulted from each of their impacts would have produced all of the results which are so well documented in Earth in Upheaval, he would have come to a different conclusion. However, it is most encouraging to know that while his theory of the Earth differs from mine, he would have encouraged the presentation of other views. Here I quote from the supplement to Earth in Upheaval. 'And should even the great ones of your age try to discourage you, think of the greatest scientist of antiquity, Archimedes, who jeered at the theory of Aristarchus, twenty-five years his senior, that the Earth revolves around the Sun. Untruth in science may live for centuries, and you may not see yourself vindicated, but dare. (i.e. Archimedes was wrong and Aristarchus was right, Aristotle's teaching supported by Archimedes persisted all the way to the time of Copernicus) Don't persist in your idea if the facts are against it; but do persist if you see the facts gathering on your side. ... The pleasure you experience in discovering truth will repay you for your work; don't expect other compensation, because it may not come. Yet, dare.' (Both this quotation and the one above are taken from the supplement to Earth in Upheaval pages 267 and 279 all of which are used with the kind permission of Rafe Sharon, his grandson.)

A considerable quantity of evidence (facts) has been gathered in support of the following discussion and so persistence would appear to be justified. Further the author can testify to the satisfaction which has resulted from sorting through this evidence to come up with a credible theory of the Earth. Whether or not, as Velikovsky understood, the major planets were involved in the unspeakable chaos that the Earth experienced is difficult to determine. However, the involvement of numerous minor planets (i.e. asteroids) is absolutely certain because evidence of their impacts has been confirmed all over the Earth. In recognition of the recent finding of several more large impact sites, no current commentator can ignore the effects that those monsters would have had. Asteroid impact sites are difficult to find because they are usually buried under thousands of meters of earth material. With this in mind there is no doubt whatsoever that even more sites will be found in the future. If the impact of asteroids with the Earth is not recognized, no modern theory of the Earth will be valid.

1. Introduction

Life processes are incredibly complex. Every organ of the human body is so complex in its function that even after intense and prolonged investigation, our most skilled and knowledgeable researchers will, while answering some questions, have produced a list of additional questions, which remain unanswered. Nor does complexity diminish at the cellular level. The functions and processes within each individual cell are the subject of many ongoing research projects. Even the cell membrane, which was once shown as a line on sketches and diagrams, contains a proliferation of valves and passageways and is under intense study by numerous investigators.

As an example of the complexity of life and the extreme unlikelihood that it could have happened by chance, we may consider the protein. Protein is a type of molecule, which is present in all types of animals as well as humans. Protein molecules are constructed from smaller parts, which are called amino acids. There are a large number of different types of protein in a human body and many of them are constructed from about 300 amino acids. While only twenty different types of amino acid are used, in order to properly form any particular protein, it is a necessary condition that they all be connected in exactly the right sequence. Even if only one of them is missing or in the wrong position, the protein will not be useful. It is, of course, also necessary that numerous factors external to the protein be present in order for the connection process to proceed. A meaningful probability calculation can still be carried out if we simplify the situation and ignore the protein formation support factors and only consider the probability of forming the protein with the correct sequence of amino acids. For simplicity we will assume that there are three hundred positions to be filled and further assume that all twenty amino acids are available in abundant supply. For the first position there are twenty possibilities. Also for the second position there are twenty possibilities. In fact for all three hundred positions in the protein there are twenty possibilities. Of course in a real protein, very particular amino acids are required at each position. For the first position there is one chance in twenty that the correct amino acid will be chosen. A similar probability exists for the second position and the third and so on. Therefore, the probability of forming one particular protein is found by multiplying 1/20 x 1/20 three hundred times. (It is understood that only very specific proteins are of interest. Even the proteins that are found in one animal body are of no use to another similar animal – even when they have been identified as the same type of protein.) The result of this calculation is a number so small that it is comparable to one divided by a number, which is similar in magnitude to the quantity of atoms in the visible universe. In plain language, it would never occur randomly or by chance. The situation is simply much too complex for this to happen. In fact even under the controlled conditions of the laboratory, it has only been possible to get an amino acid chain length of six. (5)

While proteins with 300 or even more amino-acid positions are quite common, the probability of forming a much smaller unit of only 15 positions is similarly very improbable. 'Even if we had a mixture of the correct left-handed amino acids, the chance of getting one 15-unit peptide (small protein) right is one in 3 x 10(19). (10 with 19 zeros after it)' (4)

While the complexity of a protein makes its formation by chance extremely improbable, the DNA double helix is several orders of magnitude more complex than any protein with correspondingly smaller probabilities of having been formed by any chance-time scenario. In fact, one of the co-discoverers of the DNA molecule, Francis Crick commented that it was so frustrating and virtually impossible to give any numerical value to the probability of such an unlikely sequence of events. (1) (i.e. the formation of the DNA molecule) The information content of the DNA molecule is enormous. One commentator suggested that it would be comparable to the information content of all thirty volumes of the Encyclopaedia Britannia and then some. (2) Complexity leads to improbability because the physical indeterminacy of the sequence generates the improbability of any particular sequence and this is actually what gives it meaning. (3)

In fact, the DNA molecule is said to code for protein. That means that it has the information to manufacture all of the proteins in a body. There is enough information in the DNA molecule to manufacture all 100,000 of the proteins found in a typical animal body while using less than 10% of the total information available! Then just to make matters worse we must consider the complexity of a complete life-form. Even the simplest life-form requires many billions of properly assembled in order for life to occur. (5)

The support conditions are equally complicated causing one commentator to conclude that the origin of life appears to be a miracle because there are so many conditions needed to get it going. (1)

In order for these complex life processes to carry on and for the numerous forms of life on the Earth to exist and thrive, an extensive set of support conditions or factors must be simultaneously present and each factor must be within a very narrow range of all the possibilities. If any one of these necessary factors drifts outside of its particular acceptable range, life will terminate.

Each necessary life-enabling factor may be likened to a window. A window is a relatively narrow opening, which allows light to shine through. On either side of the window, the light does not get through. Similarly, each of the windows of life has a narrow range within which it must operate in order to retain its life-enabling status. More importantly, all of the windows of life must simultaneously and collectively be within their respective life-enabling ranges.

This type of situation may be mathematically described as an 'and' function. If A and B and C = 1, life may exist. If any single entry is either missing or out of its proper range, A and B and C = 0 and life cannot exist. The more factors there are in this equation of life, the more likely it is that one of them will drift out of range, resulting in a 0 for the entire function. This would be equivalent to the termination of life.

In any real-life situation, it is probably necessary for at least several thousand factors to simultaneously be within their respective, narrow, life-enabling ranges. The windows must all line up. Any one, which drifts to the side, will be sufficient to interrupt the light and it will not be able to get through. In the mathematical realm, this would be labeled as 'improbable'. In order to estimate the probability of life being able to exist, the probability of each factor existing within its respectively narrow range, must be multiplied by the probabilities of all of the other factors. In reality, this cannot be done (because the total number of necessary factors is not known and because of the difficulty in identifying the actual probability of any particular factor) but the idea can be explored further. A reasonable estimate for each of the more obvious factors can first be made, and then all of the estimates can be multiplied together. It might be determined, for example that the first twenty factors in the list might each have a probability of 1 in a 1000 and the next five have a probability of 1 in 10,000. The overall probability would therefore be 1/1000 multiplied twenty times and then multiplied by 1/10,000 five times. The result would be 1 divided by 10 followed by 79 zeros. This would be a number, which was so small that it could not be imagined at all and in everyday terms, it would properly be concluded that whatever was being calculated would never happen.

In spite of the reality of improbability, life on the Earth does exist. Life on Earth not only exists but it is very diverse and vigorous. Even a spoonful of soil contains numerous different forms of life and the smallest of them are constructed from several hundred different types of protein. Therefore each of them is extremely improbable.

The necessary support conditions for each of these life-forms to thrive must always be present. If any one of them drifts too far one way or the other, that particular form of life will not have the support, which it needs, and it will not be able to continue.

The Earth is not alone in space. Numerous other objects exist in the solar system not to mention the rest of the universe. Included among these objects are other planets, moons, comets and, of course, the Sun. Also included are a large number of objects called asteroids. These objects are sometimes referred to as the minor planets because, like the major planets, they also orbit the Sun. They are called minor simply because they are much smaller than the major planets.

All is well as long as these asteroids continue to orbit far from the Earth. However, compared to the planets, the asteroids are not very large. It is therefore comparatively easy to disturb their orbits and cause them to travel through space on different orbits. While most asteroids orbit the Sun between Mars and Jupiter, many of them have orbits which cross the Earth's orbit and they are therefore a threat to life on Earth. (Size is of great consequence in this discussion. A house-size asteroid would be large enough to endanger life in an entire province while one ten km. in diameter would terminate life in an entire hemisphere. Unfortunately, there are millions of asteroids larger than a house including a million one kilometer across or larger.) If one of these asteroids collided with the Earth, numerous life-enabling conditions on Earth would be disturbed and forced outside of their respective narrow life-enabling ranges. This could happen at any time. Indeed, an asteroid passes close to the Earth at least once a week and so the possibility that a collision could occur is quite real.

There is no way to establish the probability of a collision even though this is claimed to have been done. The reason is that we do not have any appropriate data on which to base a calculation. The reality remains that there are frequent sightings and the fear remains that a collision is quite possible.

In fact, collisions have already taken place. The Earth includes numerous features, which are considered to be evidence of impacts by asteroids. This fact has never been in dispute. Disagreement exists over how often this may have happened but any declaration, which claims to identify how often an asteroid impact will occur (or has occurred), is nothing but speculation. It has been stated, for example, that a major impact would only occur every few million years, but this type of statement has nothing to support it. Neither is it the least bit scientific simply because there isn't any scientific data to indicate any particular frequency of impact.

The devastation that would be produced by a large asteroid would be completely overwhelming. The topography and environment of the entire Earth would be changed. In the course of such upheaval, numerous windows of life would be forced outside of their respective narrow life-enabling ranges and life would therefore terminate.

Through the ages, people have developed and offered explanations of how the world operates. As time went by these explanations have become more sophisticated because our collective understanding of nature has become more comprehensive. Several explanations of the Earth's functions have been offered relatively recently and have become very popular. Popularity, however, is not valid criteria for a theory. A proper theory should meet a set of criteria such as the following:

a. It should not contradict known and well-established physics.
b. A theory is a hypothesis confirmed by evidence and not contradicted by evidence. Hence a theory should be able to accommodate all relevant evidence. (ref. Chemistry by Quagliano page 31) As a corollary to these basic criteria a theory must avoid artificial construct. (i.e. the hypothesized arrangement of factors which would not normally occur in nature.)
c. It should be comprehensive.
d. A theory should be internally consistent.

Unfortunately, several of the most popular theories of the present age do not meet these criteria and therefore should be set aside.

The theory which has been developed and offered herein, does meet the above criteria while addressing all of the following factors. At one time, the Earth appears to have been universally warm. There also appears to have been a time when much of the northern hemisphere was covered by a thick layer of ice. There is so much carbon in the coal deposits of the Earth that they may contain 50 times as much as currently exists in the entire biosphere. Some regions of the crust of the Earth have undergone unbelievable distortion and bending and are currently found as ridges of mountains, which extend for thousands of miles. The crust of the Earth has been cracked and cracks have been found all over the Earth. Material from the interior of the Earth has been pushed up through the crust and is currently found in great mounds on the surface as well as in features called Igneous Provinces (which occur both above and below water). There are several features on the surface of the Earth, which appear to have been made by large objects from space. The crust of the Earth includes innumerable features, which appear to represent creatures, which were once alive. The ocean basins include a very great quantity of material, which appears to have been washed off the land. The islands of the high Arctic were once covered by trees and the Sahara Desert once included rich pasture land. With respect to the most ancient human records, which are available to us, life-spans during ancient times were much longer than they are now. These records also reveal that the Earth has suffered through an unbelievable calamity. Certain evidence indicates that the atmospheric pressure of the Earth was higher in the past.

The present theory recognizes all of this evidence and includes a system of ideas, which meets the necessary criteria outlined above. It is comprehensive, respects well-known physics, accommodates all known and relevant evidence (does not employ any artificial construct), and is internally consistent.

The present theory supersedes the notion of multiple mass extinctions as well as the common Ice-Age Theory, the Theory of Plate Tectonics, and The General Theory of Mountain Building, while it explains the death of the Woolly Mammoths, Submarine Canyons, Erratic Boulders, Earth Wobble and the trees of Axel Heiberg Island. It is a theory of the Earth based on asteroid impact!

2. Whoppers or Fairy Tales for Adults

The fact that opinion is widely held is no evidencewhatever that it is not utterly absurd. Indeed in view of the silliness of the majority of mankind, a widespreadbelief is more likely to be foolish than sensible.
Bertrand Russel

2.1 Overview

A theory is 'a system of ideas explaining something' (1) or a philosophical explanation of phenomena. It is a hypothesis confirmed by evidence and not contradicted by evidence (2). Through the ages, people have developed systems of ideas, which helped them to explain the world around them. Frequently, succeeding generations find previously held ideas invalid or even silly. Included among discarded theories is the idea that the Earth is the center of the solar system as well as the center of the universe.

The development and offering of theories is a consequence of the never-ending struggle to understand and make sense of the world even though only a very small portion of the actual evidence may be available to us at any one time. It also happens that good solid evidence is often simply ignored. There may be a desire to believe something, which desire may overwhelm reliance on evidence. People may desire to believe in something so much, that they ignore the evidence that is right in front of them.

Theories may be criticized or theories may be offered. In either case there should be criteria against which the theory may be evaluated. Therefore, all theories, which are either criticized or offered in this discussion, will be referenced to the following criteria.

2.1.1 Known Physics

A theory should not contradict known and well-established physics (or chemistry or any other particular scientific discipline). As an example, with the Theory of Quantum Electrodynamics there is agreement between theory and observation to within a small fraction of one percent. Similarly, the Theory of Gravity as offered by Newton and later upgraded by Einstein, has been verified by many observations and has been useful in predicting many events. For the present discussion, this type of situation will be referred to as 'well-known physics'. By contrast, the graviton particle explanation of gravity has no observations, which verify and support it. String Theory is in a similar situation. (String Theory is the idea that fundamental particles such as electrons, protons, neutrons and quarks and a host of other even more 'fundamental' particles are actually little strings. Their vibration patterns explain the observation that all particles have an associated frequency.(179)) As attractive as it may appear, String Theory has no evidence whatsoever to support it. In fact, even after more than thirty years of intelligent effort, a positive evidence-producing experiment has not been devised. It would be better to refer to it as the string hypothesis rather than the String Theory.

A theory has two basic functions. The first one is to explain and correlate observations, which have already been made. The second is to predict observations, which might be made in the future. For example, by offering a theory, which stated that shouting would be sufficient to lift an airplane, all of our physical observations made up to that time would be contradicted. The first basic function of a theory would thereby not be fulfilled. Neither would the second. In this case, predictions for the future may be readily tested and any such test would fail. A theory such as this would therefore contradict well-known physics and it should be regarded as invalid and set aside.

2.1.2   Evidence Accommodation

A theory should acknowledge and accommodate all of the known and relevant evidence, which is available. In the case of the Theory of Gravity, the ideas, which Newton offered did acknowledge the evidence, which was available at the time. As astronomical observations progressed, certain measurements indicated that the basic theory should be upgraded, and Einstein accomplished the upgrade. But what Newton had offered was very useful and enabled him to predict and describe both comet trajectories and planetary orbits.

Newton's Theory of Gravity was properly accepted during his time and is properly accepted at the present time as long as its limitations are recognized.

A theory cannot be supported by evidence which does not exist. The notion of a theory is to explain the available evidence and thereby anticipate explaining as-yet-undiscovered evidence, but evidence which would, when discovered, be conformal. The theory is therefore predictive of where to look and the search may be focused in that direction. On the other hand, if there isn't any real evidence in the first place, then there isn't anything to explain. If one should conjecture that a certain type of evidence should be there, just waiting to be found, until it is found, it cannot be used as support for a theory.

Within the realm of science, numerous consistent and logical relationships involving evidence from the natural world have been identified. Further investigations will either support these well-known relationships or extend, upgrade and refine them to enable the prediction and explanation of more observations from the world around us.

2.1.3   Comprehensive

A theory should be comprehensive. It should be able to deal with the matter of interest from beginning to end. It should not start half of the way through nor leave some great gap part of the way through. It should offer an explanatory overview. It should indicate the direction for further experimentation or study. All necessary factors, which are required for the theory to provide cohesion and completeness, should be included.

If the theory requires a force, the force should be identified. If the theory requires a globule of water to float in the air, the flotation means should be identified. If the theory requires that a gradual change from one life-form to another has happened, both an investigative procedure and observable evidence should be offered to support this idea.

Since a theory is a system of ideas explaining something, the explanation is not really complete if there aren't enough ideas included to reasonably complete the explanation. For example, a theory may require energy. If it does, the source of energy should be identified.

When a theory of the natural world is offered, there should not be appeal to the supernatural to fill in the gaps and smooth over the otherwise missing portions. If a discussion of the supernatural is the topic of the day, this should be made plain, but using the supernatural to make a theory of the natural world complete is inappropriate. This, by no means, suggests that appeal to the supernatural is wrong. It may actually be most appropriate, but it is most inappropriate to offer a theory of nature which will not work unless the supernatural is invoked several times in the course of the explanation. When this type of window is opened, any unusual and even bizarre suggestion can be made and explained by simply saying that "God did it'. Perhaps He did, but leaping into this realm may very well truncate a reasonable and justified investigation. In other words, God cannot be used as scientific evidence. God cannot be used to fill in gaps in a scientific explanation.

The evolutionist biochemist Richard Dickerson declared that we should see how far we can go in explaining the behavior of the physical and material universe without invoking the supernatural. (172) While this approach is perfectly valid for dealing with the natural world, it would not be sufficient if the supernatural were indeed involved. In such a case we have simply stepped out of the scientific realm but such excursions should only occur with the full understanding that this is what is happening.

2.1.4   Artificial Construct

A theory should avoid being setup with relationships which would clearly not occur in the natural world. If some of the factors in a hypothesis are arranged in an artificial way which contradicts readily repeatable and known relationships of science, that hypothesis or theory is being artificially propped up and any potency which it may have had, will be lost. For example, if a hypothesis requires that heavy material floated on light material, such an arrangement is artificial and hence should be avoided.

2.1.5   Internal Consistency

The ideas as well as the evidence and data which are offered in a theory should reinforce, co-relate and complement each other. Contradiction is not acceptable. If a theory is offered to explain a set of data taken from one area of the Earth but the data from another area requires an inverse explanation, such a theory is internally inconsistent. As an example, the Theory of Plate Tectonics requires that areas of expansion of the Earth's crust be offset by subduction zones where crust is swallowed to keep the surface of the Earth constant in area. The Pacific plate is declared to be moving north-west over the Hawaiian hotspot. Such movement necessitates the production of crust someplace. However, the Pacific plate is completely surrounded by subduction zones, referred to as the "ring of fire. Hence the theory is internally inconsistent. It may be excusable to believe in an idea when certain evidence supports it, but when applicable evidence requires a totally different explanation, such a theory should be set aside.

The following theories are very popular but they do not meet the above criteria.

- Swamp Theory of Coal Formation
- Ice Age
- LaPlace Nebular Hypothesis of Solar System Formation
- Continental Drift or Plate Tectonics
- Catastrophic Plate Tectonics
- Evolution
- String Theory

2.2 Swamp Theory of Coal Formation

2.2.1 The Basic Theory

The Swamp Theory of Coal Formation declares that coal has been formed from plants. This recognizes the observation that the outlines of numerous different species of plants are often observed in coal. It also recognizes that the structural component of all plants is carbon. Coal is carbon, which appears to have been pressed down and compacted into very tight and hard rock-like formations. Therefore, to declare that coal is formed from plant material is in keeping with both well-known physics as well as certain direct observations.

2.2.2 Problems with the Theory

2.2.2.1 In-situ Declaration

Another aspect of the Swamp Theory of Coal Formation declares that the plants, which formed the coal, grew in situ where the coal is presently found. However evidence has not been offered to support this declaration. In fact a large boulder has been found in a seam of coal. Was the boulder brought in and placed among the dead trees or was the entire mass moved into position and just happened to include a large boulder? (32) Appeal to reason is not a substitute for evidence. This portion of the Swamp Theory of Coal Formation is therefore not substantiated and the available evidence contradicts the declaration because some coal beds are very thick and a lot of plants were required for their formation.

2.2.2.2 Plant Rotting is Forbidden

The theory also declares that the plants grew and died in such a way that they did not rot. New plants kept on growing and piling up on top of the old ones without rotting until a huge mass of dead, preserved plants had been accumulated. This aspect of the theory violates well-known physics. Current and readily repeatable observation indicates that when trees die and fall over, they rot. Much of the material of the dead tree is thereby oxidized into carbon dioxide and becomes part of the atmosphere. Any moisture, which was part of the tree, will simply evaporate. The rest of the tree will become part of the soil and in fact will form the soil. This is what is always observed. Therefore to declare that a tree can die and fall down without rotting, violates observation. In order to circumvent this violation, it is declared that the plants, which would form the coal, grew and died in a swamp. When they fell into the water, the water kept them from rotting thereby keeping the carbon, from which they were formed, from leaving. Therefore, it is a necessary condition of the theory of coal formation that all of the plants, which formed the coal, grew in a swamp, died, fell over and were covered by the water of the swamp. This aspect of the theory violates neither well-known physics (water will keep a plant from rotting for a considerable time) nor observation (plants do die and fall over in swamps every day) as far as dying and falling over are concerned. However, the claim that the plants grew in the swamp is not valid.

2.2.2.3   Wrong Trees are Involved

The problem with the swamp origin of coal relates to the types of plants found in coal. Pine, spruce, hemlock, sequoia and other dry land conifers are found in European and North American lignites. Palms, birch, beech, magnolia, cinnamon and others are found in Cretaceous coals (4). None of these trees grow in swamps.

2.2.2.4   Swamp Stability Declaration

The next aspect of the theory declares that the plants lived and died in the swamp and were covered by the water of the swamp but the swamp never became filled up with plants. In order to produce even a modest thickness of coal, a lot of plants are required. Some coal beds are many feet thick. It has been estimated that it requires ten feet of plant material to form one foot of peat and twelve feet of peat to form one foot of coal.(117) One of the thickest coal beds found to date is thirty feet deep. (31) If this estimate of plant volume is correct, 3600 feet of plants would have been required. (Of course coal is formed from trees as well as other types of plants. 3600 feet of compressed ferns is hard to imagine but 360 trees, ten feet in diameter is a little more comprehensible.) This means that many plants grew successively in the swamp and that the water level of the swamp kept rising at exactly the necessary rate to keep the area as a swamp but not make the water too deep to terminate the growth process. A great amount of time would be required to grow all of the plants required to form a seam of coal which was thirty feet deep. Therefore, the swamp water must have kept rising at just the right rate for hundreds and even thousands of years. This type of swamp has never been observed. In recognition of the difficulty of maintaining a swamp of this nature it was declared that the swamp must have sunk instead. The swamp sank at just the right rate to allow the water to keep the dead plants covered but still allowed the new plants to grow properly. Instead of rising for hundreds or thousands of years the swamp sank for hundreds or thousands of years. A swamp of this nature has never been observed either.

2.2.2.5   Extended Times are not Involved

The final aspect of the Swamp Theory of Coal Formation declares that the plants went through their normal life cycle of living and dying for an extended period of time, which involved millions of years. According to the theory, succeeding generations of plants grew and died and added to the accumulation of material, which was forming a future coal bed. New plants grew on top of old plant material and the entire mass of material kept building up. Entire forests of plants grew and died thereby accumulating the carbon for the future coal deposits. This extended period of time has been called the Carboniferous Period and is declared to have lasted 60,000,000 years. However, the idea that a vast amount of time was required to form the coal-producing plants requires artificial construct as will be shown in the following discussion.

Certain coal beds are observed in layers. There are layers of coal interspersed with layers of rock. This necessitates that after the swamp had sunk at just the right rate for an extended period of time, it became covered with a layer of material which would later turn into rock. Then a new swamp formed on top of the layer of non-swamp material. Then the whole thing sank at just the right rate and another layer of coal-forming material was deposited. Next, another layer of rock-forming material was deposited and the sequence repeated. Of course, a lot of time would be required. Geologists have suggested that it would require 1000 years to form one inch of coal. (91) In the process, any portion of any tree, which lived and died and did not fall down properly, would be subject to rot and disappear. However, the theory is now in direct violation of observation. It has been repeatedly observed that the fossil remains of trees project right up through several layers of coal as well as the intervening rock material, which separates the layers of coal.(4) How could these things be? During the extensive times, which are required by the theory, the trees would have rotted. In fact they would have had time to rot a thousand times. This aspect of the theory is therefore not valid because it violates well-known physics. The violation is so severe that there is simply no way to adjust the theory to evade the violation.

If the trees and other plants, which were the source of material for the future coal beds, were not covered and isolated from the atmosphere quickly, they would have rotted and hence become unavailable for coal formation. In particular, the trees, which extended up through several layers of material must have been buried quickly or they also would have rotted (assuming that they did not continue living for the millions of years required). It may therefore be concluded that the material, which would become coal was gathered, placed and covered within a short amount of time (i.e. at least within the 'rot' time).

2.2.2.6   Extended Times are not Needed

Finally, it is also declared that a great deal of time was required to form the coal from the plants. This means that a great deal of time was required for the volatiles to leave the plant so that only the carbon remained. Such a declaration contradicts the following three repeatable observations.

Charcoal is made from hardwood using a process which only requires a few hours. The wood is placed in a furnace and set on fire. The furnace has a restricted oxygen supply. A small portion of the wood burns and heats the rest of the wood. The heat drives away the parts of the wood, which will evaporate, leaving only the carbon.

There is a second way that wood has been reduced to carbon in just a few minutes. Construction procedures occasionally require that piles be driven into the ground around the perimeter of a site. The soil is thereby stabilized and construction may proceed in safety. When these wooden soldier piles are driven into the ground, they may overheat. It has been observed that when a recently-driven pile was cut, only carbon remained in the interior of the pile. Since the pile had been driven into the ground within the space of a few minutes, very little time was required for the interior of the pile to be changed from wooden material to carbon material.

Another example illustrating how wood can be reduced to carbon in a short time also involves heating the wood and driving away the volatiles. This process is referred to as pyrolysis (i.e. chemical decomposition by heat action) and occurs if wood is slowly heated. As this happens, the wood will become more and more like carbon and its ignition temperature will be reduced. This may occur accidentally if wood is enclosed in a wall behind a hot stove. After a period of time, which may involve several years, the wood, (which in this case is completely out of site), will catch on fire. The gradual loss of the volatiles (or parts of the wood, which can evaporate), leaves only the carbon. Carbon will ignite at a much lower temperature than wood. The second way that this same drying process can happen involves wood, which is near an open fire. If a log is near a fire, it may gradually dry out. Then it may ignite even though it is several feet from the fire. The time required for this to happen, is measurable in hours, not years.

Fourthly, meteorite material is never found in coal. Well over 50 billion tons of coal, have been mined and never once has there been a report of meteorite material being found. This has even led to the speculation that meteorites did not fall during the 'millions' of years that the coal was being deposited. (37) Alternatively, since meteorites are observed to be continually impacting the Earth (147), the period of time involved in forming the coal and covering it with overburden, must have been less than the interval between impacts.

There is a fifth type of evidence which indicates that the coal-forming process only occurred a few thousand years ago. Since coal has apparently been in place for thousands of years, it follows that it could not have taken very long for it to form. There are three elements to this evidence. Firstly, coal has a carbon14 count, indicating that it is about 50,000 years old. (36) If it is as old as the carboniferous period, which is understood to have been more than 100,000,000 years ago, there would not be any carbon14 count at all. Secondly, the uranium238/lead206 ratios identified in the inclusions found in coal are very high. If the process had been ongoing for hundreds of thousands of years, this ratio would not be high. The high ratio indicates that the uranium has only been decaying for a few thousand years and that coal has been formed recently. (34, 35)

There is a sixth type of evidence which indicates that the trees which formed the coal were themselves formed almost instantly. This evidence has been recovered from the study of very tiny colored circles found in both rock and coalified wood. At the centers of these circles is a very tiny inclusion of lead, the last step in the radioactive decay of uranium. In all of these cases however, there is no evidence for any decay product before polonium which suggests that polonium was present and decayed but that it did not have any parent material. Hence it is referred to as parentless polonium. (Normally if uranium is decaying, polonium will occur part of the way through the decay process right after radon222.(174)) The circles that are usually found are indicative of polonium218 (1/2 life = 3 minutes), polonium214 (1/2 life = 164 microseconds), polonium210 (1/2 life = 138 days). When these circles are found in rock they require instant crystallization of the rocks at the same time that the polonium atoms are created. (175) The first two of these rings are not found in coal. This indicates that the wood that would become coal, had not yet formed around the polonium and that it did not form for at least five half-lives (or about 15 minutes) after the polonium218 started to decay. However, polonium210 is included indicating that the wood had enclosed the polonium within days of its formation. (i.e. well within 138 days x 5 = 690 days. Five half-lives are included because after this much time has passed, there is very little radiation activity taking place.)

A further complication regarding time relates to finding a fossilized forest in the ceiling of a coal mine. Mangrove-like plants were included but this type of plant was not supposed to appear until much later than the carboniferous period (i.e. at least 200,000,000 years later). (111)

All of these examples indicate that a great deal of time was not required to form coal. Therefore, the declaration that a great amount of time passed, during the formation of coal, is not valid. Two questions remain, the first of which has already been addressed. How long did it take to grow the coal-source plants? How did the CO_2 (i.e. the source of carbon for the coal-source plants) get into the air in the first place? These questions will be discussed in paragraph 2.2.3 below.

2.2.2.7   Layers Indicate Marine Environment

In some locations, numerous layers of coal are found interspersed with layers of rocky material containing sea shells. (66) 'The plants that went into the formation of ancient (coal) beds include chiefly ferns and cycads; layers of later ages are composed of sassafras, laurel, tulip tree, magnolia, cinnamon, sequoia, poplar, willow, maple, birch, chestnut, alder, beech, elm, palm, fig tree, cypress, oak, rose, plum, almond, myrtle, acacia, and many other species. …………It is said that the plants fall, but before they decompose in the air they are covered by the water of the swamps. A layer of sand is deposited over them, forming the soil for new plants, and thus the process repeats itself. In order that the sand be deposited, it is necessary that these marshy areas be covered by water in motion. Since almost regularly marine shells and fossils are found on top of coal beds, the sea must have covered the swamps at one time; then, for new plants to grow there, the sea must have retreated. There are places where sixty, eighty, and a hundred and more successive beds of coal have formed; …….many times the sea trespassed………and as many times retreated. Fossils of marine clams, snails…are abundant in the shales just above each seam of coal. Later, with fluctuating sea level, the salt water withdrew and another freshwater marsh came into being, giving rise to another bed of coal. ………Ohio displays more than forty such cycles and in Wales more than a hundred separate seams of coal have been discovered.' (67) This type of evidence is not supportive of the notion that sea and land rose and fell in unison over the extended times required to grow multiple forests on top of each other. The swamp theory of coal formation is stressed further by the split seams of coal. '…a coal bed, undivided on one side, sometimes splits on the other side into numerous beds, with layers of limestone or other formations between.' (67, 173)

2.2.3   The Carbon Cycle Theory

The carbon cycle theory is a system of ideas, which identify the several ways in which carbon is transferred in nature from one place to another. The carbon cycle theory may be validly thought of as being well established. It is supported by a host of observations. The following discussion of the carbon cycle theory will explain how carbon is transferred from plants to animals and back again to plants. Also, the way in which carbon is removed from the cycle and trapped away so it cannot be recycled will be outlined.

The structural component of all plants is carbon, and there are three ways by which this structural carbon is introduced into the atmosphere as carbon dioxide. First, as animals eat plants, their digestive systems convert part of the plant material into sugar. Sugar is a molecule, which is an assembly of carbon atoms. The carbon, which was the structural component of the plants, is converted, by the digestive system of the animal, into sugar, which is able to circulate through the circulation system of the animal. In each cell of the animal, the sugar molecule is brought into close contact with oxygen, which was also brought to each cell by the circulation system. As the carbon combines with the oxygen, heat is produced. This process of combining the carbon and the oxygen is a chemical reaction and the heat, which is produced by this reaction, keeps the animal warm. It follows that if an animal or a person does not have enough food, the heat-producing reaction cannot occur and the animal will chill and may die. For example, if a person exercises or works to excess, the amount of sugar in the bloodstream will diminish. The resulting inability to produce heat may cause hypothermia or chilling. It is unfortunate that people have died from hypothermia even when the temperature was well above freezing. A sign that hypothermia is developing is excessive convulsive shaking. The cure is warming and supplying food, which will resupply the circulation system with sugar. The carbon has therefore served a useful purpose and it is absolutely essential that an animal bring in food to keep warm and continue living. As the carbon and oxygen are combined, carbon dioxide is produced and released into the atmosphere when the animal breathes out. This is one way by which carbon is circulated from plants through animals to the atmosphere.

There are two other ways in which the carbon in plants is converted into carbon dioxide and released into the air. These two ways are rotting and burning. When plants either rot or are burned, the carbon, which is in these plants, combines with oxygen, which is in the air. Carbon dioxide is thereby created and becomes part of the atmosphere.

In summary, there are three ways for plant carbon to enter the atmosphere. The plant may be eaten by an animal which will subsequently exhale the carbon as carbon dioxide. Secondly, the plant may rot during which process the plant's carbon combines with atmospheric oxygen to form carbon dioxide. Thirdly, the plant may be burned which is a heat-producing process combining the plant carbon with atmospheric oxygen.

While there are three ways to return carbon from plants to the air, there is only one way to get the carbon from the air to the plants. The plants must grow. As plants grow, they take carbon dioxide from the air and form their respective structures. The carbon therefore becomes locked up as part of the plant and will remain as the structural portion of the plant until the plant is eaten, burned or simply rots.

As shown in the diagram, 3.13 The Carbon Cycle, the carbon cycle has two main branches, both of which are required to form the complete cycle. As plants grow, carbon from the atmosphere enters the plant and becomes its structural component. Then when the plant is either eaten, burned or rots, its carbon re-enters the atmosphere and the carbon cycle is complete.

2.2.4   The Interrupted Carbon Cycle

As coal beds are formed, (or the great peat bogs, i.e. peat is also carbon and may be on its way to becoming coal - if there is enough of it and it is properly packed down) the carbon cycle is interrupted. We understand that all of the plants, which became part of the coal beds, were formed from carbon obtained from atmospheric carbon dioxide. However as the coal-bed carbon accumulates, it is effectively trapped and is no longer available to circulate as part of the carbon cycle. It is being trapped off into a great carbon storehouse. In fact if such a situation were allowed to continue, more and more carbon would become unavailable for recirculation and the great carbon cycle would have less and less carbon in circulation. Since both plant and animal life need the carbon to keep circulating, life would consequently become less and less viable. This process could have led to a carbon starvation death for the Earth. In recognition of how much carbon is presently stored in the coal beds of the Earth, in comparison to the amount in the biosphere, it is a wonder that this has not already happened. Indeed it might have happened, except that the industrial revolution reintroduced great quantities of carbon back into the atmosphere. If there had not been an industrial revolution and vast quantities of coal had not been burned, carbon dioxide levels today would be much lower than they were prior to the industrial revolution and life in general would be less viable.

Now we are in a position to recognize the great problem which exists in trying to explain the coal beds. The carbon from the atmosphere must have formed the plants for the coal beds, but the carbon in these particular plants has not been allowed to circulate back into the atmosphere. It is still in the coal. Therefore, the carbon, which is in the coal beds, has been diverted from the carbon cycle, and has become trapped out of circulation. It is therefore appropriate to ask where the carbon in the coal came from in the first place. It is obvious that it was not exhaled by any animals, which had eaten plants, because the carbon from the plants, which formed the coal beds is still in the coal, which hasn't been eaten at all. Neither did it come from any plants, which were burned, because if they had been burned, their carbon would have combined with atmospheric oxygen to form CO2 and would consequently not be in the coal beds either. The same carbon cannot be in two places at the same time.

Once coal is formed, its carbon is tucked away in the coal storehouse and it is out of circulation. Therefore, all of the carbon in these coal formations has become completely unavailable for carbon dioxide formation and the possible production of more plants. Hence it is appropriate to seek an explanation for the source of the carbon dioxide, which supplied the carbon for the great coal deposits of the world.

The coal beds do contain a very great amount of carbon. Various estimates have been made and compared to the amount which exists in the biosphere (the total of all living things). The coal beds may contain 50 times as much carbon as the biosphere. (33) All of this carbon has been trapped away from the carbon cycle and has not been available to recirculate since it was trapped and because of this trapping, the carbon, which formed the coal bed plants, came from some source other than the metabolization process of animals. Neither did it come from the burning or rotting of plants. Of course, it came from the air because that is the source of all plant-forming carbon, but how did it get into the air?

Three possibilities present themselves.

1. Prior to the formation of the coal beds, the ancient biosphere was 50 times more extensive than it is at the present time. Forests, swamps and meadowlands were filled with an abundance of all kinds of plants. In addition a greater area of the Earth was involved including the high arctic lands, Antarctica and areas now below sea level. Then suddenly this ancient biosphere was annihilated and its carbon is now found in coal. This explanation basically shoves the question back because now we must ask where the carbon came from to form this massive assembly of ancient plants. Did they just suddenly appear? Were they created?

2. Is it possible that there could have been enough carbon dioxide in the air at some ancient time to enable the coal-forming plants to develop simply by depleting this CO_2? The amount of carbon dioxide which is in the atmosphere at the present time is more than 380 parts per million (38) which means that 380 out of every million molecules in our atmosphere at the present time are carbon dioxide molecules. If all of the carbon in this carbon dioxide were assembled together to make coal, about $6 \times 10(11)$ metric tons of coal would result. (Please refer to the appendix for this calculation.) Current estimates of the world's coal reserves are $15 \times 10(12)$ metric tons. (33) Atmospheric carbon is therefore equivalent to 6/150 x 100 or about 4% of the world's coal reserves. Therefore, if, prior to formation of the coal beds, the amount of atmospheric CO_2 had been about 25 times as great as it is now, the coal beds could have been formed by growing plants and depleting this higher level of CO_2 down to its present level. Therefore, one possibility for coal formation is that an ancient CO_2 rich atmosphere could have been depleted to an atmosphere with much less CO_2. (Please refer to the appendix for the calculation.)

3. The third possibility is that the required CO_2 was formed by a burning process, which used primeval or virgin carbon as a source. This burning process introduced the virgin carbon into the carbon cycle at just the right rate to enable the trees and other plants, which were forming the coal beds, to grow. In order to be internally consistent, it must be recognized that a vast amount of virgin carbon was required. In fact exactly the same amount of ancient virgin carbon was required as is presently found in the coal beds. This type of arrangement is recognized as an artificial construct because nature has been conveniently arranged to bring about a result, which isn't otherwise credible.

In summary, there appear to be three possibilities for the formation of the coal-forming plants.

1. A massive ancient biosphere was created and the plants from it were available to form coal.
2. The ancient biosphere was formed by depleting even more ancient atmospheric CO2 from a level approximately 25 times as high as the present level of 380 ppm.
3. The ancient biosphere was formed from CO2 which was produced by an unknown, virgin-carbon burning process.

However, all of these possibilities come with attachments. If the ancient atmosphere had 25 times as much CO2, the average temperature of the world would have been much higher and the trees would not have been able to grow properly because they would sweat too much and dry out. Also, it would have been above the body temperature for most types of animal life.

While the first possibility is totally unacceptable to many people, with both the second and third possibilities, it would have been necessary that none of the plants which grew during the extended times required, were burned, eaten or decayed. There were no forest fires caused by lightning (which currently strikes the Earth several thousand times every day). Also it was a rot-free forest wherein no significant quantity of material was eaten.

2.2.5   Conclusion

The Swamp Theory of Coal Formation contradicts both available evidence and well-known physics. Neither is it comprehensive. There is no explanation why certain swamps sank for hundreds of feet, at the rate to keep the water level just right. Further unrealistic artificial constructs must be employed to explain the origin of the carbon dioxide used to construct the plants. The widely-accepted Swamp Theory of Coal Formation should therefore be set aside.

2.3   Ice Age

The Ice Age Theory declares that for a period of about two million years, a period of several great ice ages created huge sheets of ice that built up at both poles of the Earth and moved outward under their own weight. Glaciers more than a mile thick spread over the Northern Hemisphere gouging valleys and shaving hills. (62).

2.3.1   Ice Buildup

If vast and massive glaciers are to be built where none existed before great quantities of atmospheric moisture must be precipitated in the form of snow. Since the moisture must come from the sea, sea level will be lowered. (122)

The requirement therefore, is that the climate over the continents be colder but that atmospheric moisture be much more available. Unfortunately, these two criteria are counterproductive. In order for ice to accumulate on the ground, it must precipitate. Atmospheric air currents pass over open water, which evaporates and fills the air with moisture. When this moisture-laden air passes over land, which is cooler than the open water, snow falls. However, if the climate becomes too cold, open water will crust over with ice thereby inhibiting evaporation and subsequent precipitation. There is an optimal temperature range where snow will fall. If it is too warm (i.e. just above freezing), rain falls. If it is too cold, everything freezes over and there won't be any precipitation at all. The maximum-possible snowfall temperature occurs just below freezing. Well-known physics teaches that the warmer the air is, the more moisture it can hold. (Please refer to the diagram, 7.3.1 Vapor Capacity of Air in chapter seven.) Since in this case snow is required, when the temperature is just cold enough to enable snow to fall instead of rain, maximum snowfall will occur. When it gets colder, and the lakes freeze over, snowfall will drop off dramatically because there won't be any source of moisture to form the snow. It is therefore logical to ask how the ice-age snow accumulated if the oceans were frozen and the air was deprived of the necessary moisture. If the entire world had been chilled large areas of ocean would have been covered with pack ice. Hence they would have been unable to contribute moisture for the ice build-up. Further the water that wasn't covered by ice was cold and not a good source of moisture either. At the present time major snowfalls only occur when the open-water sources of moisture are 'warm'. In this case, 'warm' is only a few degrees above freezing but the difference in moisture availability is tremendous. In a recent major snowstorm in Southern Ontario, about five feet of snow fell within a few days and was enabled to fall because Lake Huron was still 'warm'. (i.e. several degrees above freezing) Therefore, the assertion that ice could build up under the constraints of a cold climate, is in direct contradiction with well-known physics.

The further inhibition inherent in the 'cold Earth' (Milankovitch) approach relates to the greenhouse gases. Water vapor is a major greenhouse gas so if the entire world became chilled, the warming effect of the water vapor would be lost. This means that the Earth would become even colder and only stop getting colder when any further reduction in temperature did not result in any further reduction in water vapor. The end result is that the entire world would freeze solid and simply stay that way indefinitely.

It is declared that the ice moved south after it accumulated in the north. Diagrams are given to show the movement of ice from the far north. (Please refer to the diagram, 2.3.1 Ice Movement) Farther south, the sheet of ice is declared to have been two miles thick. It was presumably much thicker than this in the far north prior to spreading out. How thick could it possibly have been and how high could it possibly have built up?

On Mount Everest, which is between five and six miles high, there is very little snowfall. The reason for this is because there is very little air and even less moisture at this elevation. High up in the mountains, snow accumulation is very slow. However, even if ice were to build up this high, it is still only two or three times as high as the thickness of the postulated spreading glacier. This would not be high enough to cause it to spread very far. In order for it to spread out twenty miles for example, the ice would have to build up for some significant fraction of twenty miles. However, twenty miles above the Earth there is virtually no atmosphere at all. Therefore the declaration that the ice spread out under its own weight from locations in the far north for a thousand miles to the south is in direct contradiction with well-known physics. It did not spread out because it could not possibly have built up high enough to enable it to

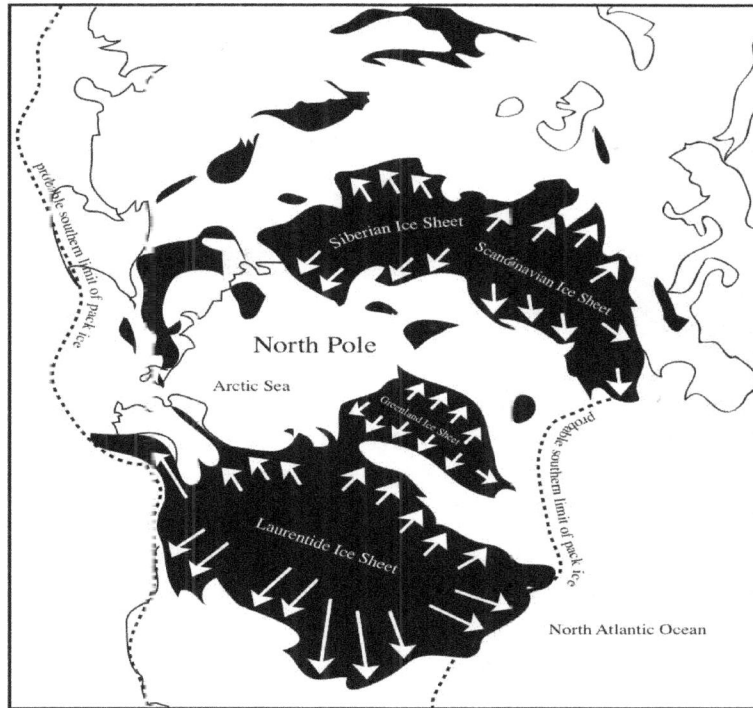

2.3.1 Ice Movement

Ice movement is declared to have been from Hudson Bay all the way to the central
United States. However, from central Quebec it was declared also to have moved toward
Hudson Bay, and then from Baffin Island to Hudson Bay as well. In Europe, it was said
to have moved from Sweden across the North Sea to cover the British Isles.

spread out. There is simply not enough moisture available at such a great height to enable any significant accumulation of snow to develop.

2.3.2   Movement Force

It is declared that the ice moved south for more than a thousand miles. It moved across all kinds of land. It moved down through valleys and up the other side. It moved right down through the Great Lakes region. As it moved, it carried a great load of scrapings in front of itself. It even moved uphill. In fact it formed its own hill to move up. As the ice increased in thickness, it depressed the ground. Hudson Bay is understood to be the result of a very heavy load of ice depressing the crust of the Earth and is expected to eventually disappear as the ground regains its original elevation. Therefore, during the Ice Age, any ice coming from Hudson Bay had to have travelled uphill. Not only that but it had to move across the Canadian Shield, the hardest place on Earth to move anything. Now the question is properly asked: what great force was causing all of this ice to move? In particular, what caused it to move uphill? Unfortunately, no explanation is given.

It is understood from observation that there are several ways for a glacier to move a short distance (i.e. from a few hundred feet to a few miles) but there is no evidence of a glacier ever having moved a thousand miles. First, it can move if it is on a mountain slope. In this case, gravity is given as an

22

explanation, and the explanation is therefore in agreement with well-known physics. (e.g. Glaciers on both Antarctica and parts of Greenland are sliding down mountain slopes.) A glacier can move more easily if it is well lubricated underneath with water. In this case less movement force is required because friction with the ground is reduced but a movement force is still required. A glacier can also move to a certain extent under its own weight. However, none of these circumstances will enable a glacier to move for more than a short distance. It may therefore be stated that because the Ice Age ice is declared to have moved for great distances, but since no explanation for the cause of this motion is given, the theory is incomplete because a necessary part of the explanation is missing.

2.3.3   Ice Strength

It is also declared that as the ice moved along it carried a great load of material in front of itself. (176) When the glacier stopped moving, the material also stopped and simply remained in place as the glacier melted. These massive accumulations of material including stones, sand, gravel and clay are now called Terminal Moraines. However, in order to move material horizontally across the land, great force is required. For example, when there is a requirement to move earth, large machinery is employed. In order to move even a small pile of soil will require a powerful machine. The forward push, which is required to move a thousand pounds of sand or gravel, may be several hundred pounds. To move the great accumulations of material which are often referred to as terminal moraines, would require an enormous force. Could a glacier provide such a force? No it could not. Ice is not strong enough. If a moving sheet of ice should encounter this much material, it would deform and ride up over it just as it does when a glacier encounters an obstacle on the way down a mountain. Mountainous glaciers ride over obstacles, which are too difficult to move, and leave them behind. Terminal moraines in mountainous areas therefore do not provide an explanation for the great accumulations of gravel in flat areas remote from mountains because glacier ice would simply have not been strong enough to move these accumulations into place. Mountain moraine material, like a mountain glacier is moved by gravity. Gravity moves the glacier and the glacier moves the material (which is also under the influence of gravity), which in turn will become the terminal moraine. The mountain glacier is therefore only providing a portion of the force required to move the moraine material. Further, terminal moraines in mountainous regions are not very large compared to the massive accumulations of gravel currently referred to as terminal moraines in various locations across the country (e.g. the Oak Ridges Moraine). It is therefore inappropriate to extrapolate from mountain situations to those remote from mountains.

The ice is said to have moved down through the St. Lawrence River valley and up the other side. (176) It is also declared to have moved down through the Great Lakes basins and up the other sides. Unfortunately, any sheet of material, which was flexible enough to follow these contours of the Earth, would have been too flexible to have been pushed. Any pushing would have resulted in folding and piling up because local irregularities in the surface of the ground can easily deflect plastic, slowly moving ice. (3) The difficulty that the flexible, viscous ice would encounter if it was forced to move forward has also been noted by others. Glacial ice movement as proposed in the glacial theory, does not seem to have obeyed the normal rules, as Erratic boulders are found in areas hundreds of miles from their supposed sources. This would require movement of the base of the ice sheets over irregular country for these great distances (124) A more important and far-reaching difficulty which the glacial champions have to face is the proved incapacity of glacial ice, as of any other viscous body, to travel over enormous stretches of level country, and up and down hills, as it must have done without failing in either compression or bending or both. (125)

If ice encounters some obstacle that it cannot move, and if the ice cannot ride up over such an obstacle, as soon as the pushing force exceeds the strength of the ice, the ice will then collapse and buckle. An example of this type of ice failure may be observed in pack ice. Pack ice is floating ice, which is found in the far north. When the wind blows, pack ice is pushed forward toward other pack ice and compressive stress builds up. The harder, longer and further the wind blows, the greater this compressive stress becomes. Frequently the ice will not be strong enough to withstand these compressive forces and it will fracture and pile up in pressure ridges. In such cases the pushing force was excessive and the ice was stressed into failure. Any more pushing would only result in more ice piling up. (41) There would be no further motion beyond the fracture zone.

If a great sheet of ice were somehow forced to move over level ground, it will become stressed in compression. This means that the ice at any particular location will be squeezed toward or pressed against the ice just ahead. Ice is pushing ice and everything is being pressed together (or compressed together). The basic reason that the ice is being compressed is because of friction with the ground. The movement force will directly depend on the difficulty encountered in sliding the ice over the ground. The thicker the ice is, the heavier it is and the more force will be required to overcome the friction with the ground. Conversely, if there wasn't any friction with the ground, the ice could slide along without any force being required.

In some ways the situation may be compared to moving a car. If a car is to be towed, a rope is attached and the car will move forward quite readily. However if the brakes are applied, the force required to move the car will be greatly increased. The increase in force is due to the friction of the tires with the ground. If it is a big car, more force will be required. If ten cars in a row are being pulled with their brakes on, an even greater force will be needed. If additional braked cars are added, the rope might break because the pulling force required will eventually exceed the strength of the rope.

A very similar situation would develop if a thick slab of ice were pushed along the ground. The longer and thicker the slab was, the greater the friction with the ground and the greater the compressive stress would be in the ice. If such a slab were much longer than it was thick, the pushing force would eventually be too much and the ice would fail in compressive stress exactly the same as pack ice. If the slab of ice were two miles thick, it would be very heavy and the friction with the ground would be very great. Consequently the pushing force would also be very great. Rather than compare this situation to pulling a string of braked cars it should be compared to pushing a freight train without wheels along ground without a track. Even a short train would not push very well but simply collapse in compressive stress failure with cars hopelessly piled up. Similarly, if a slab of ice were more than a few miles long, it would not move forward but simply break and pile up in compressive stress failure.

It is instructive at this point to consider what would happen when the slab of ice was forced up an incline. As with the level ground case, there would be basic friction stress. Added to this would be bending stress as the slab bends to go up the slope. If the slab were two miles thick, the bending force would be significant. However, a third component of stress would also be involved. The ice would be gaining elevation. The pushing force would have to be strong enough to overcome the surface friction, bend the ice to go up the hill and then lift it up the hill. If the pushing force was strong enough and the ice was strong enough, the ice would slide up the hill. However, when it came to the top, a new situation would develop.

If the crest of the hill was very smooth and the far slope was very gentle, the ice would bend over the top and slide down the other side. Unfortunately, most mountain tops are not very smooth and mountain slopes are not very gentle. But suppose that instead of a mountain, a crag and tail was involved. (A crag and tail is a situation where a long gentle upslope is followed by a sudden drop-off.) In this case the ice would come to the top but would not be able to bend over the top. So it would project out into the air. As motion continued and the slab projected farther and farther out over the drop-off, the bending stress would continue to increase. Soon the slab would break off and the broken piece would fall down the other side. If the slab still kept on coming, another piece would break off and then another and another. The drop-off area would fill up with broken pieces of the slab. Consequently the ice would no longer be moving forward but would simply be breaking up instead. If the valley were to fill with broken pieces of ice, the slab might then have something to slide on. However trying to visualize it sliding over a great field of broken pieces of itself doesn't seem very likely. The slab would just keep on breaking up as it moved over the rough field of broken slabs of ice. So far in this example, only one drop-off has been encountered. However, numerous drop-offs would be encountered by a slab of ice coming down from the far north. In addition, peaked structures may be encountered. How would the moving slab fold itself around a sequence of peaks without breaking up completely?

It is clear from this discussion that even if some great force were available to move a glacier on a horizontal surface, it still would not move. The ice would just pile up. Therefore, to avoid buckling and breakage, the ice that would be required for the general Ice Age Theory must have been a special type of very strong ice, which doesn't occur in nature. Such a requirement would be artificial construct. In other words, in order for the general Ice Age Theory to work, the circumstances would have to be artificially arranged. (i.e. special, extraordinarily strong ice would be required.) Any theory, which includes such a necessity, should be set aside.

2.3.4 Vertical Leading Edge

It appears as part of the theory that the ice formed into a sheet about two miles thick and moved across the country with a vertical leading edge as thick as the sheet. Occasionally this vertical wall of ice is depicted adjacent to grazing animals or referred to as a 'cliff-like edge' in glacier descriptions. (3) The ice is usually shown similar to a vertical cliff face and because the sheet was so thick, the upper surface is not shown. This truly would have been an incredible scene. The vertical wall of ice advancing slowly across the countryside would have been most impressive. Piled up in front would be the material, which would form the terminal moraine when the advance ground to a halt. But how did the vertical face of the advancing glacier form? What caused the falling snow or whatever formed the glacier, to form itself into a vertical wall? Supposedly, the leading edge would have appeared similar to the terminus of the Greenland glacier where it meets the sea. As this great glacier slides into the sea, the front edge keeps breaking off. This happens because the ground slopes away at the coastline and the ice cannot support itself when it projects out into the water. A vertical or nearly vertical wall is thereby formed. This arrangement, however, does not translate into a glacier moving across the flat countryside. Why would the leading edge of a land-bound glacier be vertical? At the very minimum, one would expect that there would be a great jumbled mass of broken ice and whatever other debris had been encountered along the way. The glacier itself would probably not even be noticeable in the confused assembly of material in front of the leading edge.

2.3.5   Gravel Accumulation

If it could be allowed for a moment that a great sheet of ice was able to move forward and on the way this ice encountered a horizontal layer of gravel, what would be the effect of the moving ice on the gravel? The situation would be similar to someone pushing an ice cube through a sandbox.

Gravel is understood to be a solid. However, it also has some of the characteristics of a fluid. Even a viscous fluid can be moved forward if it is confined to a pressure vessel (i.e. a pipe). The force which is required to effect motion will depend on the viscosity of the fluid and the friction of the vessel walls. While a considerable force may be required, viscous fluids are certainly movable in this way.

However if the fluid was not confined to a pressure vessel, it would not move forward. It may move but it would not move forward. A similar situation is expected with gravel. If a horizontal force is applied to a quantity of nearby gravel, why would anyone expect the gravel in the next county to move? Instead of moving forward in the direction of the cause of the movement, the nearby gravel will only move to get out of the way of the object, which is causing the movement.

Suppose that a horizontal layer of gravel one thousand feet deep is encountered by a moving block of ice, one thousand feet thick. As the ice moved forward into this gravel, there would be a pile up at the interface. As the pile got higher, it would soon be several times as high as the block of ice. Gravel, which was a few hundred feet in front of the block of ice would not be moving at all. However, the disturbed gravel would pile up and spill over on top of the ice. It would also spill over on top of adjacent gravel, which was downstream of the movement. The height of the pile would depend on the natural-angle of the surface of the pile. This means that it will only pile up until the surface angle was so steep that any attempts to pile it higher would only result in more spillage forward and backward. If the ice moved forward for ten miles, the height of the pile may reach five miles. Further, the slope of this pile will be very steep. It will be at the natural-angle (When a pile of material reaches the natural-angle, any small displacement of any part of the pile will result in the displaced portion rolling all the way down the pile.)

The general Ice Age Theory declares that ice moved forward for more than one thousand miles. In such a case it would be reasonable to expect a pile of gravel of incredible height to have formed and for it to have very steep sides. By contrast, the deposits of gravel, which are found as part of so-called moraines, do not have steep sides but instead occur in well-rounded formations where steep, natural-angle slopes are very seldom seen. Also, the depth of certain of these terminal moraines (e.g. Oak Ridges in Southern Ontario) is only a few hundred feet.

While the absence of miles-high, steep-sided, mounds of material is difficult to explain, the question of how the material became gravel is also unanswered. Since gravel consists of rounded stones, it is believed to have been formed by moving water. The common ice age theory should therefore include an explanation for the enormous water movement, which would have been required to form all of the gravel of the world prior to the ice age. Since neither of these problems are addressed, the ice age theory is incomplete.

2.3.6 Drumlins

Evidence for the notion that moving water rather than moving ice placed the gravel is provided by the great drumlin fields near Alberta, (39) Ontario and many other places in Canada. Hundreds of drumlins have been formed. They are all lined up in a north-east to south-west direction. They all have a head and a tail and appear very similar to the types of structures, which commonly form down-stream of pebbles, boulders or other eddy-generating objects in streams. The drumlins of Alberta consist of unconsolidated material. Therefore, they were not formed by a great moving sheet of ice at all as it would have scraped them all away.

2.3.7 Manson Crater

Buried beneath the town of Manson, Iowa, is an impact crater. The general topography in the region is flat. The crater was initially suspected from well drilling operations and was subsequently clearly identified by a directed exploration project. The crater is filled with gravel, boulders and other material, none of which is the least bit surprising. The fill material is declared to have been placed by glacial activity. 'Now glacial deposits of the Pleistocene Age cover the crater, which has no surface expression at all.' (126) However, this explanation is not credible. How was the glacier able to evenly spread the fill material into the crater and then carry on to the south to the end of its journey? This crater is about 38 km. in diameter. The upper layer of fill material is almost two km. deep. (126) Why didn't the glacier scrape this material away? If the glacier was able to scrape out the Great Lakes basins, which are much larger than the Manson structure, it should have scraped the Manson crater out as well instead of filling it in level. 'No surface evidence exists due to coverage by glacial till and the site where the crater lies buried is now a flat landscape.' (127) The notion that a glacier placed the gravel in the Manson Crater necessitates that it rode on top of a thick layer of gravel (i.e. the gravel in the crater) and continued spreading gravel for many miles to the south. In order to achieve this, it had to be continually pushing an enormous pile of material while gradually spreading it out in a layer more than one kilometer thick. This means that the original unleveled pile of material must have been hundreds of miles high! Alternately, the Manson Crater wasn't filled by glacial action at all.

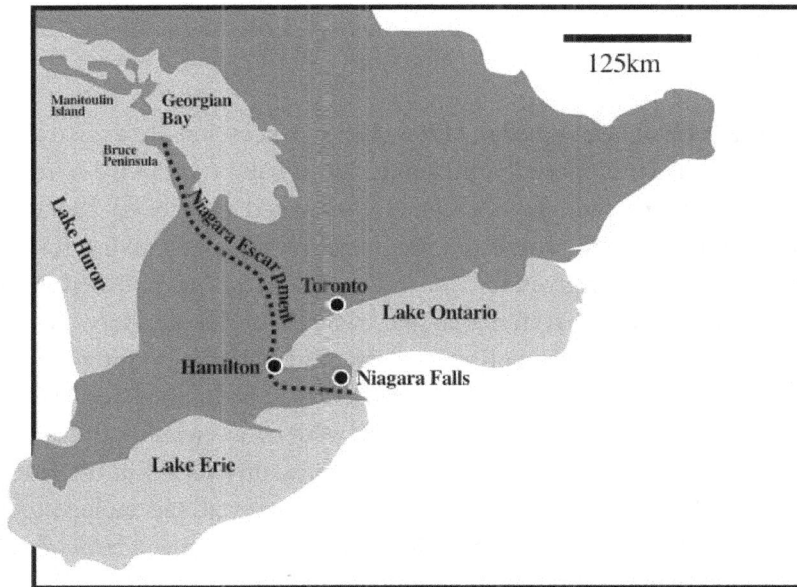

2.3.7 Niagara Escarpment

In several places along the indicated line, the Niagara Escarpment is very inconsistent, with its peaks and valleys often occuring side-by-side. However, in some areas there are no elevated or recessed areas whatsoever. The ninety degree turn the escarpment takes at the City of Hamilton is particularly curious.

2.3.8 Niagara Escarpment

The Niagara Escarpment is recognized as a sharply-elevated formation occurring from the Niagara River at Queenston (which is about half-way between Niagara Falls and Lake Ontario), all the way to Toberrmory at the tip of the Bruce Peninsula in Ontario. The formation is discontinuous and usually characterized by vertical cliffs rising up to several hundred feet above adjacent land. Occasionally, such as at Mono Cliffs Park north-east of Orangeville, there are several vertical formations near each other. How did the moving ice squeeze its way between such structures?

A more important question however relates to the general plan-view shape of the overall structure. From above, the Niagara Escarpment has the general shape of a hockey stick. (Please refer to the diagram, 2.3.7 Niagara Escarpment) The top of the handle is at Tobermory. The shaft runs down across Ontario to the west end of Lake Ontario. Then there is a sharp turn to the east and it continues to the Niagara River.

What did the great glacier do when it encountered this (almost) ninety-degree turn? One can only speculate as to what occurred. Or it could be realized that the great glacier did not do anything because it wasn't moving. (An alternative explanation for the Niagara escarpment as well as the numerous other escarpments, which randomly occur all over Southern Ontario, is that it is simply part of the Laurentian Plateau which massive formation is discussed in several different paragraphs below.)

2.3.9   Retreat of the Glacier

As previously mentioned, the vertical, leading edge of the glacier is occasionally depicted as a wall and as a backdrop for grazing animals. This, supposedly, illustrates the situation as the ice retreated. It seems to be implied from these presentations that the ice actually retreated or moved backward. Instead of continuing to slide forward as the ice age progressed, the ice sheet stopped moving forward and moved backward instead. No explanation is ever given to explain how this was accomplished. It may be that the glacier included a reverse gear. The great force-engine, which caused the forward motion became reversed and caused the reverse motion of the entire ice sheet. Did it pile up in the north? Where did the ice go when it retreated? The implication of retreat certainly seems to be implied because the vertical wall of ice is often shown. If the ice had simply melted, a vertical wall of ice would not have been involved. Instead we would expect the same types of formations as are presently found in the Rocky Mountains. Here the great ice sheets are melting and have now become very uneven and include great gouges and ridges. The idea of a vertical wall of ice moving either forward or backward and retaining any semblance of a sharp vertical structure is unrealistic and should simply be set aside.

2.3.10   Absence of Vegetation

Material which has been identified as terminal moraine is completely free of vegetation remains. If the ice had moved across hundreds of miles of land, it should have gathered up everything that was movable and carried it along to the end of its journey. No trees are found in terminal moraines. No leaves are found, nor are any other samples of former plant life. Organic material is so rare in terminal moraines that the gravel is simply scooped out, crushed and placed on roads. If it were contaminated with tree trunks or branches, it would be useless for roads. (14)

2.3.11   Greenhouse Gases

Perhaps the greatest disappointment with most ice age theories is the lack of recognition of the importance of the greenhouse gases. As discussed further in following chapters, if the Earth became chilled for any reason, the water vapor in the air (the most important greenhouse gas) would be diminished. When the water vapor is diminished the Earth would chill even more. This viscous cycle would continue until the average surface temperature was about -18C instead of +15C as it is at present. If most of the water on the Earth was crusted over with ice there would not be any moisture available for ice-field buildup and there would not be any ice age.

2.3.12   Summary

The General Ice Age Theory makes unrealistic declarations without providing any explanations consistent with well-known physics. Even super-ice could not have been pushed forward for one thousand miles (i.e. unless it is sliding down a mountain slope as in Antarctica and Greenland) while carrying a great load of material in front of itself and leaving a polished rock pathway to indicate it's movement. But more importantly, there has never been any hint given to explain what caused the ocean to evaporate 1500 to 2000 feet or more of water at a rate which was fast enough to cause glacial ice to accumulate. While the General Ice Age Theory really is a preposterous tale, never-the-less the ice was here. An explanation for the very unusual requirements of the Ice Age will be offered in a later chapter entitled; Post-Impact Developments.

2.4   LaPlace Nebular Hypothesis of Solar System Formation

The Sun and the planets, and the other parts of the solar system such as comets and asteroids all formed from the gas and dust residing in a flat, spinning disk called the solar nebula. The Sun formed at the center of the nebula and pulled in so much material that eventually the pressure and temperature in its interior rose so high that nuclear reactions could begin. Meanwhile other material duplicated the process on a smaller scale and formed the planets. Their gravity attracted the nebula's abundant hydrogen and helium gases, building up the huge gas giant worlds of the outer solar system. Eventually the Sun turned on and blew away the leftover material, resulting in the solar system as we see it today. (42) In other words, a large cloud of space gas collapsed. Most of the gas collapsed into the center and formed the Sun while the rest formed the planets. As it collapsed, it began to rotate and this accounts for the fact that the Sun rotates and that the planets rotate around the Sun.

Unfortunately, this popular science version is contradicted by authorities such as 'G. R. Burbidge, a recognized authority on the "evolution of elements" in stars: He complained that the problem simply is that the condensation of a star from interstellar material would violate what we know about the laws of nature. (113)

2.4.1   Gas Cloud Collapse

Unfortunately, it is well understood that gas clouds in free space do not collapse at all unless they are already relatively small. Usually they expand. There is a missing explanation for the initial formation of stars because stars will not spontaneously form in space since the dominant outward gas force will forbid collapse. Instead, gas clouds dissipate outward. (101) Even if the cloud of gas is relatively cool, it is still expected to expand because the outward push due to thermal motion of the molecules, even at 100 Kelvin, is greater than the gravitational pull inward. (114) Therefore in order to make this theory work, the gas cloud must have been forced to collapse down to its critical size, which is the size from which it could continue to collapse due to its own gravity. The agent, which has been specified in the literature for this task, is another gas cloud. (40) In fact you must surround the cloud you wish to compress with a hotter cloud, so that the molecules at the surface of the inner cloud will be bombarded by the faster moving molecules of the outer cloud and pushed inward. (40) This outer, hotter gas cloud would then put pressure on the inner gas cloud (i.e. the future solar system) to collapse it. The inner gas cloud must be condensed down to a critical size (called the Jean's length, after Sir James Jeans who initially developed the mathematics) by some other mechanism before gravitational collapse will work. (115,123)

In order for the outer gas cloud to put pressure on the inner gas cloud, it too must have already been down to its critical size. Otherwise it would have been expanding out into space and could not have put any pressure on anything (What caused the mystery gas cloud to start collapsing?). Somehow after the inner gas cloud was down to the right size, the outer gas cloud disappeared. (Why didn't the outer mystery gas cloud continue to collapse also?)

Therefore, it is clear that three artificially arranged conditions are necessary for the theory to work. An outer gas cloud must be present, it must already be at a certain critical size and temperature and it must disappear after the inner gas cloud is down to its critical size and the solar system is well on its way to being formed.

2.4.2 Angular Momentum

Momentum is the idea that when something is moving, it will keep on moving unless it is somehow forced to stop. For example, if a stone rolls down a hill, it will not stop at the bottom. It will keep right on rolling past the bottom. Similarly, if a wheel is turning and the turning force is removed, the wheel will keep on turning until friction stops it. The wheel is understood to have angular momentum or rotational momentum, whereas the stone has linear momentum.

The main problem with the idea that a gas cloud can collapse and form a star is that the angular momentum of the finished product(s) does not match the angular momentum of the original cloud. According to observation, interstellar clouds have up to 100,000 times as much angular momentum per unit of mass as their progeny stars. Any theory of star formation must therefore describe how a cloud disposes of all of this angular momentum before it collapses to form a solar system.

The Sun does not have angular momentum, which is in proportion to its size. While the Sun has most of the material of the solar system (i.e. 99% (15)), it has hardly any of the rotational factor of the solar system (i.e. It only has 1%), which factor is called angular momentum. In order for the Sun to have angular momentum, which is in proportion to its size, it would have to be spinning one hundred times faster. It is not rotating nearly fast enough to agree with the theory. The outer planets including Jupiter, Saturn, Uranus and Neptune have most of the angular momentum of the solar system while the Sun has most of the material. Therefore, the declaration that a large gas cloud collapsed to form the solar system and rotated as it collapsed does not coincide with observation of the solar system and is in direct contradiction with the principles of well-known physics.

2.4.3   Deep Time Problems (Faint Young Sun Paradox)

It is well recognized that the Earth must have a very stable temperature because if it wandered from its present orbit or if the energy output of the Sun changed, the window of life on Earth would close. With these necessities in mind, it is of interest to note that the Earth is declared to have been formed some five billion years ago subsequent to the collapse of a massive gas cloud. (42) During the extended period of time since the Earth was formed, the Sun was shining on the Earth and life is declared to have somehow developed. At the same time, theories of solar function tell us that the output of the Sun would increase over a five-billion-year timeframe by a factor of 25%. A serious consequence of the solar evolutionary model is the increase in theoretical luminosity. Over the supposed 4.5 billion-year history of the Sun its luminosity is thought to have increased about 25 percent. Present climate models cannot tolerate a 25 percent decrease in solar luminosity without leading to an ice-covered Earth. This problem is called the Faint Young Sun Paradox, and contrasts sharply with the presumed warmth of much of the geological history of the Earth. (43)

Therefore since we appear to be at just the right temperature now, it must have been a lot colder on the Earth through much of the previous five billion years until just a short time ago. How then did life develop? Twenty-five percent less energy is equivalent to being about 12% farther from the Sun. This is well past the 5%, which is recognized as being the limit for life on Earth. (44) It may therefore be concluded that either the theory of solar system formation is invalid or that the theory of solar function is invalid or both of them are invalid.

Current understanding of the warming effects of greenhouse gases further invalidates the notion that the Earth is several billion years old. The present average temperature on the surface of the Earth is about 15C. (Please refer to Climate and the Oceans by Vallis p. 165) This is due to incoming solar energy AND the characteristic of the greenhouse gases (mostly water vapor and CO_2) to retain some of this energy to keep the surface temperature at its present level. The energy from the Sun by itself is currently not sufficient to keep the surface temperature at the proper level. This means that if the Earth had been frozen solid at some time in its past, the Sun would not be hot enough yet to thaw it out. Of course water vapor would not appear until it was thawed out. Therefore it would require another one or two billion years into the future before the Sun would be hot enough on its own to raise the surface temperature of the Earth until it was above the freezing point. Once this happened, water vapor would appear in the atmosphere further helping to raise the temperature. However the Sun would now be considerably hotter than it is at present and together with the heating effect of the added water vapor the temperature would continue rising until it was too high for life to exist. The transition from 'starting to melt' to 'too hot' would only take a few years – not nearly long enough by any theory of life.

2.4.4   Retrograde Rotation

If all of the planets of the solar system formed as a result of a collapsing gas cloud, they would be expected to be rotating in the same direction. However they are not. Venus has so-called retrograde rotation because it turns in the opposite direction to the other planets. Also three of Jupiter's moons and one of Saturn's moons have retrograde rotation. 'A rotating nebula could not produce satellites revolving in two directions.' (68) The large moon, Triton, of Neptune, has a retrograde orbit. (100) The theory cannot explain exceptions such as this. In addition to a planet with retrograde rotation, there is a planet with an axis which points toward the Sun every time it comes around. Uranus does not spin with an axis basically pointed at the pole star but with an axis that is close to the plane of its orbit around the Sun. Also the moons of Uranus 'revolve in a plane almost perpendicular to the orbital plane of their planet.' (68) This is not good news for the Nebular Hypothesis of Solar System Formation because it does not fit into the expectation that rotation should be the same way for all of the planets.

2.4.5   Planetary Rotation Periods

The planets do not rotate with the same speed and the deviations from any suggested average are enormous. The rotational speeds of the planets are: Mercury 58.6 days, Venus 243 days, Earth 1 day, Mars 1.03 days, Jupiter 0.41 days, Saturn 0.45 days, Uranus 0.72 days, Neptune 0.67 days and Pluto 6.39 days. (60) The Nebular Hypothesis for Solar System Formation should have addressed this problem. Since it hasn't, it is incomplete.

2.4.6   Gravity Insufficient

Gravity is the basis for the formation of the Sun at the center of the rotating gas cloud as well as for the formation of the various planets. However, if the central majority of gas separated to form the Sun, there would not have been sufficient self-gravity in the remaining portions of gas to cause them to coalesce to form the planets. '....but even if they had broken away, they would not have balled into globes.' (69) This problem is compounded with respect to the moons. If there wasn't enough gravity to form the planets, there certainly wasn't enough to form the moons.

2.4.7   Patterns Nonexistent

Neither Mercury nor Venus has a moon. Earth has a very large moon. Mars has two extremely small moons. Jupiter has sixteen moons. Saturn has twenty-one moons. Uranus has fifteen moons. Neptune has eight moons. Pluto has one moon. Mercury is quite small. Venus is larger and Earth is larger still. Jupiter is very large but Mars which is located between Jupiter and Earth, is quite small. Between Mars and Jupiter there may have been another small body which disintegrated to form the asteroid belt. Saturn is smaller than Jupiter and Uranus and Neptune are similar in size. 'No orbit is an exact circle; there is no regularity in the eccentrical shapes of the planetary orbits; each elliptical curve verges in a different direction.' (70) 'The celestial harmony is composed of bodies different in size, different in form, different in velocity of rotation, with differently directed axes of rotation, with different directions of rotation, with differently composed atmospheres or without atmospheres, with a varying number of moons or without moons, and with satellites revolving in either direction.' (71) It appears to be by chance or some other reason that our planet has oceans, atmosphere, oxygen and one moon and that it is located between Venus and Mars with a nearly circular orbit but the Laplace Nebular Hypothesis of Solar System Formation offers no explanation for any of these circumstances.

2.4.8   Iron Distribution

If an Earth-forming gas cloud should contract, it is reasonable to expect that heavier elements would settle toward the core. While it is commonly held that the core of the Earth may be iron, there is no explanation why iron is found throughout the crust right to the very surface. Even a violent explosion cannot explain this because the iron is well mixed with other elements over untold thousands of square miles of the Earth's surface. The theories explaining gas cloud collapse are unable to explain iron distribution throughout the crust of the Earth.

2.4.9   Multiple Star Systems

Most of the stars in the sky do not occur as single entities but as doubles or triples. (133) In some cases, two stars orbit around each other and a third one orbits further away around the first two. If these stars formed at nearly the same time, they should be nearly the same age. This however, is apparently not the case. There is a class of double stars known as semi-detached systems where the stars almost touch each other. These systems have a large secondary of low mass, and a small primary of high mass. The larger secondary appears to be at a later stage of evolution than the primary. This presents a problem, for computer models because they predict that the more massive star should have evolved quicker than the less massive. Therefore, why is the star of lower mass, larger than the other? (134) It is common for binary star systems to include a white dwarf. Such stars are about the size of the Earth and about as massive as the Sun, while the companion is much larger and not nearly as massive. (135) The Laplace Nebular Hypothesis of Solar System Formation cannot account for these discrepancies.

2.4.10   Asteroid Shape

An integral aspect of the Nebular Hypothesis is that small objects like asteroids would never be found with a spherical shape. A planet's ' … mass and gravity must be large enough to have squashed it into a sphere through its own rotation and interaction with other bodies. Asteroids and comets do not possess enough mass for this to occur, and thus they are often seen to have irregular shapes …' (180) Unfortunately for the theory the asteroid 2005 YU55, scheduled to pass the Earth in November 2011, is only 400 meters in diameter but it clearly has a spherical shape. In fact it has been referred to as ' … a round mini-world … ' (181)

2.4.11   The Death Knell

The Nebular Hypothesis was held to be the explanation of how the solar system came about up until other solar systems were discovered. Somehow, the Nebular Hypothesis anticipated that rocky planets would form closest to the Sun and gaseous planets would form much further out. Since this was indeed the case with the Solar System, the hypothesis seemed to be correct. Then, other solar systems were discovered. These other systems are all constructed contrary to how a solar system should be constructed and they have large gaseous planets near their host sun. In fact, their variant construction enabled them to be discovered. It is, needless to say, difficult to 'see' a planet near a faraway star. The light from a star is so bright when compared to a non-glowing object like a planet, that visual detection is impossible. However, if a sun has a planet nearby and if it is big enough, it should put a slight wobble into the host star's pathway. Then it would not be necessary to actually see the planet but only the wobble. The Sun has a wobble. The Sun and Jupiter both orbit around a common center which is located 50,000 km above the surface of the Sun. (42) Therefore, if the Sun were observed from far away, its wobble would enable an observer to deduce that Jupiter existed. The presence of very large objects, in some cases, very near their host sun, effectively disabled the Nebular Hypothesis. Unfortunately, its replacement is slow in arriving as the following illustrates. The presence of such huge bodies so close to their stars challenged prevailing theory. How could gas giants form so close to their suns? Could they have formed elsewhere? If so, how did they get to their present locations? Are their orbits stable? And what does all this say about our solar system? Is it a freak whose apparent orderliness is illusory? The discovery of more than a dozen extrasolar planets (the number is steadily rising) has forced a serious rethinking of many details of the solar nebular theory and, in particular, the subtleties of orbital dynamics. One cornerstone of the standard theory has been that the planets first formed at or near their present locations relative to the Sun. But the news from afar, combined with more sophisticated computer modeling techniques, has suggested a more complicated scenario for planet formation. (42) In other words, the old theory is dead and there isn't a new one.

2.5 Continental Drift or Plate Tectonics

Continental Drift is the theory that the continents of the Earth are drifting around on the surface of the Earth. If a map of the Earth were drawn to illustrate how the Earth appeared a long time ago, it would appear very different from the way it is at the present time. It has been conjectured that, at some time in the distant past, North America was touching Eurasia (The entire area was called Laurasia.) and South America was touching Africa. (This portion was called Gondwana.) (45) Later this idea was modified and refined to identify that large crustal plates of the Earth were drifting, not whole continents. This improved version is called the Theory of Plate Tectonics.

The theory was developed to explain numerous observations, which had been made around the world including: the Mid-Atlantic ridge, submarine trenches, surface discontinuities and mountains as well as volcanoes and earthquakes.

2.5.1 Plate Movement

It has been observed that large portions of continents are moving with respect to each other. They are sliding along each other. However, no force function is offered to explain how or why these movements are taking place. In order to move anything as large as a mountain or a portion of a continent, a very large force would be required. Supposedly the underlying fluid layer is moving in a manner which drags the continental portions of crust along as it goes. But why is the fluid layer moving in such an odd manner? A coastal portion of California is moving northward but a little further inland, the continent is not moving. What incredible force is causing this movement? There has never been any suggestion to account for this mysterious force or how or why it is acting. If a system of ideas is offered to explain something, which requires very great forces and incredibly high amounts of energy, then included among the ideas there should be a discussion of the source of both the force and the energy. Since the Theory of Plate Tectonics does not include any such discussion, it is incomplete.

2.5.2   Ocean Floor Area Problem

The Theory of Plate Tectonics declares that the Atlantic Ocean is growing wider and has been growing wider ever since the original supercontinent cracked. Europe is separating from North America. (63) The source of this separation is the mid-Atlantic ridge, which is supplying new ocean floor material which could make the surface of the Earth larger. In particular, the Atlantic Ocean must be getting larger unless the extra material can be accounted for some other way. If an equivalent area of ocean floor was being consumed elsewhere around the Atlantic, it would not be getting any wider and there would consequently be no influence on the size of the Pacific Ocean. Subduction (i.e. movement of ocean floor down into the interior of the Earth) could account for any new ocean floor being produced. This is thought to be occurring where there are deep ocean trenches near continental shores. However, the North American plate continues out across the Atlantic Ocean as far as the mid-Atlantic ridge, making the idea of subduction around the Atlantic Ocean untenable. Since the floor of the Atlantic Ocean is not being consumed by diving into the interior, and since the mid-Atlantic Ridge is producing more ocean floor every day, the Pacific Ocean floor must be getting smaller and it must be diving into the fluid interior of the Earth. Up until it started to dive, this material was floating on the molten interior material. It was floating supposedly because it was lighter or at least not heavier. The Theory of Plate tectonics is therefore declaring that, an area of Pacific Ocean floor as large as the entire Atlantic Ocean floor has been unnaturally forced down into the interior of the Earth. However, this is in direct contradiction with well-known physics. It is very difficult to force an object that is floating, down into the material that it is floating on. A situation like this is an artificial construct. Nature is declared to have done something extraordinary which it normally or naturally would be unable to do.

2.5.3   Rate of Separation Hypothesis

The rate of separation of the Americas from Europe and Africa has been declared to be one to two inches per year. (46) This is not a measured amount but rather a deduced amount, which is based on a calculation. The distance from the center of the mid-Atlantic ridge to magnetic anomalies (approximately 500 miles) on either side of the ridge were used for the calculation together with the 'known' ages of the rocks at these anomalies. Unfortunately the ages of the rocks are not known but only surmised from other theories. It is also unfortunate that older rocks have been found closer to the ridge using these same theories, all of which places the separation deduction in the dubious category.

Just to make matters worse, satellite measurements do not show any widening of the Atlantic Ocean at all. (47) Neither do other observations. 'If the theory is correct, the motion of the continents should be observable at present; though Wegener claimed, on the basis of certain reports, that Greenland and an island near its western coast still move, repeated observations and triangulations do not support this claim.

… The land masses of today do not change their latitudes; the motive force claimed is insufficient by far. Coal beds in Antarctica and recent glaciations in temperate latitudes of the Southern Hemisphere all conspire to invalidate the theory of wandering continents.' (142)

The entire concept of plate tectonics relies heavily on the magnetic data from the ocean floor and without it, the Theory of Plate Tectonics would not have been developed in the first place. In fact the so-called spreading zones are classified as either fast or slow as determined by calculations based on the magnetic data. (94)   However the data are not convincing to everyone. 'The theories of continental drift and seafloor spreading are highly conjectural, but it is hard to stop anything as big as the floor of the ocean once it has been put into motion.' (95) Other commentators have similar reservations. 'The foregoing discoveries led the author to one conclusion only, that paleomagnetic data are still so unreliable and contradictory that they cannot be used as evidence either for or against the hypothesis of the relative drift of continents or their parts.' (96) Magnetic reversals in a horizontal direction are difficult enough to explain with any hypothesis but their occurrence in both vertical and horizontal directions places the continental drift notion in utter jeopardy. '…these several vertically alternating layers of opposing magnetic polarization directions found in cored oceanic crust disproves one of the basic parameters of sea-floor spreading theory, namely that the oceanic crust was magnetized entirely as it spread laterally from the magnetic center.' (97)

'An even more puzzling fact is that the rocks with inverted polarity are much more strongly magnetized than can be accounted for by the Earth's magnetic field. Lava or igneous rock, on cooling below the Curie point, acquires a magnetic charge stronger than the charge this rock would acquire in the same magnetic field at outdoor temperature, but only doubly so. The rocks with inverted polarity, however, are magnetically charged ten times and often up to one hundred times stronger than they could have been by terrestrial magnetism. "This is one of the most astonishing problems of paleomagnetism, and is not yet fully explained, although the facts are well attested."' (132)

The case for using magnetic reversals as indicators of time and hence of the rate of ocean floor formation, is weakened further by the recent observations that reversals can occur within a few days. A team of geoscientists investigated the Miocene lava flows at Steens Mountain, Oregon. They observed that; The seven lava flows stratigraphically above B51 are of normal polarity and the ten below it are of reversed polarity. Numerous samples taken through the several-meter thickness of flow B51 show a continuous transition from the reversed polarity below to the normal polarity above. The flow would have cooled to 500c or below in about 15 days. The investigators thought that such a rapid change was unbelievable. The rapidity and large amplitude of geomagnetic variation that we infer from the remanence directions in flow B51, even when regarded as an impulse during a polarity transition, truly strains the imagination. (99)

2.5.4   Crust Formation Problem

Supposedly, the source of material for the new Atlantic Ocean floor is the mid-Atlantic ridge, which means that the mid-Atlantic ridge is somehow deforming from a mountain ridge down into ocean floor. Unfortunately, no evidence is offered to show that this is happening. Instead, the ridge material appears to have piled up on the ocean floor and is just sitting there. The theory is therefore in direct violation of the evidence.

2.5.5   Force Direction Problem

In order to force the Americas away from Europe and Africa, a tremendous force would be required. This force must act outwards both ways from the central region and it must be directed horizontally without having any significant vertical component. Otherwise the central ridge would just get higher and higher and pretty soon it would stick up through the ocean surface. However there is no evidence that the central ridge is getting any higher at all. In order to have formed in the first place there was certainly a vertical force involved because the ridge material is piled up quite high. Therefore we must conclude that the force that pushed up the ridge material somehow lost its vertical factor and is now only pushing horizontally. In order to achieve the stated objective of widening the ocean floor, it must be pushing outwards in two directions at once. It would therefore have to be a type of wedge developing an outward push without piling the ridge material any higher. However, there is no evidence that any such action is occurring.

2.5.6   Expected Ocean Floor Failure

Why isn't the ocean floor in a state of buckling failure? The theory of Plate Tectonics requires that the outward force from the center of the ocean be strong enough to either force the continents apart or at least force ocean floor material down into the interior of the Earth. The magnitude of any such force would be astronomical. Why is it preferential for the continents to separate rather than for the ocean floor to buckle and crumble and just pile up? This is certainly what we expect in the arctic. When the compressive forces become too great, sea ice crumbles and buckles up into great pressure ridges. The ocean floor is understood to be relatively thin (approximately three km. thick, (49)) in comparison with the distance from the mid-Atlantic Ridge to any shore. Long thin objects such as this cannot carry compressive stress. They will, like pack ice, buckle and fold up. How then would the ocean floor, without buckling, be able to carry the forces required to separate the continents around the Atlantic Ocean or to force floor material to dive into the interior of the Earth?

2.5.7   Trench Sediment Missing

It is declared that the ocean floor has been forming for a considerable length of time and that as it formed, it has been moving towards the continents. Where the ocean floor meets the continents, it is understood to slide underneath, or subduct. Trenches are understood to be the evidence of subduction taking place. If this were the case, the trenches, should be full of the ocean floor sediment, which had accumulated over long eons of time and been dragged along towards the continents for those many years. They are not full. (32) Actually many of them are empty with almost no material in them at all, which makes them quite easy to identify. In particular the Kermadee Trench north of New Zealand, the Chile Trench, the Middle America Trench and the Tonga Trench are rock bare. (82) Most ocean floor has a sediment layer but most of the trenches are virtually empty. Even though a lot of ocean floor is almost void of sediment in many places, (which itself is inexplicable) if the trenches indicate where ocean floor is being subducted, they should be full of sediment. Being empty, they look like they formed quite recently by a rapid displacement followed by a slumping back. (48) A sudden one-time force could account for their appearance. However, without being full of ocean floor sediment, the ocean trenches testify to an event that happened recently and contradict the expectation that they have traveled one-half way across the ocean over long eons of time. The theory is therefore in direct violation of the evidence.

2.5.8 Missing Trenches

Alternately, if the ocean floor is expanding towards the continents, and the continents do not move apart, an ocean trench must be found next to the continents where the ocean floor is being consumed. However, neither Africa nor most of the rest of the rim of the Atlantic Ocean, have any adjacent trenches. The theory is therefore in further violation of the available evidence.

2.5.9 Magma Pressure Problem

It is asserted that molten fluid magma is oozing up through a fault and forcing the ocean floor on either side to move away. 'The Mid-Atlantic Ridge is being built by magma oozing between two plates, forcing them apart and continuing the separation of North America and Africa at the speed of a growing fingernail.'(64) As mentioned above, the force which would be required to cause the plates to separate would be very very large. How much force does it take to move a continent? While it would be difficult to specifically identify the magnitude of such a force, suggesting that the required pressure must be hundreds of thousands or even millions of pounds per square inch would be quite realistic. This high pressure magma is supposedly being contained in a fault between two slabs of the Earth's crust without blowing out into the ocean and causing continual volcanic activity. Further, these extreme pressures would hardly be confined to the fissure between the separating plates, as if they were in some kind of pressure vessel but they would also exist in all of the magma in the region. The pressure required to separate the ocean floor would be many orders of magnitude higher than the pressure that magma would normally develop due to its depth below the surface. If somehow the magma pressure became much higher than normal, basic physics predicts that the entire ocean floor would be pushed up until it was well above sea level. It is inexplicable how the weight of the ocean floor is keeping such pressure from elevating the whole ocean. It is inexplicable how the fault is containing the extreme pressure without allowing it to escape at high speed up through the ocean and into the air, and it is also inexplicable how such enormous pressure is being generated. Therefore the above statement explaining how ocean plates are being separated stands in total disregard for well-known basic physics. Consequently, the above quotation is neither a statement of fact nor a statement with any validity. It is rather a tale that has just been made up.

2.5.10 The Hot Spot Dilemma

A hotspot is a location on the Earth where volcanic activity is happening remote from faults where two crustal plates meet. One expert has defined a hotspot as 'persistent volcanism in a location that is relatively independent of plate motions and moves only slowly relative to other hotspots, often with an associated topographic swell. (118) It has been estimated that there are between 40 and 100 hotspots on the Earth. The wide spread in this estimate is due to disagreement among the experts as to exactly what constitutes a hotspot. However, there does seem to be agreement that the three best known are; Iceland, Yellowstone and Hawaii. Of particular interest at the present time is the fact that a new volcanic hotspot, named Loihi, is developing south of the main island of Hawaii.

One of the most important concepts in Plate Tectonics Theory is that hotspots provide a record of actual plate movement. By measuring the distance between rocks along the hotspot trail, as well as the ages of these rocks, a record of plate movement over many millions of years should be provided. The following quotation focuses on the Hawaiian Islands as evidence of plate movement. 'Each of the Hawaiian Islands formed over a hotspot. A hotspot is the result of a persistent region of molten or melted rock known as magma. There are at least 50 hotspots worldwide and they occur at several places beneath the moving plates of the Earth's crust. The Earth's crust is divided into a dozen or so moving plates, called tectonic plates. The plate that is over the Hawaiian hotspot is called the Pacific plate. These tectonic plates are adrift on the molten magma beneath. The Pacific Plate is currently moving northwest about 3.5 inches per year. Thus when a volcano forms over the Hawaiian hotspot, it is eventually pulled northwest and becomes extinct. In this way a chain of volcanoes is produced.' (119) The Hawaii Center for Volcanology offers a similar scenario. 'The Hawaiian Islands are volcanic in origin. Each island is made up of at least one primary volcano…..In fact even beyond Kure the Hawaiian chain continues as a series of now-submerged former islands known as the Emperor seamounts.' (120) The Kure Atoll is located about 1500 miles northwest of the big island of Hawaii and is the furthermost feature included as part of the Hawaiian Islands. While the Hawaiian Island chain runs generally northwesterly, the Emperor Seamount chain continues from the end of the Hawaiian Island chain and runs almost strait north for another 2500 miles to within 600 miles of the Kamchatka Peninsula of Russia.

The claim being made for the Plate Tectonics Theory is therefore that the entire Pacific Plate moved strait north over a hotspot for a distance of about 2500 miles. During that time, seafloor was being consumed by subduction zones to the north of the plate, as the plate was sliding along other plates on both sides. At the south end of the Pacific plate, new seafloor was being produced. Then the entire Pacific Plate changed direction. Instead of moving strait north, it moved northwest. A reason for the change in direction is not given. Subduction must then have commenced to the northwest of the plate while all along the trailing edge new seafloor was produced. However, the Pacific plate is completely surrounded by subduction zones which are referred to as the Pacific 'ring of fire'. (119) The theory is therefore internally inconsistent and does not agree with itself.

The force required to move the entire Pacific Plate would be incredibly high. A plate that large could not be pushed because it would just buckle up. Further, whether the motion was north or northwest, the perimeters of the plate (which would be the sides in either case), are far from being strait. One therefore wonders how the edges of the plate slid in and out of the various recesses along the way. Then there is the problem of magma pressure. There is no pressure of magma that could conceivably have been nearly great enough to move the plate without simply blowing out and no mechanism has ever been identified that could produce the required incredibly high pressures. In other words the entire idea is totally lacking of any realistic sense of basic physics and consequently isn't credible. Recently, more advanced study of the Hawaiian hotspot suggests that the hotspot may be moving and not the seafloor. If this is the case, the entire Plate Tectonics Theory would be rendered moot. (121)

2.5.11   Sulfide Problem

'A question that does not appear to have been asked by researchers is; if the ocean crust were formed by conveyor belt type spreading away from the ridge crest, why is not the ocean floor as enriched in sulfides as the ridge crests? If the sea floor spreading really has been occurring at a slow rate of cm/yr over millions of years, the surface of the ocean crust should be dotted with sulfide chimneys, sulfide mounds, and fossil vent communities, under a blanket of sediment of course.' (140)

2.5.12   Magnetic Anomalies

Magnetic reversals of the ocean floor have been used as evidence of seafloor spreading. However, this information is not as supportive as its proponents declare. 'It is not true that linear magnetic anomalies can be correlated from the North Atlantic via the Indian Ocean to the northeastern Pacific. The magnetic signatures are very similar in limited areas, but are very different among different areas. Moreover, magnetic stripes need not be caused solely by alternate bands of 'normal' and 'reversed' polarization, differences in magnetic susceptibility values of adjacent rock types can produce the same. ... The so-called magnetic anomalies are not what they are purported to be – a 'taped record' of magnetic events during the creation of new ocean floor between continents.' (143) Unfortunately the rocks include erratic magnetic patterns. It is known that 'lightning can magnetize rocks. ...It is probable that much of the scatter observed is the result of lightning strikes.' (144) (While it is clear that a lightning strike cannot affect the floor of the ocean, the fact that a surge of current can magnetize a rock leads one to wonder what other natural phenomenon could do the same.)

2.6   Catastrophic Plate Tectonics

The Theory of Catastrophic Plate Tectonics (CPT) is a modified form of the Theory of Plate Tectonics, (78) which declares that an ancient massive supercontinent, Pangea, broke apart but instead of requiring millions of years, the time frame for continent separation was very short. (65) The great super-continent split and moved apart to form the Atlantic Ocean in a matter of weeks instead of hundreds of thousands of years. This is basically an effort to reconcile a basic belief in the widely accepted Theory of Plate Tectonics with the much shorter times reported in the Genesis account of the great flood.

This theory is dependent on the notion of runaway subduction, which depends on the idea that the ocean floor all along its perimeter with the continents, split from the continents and suddenly started to sink. 'The flood was initiated as slabs of oceanic floor broke loose and subducted along thousands of kilometers of pre-Flood continental margins.' (78) As the ocean floor at these coastal locations dove down into the molten interior, it dragged the rest of the ocean floor along behind it. After the entire ocean floor had been dragged under and new ocean floor was in place, the movement stopped. 'When virtually all the pre-Flood oceanic floor had been replaced with new, less-dense, less-subductable rock, rapid plate motion ceased.' (80) However, in order for ocean floor to dive into the interior, it would have been necessary for the original ocean floor to have been heavier than the material underneath. If this was the case, once movement started, it would have continued, powered by gravity until the floor of the whole ocean had been replaced with fresh material from the molten interior. As the old floor slid away, it exposed the hot interior which rose, cooled, hardened and became new ocean floor.

2.6.1   Crustal Density

In order for the CPT theory to be true, heavy material (i.e. old ocean floor) had to have floated on top of lighter material further down, for an extended unknown period. This, however, is a violation of basic well-known physics from which we understand that light material floats on heavy material. The declaration that heavy material floated on top is therefore an artificial construct.

2.6.2   Movement Initiation

The theory declares that the ocean crust, where it met the continents, started to sink. No reason is given. Supposedly after an unknown period of years during which the ocean floor was stable, suddenly it started to sink along the entire perimeter of the supercontinent. Why did the ocean floor fracture away from the supercontinent? If the entire ocean floor was heavier than the material beneath, why did the sinking commence and sustain only along the continental boundary? Since no reasons are given, the theory is incomplete and hence not comprehensive.

2.6.3   Continent Separation: Cause not Given

The theory declares that the great super-continent, Pangea split and separated to form the Atlantic Ocean. There isn't any reason given for the split. Therefore the theory is incomplete on this point as well.

2.6.4   Atlantic Ocean Formation

Prior to the breakup of Pangea, there wasn't any Atlantic Ocean or Atlantic Ocean floor. Consequently, the declaration that the original ocean floor sank and new floor was formed, is not applicable to the Atlantic Ocean. It may therefore be concluded that the theory is internally inconsistent. A procedure is offered, which does not apply. Further, as the Atlantic Ocean floor was forming and increasing in size, the Pacific Ocean floor must have been reducing in size. Was this the part that sank? Or did the entire original Pacific Ocean floor sink and new floor form over what was left (i.e. the present Pacific Ocean)? Or did it only sink exactly enough to make room for the Atlantic Ocean to appear? Possibly the reason the process terminated prior to replacement of the entire Pacific Ocean floor was because part of the Pacific Ocean floor was light enough to remain on top. However, there is no doubt in the proponents mind that the entire present ocean floor is new and was formed in a short period involving only days, 'I consider …the present ocean basement to be no older than the Mesozoic portion of the continental fossil record. This requires the entire pre-Flood ocean floor, as well as any generated when the Paleozoic fossils were being deposited, to have vanished from the Earth's surface.' (79)

While this type of scenario is unbelievable enough, the actual scenario envisioned is even worse. The hypothesized ancient landmass named Pangea was thought to include all of the continents of the entire Earth. (Pangea actually means 'all lands'.) This massive continent was surrounded by one massive ocean to which the name Panthalassa was given. (45) Pangea had a very irregular shape which would be similar to crowding all of the Earth's continents together without any spaces in between. It was considered to be a single landmass. The above comment 'entire pre-Flood ocean floor' is therefore referring to Panthalassa. So the entire ocean floor under Panthalassa was spontaneously subducted. It is not mentioned where this subduction occurred and neither is it obvious with respect to the shape of Pangea. The irregular shape of the coastline would have precluded any general horizontal movement of ocean floor in any direction. Even if it had been divided into strips it would not have been able to move towards any shoreline because the Earth, after all, is a sphere not a flat plate. Widespread ocean floor movement either toward or away from any shoreline would have been impossible.

Not only was the entire Panthalassa Ocean provided with new floor but it all happened in a matter of a few days. The rate of separation is given as meters per second. (79) This means that Pangea broke up and separated into all of the continents of the world at astonishing speed and they continued to move at this high speed until they arrived where they are presently located. Then they stopped. Why? Momentum would surely have carried them further. There is never any mention of an agent causing any of the movement. Needless to say the forces required to move a continent and move it at top speed would be astronomical. What agent provided this force? Why did it suddenly start and just as suddenly stop? Such information is necessary for completion of the theory.

2.6.5 Mid-Atlantic Fault

The Atlantic Ocean floor is observed to have a fissure together with an accompanying accumulation of material on top of the basic ocean floor. This feature exists all the way from Iceland in the north right to the southern regions of the South Atlantic Ocean and it occurs right near the middle of the ocean. Why is it in the middle? The Theory of Catastrophic Plate Tectonics does not identify why this is the case.

2.6.6 Pacific Ocean Floor Fractures

The floor of the Pacific Ocean includes numerous fractures, with many in the eastern regions being parallel to each other. Most of these fractures, like the mid-Atlantic Ridge, have accompanying accumulations of material. The theory of Catastrophic Plate tectonics does not address any of these factors.

2.6.7   Subduction Questionable

Both the Theory of Plate Tectonics and the Theory of Catastrophic Plate Tectonics depend on one particular interpretation of a certain geological feature. The notion of subduction is absolutely necessary for both theories. The long, narrow, deep valleys found at numerous locations in the ocean floor must be subduction zones in order for either theory to have any validity at all. However, it is not known for certain that these particular valleys are indicative of subduction taking place. If they represent ocean floor that has travelled one-half way across the ocean, at the very minimum, they should be full of the sediment which had been gathered along the way. They are not full. Indeed, if they were full, they would be harder to locate and would only be found with the aid of instruments. As it is, they have no more sediment in them than may be found on the ocean floor much further from shore. On the other hand, next to the mid-Atlantic Ridge there is plenty of sediment. Why isn't there any in the trenches? Occasionally, it has been suggested that material is scraping off the ocean floor as it slides underneath of a continent. 'During collisions of oceanic and continental crust, or oceanic and oceanic crust, some of the sediment and rock on the down-going slab may be scraped off and pasted onto the overriding plate. The island of Barbados is built on a wedge of material scraped off the Caribbean plate as it dives beneath the South American plate.' (81) Why is this happening only at Barbados? If a great slab of rock several miles thick was forced to bend and scrape down under another slab of rock of even greater thickness, a great deal of resistance would develop at the interface of the two slabs. One would certainly expect a portion of each slab to be damaged and material to accumulate. The force required to push and bend one slab of rock down under another would reasonably be expected to result in damage to the slabs. Why doesn't this type of development occur at every crustal interface? Are continental plates nicely tapered at the edges to allow ocean plates to slide under? If they aren't, the ocean plate would have to bend down quite sharply. This would require more pushing force. If it could not bend down enough, it would break and if the pushing kept happening, there would be a great accumulation at the interface. Islands and mountains would be expected. Trenches would not be expected. At the very minimum, subduction trenches should be full of material acquired on the way but more reasonably they should be full of the debris from the colliding activity. Since the expected material is completely missing, it may be concluded that trenches do not represent subduction activity at all but some other type of event. The notion of subduction was actually an invention in the first place. 'The concept of subduction in the framework of plate tectonics was introduced more or less as a logical consequence of sea floor spreading to keep the surface area of the Earth constant.' (110) In other words, subduction was needed so it was invented.

2.6.8   Magnetic Reversals

It has occasionally been suggested that the only explanation for magnetic reversals in the ocean floor is Catastrophic Plate tectonics. However, there are other explanations for the magnetism found in rocks. As mentioned in par. 2.5.3 above, the magnetism of rocks is often many times stronger than the magnetism of the Earth. These stronger levels may be explained by magnetostriction. 'A mechanism for changing the direction of magnetism in magnetized igneous rocks is magnetostriction. This is caused by distortions of the rock due to shock or stress. Since there has been considerable earth movement in the past, magnetostriction may possibly explain many reversed rocks. Vestine says; "Perhaps some of the interesting magnetostriction of rocks stem from the effect of geologic stresses, rather than from changes in the geologic field." …Shock induced NRM (natural remanent magnetism) from meteorite impact has been invoked to explain NRM in Precambrian rocks from Lake Superior. Shock induced magnetization has even been demonstrated in the laboratory by simulating meteorite bombardments.' (137)

'Magnetostriction can cause "remanent magnetizations of rocks to be in directions that are not those of the fields acting when remanent magnetization was originally acquired". …great upheavals and folds in the Earth beneath the oceans must certainly have yielded profound magnetostrictive effects, altering the orientation of the magnetization in the rocks.' (145)

There is no doubt that the Earth has experienced many asteroid impacts. The shock waves propagating from all of these impacts as well as from all of the major Earth faults (as they formed) would have been sufficient to cause magnetism of varying orientations at many places around the world. Such developments would have occurred whether the Earth had a magnetic field or not.

2.7   Evolution

The General Theory of Evolution is widely accepted and is defined as 'the theory that all the living forms in the world have arisen from a single source which itself came from an inorganic form.' (169) Since the theory embraces all living things and includes everything which is alive now as well as everything which lived in the past, the theory may be considered comprehensive. This is the basic reason why it is so widely embraced and why it has survived for so long.

2.7.1   Fossil Evidence Contradictory

The theory declares that the earliest forms of life were very simple and that over great periods of time they became more complex. Since these things took place in the distant past, the only place to obtain evidence is from fossils. 'Direct study of macro-evolutionary patterns is only possible in the fossil record, ... and this fossil record forms our most direct evidence... of the course of biological evolution' (50) However the fossil record contradicts the assertion of gradual progression from one type to another. Instead, numerous different types of complex life forms appear 'suddenly' without any predecessors. In particular consider the "Cambrian Explosion". 'Prior to this time (i.e. the Cambrian) the only forms of life that show us fossils are simple single-celled types such as algae and bacteria. Then suddenly a gorgeous array of shelly invertebrates appeared (as paleontologist Niles Eldridge said). As he put it, "Indeed, the sudden appearance of a varied, well-preserved array of fossils … does pose a fascinating intellectual challenge." These include more than 5000 species, including sponges, jellyfish, corals, worms, mollusks, trilobites and crustaceans. There is no apparent explanation for this. We find traces of the algae and other simple forms that preceded these invertebrates and millions of Cambrian fossils have been found. But surely, if evolution is true, there should be a great many in-between transitional fossils.' (102) Another major discontinuity occurs with fish. 'There's another cavernous gap in the fossil record, between Cambrian invertebrates and fish. ... There are absolutely no intermediate forms found in the rocks, it's as if fish were just suddenly there, having no ancestors.' (102)   Further comment reinforces these observations. 'Yet the fossil record is characterized by systematic abrupt appearances of groups of organisms and systematic gaps between groups, rather than by a generally progressive emergence of one group from another. The attempt to explain all living forms in terms of evolution from a unique source, though a brave and valid attempt, is one that is premature and not satisfactorily supported by present-day evidence.' (50) With this evidence in mind, the declaration opening this paragraph must be set aside.

2.7.2   Required Information Not Explained

The Theory of Evolution declares that simple things became more complex but this violates basic well-known physics. Instead of becoming more complex, things wear out, rust, fall down, rot or disintegrate. This running down and cooling off trend dominates all of nature and has been mathematically described by the Second Law of Thermodynamics. In this world, things do not become more complex. When they change, they drop to a lower energy state or lose information, rather than climb to a higher energy state and gain information.

In nature, as plants grow and animals grow, some of the material of the Earth is brought into higher order. This is made possible and enabled to come about because of the information which is already resident in the cell of the plant or animal. It is this resident pre-existing information that enables the plant or animal to grow. During the growth process this resident information does not increase.

Neither do growth processes violate the laws of thermodynamics because during all transactions that involve energy, some energy is lost. The energy to carry out all growth processes comes from the Sun. The Sun provides the energy to organize some of the material of the Earth into a higher thermodynamic state (i.e. grow). During the growth process, a certain portion of the energy involved is lost. The fact that local higher thermodynamic order is achieved is sometimes identified as the means to increase the information of the universe and ultimately produce a greater variety of life forms. In other words, solar energy is credited by some commentators as being the agent that drives evolution resulting in a great diversity of life forms. This is, however, a false assertion.

'The natural economy is solar powered. Photons from the Sun rain down upon the entire daytime surface of the planet.' (164) Comments such as this are perfectly valid. However, they cannot be used as a launching point for the following. 'The whole system … is … dependent on the steady flow of energy from the Sun … Life evolves greater complexity … because of the ceaseless supply of energy from the Sun.' (165) This type of statement is gravely misleading, While the energy from the Sun is required to drive photosynthesis as well as other life processes, none of these processes result in an increase in information. It is the information already resident in the plant that directs the process to utilize the incoming solar energy and achieve the desired result (e.g. larger plants). Without the previous existence of this information, the energy from the Sun would result in disaster.

The situation could be compared to a race car speeding around a track. It is readily acknowledged that the engine is converting the chemical potential energy of the fuel into mechanical energy to move the car but what would happen if there wasn't any driver? Intelligence (i.e. information) is necessary to keep the car going in the right direction. Without it, even one lap would not be completed.

In order to bake a cake, build a house or create a rabbit, both information and energy are required. The information must be very particular to obtain the required very particular results. For example, if a cake recipe for a particular cake requires three cups of white flour, substituting whole wheat flour will not do because a different cake would result. Neither can we substitute three cups of one inch nails because then we would not have a cake at all.

The information to bake a cake is contained in the recipe and the cook's memory. The information to build a house is contained in the architect's plans and in the builder's memory. The information to create a rabbit is contained in the cell which was formed from its mother and father. It is a marvel of nature that the information in each individual cell is not only sufficient (i.e. that is along with the required energy) to create the particular entity involved but is also sufficient to create more seeds for the next generation.

When the Sun shines on a patch of ground nothing useful is produced. Even if it should shine on the ground for an extended time, information would not be created. The Sun does not create information when it shines on something.

In fact, unless solar energy is controlled and directed properly, damage to most forms of life will result. As an example, solar energy is devastating to microscopic life forms like bacteria. The situation may be likened to a hammer being brought down on a string of beads. The beads will be damaged every time the hammer comes down. In the case of bacteria, the energy in the violet and ultraviolet portion of the solar spectrum will destroy some of the molecular bonds in the bacteria and thereby render them harmless. In fact, this characteristic of ultraviolet light is commonly used to treat sewage (the treatment in this case involves shining ultraviolet light on the sewage stream to destroy the bacteria).

It is therefore not valid to claim that the energy of the Sun has produced the great variety of life forms on the Earth all by itself. The energy from the Sun never produces any new information. It always requires BOTH ENERGY AND INFORMATION to build an organized structure from the Earth's raw materials and this is true whether the structure is plant or animal. Since the Sun cannot account for the necessary information, some other explanation is required.

Further, the amount of information required is prodigious. 'The average human brain packs a hundred billion or so neurons – connected by a quadrillion 10(15) constantly changing synapses.' (182) Or consider the following quotation. 'There is enough information capacity in a single human cell to store the Encyclopaedia Britannica, all 30 volumes of it, three or four times over.' (162) (The information in the DNA molecule, which is in the cell, is sufficient to manufacture all 100,000, of a typical body's proteins, while using less than 10% of the information available. It also has the information required to make a human brain with a hundred billion neurons.) Plant cells and animal cells are no less complicated than human cells and science is totally silent concerning how such a complicated situation happened. Since there is no evidence that the Sun somehow caused the necessary information to appear, simply declaring that it did so is wishful thinking.

In the laboratory, decreasing complexity instead of increasing complexity has been observed. 'In 1997, Sol Spiegelman was working with strands of viral RNA in the lab. Even though no cells were present, the RNA fragments readily reproduced. As they did so, strands that started out 4,500 nucleotide bases long started slimming down, sloughing off parts of the genome that weren't needed. Smaller strands reproduced more rapidly and the end point was a tiny genome fifty base pairs long that out-reproduced everything else. ...In the conventional wisdom of biologists, this "survival of the smallest" is a problem for evolution because it prevents larger and more complex organisms from emerging.' (159)

It requires a lot of information to make a dog. It also requires a lot of information to make a seagull. The amount of information in a typical animal genome is about a gigabyte. When plants, insects and animals are considered the total amount of information required is several megabytes of gigabytes. The origin of all of the required information is not explained by sunshine (or natural selection as discussed below) and therefore has not been explained by promoters of the Theory of Evolution.

2.7.3   Mutations are Disadvantageous

The Theory of Evolution declares that mutations offered an advantage to the creature and thereby enabled it better able to survive. This declaration is in direct violation of the evidence. "The topic of mutations is of major concern in medicine because mutations are a major cause of disease. The vast majority of all non-neutral mutations are harmful in most environments, partly because they usually result in loss of information. ... no documented case of a mistake or error type of mutation that has produced a phenotype change that is beneficial in the wild exists. In the medical field, non-neutral mutations are assumed to be harmful without exception.' (5) Examples of harmful mutations include; cystic fibrosis, sickle cell anemia, hemophilia, muscular dystrophy, and Tay-Sachs disease. (5) When mutations occur, the creature or plant is invariably disadvantaged. It must be noted that an advantage for man is not necessarily an advantage for the plant or animal of interest. As an example, consider the Angora Rabbit. While white fur and pink eyes may be functional in captivity, an animal with these characteristics in the wild is disadvantaged. Survival is more difficult and in many cases it is not even possible. Many mutants are disadvantaged so much that they expire even before birth. 'Many of these mutations are lethal and result in a miscarriage or a spontaneous abortion.' (5) Therefore, the evidence contradicts the declaration and it must be artificially arranged in order to be supportive of the theory.

2.7.4   Mutation Accumulation Deleterious

It is a necessary condition of the Theory of Evolution that animals changed into other animals by the accumulation of mutations. However the accumulation of mutations has caused the defective gene load to increase until now an estimated 5,000 mutation-caused diseases exist in humans. As the mutational load increases, more and more mutation-caused diseases enter the human gene pool. While this unfortunate development has resulted from the marvels of modern medicine, it never-the-less illustrates that accumulating mutations are increasingly detrimental and not beneficial as the Theory of Evolution would have us believe. 'Spetner noted that: all mutations studied destroy information. None of them can serve as an example of a mutation that can lead to the large changes of macroevolution. The neo-Darwinians would like us to believe that large evolutionary changes can result from a series of small events if there are enough of them. But if these events all lose information they can't be the steps in the kind of evolution the neo-Darwinian Theory is supposed to explain, no matter how many mutations there are. Whoever thinks macroevolution can be made by mutations that lose information is like a merchant who lost a little money on every sale but thought he could make it up on volume.' (5)

In all of the literature that discusses evolution, an example of the beneficial accumulation of mutations has never been shown. It is simply an assertion. In fact, 'Random mutation … is incoherent: that is, any given evolutionary steps taken by a population of organisms is unlikely to be connected to its predecessor.' (161)

Mutations invariably represent trouble, which may even lead to an early death. What the evidence actually indicates is that accumulating mutations would not be an advantage at all. It may therefore be concluded that the assertion contradicts the evidence. This is another artificial construct because the trends, which are observed in the evidence, must be artificially arranged in order to be supportive of the theory.

2.7.5   Logical Disadvantage of Mutations

While the evidence shows that mutations are a disadvantage, logic may also be employed to understand the disadvantage factor. In order for a reptile to become a bird, a wing must be formed. A partially-formed dysfunctional limb such as a wing would be a disadvantage to the creature. What advantage would it be for a reptile to drag a partially-formed wing around for a million years while it was developing into something useful? Survival would be more difficult instead of easier. Or what would be the use of a partially formed eye? An eye is a very complex organ and it must be 100% developed and operating properly to be of any use at all. In this manner, practical reality must be contradicted or artificially arranged to be supportive of the theory.

2.7.6   Fossil Time Factors Questionable (soft tissue found)

The Theory of Evolution declares that the fossils were deposited over great spans of time. How is this known? The fact is, it is NOT known. The evidence shows that fossils are not usually laid out in layers to which times may be applied. They are often found in great jumbled heaps all mixed up together. 'Ancient' creatures are found with 'modern' creatures. Some are found nose down. Some are found nose up. Mostly they are so mixed and pressed together that even getting them apart is a major undertaking. (51) How could time factors be applied to this type of chaos?

It has been declared that the dinosaurs died 68 million years ago. (52) However, recently dinosaur fossils which include soft tissue, have been found in Montana. 'There were the remains of blood vessels as well as 'osteocytes with internal cellular contents and intact, supple filipodia that float freely in solution' (6).

Another even more interesting find was made in North Dakota in 1999. A complete Hadrosaur found in a mummified state 'fully articulated and properly proportioned with skin andsoft parts' has allowed paleontologists to understand more details about the skin patterns, muscle mass, and body proportions of Hadrosaurs. (109) In 1995 from the Dominican Republic bacterial endospores (Bacillus sphaericus), as well as some other species of bacteria, yeasts, and even fungi, were recovered from the abdomens of fossil bees enclosed in amber. Similar reports came from drilling operations in Antarctica where living bacteria were found. Bacterial endospores respire very slowly (i.e. they breath slowly) but if they completely run out of air, they die. (116) How long could the miniscule air supply for these bacteria have lasted?

Even more examples of soft tissue have been found. Recently in Alberta while a fossilized bone was being extracted the impression of skin was observed. Then the skin fell off of the bone and it turned out to be "real skin". 'The rare fossil, only the third such 3-dimensional dinosaur skin fossil ever discovered was found in an area described as a "robust bone bed."' (183) The sample was so well preserved that it could be analyzed for "proteins, sugars and fats". Another report of dinosaur skin was included in the May 2014 issue of National Geographic. This article was somewhat confusing because it only barely mentioned the skin which would normally be the find of the season but then later it conjectured that the dinosaur was probably "feathered". One would expect that if a sample of skin had been found that the question of feathers could have been answered quite definitely.

Ground is not a long-term storage medium. If soft animal remains are to be preserved, they cannot be stored directly in the ground. For example, if it is desirable to bury the remains of a person and keep these remains intact, they must be protected from the ground with a watertight enclosure. Even in dense clay there is movement of moisture. As it moves it carries with it dissolved minerals, salts and so on. As time goes on, any material which has been buried directly in the ground will gradually dissolve and be carried away. If animal remains are buried in material which is becoming rock, the mineral in the rock may gradually replace the buried material. In this manner the exact shape of the original object may be preserved and identified at a later time. This is commonly seen with trees. The exact shape of the tree may be clearly identified but the material is no longer wood but mineral (i.e. stone). This in fact is what constitutes a fossil. A shape is found but it isn't bone or wood, only mineral.

Time is an uncertain parameter in cases like this but one would never expect to find soft tissue intact after being buried in the ground for one thousand years. Soft tissue, skin and clothing will be leached away completely within a few years. Even if the specimen becomes entombed in material which will become rock, it may disappear and only leave a cavity in the rock. This was the case at Pompeii. Cavities were found and when they were filled with casting material, the shapes of people could be identified. Therefore when the difficulty in preserving animal remains by direct burial is realized, the discovery of soft tissue immediately suggests that burial was recent. How recent is impossible to determine but to suggest one hundred thousand years is a serious stretch. All of the evidence available to us shows that animal remains will not be preserved if they are buried directly in the ground. Consequently it is contrary to the available evidence to suggest that soft tissue could remain after 68 million years so it must be concluded that these remains were from animals which were alive much more recently.

2.7.7 Fossil Formation

Actually, it is very difficult to make a fossil. 'Viewing the great quantities of petrified wood and other fossils exposed at Petrified Forest may make it difficult to believe that plants and animals rarely are fossilized... Plants and animals normally decay and completely disappear relatively soon after death.' (138) After a creature dies, its remains will usually be consumed by other creatures. Even the bones will be consumed. This explains why the antlers of moose and deer are rarely found in the forest. Other creatures eat them and thereby recycle the material. Even if an animal becomes mired in the mud and dies, it will probably not become a fossil. Either it will be eaten or it will rot. Buffalo remains are seldom found on the plains of North America even though the great buffalo herd numbered forty million within the last few hundred years, none of these buffalo remains are becoming fossils because there are no buffalo remains. (146) Further, while the oceans teem with life of many different kinds, the 'ocean bottom is never found littered with dead bodies waiting to be fossilized.' (98)

'Even after fossils form, they may later be destroyed by natural events or they may simply be deeply buried and remain unknown to man.' (138)

It is a necessary condition of the Theory of Evolution that fossils will form from most creatures after they die. As the years go by, and creatures continue to die, more and more fossils would be formed. In this way fossils would represent a continuous record of life on Earth over the long ages of the past. In order for this to happen, the animals must basically be allowed to live their normal lives and expire in a reasonably expectable manner. Unfortunately, the evidence contradicts this expectation.

In order for fossils to be formed from plant or animal remains, the material must be immediately isolated from microbial activity (which is enabled by the presence of air). 'Before fossilization can occur, the evidence must be buried naturally and rapidly, thus preventing weathering and decay.' (139) It is not known how long the fossilization process takes. (Partial fossilization of trees, believed to have been buried for only a few hundred years, has been observed. (157)) Whether it takes many years or just a few hours, is of secondary importance to the fact, that a creature must be isolated from the atmosphere, quickly and completely, in order for there to be any hope of it becoming a fossil. With this knowledge in mind and without any further understanding being necessary, it is immediately appropriate to conclude that the extensive fossil record is indicative of sudden isolation from the atmosphere. However, this violates the above-mentioned expectation that the record shows animals that had lived a reasonably normal life and simply died in due course. While rapid burial is necessary, it seriously violates a basic requirement of the theory. Also, many fossils bear direct testimony to their destruction being violent as well as sudden. The following examples from numerous places around the world all indicate that both suddenness and violence were involved.

First we consider the evidence from the islands north of Russia. 'Other hills on the same island, (i.e. New Siberian Island) and on Kotelnoi, which lies further to the west, are heaped up to an equal height with skeletons of pachyderms (elephants, rhinoceroses), bisons, etc.. On Moloi, one of the group of Liakhov

Islands, Toll (Eduard von) found bones of mammoths and other animals together with the trunks of fossil trees.' (74) In order to account for the catastrophic arrangement of the trees and animals, Velikovsky offered 'A hurricane, apparently, uprooted the trees of Siberia and flung them to the extreme north.' (74)

The caves of England have yielded the following assortment of creatures. 'In a cave in Kirkdale, eighty feet above the valley, ...the bones and teeth of elephants, (an extinct variety) rhinoceroses, hippopotami, horses, deer, tigers (the teeth of which were larger than those of the largest lion or Bengal tiger), bears, wolves, hyenas, foxes, hares, rabbits, ...ravens, pigeons, larks, snipe, and ducks' were found. 'Reindeer and grizzly bear lived with the hippopotamus at Cefn in Wales. Lemming and reindeer bones were found together with bones of the cave lion and hyena at Bleadon in Somerset.' (73) The unusual assortment of animals which would normally only be expected from very diverse climatological areas speaks of violence being involved in their assemblage into the same cave.

In the Old Red Sandstone in Scotland, there is 'a wonderful record of violent death falling at once, not on a few individuals but on whole tribes. ...The animals are in disturbed positions. ...which exhibit unequivocally the marks of violent death. The figures are contorted, contracted, curved; the tail in many instances is bent around to the head; the spines stick out; the fins are spread to the full as in fish that die in convulsions. ...attitudes of fear, anger and pain.' (72)

At Rancho La Brea near Los Angeles, 'bones of extinct animals and of still living species are found in abundance...most remarkable mass of skeletal material...a bed of bones was encountered in which the number of saber-tooth and wolf skulls together averaged twenty per cubic yard. ... among other animals unearthed in this pit were bison, horses, camels, sloths, mammoths, mastodons, and...birds, including peacocks.' The majority of animals found were carnivorous whereas in a normal population herbivorous species would dominate the numbers. Further evidence of violence follows. 'Contrary to expectations, connected skeletons are not common. The bones are 'splendidly' preserved in the asphalt, but they are broken, mashed, contorted and mixed in a most heterogeneous mass, such as could never have resulted from the chance trapping of a few stragglers.' (75)

In Agate Spring Quarry is a fossil-bearing deposit up to twenty inches thick. 'The fossils are in such remarkable profusion of interlacing bones, very few of which are in their natural articulation with one another...block contains about 100 bones to the square foot. There is no way of explaining such an aggregation of fossils as a natural death retreat of animals of various genera.' (76)

2.7.8   Chirality

The Greek word for hand is cheir, thus chiral simply means handed. Biological molecules can exist in either left-handed mode or right-handed mode. However, life only uses the left-handed type. All of the amino acids (the small molecules that form protein) in every form of life on the Earth, are the left-handed type. If, for any reason whatsoever, a right-handed molecule should be included or if a left-handed molecule spontaneously flips into a right-handed one, the protein would be useless. If this should happen on a wider scale, the host cell would die.

The initial activity in the formation of a protein is the proper coupling together of all of the amino acids that the protein requires. Then the protein must be folded up in a very particular manner so exactly the correct atoms appear on the surface. If the folding activity does not proceed properly, the protein will be of no use. This, unfortunately, is what happens with the so-called mad-cow disease. All of the required parts of the protein are in their respectively-correct positions but the folding is wrong so the final product is detrimental to its host. If improper folding keeps happening, the cow will die. This malady is also called spongy brain disease because the brain will form openings and appear like a sponge. (130)

Unfortunately, the state of handedness does not have long-term stability. Certain amino acids spontaneously flip from left-handed to right-handed. Any proteins in which this happens must be removed from the cell. Failure to attend to keeping the cell free of useless proteins will lead to the cell becoming useless. This appears to be what happens with Alzheimer's patients. A residue of useless protein called B-amyloid protein builds up in the cell and forms a mass of tar which will eventually kill the cell. (131)

The problem with all of this from an evolutionary point of view is that the lack of long-term amino-acid stability means that there would never have been enough time for any complexity to evolve. An animal body is incredibly complicated. In particular the inner workings of each individual cell are continually demanding the best minds we have to identify it's various operating procedures. If the basic building blocks of proteins at work in the cell had not been stable long enough in the first place for these procedures to be set up, life would never have evolved. How long would it have taken for the primordial cell to evolve – a million years? 'This is significant because one out of 1200 aspartic acid molecules spontaneously flips chirality every year.' (131) If the cell was not already organized and able to deal with such developments, it will die. With Alzheimer's disease, both aspartic acid and serine form and the cell dies. (131)

Without offering a solution to the problem of spontaneously-flipping amino acid molecules, these most basic of life's building blocks, will, rather than provide basic structures for the cell, only provide useless material which a normal, operating, healthy cell would simply discard. Since the Theory of Evolution has not addressed this problem, it is incomplete.

2.7.9   The Mass Extinction Problem

Whenever a large asteroid hits the Earth, mass extinction follows. The numerous forms of devastation resulting from the impact of the asteroid with the Earth are completely overpowering to virtually all forms of life. The atmospheric shock wave would instantly kill entire herds of animals, such as the Woolly Mammoth all of which would then be completely swept away. The gigantic waves spreading out from both the impact site as well as any coastlines within several thousand miles of ground zero would wash completely over continents. The arrival of erratic boulders would ensure that any animals left standing would be killed. All of these miseries are discussed in detail in later chapters as well as the post-impact developments which include complete world-wide darkness and the death that this would cause. If the plants do not have light they die. If the animals do not have plants to eat, they die.

In spite of the universal nature of these happenings, most commentators still insist that a few animals would survive with preference being given to very small ones. For example it is repeatedly stated that all of the dinosaurs were killed by the Chicxulub impact event. Not only all of the dinosaurs but also seventy to ninety percent of everything else. Such comments might be credible except that the Earth has experienced numerous impacts, all of which were capable of achieving similar results. This presents a serious problem for The Theory of Evolution because a diverse cohort of life forms must evolve after every impact. If the first major impact resulted in seventy percent extinction, the remaining thirty percent would be the starting point for the next phase of evolution. However the Earth has been hit more than one hundred times. Many more hits are suspected but not yet discovered because the Earth is seventy percent covered by water. This means that after every hit, evolution had to start all over again but each time it would have to start with a different cohort (which was always dominated by small animals). It also means that the present cohort has evolved from the animals which were left after the last major impact. Such a scenario is not credible. Neither is it explained which leaves the Theory of Evolution incomplete.

2.7.10 Natural Selection

Natural selection is offered as the mechanism that drives evolutionary development. Natural selection is an improbability pump It systematically seizes the minority of random changes that have what it takes to survive, and accumulates them. The improbability pump of natural selection is a kind of statistical equivalent of the Sun's energy raising water to the top of a conventional mountain. Life evolves greater complexity only because natural selection drives it …'(160)

'Natural selection' and 'artificial selection' are phrases used when referring respectively to nature-produced and human-produced changes in a plant or animal life. The natural selection phrase is used when the change only involves nature and nature is somehow bringing about the change. The artificial selection phrase is used when man intentionally makes the change.

The classic example of artificial selection is dog breeding. All dogs have been bred for some particular characteristic that somebody wanted. Hounds with very short legs would be useful for going down animal holes in the ground. Very small dogs would be suitable for household pets. Every breed of dog has come about because somebody desired a particular characteristic. A similar situation has occurred with plants. If a desirable characteristic is noticed by a plant breeder, he may keep breeding the plant to emphasize the characteristic of interest. For example, strawberries have been produced from a wild variety which was much smaller. Similarly, certain varieties of corn have been cultivated for sweetness whereas the original plants were not sweet at all.

It has been declared that nature operates the same way through natural selection wherein some entity of nature "breeds" for some desirable characteristic in another entity of nature. For example, Richard Dawkins declares; 'Insects breed flowers to be bright and showy …Insects made roses pleasantly fragrant …The insects, by choosing the most attractive flowers to visit, inadvertently "breed for" floral beauty.

...Insects have ... bred (flowers) for increased yield (i.e. nectar) ... Bright colors ... attract female pheasants. The same thing happened with peahens selectively breeding peacocks.' (163)

Even fish have gotten into the act. 'Endler ... noticed ... the adult males (guppies) were rainbow-colored (and) surmised that their ancestors had been selected for their bright colors by female guppies, in the same manner as cock pheasants are selected by hens.' (167) He noticed that 'where the males were less bright were also the streams where predation was heavy. In streams with only weak predation, males were more brightly colored ... here the males were able to evolve bright colors to appeal to females.' (167)

Examples such as these have been offered to show that natural selection is continually operating throughout nature to direct the breeding process. From these observations, a great leap is made by declaring that this mysterious mechanism of natural selection has resulted in the great diversity of plant and animal life on the Earth. 'The difference between life and non-life is a matter not of substance but of information. ... (Through natural selection) the information content builds up over geological time.... All living creatures on this planet are descended from a single ancestor.' (165)

While the variation in the coloring of guppies is interesting, it is unjustified to leap to the above conclusion. One of the basic declarations of evolution is that 'The first limbed land dwellers, amphibians, evolved from fishes in the Devonian period about 340 million years ago.' (168) Since natural selection (and hence evolution) is currently in operation, we should expect to find some evidence to support such a basic declaration. For example, if some of the guppies had been observed with frog's legs protruding from their sides, there would be something to get excited about. Unfortunately, even though countless millions of fish have been caught – even since the time of Darwin, - not one has shown any evidence of legs appearing! The leap from observable natural selection to "every living creature" is therefore not substantiated.

Even more critical however is the fact that natural selection requires pre-existing life. The evolutionist Theodosius Dobzhansky recognized this problem clearly and stated that he was afraid that not all authors have used the term (natural selection) carefully. Natural selection is organism reproduction. In order to have natural selection, you have to have self-reproduction or self-replication. Therefore prebiological natural selection is a contradiction of terms. (170)

2.7.11   The Start-up Problem

It all had to begin. However, there has never been any explanation of how this happened. There is no evidence supporting the great event that was the start of evolution. The truth is that there is no evidence. The first step in making life is missing. (160) In fact, in recognition of the great difficulty in postulating a start-up scenario, a prize of $1,000,000 has been offered to anyone who can outline a plausible solution to the problem. 'The Origin of Life Prize (hereinafter called the Prize) will be awarded for proposing a highly plausible mechanism for the spontaneous rise of genetic instructions in nature sufficient to give rise to life. To win, the explanation must be consistent with empirical, biochemical, kinetic, and thermodynamic concepts as further delineated herein, and be published in a well-respected peer-reviewed science journal(s).' (171) Unfortunately without an explanation of how it all began, the Theory of Evolution is incomplete.(To date no one, including Dawkins has applied for this prize.)

2.7.12　The Requirement for Time

The requirement for long periods of time does not disappear with the formation of the primordial cell. It is thought that enormous amounts of time would be required to proceed from a single cell to the first creature that could be identified as the ancestor of all of the creatures of the Earth. If the Earth was the initial home for life, the time between the origin of the Earth and the existence of the common ancestor of all creatures, appears very short. In view of the great and necessary complexity of the 'common ancestor', it would be reasonable to allow 4 billion years for its development from the initial cell. (172)

2.7.13　Evolutionary Offerings of Evidence

While it is herein being pointed out, that there is no evidence at all for the theory of evolution, it would be fair to mention that over the years several items have been offered as evidence.

2.7.13.1　Missing Links

Several times since the Theory of Evolution was outlined by Charles Darwin, it has been recognized that intermediate life forms would be required to substantiate the theory. In fact Darwin himself recognized the need but anticipated that as more fossils were found, the evidence would be forthcoming. 'The evolution model predicts that birds have evolved either directly from dinosaurs or from crocodile-like creatures that were the ancestors of both dinosaurs and the birds. For such a significant transformation to happen there would have had to be millions and millions of intermediates. As Darwin wrote "…the number of intermediate varieties, which have formerly existed, must be truly enormous (Darwin, 1967, p 292). Darwin's explanation for the lack of transitional fossils known during his time was the imperfection of the geological record. But if it was a valid argument at that time, this is no longer the case. As Gish (1993, p 111-12) stated; "In the natural history museums of the world are to be found more than 250,000 different fossil species, represented by tens of millions of catalogued fossils. These have been taken from every one of the so-called geological periods. Thus, the fossil record is almost immeasurably rich. An appeal to the "poverty of the fossil record" is no longer available.' (104) However, over the years since the theory was introduced, from time-to-time "missing link" creatures have been promoted and in no case – not even in the case of man - has more than a bare minimum of candidates been advanced as intermediate between two other completely different types of creatures.

While it may be claimed that all creatures have a great amount of basic life material - including DNA - in common, it is also true that the collective number of incremental differences between any two creatures, from different species, is incredibly huge. As well as DNA there are, in every living creature, a large number of unique proteins. If only the protein and DNA differences were counted, the number of differences between similar creatures would still be incredibly high. 'The DNA of a person and a chimpanzee, for example, are about 99 percent identical – meaning that … less than one percent of the three billion letters in the human genome have changed.' (141) According to this comment there have been thirty million changes in DNA to evolve from a chimp to a human. It follows that any theory, which is attempting to show progression from one type of creature to another, should provide, as evidence, some reasonable fraction of the intermediate possibilities. In such a case, would it be unreasonable to expect a thousand intermediate species or variations as evidence to substantiate such a hypothetical change? Further, a hypothetical mutational sequence should be identified as a possible pathway from one type of creature to another. To use an interesting fossil as evidence for macro-evolution, should one not have a reasonable, detailed explanation on how it could have evolved, tiny step by tiny step, by the mechanism of natural selection and from which ancestors?

While there is a theoretical difficulty with the 1% claim, there is also a factual difficulty. However, the chimpanzee genome is 11.5% larger than the human genome (CSAC, 2005). Then how could there be only a 1% difference between the two genomes if the chimpanzee genome contains almost 12% more DNA? The answer is there cannot. The 12% difference is simply not counted, almost as if that extra DNA of the chimpanzee genome is unimportant (I am not sure the chimpanzee would agree). What is more, the claimed 98-99% similarity of human and chimpanzee genome is based on a comparison of less than 80% of the respective genomes (Britten, 2002, CSAC, 2005). And of this 80%, the comparison really only involves stretches of DNA where the human and chimp genome can be directly aligned.' (156) In engineering terminology this would be called 'cooking the books'.

The ancient mammoth and the modern elephant are very similar animals. One would therefore expect that there would be great similarity at the genetic level and indeed this is the case. However similarity still involves a large number of differences. In pursuit of ideas to recreate ancient creatures one research team suggested the following. 'Transforming this data (DNA from mammoth hair) into a woolly mammoth will be far trickier, though the existence of close living relatives, the African and Asian elephants, helps…..one way to make living mammoth genetic material might be to modify elephant chromosomes at each of the estimated 400,000 sites where they differ from the mammoth's…' (129)

Never-the-less hope springs eternal. 'By making a few genetic tweaks to its modern-day descendant, the bird, he wants to hatch a dinosaur straight from a chicken egg. By tripping the right genetic switches at the right time, Larsson should be able to build a dinosaur inside a chicken egg.' (141) While this incredible feat is never going to happen, if it did, it wouldn't provide evidence that something similar actually has happened. Without evidence, such comments are not compelling.

Without showing some reasonable number of intermediate forms, in light of the magnitude of the number of changes which are required, it is improper to offer evolution as a theory. If a creature is offered as an intermediate or missing link, and there are a hundred thousand distinctive factors between it and the creatures which are supposedly on either side of it, suggesting that it is evidence of progression from one to the other is not substantiated and therefore not properly offered as support for a theory. One is inclined to ask; where is the abundance of fossils for these missing links? In particular, in the horse sequence discussed below, it would be reasonable to expect some progression from the four-toed 'eohippus' to the single-hoofed small horse offered as the next stage of evolution. How many mutations does it take to change from four toes to one hoof? Several examples of suggestions for missing links have been put forward over the years and the more popular examples are listed here.

2.7.13.1.1 Piltdown Man.

During the period from 1908 to 1912, parts of a skull, most of a jaw and a tooth were found in England. The discovery was 'validated as Eoanthropus Dawsoni - meaning Dawson's dawn man (in honor of the discoverer, Charles Dawson) by several of Britain's leading scientists, including …Graham Elliot Smith, …Arthur Keith…and …Arthur Smith Woodward.' (53) All of these gentlemen were knighted within the next few years. The New York Times even joined in the revelry "Darwin Theory Proved True." (53) Objections at the time were suppressed. However, in 1953, a newly-developed fluorine test was carried out. This revealed that the bones were of recent origin. Then it seemed quite obvious that the teeth had been filed down and that the bones had been stained to make them appear old. (90) The jaw had always appeared too large for the pieces of skull and it was finally concluded that it was an orangutan jawbone planted with some human skull fragments. The Piltdown hoax was so clearly exposed that Piltdown man is, unlike Java man and Peking man, very rarely mentioned any more

2.7.13.1.2 Nebraska Man.

In Nebraska in 1922, a tooth was discovered and declared to be from a very primitive pre-historic ancestor of modern man.'… Osborn…. was shown a single tooth found in Nebraska…declared it belonged to an early ape-man… became known as Nebraska Man.' (54) Reinforcements soon arrived. '…Gregory and …Hellman, specialists in teeth, said after careful study that the tooth was from a species closer to man than ape. Wilder, a zoology professor…wrote; "Judging from the tooth alone the animal seems to be halfway between …Java Man and …Neanderthal…" Graham Elliot Smith convinced the …news to publish an artist's rendering…showed two brutish, naked ape-persons…" All of this came from one tooth! However, further excavations revealed that the tooth belonged to '…neither ape nor man, but to a peccary, a close relative of the pig.' (54)

2.7.13.1.3 Lucy.

In this case, a skeleton of a supposed ancestor of modern man was presented. It was a great find and much press was consequently obtained. However, things were not exactly as they were portrayed. In order to make this "discovery", a great number of bone fragments were gathered from a large area. Then they were placed on a table and pieces were pulled from the numerous possibilities and arranged into an animal shape. However even with the large selection of bone fragments to choose from only a little more than one-half of a skeleton could be assembled. Some estimates declare that there was barely enough to form 40% of a skeleton. (55)

Only one hip bone was found. Unfortunately, it was more ape-like than man-like and indicated that the creature walked on all fours most of the time. This was unacceptable and so the casts of the hipbone were modified to show bipedality (two-leg walking) instead. 'Dr Own Lovejoy (the primary anatomist involved in Lucy's reconstruction) …described how he proceeded to reshape a cast of the pelvis to give it a human anatomy. His conclusion was that this gave it a more accurate reconstruction. This reshaped pelvis was subsequently incorporated into the "official" Lucy reconstruction. …many of these reconstructions have (also) given her a white sclera (the white of the eye). This is almost certainly erroneous since only humans, not primates, have white sclera. Lucy clearly has no fossilized eyes to determine her sclera color.' (155)

2.7.13.1.4   Java Man

While exploring in the East Indies for evidence of a link between some primitive form of life and modern man, a person named Eugene Dubois, a student of Ernst Haeckel (see discussion below) in the late eighteen hundreds, recovered several body parts. A tooth was 'found in Sept. 1891, a skullcap in October, a thigh bone the following August, and another tooth found in October'. (85) He assumed that these bones belonged to the same creature and that they therefore constituted the looked-for missing link. It was declared that a missing link had been found and "Java Man" was recognized as a missing link for many years. In fact it received recognition until at least 1985. 'In the late 1880's… a Dutch doctor, Eugene Dubois, burned with a desire to find the missing link. With 50 convict laborers… in Java… in 1891… He called the creature Pithecanthropus erectus, "upright ape-man." To the public it would be known as Java man.' (56)

While the report was treated with considerable skepticism at the time, things only degenerated further as time went by. Apparently, Dubois was not present when the various finds were made. Then it became evident that two fully human skulls had been found in similar geological strata at Wadjak which is about 60 miles from the other finds. This was not reported until 1920 as it would have detracted from the original declaration but detraction occurred anyway and Dubois never received the recognition he sought.

2.7.13.1.5   Peking Man

During the 1920's, within a few miles of Peking, westerners found the source of fossil bones which were being recovered by locals. A search for a missing link ensued. In 1929, 'an almost complete brain case was discovered'. This was the best of a total of fourteen that were eventually found and none of them had either face, jaw, or base. Eleven jawbones, portions of seven thighbones, two upper arm bones, a wrist bone and 147 teeth were also found in the same excavation. (86) From one of the skull portions, Peking Man was recreated and replicas were placed in various museums around the world. (87)

When this story is retold during the present day, the report sounds quite credible as noted by the following. 'The original Peking Man bones, representing more than 40 individuals, mysteriously disappeared during World War II. Fortunately, we have excellent casts. Today we include both Java man and Peking man in the category Homo erectus, "upright man." (57)

The most important parts of the story have been left out of such reports. Thousands of animal bones were also found including elephant and deer and the above finds were mixed with the animal bones. Further, 'each skull had its base broken away in a definite manner suggesting that the individuals had been decapitated and the brains eaten'. (88) In 1934 parts of six truly human skeletons were found including three complete skulls. In 1931 Professor Henri Breuil found abundant evidence of a large-scale human operation. A great number of antler bones had been worked by stone tools imported from more than one mile away leaving a pile of chips 18 inches deep. There was also a furnace operation as evidenced by an ash heap 23 feet deep. (89) The apparent deliberate damage to the original skulls along with their small size and monkey-like appearance, leads to the suspicion that the furnace-operating humans had been eating Peking man and that he wasn't really a man at all.

2.7.13.2   Haeckel, Ernst Heinrich

Ernst Haeckel was a biologist during the late eighteen-hundreds. In support of the Theory of Evolution, he elaborated and promoted the biogenic law which had earlier been discussed by Kielmeyer, Mullen, Karl Von Bair and Darwin. (30) The law declares that an organism in its prebirth development is an epitome of the changes, which an organism has gone through in the course of its evolutionary development. In support of this notion, he prepared a series of drawings of the human embryo during its development. (29) These drawings showed that the human being, before it is born, goes through the same stages of development that it did as it evolved. The drawings however did not show actual human development at all but were fabrications. 'This is one of the worst cases of scientific fraud. It is shocking to find that someone once thought to be a great scientist was deliberately misleading. It makes me angry. What Haeckel did was to take a human embryo and copy it, pretending that the salamander and pig and all the others looked the same at the same stage of development. They don't! These are fakes.' (29) The fault was actually pointed out by other biologists at the time but his drawings have remained in numerous textbooks right into the present age.

2.7.13.3   Evolution of the Horse

The evolution of the horse has commonly been depicted in books which discuss evolution, by showing a sequence of four or five drawings which trace the horse's evolutionary development. The smallest creature is labeled 'eohippus', which is a word assembled from Latin, meaning 'dawn horse'. This drawing is a reconstruction from a fossil, which was found, and attempts to show what it would look like with flesh and skin. The other drawings show horses increasing in size up to a full-size horse. These drawings show clearly that the horse did evolve and that it became larger as it did. (28)

However, the eohippus had four toes on each front foot and three toes on each back foot. This makes it very similar to the present-day hyrax, which has three toes on each front foot and two toes on each back foot. (58) It lives in Africa south and west of the Sahara and extreme south-western Asia. It is about the size of a rabbit or fox with a plump pointed head, short neck, short slender legs and a squat almost tailless body. (84) The following question is therefore properly raised. Why isn't the fossil, which is labeled 'eohippus', recognized as an ancestor of the modern hyrax? It much more similar to a hyrax than it is to a horse.

The other drawings are actually current members of the horse family but the one next to eohippus, is usually drawn disproportionately small to make it appear as a better link between eohippus and a full-size horse. One wonders why they didn't simply use a drawing of a miniature horse. A miniature horse is only about three feet tall and so would seem to be an appropriate entry for this position. Actually miniature horses, like the hyrax, still exist today. Therefore, they did not evolve at all and, like the hyrax, did not need to evolve to survive.

There was another ancient creature, which is similar in appearance to a horse but which is never shown in the horse sequence. The Baluchitherium looked very much like a horse but probably due to its large size, was never included in the horse sequence. The Baluchitherium was about eighteen feet high at the shoulder, which is approximately three times as high as present-day horses. If horses evolved, how do we know that they didn't descend from a very large creature instead of from a very small creature? It isn't very likely that the Baluchitherium was an ancient horse but it looked a lot more like a horse than eohippus. (59)

Or perhaps we should consider Moropus. This animal had a head like a horse but it also had heavy legs and claws on its feet. (77) In this case all that needs to be changed to make a modern horse is the legs and feet. Interestingly, to change a hyrax-like eohippus into a horse, the main changes required are also the legs and feet.

Just when things could not get worse, 'Fossils of three-toed horses were found in layers above (and thus younger than) those containing the one-toed horses …' (178)

All of these facts are well known and have resulted in the horse sequence not being used very often at the present time. 'The construction of the family tree of the horse is therefore a very artificial one, since it is put together from non-equivalent parts, and cannot therefore be a continuous transformation series. Its evolutionary value is therefore made totally untenable through the new research.' (83)

2.7.13.4 Geologic Column

The concepts and ideas which are involved with the geologic column are so closely related to the Theory of Evolution that the two topics are inseparable. The geologic column shows a sequence of layers of Earth material, each of which is said to represent an age of the Earth. (Please refer to the diagram, 2.7.13.4 Geologic Column. also called Geologic Time Scale) In this diagram, each layer includes a certain assortment of fossils which are said to represent animals which lived during the age which the layer represents. These layers are always shown to be very neat and each one is declared to represent a period of time on the Earth, which was several million years in length. The reason we know that the layers are very old and represent a certain period of Earth history is because they contain the fossils, which represent the animals, which lived during that time. The reason we know that the fossils lived during that time is because they are included in the layer, which shows that time. Ordinarily this would be called circular reasoning but since it works so well for the Theory of Evolution, an exception is apparently in order.

Unfortunately, the complete geologic column is not found in nature with the exception of several very small areas comprising much less than 0.01% of the Earth's continental surface. Over most of the Earth, only two, three or four layers are actually found. For example, in Southern Ontario only Ordovician and Silurian layers are found. (17) The others which are expected, from the geologic column presentations, are simply not there. The situation is even worse over more than one-half of the rest of Canada. The Canadian Shield (also known as the Laurentian Plateau), involves virtually all of Quebec north of the St. Lawrence River, Northern Ontario, most of Southern Ontario north of Lake Ontario, a northern portion of Wisconsin and Minnesota, most of Manitoba, northern Saskatchewan, most of Michigan, most of northern mainland Canada, the Arctic Islands and Greenland. The total area involved is more than 3,000,000 sq. mi. (The total area of Canada is only 3,800,000 sq. mi.) Only Precambrian rock is found over this vast area, meaning that there are no layers at all!

2.7.13.4 Geologic Column

ERA	PERIOD		EPOCH	APPROXIMATE TIME BOUNDARIES	LIFE FORMS ORIGINATING
CENZOIC	QUATERNARY		RECENT		MAN
			PLEISTOCENE	2,500,000 YRS.	
	TERTIARY		PLIOCENE	12,000,000	
			MIOCENE	26,000,000	
			OLIGOCENE	38,000,000	
			EOCENE	54,000,000	
			PALEOCENE	65,000,000	
MEZOZOIC	CRETACEOUS			136,000,000	PRIMATES - FLOWERING PLANTS
	JURASSIC			195,000,000	DINOSAURS - BIRDS
	TRIASSIC			225,000,000	MAMMALS
PALEOZOIC	PERMIAN			280,000,000	
	CARBON-IFEROUS	PENNSYLVANIAN		320,000,000	REPTILES
		MISSISSIPPIAN		345,000,000	
	DEVONIAN			395,000,000	AMPHIBIANS
	SILURIAN			430,000,000	VASCULAR LAND PLANT
	ORDOVICIAN			500,000,000	
	CAMBRIAN			570,000,000	FISH - CHORDATES
PRECAMERIAN				(700,000,000) (1,500,000,000) (3,500,000,000) 4,650,000,000+ (FORMATION OF THE EARTH)	INVERTEBRATES EUCARYOTIC CELLS PROPCARYOTIC CELLS

The evidence therefore does not support the theory. So explanations are offered instead of evidence. For example, "the missing layers all washed away". However, there is no explanation of how they were washed away or where they went. Unfortunately, explanations are not a valid substitute for evidence. By using an obviously made-up explanation, resort is being made to artificial construct. In the absence of observational evidence, an explanation is offered instead.

It is also unfortunate for the theory that the evidence not only does not support the theory but actually contradicts the theory. In nature, fossils are seldom found in neat layers but are more often found in chaotic piles. They are often found broken or smashed and often found with entire body sections missing. Sometimes they are found in great deposits with ancient and modern creatures all pressed and rolled together in a great jumbled mass. (18)

It is also the case that where a particular order is required by the theory, the layers are actually found reversed with the so-called older fossils on the top of the younger fossils. (19) In order to force this type of evidence into the theory, it is explained that the rock containing the 'older' fossils was forced to move around and slide over the 'younger' rock. Thus an explanation is offered to contradict the evidence in order for the evidence to comply with the theory. A similar type of explanation is required when 'old' fossils are found mixed in with 'young' fossils. Whenever this type of contradictory evidence is found, artificial construct is absolutely necessary to make the evidence support the theory.

2.7.13.5 Bacterial Resistance

Certain bacteria acquire resistance to drugs. This is a major cause for concern because if the bacteria are damaging to human life, they should be destroyed. If they are not destroyed, the damage that they cause will continue. The acquisition of the drug-resistant characteristic is offered as evidence of evolution. However, it turns out to be a type of change, which is called horizontal transfer. The genes which provide the beneficial result for the bacteria were already in existence and were acquired by the bacteria from an outside source. However, as a result of this acquisition, the bacteria lose a portion of their original function. The loss varies from one type of bacteria to another but there is always a loss. These losses include 'the reduction of the function of transport proteins, protein binding affinities, enzyme activities and regulatory control systems' (7). Further, One means by which bacteria can acquire antibiotic resistance is via horizontal transfer of antibiotic resistant genes. Such transfer of resistant genes is common (Gomez, 1998; Top et al., 2000), accounting for many examples of resistant bacteria. But, horizontal transfer merely involves the transfer of resistance genes already present in the bacterial world. While horizontal acquisition of resistant genes is "beneficial" to those bacteria exposed to a given antibiotic, such gene transfer does not account for the origin or the diverse variety of these genes. As such, it fails to provide a genetic mechanism for the origin of any antibiotic resistant genes in the biological world. Evolution, through the process of common "descent with modification", predicts it can account for the origin and diversity of life on Earth; however the mere shuffling of pre-existing genes between organisms via gene transfer does not provide the necessary genetic mechanism to satisfy this prediction. Nor can it readily account for the simultaneous development of both the antibiotic biosynthesis and resistance genes – an evolutionary enigma (Pemrose, 1998). Thus horizontal transfer of resistant genes cannot be offered as an appropriate example of "evolution in a petri dish." (92)

The situation is similar to attaching an eraser to the end of a pencil and proclaiming that the pencil has evolved. The pencil didn't really evolve because the eraser was already in existence. On the other hand pencils do not evolve anyway. Neither do bacteria.

2.7.13.6   Reptiles into Birds

Proponents of the Theory of Evolution declare that reptiles evolved into birds and that Archaeopteryx is the evidence to support this declaration. 'The five known specimens of Archaeopteryx preserve the only solid physical evidence of the earliest recognizable stage of bird evolution and therefore provide the most compelling evidence about bird origins – which all point to a coelurosaurian ancestry – not crocodilian and not the condontian.' (103) However, this declaration is not substantiated by evidence because the features of Archaeopteryx reveal it to simply be a bird. 'Archaeopteryx is not intermediate at all because, as paleontologists acknowledge, Archaeopteryx was a true bird – it had wings, it was completely feathered, it flew. It was not a half-way bird, it was a bird. Martin (1985, p 182) concluded that Archaeopteryx is a genuine bird.' (103). It not only had feathers, the feathers were asymmetrical, which is indicative of a superior flyer like a swallow or falcon. (14) Alan Feduccia, an evolutionist, confirmed that '… the fossil's (i.e. Archaeopteryx) feathers had leading edges significantly shorter than their trailing edges, which is characteristic of all modern flying birds.' (106) Great excitement followed the discovery of Archaeopteryx because it was initially seen as an intermediate creature. As stated above, it became the only evidence available but soon it was recognized as a bird, which happened to have small claws on its wings and a peculiar tail. The claws on the wings were similar to the claws on the wings of the present-day hoatzin, which is a bird of South America. The juvenile of this species uses the claws to cling to branches in the trees. (20)

While the archaeopteryx was abandoned, another similar type of creature was soon offered to take its place. The Archaeoraptor was reconstructed and placed on display by the National Geographic Society at their headquarters. The reconstruction appears like a bird with undersized wings and oversized legs. The tail appeared cumbersome. Then it was pointed out that the tail was probably not even properly associated with the creature at all but actually belonged to a small dromaeosaur fossil found in China. '… it has been suggested by Chinese paleontologist Xu Xing that Archaeoraptor's alleged tail may not belong to the main body of the fossil and may instead be the counterpart to a small dromaeosaur fossil back in China. National Geographic may be…presenting a specimen of questionable pedigree.' (107) This commentator acknowledges questionable conduct but then goes on to assert that there was a dinosaur which unquestionably had feathers. 'While unmistakably a dinosaur and possibly ancestor to Velociraptor, Sinornithosaurus clearly was covered with feathers.' (107) Hope does spring eternal. The 'clearly' was not so clear to others interested in the subject. 'The much-touted feathers on certain dinosaurs may be nothing more than collagen fibers. An article on ABC Science Online says "Dinosaur 'feathers' are no such thing." Instead it's just decayed collagen, like that found on sharks and reptiles.' (22) Other commentators were even blunter. 'Another recent case of fraud in evolution is that of Archaeoraptor, the "evolutionary find of the century"… It was the Piltdown Man story all over again.' (21) While there appears to be a full share of fraud and humbug surrounding the debate, if a feather is ever found on a dinosaur, plenty of comments will be sure to follow. In the meantime, on behalf of those of us who cannot do primary research, it would be appreciated that any and all evidence be made manifest so people can consider it and come to whatever conclusion they decide.

Just when things couldn't get worse, Protoavis was found. This creature 'is thought to have lived 75 million years before Archaeopteryx, also had advanced avian features. Protoavis is so much older than Archaeopteryz that it would have existed long before the dinosaurs that are supposed to have been the ancestors of Archaeopteryz.' (12)

The other main offering to support the claim that reptiles evolved into birds relates to the claim that some dinosaurs have been found with 'protofeathers'.(22) The prefix 'proto' means original or primitive. The clear implication is therefore that these dinosaurs had primitive feathers and were therefore on their way to becoming birds. However when these articles are read more carefully, it is clear that the protofeather may actually be a hair which was hollow. The question is then reasonably asked: how does a hollow hair become a feather? A feather is a very complex aerodynamic device with a very particular asymmetrical cross-section. When it is moved through the air at the appropriate angle, lift is generated. Comparing this structurally ingenious device with a hair is grossly misleading. It is also appropriate to point out that polar bears have hollow hair. Also they are white. Would it be reasonable therefore to declare that polar bears are evolving into swans? The fossil marks declared to be protofeathers may have been part of the skin. In response to the wide-spread claims that "feathered dinosaurs" had been found, the noted paleontologist Alan Feduccia stated; 'Yet no one ever bothered to provide evidence – either structural or biological – that these structures had anything to do with feathers. In our new work, we show that these and other filamentous structures were not protofeathers, but rather the remains of collagenous fiber mesh-works that reinforced the skin.' (106) The simple fact is, that the marks or features under discussion are not well enough differentiated to conclusively determine that they were feathers, so advancing a claim that they were feathers is definitely not warranted.

2.7.13.7   Peppered Moths

The Peppered Moth of England has been offered as evidence of evolution and discussions have included a photograph of moths on a tree trunk. There are two color phases of the Peppered Moth, one being black and the other gray. During a period of increased air pollution in England in the 1800's, tree trunks became darkened. As the tree trunks darkened, the proportion of gray moths was reduced from 98 percent in 1848 to 5 percent in 1900. Then as the air quality was improved, the tree trunks became lighter again and the proportion of gray moths again increased. Since the black phase was less obvious against the black tree trunks, the black moths enjoyed an improved survival advantage compared to the gray moths.

However, with respect to evolution, the argument is not compelling because the moths have always remained as moths. This is an example of change in gene frequency rather than gene modification. There is no evidence that the moths were changing into anything else and when the industrial pollution problem was addressed, the frequency of gray moths increased. The more disappointing aspect of the story relates to the photograph, which is usually shown, of the moths on a tree trunk. Apparently these moths do not usually land on tree trunks. Therefore, in order to get such a photograph they must have been placed there and attached to the trunk. (23) If this is what actually happened, the report which included the photograph cannot be considered good science. It would be deliberately deceiving to stick moths to a tree trunk in order to get a picture to support an argument that they were evolving.

2.7.13.8   Body Fluid Saltiness

The oceans of the world are credited with being the location where life developed. In support of this hypothesis, it has been declared that the salts in the oceans occur in the same proportion as they do in the human body.

There are two aspects to this declaration, which render it inappropriate.

2.7.13.8.1   Ocean Salts and Minerals

Neither the mineral nor the salt content of the oceans is static. They increase continually. Salts and minerals are being brought into the oceans by the rivers so the concentrations are continually increasing. (24) Therefore, even if a particular concentration of interest were to occur, it would only be valid for one particular time. The next day it would have changed. It follows that looking backward in time, the concentrations would all have been lower. The situation is even more exaggerated if the extrapolation is extended backward for several thousand years. If fact, from the information, which is presently available (Please refer to the table entitled 'Residency Periods for Chemicals in the Ocean', which is included in the appendix.), the concentrations of many of the minerals, which are presently in the oceans, would have been virtually zero only a few hundred years ago. It should be observed that extrapolation is always risky but in this case the time frames are so short that some credibility will remain with such extrapolation. It further follows that if this extrapolation were carried into the distant past (i.e. several million years), the oceans would have been free of minerals altogether and would also have had much less salt in them.

2.7.13.8.2   Element Concentration Comparison

The respective concentrations of the following three elements in blood plasma in a human body are: (25)
  i.   Sodium   0.762%
  ii.   Potassium   0.03%
  iii. Calcium   0.03%

The present concentrations of the same elements in ocean water are: (24)
  i.   Sodium   2.6%
  ii.   Potassium   0.07%
  iii. Calcium   0.31%

The respective concentrations and amounts in blood plasma are in a very particular and delicate balance and the concentrations and amounts in ocean water are continually increasing. The total of the above three items in plasma is only 0.8% whereas in sea water it is about 3% Therefore the relationships are not very close at this particular time and were probably not in the right proportion years ago either when sea salt concentrations were much lower. Since the ocean concentrations are continually increasing, making the above salt equivalence declaration is inappropriate.

2.7.13.9 Natural Self-Organization

While reviewing the various ideas which have been offered as evidence for evolution, it is appropriate to include an idea which actually earned its originator a Nobel Prize. Ilya Prigogine declared the following: 'Systems which are significantly out of equilibrium have a probability of spontaneously organizing themselves into temporary structures. The process would repeat resulting in increasing levels of organization. The usually expected increases in entropy would not occur and we are on our way to an increasingly organized state of affairs which would lead directly to the development of living self-reproducing creatures of all kinds'. (26)

The two difficulties which must be sidestepped to accept this idea are:

a. The second law of thermodynamics which declares that every process results in a less organized universe (i.e. every process results in an increase in entropy). Every activity that we know of whether it is climbing the stairs, painting the house, driving the car or stamping out license plates results in the production of a little bit of heat. This heat always radiates away into the universe and can never be recovered. This means that no matter how careful we are with any activity, some portion of the energy required, will not be accounted for in the end result. A portion will be lost every time. This basic fact of nature precludes the construction of any perpetual motion machines. This is one of the best examples of theory and observation being in very tight agreement. This, however, was not a constraint for Prigogrine. '.....spontaneous self-organization, not decay, is the natural order of things. Evolution is the natural outworking of this principle. Systems significantly out of equilibrium....tend to spontaneously organize themselves. ...if the system continues out of equilibrium, a current dissipative structure becomes the starting point for a second, more organized one.....organization continually increases. etc.etc. (26)

b. The second difficulty is the complete lack of observation of the anticipated ' increasing levels of organization'. Two of the offerings for 'increasing levels of organization' are thunderstorms and tornados. While such examples are questionable to begin with, these types of structures do not last long enough to lead to anything else. Temporary questionable examples are not sufficient to support the magnificent far-reaching claims that were made. '.....dissipative structures provide an adequate means for a natural formation of life (i.e. evolution)....these concepts are now applied to a wide group of problems in biology, sociology, and economics.' (26)

2.7.13.10 Evidence Summation

These examples illustrate that the available evidence doesn't really support the Theory of Evolution. Therefore it does not meet the criteria for a theory. Other less-demanding terms would be more appropriate including, speculation, hypothesis and conjecture. In order for the Theory of Evolution to have any validity it must provide evidence to support; a. the complexity of DNA, b. the vast differences in DNA between similar-looking creatures and c. the startup problem. Since these factors are not addressed the conclusion that must be reached is that the Theory of Evolution is not a scientific theory at all. Further, in the light of the unthinkable complexity of living organisms, (e.g. A pinhead of DNA could hold 40 million times as much information as a 100 gigabyte hard drive. (177)) the explanations offered by proponents of the Theory of Evolution are simply childish.

What we are being asked to believe is that millions upon millions of creatures ((from the virtually-microscope size to the great dinosaurs and monsters of the seas) some with two wings and some with four wings; some with two eyes and some with many eyes; some able to swim on top of the water as well as dive down and swim under water before emerging to fly to some other location; some able to move about on two legs and some with many legs; some able to fly across country, then dive into a pond and live out their lives underwater; some able to soar to enormous heights and some able to fly from one pole to the other and then make the return trip a few months later) developed, over enormous stretches of time, without any direction being applied whatsoever and without leaving any trace or evidence of that development. It really is a lot more sensible to simply believe in Creation.

2.8 String Theory

One of the most interesting theories to be offered in recent years is String Theory. In an attempt to identify what fundamental particles such as electrons, quarks (parts of protons) and all other so-called fundamental particles actually look like - that is if we could see them - it was suggested that they may actually be various configurations of little strings. The characteristics of the particle would them be explainable in terms of the mass of the string, its tension, length and so on. It really was an exciting idea and has spawned a host of projects at the PhD level and beyond over the thirty-plus-years since the theory was first proposed. It is very exciting to have a fundamental all-encompassing idea such as this from time to time and a lot of physicists became deeply involved in studying it. The problem remains, however, that after all these years and after all the research money has been spent, there isn't any evidence to support it. In fact the problem is worse than a lack of evidence. There hasn't even been an experiment proposed which might recover some evidence. (27) (More recently an experiment was carried out at the large European accelerator but the results were negative.) Without evidence, all that we have is a hunch. Conjecture has a place. Hunches have value. Hypothesis is often a starting place to pursue an investigation. However, without evidence this is all we have. To suggest that we have a valuable theory is inappropriate. We may actually have a whopper. After such an extended period of intense investigation has produced no evidence, it is increasingly probable that there is no evidence. If this is the case we would indeed have a whopper where we thought we had a theory.

2.9 Ice Age Mammals

2.9.1 The Species

It has frequently been declared that mammals roamed the Earth during the great Ice Age. In fact some animals are referred to as Ice Age animals. 'Harrington, one of the world's experts on the fauna of the last glaciations, has hauled thousands of ice-age oddities from the northern mud. ...an ice-age elephant –

better known as a woolly mammoth… ground sloths. …beavers… bears…moose.. etc.' (149) Other commentators concur with the basic assumption. 'This "muck" (i.e. the muck of Alaska) contains enormous numbers of frozen bones of extinct animals such as the mammoth, mastodon, super-bison and horse. These animals perished… at the end of the Ice Age or in early post-glacial times.' (150) 'Few controversies rage more fiercely in paleontology than why the megafauna vanished – not just in Australia but in North America, where mammoths, horses, camels, and dozens of other large Ice Age mammals all vanished … the giant mammals of the Ice Age.' (151)

2.9.2 Environment

Other commentators offer even more graphic information, not only about the animals but about the environment where these animals lived. In particular, the north-western portion of North America and the eastern portion of Russia together with land which would have been above sea level at Ice Age maximum, has been given the name, Beringia. The secondary name given is "An Ice-Age Serengeti". (149) The Serengeti is a large area of Central Africa where great numbers of animals live. Several rivers cross the Serengeti, it covers several thousand square miles, and includes portions of several countries. The herd of Blue Wildebeest (or Gnu – a species of large antelope) alone numbers many thousands. Other species included are giraffes, hyenas and lions. These animals live on the Serengeti Plain all year and migrate from place to place to take advantage of the best grass. Generally speaking, the area was considered to be an ideal setting for a great diversity of species which infers that the food supply is at least adequate and usually much better than adequate.

2.9.3 Glacial Presence

During the "Ice Age Serengeti" period, the great glaciers of the Ice Age were nearby. In fact they were distributed throughout Beringia. In particular they were understood to cover the southern third of Alaska as well as numerous other locations throughout Alaska and Eastern Siberia. These glaciers would have been typical of glaciers of the Ice Age wherein the ice accumulated until it was hundreds or even thousands of feet deep.

2.9.4 Snow or Grass

It may be recalled that it requires between ten and twelve feet of snow to form one foot of glacial ice. If this was indeed the case, ten thousand feet of snow would be required to form one thousand feet of ice. Therefore the situation which is presented is that grass and other vegetation grew in abundance while interspersed throughout the area the ice was accumulating until it was thousands of feet deep. The snow was falling in the next field but it wasn't falling in this field. It was cold enough in the next field to keep the snow from melting but in this field it was warm enough to enable abundant vegetation to grow so a diversity of fauna could thrive. It was warm here but cold over there. Over there the snow was piling up to incredible depths but over here there wasn't any snow at all. Summer came and the grass grew. Then winter came and the animals accessed this grass through a few inches of snow. With this arrangement it is declared that the great diversity of Ice Age animals thrived and grew to be very large.

2.9.5   Gross Contradiction

Unfortunately, such presentations are gross contradictions. It cannot be snowing profusely a few miles away and not snowing here at all. Wasn't there any wind blowing? If not, what air movement brought the snow-laden storm clouds from the ocean? It is contradictory to assert that it was warm enough to enable grass to grow in apparent abundance throughout the area while at the same time it was cold enough for snow to fall and remain frozen. We cannot have major buildups of snow right next to areas that are snow-free. It cannot be warm and cold at the same time in virtually the same place. If it is snowing so heavily only a few miles away in every direction, it must surely be snowing here as well.

Arctic Islands and continental areas in both Siberia and North America are declared to have been glacial-free during the Ice Age. (152)While this may have been the case it does not mean that these areas were snow-free. It is more likely that there wasn't enough accumulation of snow to form significant thicknesses of ice. There still could have been an overwhelming abundance of snow. After all, it requires at least eighty feet of snow to just start to form ice. If two hundred feet of snow accumulated, 120 feet of it would compress into about ten or twelve feet of blue ice. The remaining eighty feet would remain as snow. Years later when the area was investigated for glacial evidence, would ten feet of ice have left any sign of its existence? Glacial ice in a flat area does not move. There would not be any signs of movement. How then could it be determined that it had or had not been there? It is even likely that several hundred feet of ice, which would only have been on the ground for a few years, would not leave behind any particular indication that it had ever been there. But two hundred feet of ice represents two thousand feet of snow. An animal has never existed that could thrive in two thousand feet of snow. In fact, no animal has ever existed that could thrive in ten feet of snow.

However, all of the Ice-Age animals are presumed to have survived the trauma of the Ice Age only to die when it was all over. The various megafauna species survived the great Ice Ageand only became extinct when it was all over. Mammoths, mastodons, saber-toothed cats, horses, ground sloths, camels, glyptodonts, peccaries, mountain deer, monster beavers, antelopes, wolves, lions, and giant bears are understood to have survived the trauma and desperation of the Ice Age only to simultaneously die later! Apparently scientists have grappled with this question for hundreds of years but they still have no explanation and appear to be no closer to an explanation. (153)

Either the snow was there or it wasn't there. Either the animals were there or they were not there. If there was fifty feet of snow there could not have been any animals. If there was a great diversity of animals, there could not have been any snow. To assert that so many animals lived in the snow fields during the Ice Age is not a true story. Beringia, an Ice-Age Serengeti, is a fairy tale.

2.10   Conclusion

If there is no evidence for a system of ideas which is being offered as a theory, it should be relegated to lesser status than theory until some evidence is found. If a theory depends on artificial construct where basic and well-known physics must be overturned in order to be supportive of the theory, it should be set aside. Further, if a theory is only able to deal with some evidence but is contradicted by other evidence it is not internally consistent and is really of no use.

Unfortunately, none of the above theories stand up to serious scrutiny and none of them meet the basic criteria for a theory which was offered at the beginning of this chapter. They should therefore be relegated to lesser status and referred to as hypothesis or conjecture instead.

Further, because these theories have been so boldly promoted and so vigorously defended as the ultimate in absolute truth, one becomes increasingly suspicious that what is being offered with each one of them is not really truth at all but closer to a fairy tale. Because of the lack of evidence, offering any of the above systems of ideas as theories is similar to offering a fairy tale as a true story. People do like fairy tales. Actually, adults like them. All of the above offerings may be compared to fairy tales and they are all so brazen in their unsubstantiated declarations that they may also be referred to as 'whoppers'.

In support of this assertion, the following quotations have been recovered from discussions of these various offerings.

2.10.1   Nebular Hypothesis

Within the Milky Way Galaxy was a cloud, of gas and dust, which was not doing very much until one day it started to collapse for some unknown reason. Of course, what caused the collapse is a matter of speculation. As it collapsed it bulged in the center and flattened around the edges and then began to rotate. The bulge in the center would become our Sun. (8). 'Unknown reason, – a matter of speculation' – indeed it is. Since there isn't the least bit of evidence to support any portion of this statement, it, clearly, has just been made up.

The following comment also tries to capture the situation however somewhat pathetically. 'Contemporary opinion on star formation holds that objects called protostars are formed as condensations from interstellar gas. This condensation process is very difficult theoretically, and no essential theoretical understanding can be claimed; in fact, some theoretical evidence argues strongly against the possibility of star formation. However, we know that stars exist and we must do our best to account for them.' (112)

2.10.2   String Theory

The dilemma of String Theory is captured by the following quote. 'If we restrict our attention to the theories that are known to exist – those that allow us to do actual calculations and make predictions – we must conclude that string theory has nothing to do with nature, because every single one of these disagrees with experimental data. So the hope that string theory may describe our world rests wholly on a belief in string theories whose existence is wholly conjectural.' (179)

2.10.3 Swamp Theory of Coal Formation

Coal is the remains of plants that lived and were buried in the same general locality. It is believed that the plants were swamp plants which did not decay completely when they died. A layer of peat was formed from these plants and built up over long periods of time. Then, the swamp sank and was covered by the ocean and was buried later by other strata. Burial compressed the layer of peat and the temperature increased gradually transforming the peat into coal. (10) But, as discussed in section 2.2.2.3 above, the plant remains found in coal are not swamp plants at all.

2.10.4 Ice Age

Uncertainties abound: 'Many attempts have been made to reconstruct a chronology of Pleistocene events in terms of actual years. …Although ingenious, all these calculations have to take into account factors so variable as to render the results untrustworthy. …The resulting figures are therefore inaccurate, though probably they are of the right order of magnitude!!!' (`122)

Many methods to account for ice ages have been proposed, and likely no other geological problem has been so seriously discussed by glaciologists, meteorologists and biologists, however, there is no consensus but opinions are hopelessly in contradiction with each other. (11)

2.10.5 Continental Drift

In the beginning was Radinia. '750 million years ago Radinia split. 500 million years ago Pannotia formed. 514 million years ago a chunk of Pannotia broke off leaving Gondwana at the South Pole. 390 million years ago, Laurentia collided with Baltica and later with Avalonia. 237 million years ago Pangea formed and stretched from pole to pole. 200 million years ago Pangea broke up as soon as it formed. 94 million years ago Africa separated from South America and North America broke away from Eurasia. Gondwana broke apart with Africa, Australia, Antarctica and India all going their separate ways. 50 million years ago, drifting pieces of the former land masses of Pangea collided with each other. Africa pushed into western Eurasia raising the Alps. India rammed into Asia creating the Himalayan plateau and squeezing southeast Asia out of the way'(63) A most incredible tale none of which is the least bit verifiable.

2.10.6 Catastrophic Plate Tectonics

Baumgardner, the originator and promoter of the Catastrophic Plate Tectonics Theory, had the following comment. 'I consider…the present ocean basement to be no older than the Mesozoic portion of the continental record. This requires…the entire pre-Flood ocean floor…to have vanished from the Earth's surface…the lack of pre-Mesozoic seafloor to tell us what these motions may have been leaves everyone…with precious little to work with…beyond one's own imagination.' (13) Having a good imagination is, no doubt, occasionally an asset, but when it comes to offering a scientific theory, there is no substitute for evidence.

2.10.7   Evolution

In response to recognition of the tremendous problem that the Cambrian Explosion presents for evolutionary thought, Richard Dawkins, probably the most enthusiastic defender of Evolution at the present time stated, 'It is as though they were just planted there, without any evolutionary history.' (102) How close can a person come to recognizing the fallacy of an idea without rejecting it outright?

Further he stated; 'If evolutionary processes had to rely on single-step selection, it would never have got anywhere.' (148) (The structure of a comparatively simple bacterial ribosome requires the precise assembly of over 4,500 nucleotides and 800 amino acids.)

Carl Sagan, the popular science writer who championed the position that life exists in outer space, recognized these difficulties as well as the influence of the popular press. 'The great evolutionists have always been united against ETI. (extraterrestrial intelligence, i.e. life elsewhere in the universe) The biologists who have supported ETI have generally been biologists with the viewpoint of a physicist. (quoting Tyler, 1981, p 140) Biologists generally argue that the enormous complexity of life and the fact that the…"likelihood of the evolution of an intelligent species… (is) essentially zero…," strongly argues against the evolution of life elsewhere (Tipler, 1981, p 140; and see also Faulkes, 1991). This view though, is not held by large numbers of people, nor by many biologists who specialize in other areas. Many science discoveries which became popular, such as the claim of "canals" on Mars made by Percival Lowell, and the origin of life experiments by Stanley Miller, have fueled a belief both in lay persons and scientists that life could have evolved elsewhere in the universe (Sagan, 1980).' (128)

2.10.8   Ice Age Mammals

'…these species thrived in great numbers despite bitter winter winds and long months of darkness. The secret of their survival and the key to the giant size of so many of Beringia's animals… lay in the rich nutrients contained in the ecosystem's grass…the light winter snowfalls could be easily brushed aside for grazing.' (154) So the grass grew very well during the summer and during the winter the snow was shallow enough to allow this grass to be accessed quite readily. This type of situation seems more like central North America at the present time than it does to an ice age. The Ice Age was a time when abundant snow fell virtually continuously, it was dark all of the time and the temperature was always well below freezing. Plants could never have grown during such a time and there has never been an animal that manage to get around in a thousand feet of snow

2.11   Coherency

Science is supposed to be about logic and evidence.? However, when data is in short supply scientists speculate and when they speculate emotions creep into the thinking process. (158)This is a time when there are a lot of scientists engaged in complex, valid and very sophisticated scientific endeavors. It is therefore most disappointing that such unsubstantiated beliefs as those discussed above, are being dragged along. One might expect this type of situation years ago when there was a reduced inclination to measure and a greater inclination to try to arrive at the truth by debate. However, to persist with questionable theories such as those above, is inexcusable in this modern age.

The world view offered by the above assortment of ideas is incoherent and does not provide an integrated view of the world. The assortment of offerings discussed above, does not fit together and the lack of togetherness is another element of disappointment. Further by repeatedly introducing tangents, they collectively do nothing to really explain the very improbable circumstance which will be discussed herein as "The Window of Life".

2.12 Declaration

By contrast, the system of ideas, which is offered herein, A Theory of the Earth Based on Asteroid Impact, does meet all of the earlier-mentioned criteria for a theory and will help to explain The Window of Life in a coherent fashion. Everything fits together.

2.12.1 Known Physics

Known physics (or well-known physics) is what is used every day by engineers and architects all over the world. In order for an engineer to design a bridge he/she must recognize the characteristics of the material to be used as well as how that material is arranged. Neither did the Space Shuttle just pop into existence. Every miniscule part and the relationship of that part to other parts had to be studied very carefully. Then the completed assembly had to be tested rigorously. Will this arrangement perform as expected?

The ideas herein are in agreement with and do not violate well-known physics. For example, in order to move ocean water from the ocean onto the land, energy must be transferred to the ocean water. Similarly to build a mountain range, energy must be transferred into the sedimentary layers of the Earth to cause them to crumple up into mountains. These energy transfer mechanisms have been identified.

2.12.2 Evidence

There are two types of evidence, which have been identified. The first is evidence identifying the required source of energy. The second is evidence which identifies the results of transferring the energy from the source into the Earth.

Conversely, the requirement that all available evidence should not contradict the basic theory has also been met. There simply is no evidence on the Earth, which is not compliant with the system of ideas which is offered herein. Two chapters in particular discuss evidence which relates directly to the theory and a further list of relevant items has been included in the appendix.

2.12.3 Comprehensive

This theory is complete. The explanation does not start part of the way through. The situation from the very beginning to the end has been described and explained. A great amount of energy is required by this theory and the source of this energy has been identified. A great amount of time is not required and evidence has been offered to support this contention.

Neither has appeal been made to the supernatural to cover a weak area. This is not to say that the Deity was not involved. 'Who has understood the Spirit of the Lord'(Isaiah 40:13)? Truth seekers must accept truth wherever it is found and in whatever form it is found. It is certainly correct to observe that if any life forms were to survive the overwhelming catastrophe discussed herein, protection would have been required, but any such arrangement is beyond the scope of the present discussion. (The survival problem is discussed extensively in Part 2, Surviving the Ice Age)

2.12.4 Artificial Construct

No artificial construct has been employed. Things have not been artificially arranged in violation of how things naturally work in nature to make the idea work.

2.12.5 Internal Consistency

The system of ideas offered herein is coherent. Everything works together to provide a consistent and comprehensive world view.

3. Some Windows of Life

3.1 Introduction

Life is like a window. As long as the "light" shines through, life can exist. If the window closes, life ends. Life is like a window because there is a range for each of the parameters essential to life, which make up the window. There is a maximum and minimum for every factor, which exists as a necessary condition for life to exist. Actually, life is like a series of windows, all of which must be lined up to enable the light to get through. Even if all of the windows are lined up except one, life cannot exist because the light will not get through.

Every life form is complicated. Even the simple ones are complicated. All forms of life, right down to the simplest bacterium, require many different types of protein to be present in their structures in order to exist. It has been calculated that the minimum number of proteins required for a microscopic entity to actually be alive, is 238. '….a set of 238 proteins, the minimum number which would sustain life.' (17) This is a theoretical number. In reality, the simplest known bacteria have over 300 different types of protein. The complexity continues because each kind of protein is assembled from about 300 smaller parts (called amino acids) and some proteins have several thousand parts.

In order for any form of life to exist there must be a supportive environment. This environment must supply the necessary nutrients as well as provide the appropriate living conditions. In order to achieve the life-supporting function, environments are complicated. Numerous factors must be in place at the same time and if even one of them ceases to exist, the life form, which depended on them, will cease to exist.

The current and ongoing concern about the environment of the Earth recognizes that if one or more of the factors, which is essential for our existence, is taken away, life on the whole Earth will be in danger. An example of this is the present concern regarding endangered species. There may be many reasons why a species becomes endangered. For example, the habitat may be destroyed. If all of the swamps and other wetlands were destroyed, ducks and geese would cease to exist because the places, which they required to live, would no longer exist. Similarly if there were no more safe sandy beaches, the great sea turtles would have no place to lay their eggs. They would therefore cease to exist. If the pack ice in the northern waters all melted, the Polar Bears would not be able to catch seals – their principle food supply. Some commentators think this would cause the extinction of Polar Bears. Further, if there was no pack ice, where would the seals give birth?

The situation is far from academic. Numerous species have already been lost. (8) Many more are on the endangered list. There is great fear that these too might be driven to extinction due to the loss of an appropriate environment in which they can to live.

Neither does the situation terminate with birds and animals. People may be at risk. When animals are at risk, human life may be in danger because in many cases people depend on animals. Even if those on which we are directly dependent are not in immediate danger, the fact that others, which are similar in their requirements, are endangered raises the general awareness of the problem to a much higher level.

An example of our dependence on a species to which we do not usually pay much attention, is honey bees. How many of us worry about the bee supply? However, if there are no honey bees, there will be very little pollination and as a result there will be very little fruit. Our food supply will be jeopardized. Recently a serious problem has developed with bees where large numbers have been dying. This is not good news and we must sincerely hope that the situation will not deteriorate further. (18) 'Insects, especially bees, are also crucial for pollinating flowering plants. One in every three bites of food you eat relies on insect pollinators for its production.' (47)

Numerous aspects of nature must remain stable within a very narrow range for both animals and people to survive. If a few swamps were lost, it might be upsetting but what if the atmosphere became too polluted to breath? What if the temperature of the whole Earth increased a few degrees? What if the water in the ocean suddenly washed over the land? What if most of the honey bees died?

3.2   Water

There are several different ways in which water is necessary for life on Earth and each of them have a range wherein life is enabled. If this range is exceeded, life will not be viable. Therefore this range is like a window. If it moves too far either way, life will be eliminated.

3.2.1   Sustenance Water

Drinking water must be available to all life forms in the proper amount and with the appropriate qualities. All birds and animals require clean, unpolluted water to drink. Water is easily contaminated. Contamination is already a major concern for many municipalities. If industrial waste or crop fertilizer seeps into the drinking water supply, the entire community is at risk. If run-off from pasture-land gets into the water supply, it is not usable for drinking and neither should it be used for washing. It is, of course, common for some minerals to be present in drinking water but salt is not permitted. However, some wells produce salty water, which cannot be used for drinking. Industrial pollution can make water acidic. In this case the fish cannot live in the water. It has not been uncommon for entire lakes to become too acidic for the fish to survive. Some rivers carry a silt load. They are always brown and muddy looking. Water such as this should not be used for drinking either until the silt is removed. Some places have water, which is full of bacteria. A person who drinks such water will become ill. Water for human or animal use simply must be clean. It is unfortunate that at the present time, there are many areas, which do not have a good source of clean drinking water. This has led to the creation of a major industry, which supplies water in bottles for drinking purposes. What would happen if the source of the bottled water became polluted?

Sustenance water or drinking water may be thought of as a one-sided window. Water cannot get any purer than pure but degradation and pollution have an unending range. A certain amount of impurity can be tolerated by the human body, but when the tolerable range is exceeded that particular "window of life" closes.

3.2.2   Liquid Water

Water exists in three forms. When the temperature is low enough it exists as ice. Above freezing there is a range of temperature where water is liquid. In this range, water can be pumped and moved around as required. When the temperature is above the boiling point, water exists only as a vapor. Water is useful in all of these forms but what if the temperature was too high and it only existed as vapor or steam? Or what if the temperature was too low and water only existed as ice? Very few creatures can exist for any period of time without liquid water. Certainly people cannot exist without it. Some people live in deserts and some live in the far north. All of these people must have water to drink. Those who live in the far north must have it and if they cannot find it they must use heat to melt snow in order to get a life-sustaining drink. If liquid water did not exist extensively around the world, there would not be any in the north or the deserts either. If the usual sources of water (e.g. rivers and lakes) are absent from a particular area, desert dwellers must depend on either atmospheric moisture (i.e. rain which entered the atmosphere remote from the desert) or deep underground water reservoirs (which are replenished by rain). Most of the creatures of the Earth, especially people, require clean water in liquid form and without it life would not be possible.

3.2.3   Ocean Water

Ocean water is salty and therefore it cannot be used for drinking. Ocean water is useful for another reason, however, and it is not only useful, it is absolutely necessary. Ocean water covers more than one-half of the surface of the Earth. It covers much more than one-half in the Southern Hemisphere. As well as simply covering the Earth, the oceans have depth and so a great volume of water is involved.

An important characteristic of water is its high specific heat. It takes a lot of heat to increase the temperature of water. After it has been warmed up, water must lose a lot of heat before it will cool down. By contrast, land requires much less heat to warm up a similar amount.

The orbit of the Earth around the Sun is almost circular but not quite. While the average distance of the Earth from the Sun is 92.9 million miles, it actually drifts out about 1.5 million miles and then six months later on the other side of the Sun, it will drift in about the same amount. When it is summer in the northern hemisphere, the Earth is farther from the Sun. This is most fortunate because if the northern hemisphere was closer to the Sun at this time, summertime temperatures would be higher (because the land is easier to warm up and if the Sun were closer, it would warm up even more). By the time it is winter in the northern hemisphere, the Earth has moved closer to the Sun. This means that winters will not be as cold as they would have been if the Earth was further from the Sun at this time. (i.e. If the Earth can heat up easier, it can also cool down easier.)

In the southern hemisphere it is the other way around. While it is winter up north, it is summer down south. Therefore the great ocean areas are facing the Sun more during the time when they are closer to it.

Then, when winter comes down south, the Earth is further from the Sun but now the great ocean areas are not facing the Sun as much. (The oceans store up a lot of heat and then release it later when it is needed more.) The net result of all of these effects is that the temperature of the Earth is evened out more than it would have been with any other arrangement. The great ocean areas have worked together with the eccentricity of the Earth's orbit to even out the temperature of the entire Earth including the oceans, the land and the atmosphere. A better arrangement is hard to imagine. If things had been arranged any other way, the temperature variations on the Earth would be more extreme and survival on the Earth would be more difficult.

The oceans have this moderating effect on the temperatures of the world because they are mostly open water. When water is 'open' and not covered by ice, the Sun is able to warm it up easier. The Sun will penetrate right down into the water. This means that the heating effect is distributed throughout the top several feet of water. But if the water is frozen, most of the Sun's energy just bounces off. In this case, the water underneath will not be heated at all. Even if the open water is almost cold enough to freeze, it will still have the ability to absorb the Sun's energy. Open water, even down near the freezing point, will have an influence on the temperatures over the adjacent land because open water has a moderating effect on the temperature of the air right above it. As this air moves inland, the temperature of these adjacent land masses will be moderated. However, the proximity of the water to any given area of land will determine the degree to which the air temperature is affected or moderated. In other words, the further away from the ocean a particular land area is, the more extreme its weather will be. The greatest land area on Earth is the continent of Eurasia. Central Siberia is a land-locked area of Eurasia, having the furthest inland region from all major bodies of open water. It, therefore, has the least benefit of any region on Earth from the moderating effect of ocean water on air temperature. It is a place where temperature extremes are similar to the South Pole. It gets about as cold in central Siberia in winter than it does at the South Pole. In fact, the northern regions of Siberia are commonly warmer than the central region.

It is safe to conclude that if the Earth had greater land areas and less ocean surface, there would be greater weather extremes on land. Survival in central Siberia is very difficult at the present time. If this area of Asia was even greater, the weather extremes would be greater. The situation would be similar in North America if it was twice as wide, in which case the weather patterns would be more extreme and survival would be more difficult.

Suppose that there were no oceans. In this case, even if there was sufficient water for drinking, would it be possible to survive on the Earth? Greater land areas would heat up more in the summer and cool down more in the winter. The heat storage function of the present oceans would be missing so it is safe to conclude that all areas of the world would go outside of the acceptable temperature range for survival either in the summer or in the winter or both. It is true that many other factors are involved with temperature regulation, but the oceans of the world are a major factor. During summer in the southern hemisphere, a great amount of heat is stored in the oceans. However, if significantly more or significantly less than the expected average amount of heat is stored, the weather patterns of the entire world will be modified, Increases in ocean temperature have been dubbed El Nino and decreases have been dubbed El Nero, which terms provide an abbreviated way of referring to the weather-modifying phenomenon of ocean temperature change. What would be the effect of these changes if the oceans were smaller or non-existent?

On the other hand what would it be like if the oceans were larger? No doubt larger oceans would influence our weather patterns. Other effects besides temperature would soon become apparent. Included among these effects would be humidity, rainfall, hurricanes and waves. Waves become larger the further the wind is allowed to blow over the water. For example, on Lake Ontario, if the wind blows directly across the lake, wave amplitudes of three or four feet may develop. If the same intensity of wind were allowed to blow length-wise down the lake, much larger waves would develop. Generally speaking, a similar situation happens on the ocean. In the North Atlantic, wave amplitudes of 30 or 40 feet are not the least bit uncommon. If the North Atlantic was twice as wide, much larger waves would result. One of the worst places on Earth for large dangerous waves is the Antarctic Ocean. The Antarctic Ocean surrounds Antarctica which means that the wind is allowed to blow all the way around without being interrupted. Monstrous waves continually develop in this area. Thankfully, most of the time continents get in the way and interrupt the unlimited buildup of waves. Continents thereby prevent the development of world-wide wave patterns. If the continents were small or if there was only one, there would be nothing to stop waves from building up and traveling all the way around the world.

Similarly, the extent of ocean currents is restricted by the continents. Without the inhibiting affect of continents, ocean currents would be operating over much greater distances as well. In such a case, the moderating influence of the current-entrained ocean water would be sensed over these much greater distances. Currently, the Gulf Stream travels from the warm sub-tropic regions adjacent to Central America to Northern Europe. As result the climate of England is much warmer than similar-latitude areas of North America. The further this current travels, the further the influence of the sub-tropical temperatures of America would be felt.

As continents have an influence on the distance over which waves build up and the distance that ocean currents travel, they also have an influence on the distance that hurricanes travel. Since it is warm surface water which provides the energy for a hurricane, the farther that a hurricane travels over warm water, the more ferocious it becomes. At the present time hurricanes only travel part of the width of the Atlantic Ocean. If the Atlantic Ocean went all the way around the world, how dangerous would a hurricane become? If such a hurricane ever came ashore, the rainfall and the winds would render survival impossible.

It is appropriate to note at this time, that the hypothesized ancient landmass called Pangea, which was supposedly surrounded by a giant ocean called Panthalassa, would have been a difficult place to survive. The inland regions would have experienced climate extremes because they would have been far from the giant ocean. Coastal regions would have had to put up with over-sized waves and ocean currents from far away. The greatest nightmare would have been the hurricanes, which would have travelled for many thousands of miles over open ocean water and would have become entrained with extremely high amounts of energy. However there is no need to worry. Current theories of the Sun tell us that 4 billion years ago the Sun would have been 25% cooler than it is now. This means that Panthalassa would have been frozen solid right to the bottom and would have stayed that way right to the present time. This contradiction is called 'The Faint Young Sun Paradox' and has never been resolved.

It is safe to conclude, even from this brief consideration, that the existence of the oceans of the world are a "window of life" simply because of their respective areas compared to the areas of the land masses. Changing this ratio would affect our ability to survive. Surprisingly, by altering the area relationship of water to land either way, the window of life would get narrower.

3.2.4   The pH of Water

pH is a term used to identify the acidity or alkalinity of water. If the pH is neutral, the water is neither acidic nor basic. A pH close to neutral is desirable. It is not only desirable for life on the Earth but is a question properly raised with respect to the possibility of life elsewhere. 'Even if a planet has oceans, does the water have a pH neutral enough to permit cells to grow?' (13) A pH of 7 is neutral. Lower than 7 is acidic and higher than 7 is basic. (14) Water can be shifted into either of these less desirable regions if certain materials are added to the water. For example adding lime to water will make it more basic. In fact this has occasionally been done to offset too much acidity. Of course adding any acid to water would make it acidic and this has been unintentionally achieved as well.

During recent years the pH of numerous lakes has been modified by pollution and the pH levels have drifted away from neutral. In particular, many lakes within a few miles of the large smokestack at Sudbury, Ontario, have been subject to acid rain from this smokestack. As a result the water in these lakes became very clear. While 'very clear' is desirable for viewing the bottom of the lake, it is not at all desirable for the minute forms of life in these lakes. As a consequence they die and any creatures dependent on them die as well. This includes fish. If there are no fish there will be no fish-eating birds and the entire wildlife scene is negatively modified. Fortunately, improved technology and more relevant legislation have resulted in many of these lakes becoming murkier and once again being able to support a diversity of wildlife.

The pH of water is a two-sided window. A shift is possible either way from neutral neither of which is desirable and a serious shift either way would render such water unfit as life-support material.

3.3   Ocean Saltiness

Ocean water is salty but none of the salts have reached saturation. The same cannot be said of the Dead Sea where the water has reached saturation and simply cannot hold any more salt in solution. Therefore as more salt comes into the Dead Sea (primarily from the Jordon River), a similar amount of salt precipitates out of solution and may be found on the bottom or along the shore. These large amounts of salt render the Dead Sea useless for any form of life, which may be useful for human consumption. Neither lobsters nor fish can live in the Dead Sea. Consequently, no one goes to the Dead Sea to find food. (19)

Conversely, the oceans of the world include a great number of creatures, which are used for food by the people of the world. Billions of people eat food from the sea. In fact a significant percentage of the food requirements for the entire world are met from the oceans.

This would not be the case if the oceans were as salty as the Dead Sea. In this case the oceans would not provide any significant percentage of the food supply for the world. It is therefore appropriate to recognize ocean saltiness as a window of life. In this case, it is virtually a one-sided window. No salt is acceptable. High concentrations are unacceptable.

While ocean saltiness is a circumstances window, it is also a temporal window. As time goes by it is shifting slowly in one direction and that direction is towards closure. Since the oceans continue to become saltier (and they are becoming saltier every year because the rivers continue to bring in more salt from the continents), they will be less and less able to provide food for humanity. While the oceans are presently at about 10% saturation, long before saturation is reached, the balance of life in the oceans will have been upset. While the oceans are a source of food for humanity, they also accommodate very complex internal food chains and interactive plant and animal forms of life. These are needed as internal regulators enabling the portion that humans remove for food to be sustained. The optimal balance is increasingly difficult to maintain. For example, various groups around the world repeatedly over-fish. It is therefore suspected that long before the oceans have reached the 50% salinity saturation level, the ocean's internal food chains will have been interrupted so badly that the oceans as a source of food for humanity will be seriously reduced from their present levels. As that situation develops – which at present rates of saltiness increase is still many years away, - the oceans will cease to be a source of food and that particular window of life will have closed. When it does close, there is no conceivable way that it will ever be reopened.

3.4   Rainfall

Rainfall is a necessary element of survival. Rainfall is necessary to enable crops to grow and for animals to survive. There is a range of rainfall, which is useful and beneficial for our survival. Beyond this range there is a wider range which could be called tolerable. If the tolerable range is exceeded in either direction, life cannot continue.

The beneficial range of rainfall recognizes that a certain amount of rain is required for crops to grow and for food to be produced. Of course some crops do not require as much rain as others but there is minimum quantity of rain required for any meaningful amount of plant growth. Most crops require more than the bare minimum and even the crops, which survive with the bare minimum, will do much better with an increase. Rain is not only a basic requirement for food production. It is required at a specific time in a plant's life cycle. If the rain comes too late it may as well not come at all because the crops need it during the growing season within a fairly narrow period of time. If a particular pattern of rainfall is required, having an adequate supply at the wrong time will not be very helpful. Lots of rain in the fall, for example, it will not be useful for crops, which needed it in the spring.

The timing of the rain also relates to the activity of the honey bee. While we recognize that every third bite we eat required the action of an insect pollinator, (47) it must also be recognized that honey bees cannot function when it is raining because they will drown. Therefore if it rains too much during the part of the season when the plants must be pollinated, the honey bees will be unable to fly and many plants will not bear fruit.

There are also crops, which will thrive with an abundance of rain. Usually the pattern of rainfall is reasonably predictable and known ahead of time. Crops, which thrive with less rain, will not be planted in areas, which have rain in abundance. Only the crops, which thrive with an abundance of rain, will be planted in areas where this is expected to occur. Therefore it is recognized that there is both a beneficial range of rainfall as well as workable distribution patterns of rainfall.

Extending beyond the beneficial range, there is a wider range, which we may recognize as tolerable. Included in this range are areas, which receive very little rain such as deserts. If anyone is to live in a desert, they may require an outside source of water and food. Some animals can live where there is not enough rain for humans to exist. However, animals do need water even in those cases where they do not need very much of it. In some desert areas, the little rain, which does fall, is sometimes stored in protected pools in the rocks and thereby affords a source of water for any animals in the area. Water may also be retained in small waterholes and these waterholes provide a source of water, which, even though it might be pretty muddy looking, will be the difference between life and death for those creatures, which are dependent on it. Some plants are quite resistant to drought because either they do not need very much water or they may be able to store it within themselves for use during extremely dry spells, thereby extending the tolerable range.

Another area where rainfall might be called tolerable but not really adequate is in the far north. In these areas, there may be very little rain but some animals are still able to survive. The rain, which does come, enables grass and other pasture crops to grow. These crops may be uncovered and eaten far into the winter season, when there is no rain, thereby enabling the animals in the region to survive. In many cases, survival of the animals means survival of the people. In the far north, for example, the musk oxen feed on grass which they uncover in the winter, long after this grass has ceased to grow. In this way they are able to survive the harsh winter season. In these regions, a portion of the diet of the local people might be musk ox and the survivability of these people would thereby be enhanced by the survival of the musk ox.

While less-than-adequate rain may enable a reduced population to survive, it also occurs that a few people can survive where the rainfall is more than adequate. Some areas of rain forest might be included in this category. When torrential rain is pouring down and the river has risen almost up to the floor of the house, the local residents must go about their routines in a way that meshes with these realities. People who live in these areas must recognize that it is pouring now, but in a few weeks it will be dry again and they must plan for these extremes. They must understand the rainfall pattern and gather their food with this in mind. In such areas as these the population will not be abundant and may even be as sparse as the semi-desert areas where rainfall is much less than adequate.

There will always be room for discussion on just how wide the tolerable range of rainfall is, but the fact will remain that there is a range and this makes rainfall a "window of life". Outside of the tolerable range, agreement may be reached on the fact that some minimum amount of rain is necessary. Agreement may also be reached on the fact that too much rain is possible. The rain, which comes with a hurricane for example, is intolerable. All that anyone can do at that time is take shelter and wait for it to end. The fact that it will soon be over is all that enables people in this type of situation to survive. If rainfall continued indefinitely at the rate, which occurs in a hurricane, there would be no chance of survival. The heavy rains which occur during a hurricane not only make survival difficult but they may wash away the support system of roads, bridges and gardens as well as large fields where food crops for the entire country are grown. If such rains continued, houses and workplaces would all be washed away as well as the support networks of power lines and fuel lines. Rainfall, which is in the hurricane category, is well outside of the range which is tolerable for life to survive.

Rainfall is therefore a "window of life". Some rainfall is absolutely necessary but too much is intolerable. At either extreme, life is impossible.

3.5   Cloud Cover

Cloud cover is a "window of life". It is similar to rainfall because a certain amount of cloud cover is required but too much is detrimental.

What is the minimum amount of cloud cover, which is required for life to exist? It is probably impossible to say but it is known for sure that it is the clouds, which bring the rain and some rain is essential. It is also a fact that too much solar exposure is not a good thing. In certain areas of the world, there is very little, cloud cover. In such areas we would expect to find the problems which accompany too much exposure to the Sun. Since certain skin diseases are attributed to over-exposure, it is recommended that over-exposure be avoided. A person's skin may even be burned. Burning in childhood has been recognized as problematic later in life. Sometimes the skin breaks down later because of too much solar exposure early in life. In some of these cases, the problems that develop are life-threatening. So it is safe to assert that a certain amount of cloud cover is beneficial for our survival.

An overabundance of cloud cover is not beneficial either. For example, the clouds of Venus are much too thick and allow hardly any solar energy to reach the surface. While clouds provide protection from the Sun, a certain amount of solar energy must reach the Earth. Crops require sunshine. (Mushrooms do not require any sunshine but we must eat more than mushrooms.) The Sun must shine directly on a plant in order for the plant to carry out its various growth processes. There are a few plants, which will do well in the shade but even here, in most cases, some reflected sunlight gets through. Most crops require direct solar exposure for at least part of their growing cycle.

If there is too much cloud cover the amount of sunlight which reaches the surface of the Earth is reduced. This means that the amount of heat which reaches the surface of the Earth is reduced. If there is too much heat reduction, the temperature of the Earth will drop. The clouds will reflect the sunlight away and when it cannot get through to the surface of the Earth, the Earth will cool down. Occasionally, there is a volcanic eruption on the Earth and a great amount of material is thrown up into the atmosphere. In some cases, material may be thrown up for thousands of feet into the upper atmosphere. At these higher levels there is no rainfall and the volcanic dust may stay there for extended periods of time. It has happened that the dust from a single eruption has, after being injected into the upper atmosphere, gone completely around the Earth. This dust will not only change the appearance of sunsets but the presence of the dust will partially block out the Sun and there will be less sunlight getting through to the surface of the Earth. Temperatures world-wide will be affected. It has occurred that both unusual sunsets and temperature reductions were noticeable for a whole year from a single eruption. (Examples of volcanoes which had world-wide effects are; Mount St. Helens and Tambora.) The effect of multiple eruptions or one super-eruption (i.e. like Yellowstone Park, which is understood to be an extinct giant volcano approximately forty miles across.) would be very detrimental. The saving factor in most cases is that the dust will settle out of the atmosphere within a few months and most of it will settle out even sooner. In some ways the atmosphere is self-cleaning, which is another beneficial aspect of rainfall. If there were no rainfall, the atmosphere could become overloaded with unwanted dust. As the rain falls, it brings down a lot of material which is undesirable in the air and in some cases this material is not much more desirable when it lands. At least the air gets cleaned this way and without rainfall the dust from a single volcano could stay airborne for a very long time.

It is very fortunate that included in the material, which is ejected by most volcanoes, is a large amount of water. This water is ejected with the other material but because of its temperature, it is in vapor form. As it leaves the eruption area, the vapor will cool and then much of it will fall as rain. As it falls, it will bring down a lot of the dust. cleansing the atmosphere. Without this cleaning procedure, there would be a lot more dust to cover the Earth, which, of course, would cause problems in addition to the world-wide cooling effect. Lower world-wide temperatures caused by extensive cloud cover would also result in less evaporation. Less evaporation means less rainfall so the potential cleaning effect of the rain would be further reduced.

Cloud cover can therefore be considered as a "window of life" because some cloud cover is essential to life but too much is detrimental to life.

3.6   Snow Cover

There are some areas of the world, which are constantly covered by snow and ice. This does not really bother most of us most of the time. These snow-covered areas are in the far north, the far south and the high mountains. Greenland, in the north, is almost covered by a great ice field all year. Antarctica, in the south, is similarly covered by a great glacier all of the time. Many of the high mountains of the world are snow- capped all year, including some that are close to the equator.

Many other areas are snow-covered during the cold season. This is certainly inconvenient and inhibits many activities but is tolerable because within a few months this snow will disappear again. While these areas are snow covered, most life support factors must be obtained from elsewhere. In the high arctic where people live all year round, life would not be possible without a warm season when the snow cover is gone for several weeks.

Even in more temperate regions, if winter snow did not melt in the spring, people would be forced to crowd further south and life on Earth would be restricted to those areas, which are closer to the equator.

While snow cover is a circumstances window, it is also to a certain extent a temporal window. As the Earth warms, the great glaciers and ice-fields will melt. In fact, most of the great glaciers of the world are melting at an alarming rate and many will completely disappear within another few generations. As these glaciers shrink, less sunlight will be reflected and the Earth will warm up a little more. The glaciers of the world currently cover about 22% of land surface area. (20) When they are all gone there will be a lot more ground exposed to sunlight which is not being reflected. Since the warming trend is melting the ice and snow in the first place, any reduction of ice and snow will only accelerate the warming trend. As time goes by, more snow cover will be lost and the loss will be one more factor allowing the Earth to warm up.

Snow cover is therefore a "window of life". While life on Earth may not require any particular minimum, less snow cover will result in a warmer Earth (which is understood to be less desirable).

Conversely, when we look in the other direction and try to visualize a snow-covered world, it is clear that life-sustaining activities would be curtailed in that event as well

3.7   The Crust of the Earth

The crust of the Earth is a "window of life". There is a range of thickness of the crust, which will enable life, in all its diversity, to thrive. Beyond this range, the activities of life would be inhibited.

In comparison to the size of the Earth, the crust is very thin. It is floating on top of a molten interior. The Earth is basically a red-hot ball of molten rock, which is covered by a thin layer of cooler solid material.

Even though the crust of the Earth is very thin – when compared to the diameter – if it were thinner still, the surface of the Earth would be warmer. It is well known that the subsurface temperature of the Earth is virtually the same as the average annual air temperature. In fact, by measuring the temperature of the water from a shallow well, the average annual air temperature in that area will be known. (21) If the crust of the Earth was thinner, these near-the-surface temperatures would be higher because the hot molten rock in the interior of the Earth would be closer to the surface. An example of this effect may be observed in Yellowstone Park in the western USA. The entire area of this park is understood to be a crusted-over volcano with a crust that is not very thick. Hot springs are plentiful and mixtures of steam and hot water are continually erupting from the surface. Water from wells in the area is warmer than usual, indicating that subsurface temperatures are warmer than usual. Local weather will be slightly more moderate than areas of similar latitude in either direction.

A second example of local surface warming is the floor of the ocean near Hawaii. In this area there are several hot-spots where the ocean floor is several degrees warmer than surrounding areas. This effect is attributed to the crust of the Earth being locally thinner than average. If the entire area of North America had crustal thickness similar to these thin regions, the weather patterns of the entire continent would be more moderate than they are at the present time.

As discussed above, a slight increase in surface temperature may be tolerable or even beneficial to life on Earth but an average increase of more than a few degrees would be devastating. As the Earth warms and the subsurface temperature goes up, the temperature of ground water will go up. Also the temperature in all of the mines will be higher. Warmer soil would produce crops sooner in the spring, but warmer soil in the summer would dry out the ground. Further there may be many areas of the world that simply cannot stand warmer temperatures. A little warmer might be acceptable in the mid-latitudes and the northern regions but it might be a disaster in India and Africa. If the crust was thinner, water would evaporate much faster from all of the bodies of water around the world. This would result in an increase in rainfall. A modest increase, as with temperature, might be acceptable but a large increase would not be acceptable.

A crust that was much thinner would not tolerate tidal stress. As the Sun and the Moon tug on the Earth, tides are created. They are particularly noticeable in the oceans but also occur on land. (39) The force being exerted on the Earth by the Moon is very high (38) and if the crust of the Earth were only a fraction as thick as it is, instead of simply rising upwards for a few inches at high tide, it would rise up several feet and crack. Cracks would continually be appearing as the Moon passed overhead. Once the crust was cracked, the cracks would simply reopen the next day and material from the interior would again be allowed to escape. Under such circumstances, life on Earth would be intolerable.

Conversely, if the crust was thicker, there would be less heat loss from the interior of the Earth and the subsurface temperature of the Earth would be lower. In fact, it would be lower all year. The water in all the wells would be a little colder. The temperature deep in all of the mines would be lower. Miners may welcome this development. However they would not welcome the greater extremes of weather on the surface. The variation in temperature over the year would be greater which means that the weather patterns around the world would be more extreme. Winters would be longer. The growing season would be shorter. The situation would be self-exaggerating because with any increase in snow cover (because more of the heat of the Sun would be reflected away from the Earth) the average annual air temperature would be lower. If the crust of the Earth was thicker than it is at present, living conditions world-wide would be similar to living in Siberia.

While the Earth is currently undergoing a warming trend due to atmospheric changes, there is also a long-term change occurring due to a changing crust thickness. While the interior of the Earth is quite hot at the present time, as it ages it will cool down. (22) As this cooling proceeds and internal heat is lost, less and less of the interior of the Earth will be hot enough to remain molten. The cooler Earth will develop a thicker crust. Consequently, over the long term as the Earth cools and the crust gets thicker, all of the above-mentioned results of a thicker crust will develop.

While there are thought to be several causes of heat in the Earth's interior, any which are included as either original heat or heat resulting from some internal process will gradually dissipate. Tidal friction caused by both the Moon and the Sun will also generate heat in the Earth's interior. As the Moon recedes from the Earth, the portion caused by the Moon will also dissipate. The portion caused by the Sun will not dissipate as long as the Earth remains in its present orbit. However this portion of friction heat is quite small and will not offset the other factors.

The Earth must be a fluid body. That is it must be flexible enough to enable an equatorial bulge to develop because an equatorial bulge is a necessary factor involved with the Earth's axial tilt and consequently its seasons. The rotation of the Earth causes an equatorial bulge to develop which would not happen if the crust of the Earth was too thick. As the Sun and the Moon pull on the equatorial bulge a restoring torque develops to offset the overturning torque caused by the tidal bulge. The two forces cancel when the Earth's axial tilt is twenty-three and one-half degrees. As a result the Earth has an ideal seasonal arrangement.

The thickness of the crust of the Earth is both a circumstances window and a temporal window. If it were much thinner, the Earth would be too hot for life to thrive. As the Earth ages and cools, the crust will become thicker. As a consequence all climates will be cooler, extremes of temperature will be greater and life will become less and less viable.

3.8   Orbital Stability

The orbit of the Earth around the Sun is very very predictable. It is exactly the same every year. The orbit is almost perfectly circular and the small variation from circular always occurs the same way. During winter in the northern hemisphere, the Earth is closer to the Sun. During summer in the northern hemisphere, the Earth is further from the Sun. This situation repeats every year exactly the same way.

Most of the Earth's energy comes from the Sun and much of it is converted to heat and keeps the temperature of the Earth within a very narrow range. A host of other factors contribute to the stability of the Earth's temperature but the distance from the Sun determines just how much energy the Earth receives. If it received more energy, the other stabilizing factors would still be at work but the temperature would settle out at some higher range. This is why the distance from the Sun is so important. The Earth cannot be at any other distance than it is because if it was the temperature range would have an average which would be either too hot or too cold for us to survive.

The maximum, tolerable variation in the Earth's orbit has been calculated. If the Earth was further away from the Sun, but still in a stable orbit, the annual average temperature would be lower. It has been estimated that if the Earth were only five percent further away from the Sun, life on Earth would perish. (23) Similarly it has been estimated that if the Earth were five percent closer to the Sun, life would also perish. This means that the Earth is somehow operating right in the middle of a very narrowly-acceptable distance range from the Sun. This seems most fortuitous. It also means that the maintenance of the Earth's stable, repeatable orbit is a basic and absolute necessity to enable the great proliferation of life forms on the Earth to thrive.

It may therefore be concluded that the orbit of the Earth about the Sun is a "window of life" and that it is a very narrow window as well.

The orbit of the Earth is so stable and predictable that it is almost as if the Earth were moving through space along a track. While it is understood that this stability is absolutely necessary, there is still some mystery concerning just how it is being achieved. Computer modeling of the Earth's orbit indicates that over the long term, due to the influence of the major planets, the Earth will deviate slightly from its present orbit. Fortunately, the deviations are only in the one or two percent range. From such data, conclusions have been drawn suggesting that deviations of this nature could enable an ice age to commence. While this conclusion was hasty, and consequently has become less popular, it did recognize the importance of a stable Earth orbit. Apparently, the orbit-modeling did not include the effect of transients, such as the passing of massive comets, which does occur with frequencies, which we would prefer to ignore.

Man-made satellites may be placed in orbit around the Earth by launching them up to just the right height with the appropriate speed. Once these conditions have been achieved, the satellite should orbit around the Earth forever. However they will not orbit forever. They will come down again. In order for an Earth satellite to stay in orbit for a long time, adjustments to the orbit must be repeatedly made. Most satellites include several very small adjustment devices, which repeatedly fine-tune the orientation of the satellite as well as adjust its orbit. This type of active ongoing activity is necessary because even if the ideal orbit were achieved, there are always factors, which can cause detrimental deviation. In the case of an Earth satellite, these factors include minor variations in the gravity field of the Earth as the satellite moves along its orbital pathway. Lower satellite orbits may also be influenced by atmospheric effects. Even at satellite altitudes, space is not an absolute vacuum. There is also the influence of the Sun and the Moon. The gravity of the Sun and the Moon are responsible for the tides but they also will have an influence on any object, which is in orbit around the Earth. The achievement of a stable orbit around the Earth is not simply a one-time event but rather an ongoing activity, which requires continual attention to maintain.

It is therefore quite remarkable that the Earth has a stable orbit. If the Earth wandered from its nearly-circular orbit, it would be less surprising. Since there is no wandering, further investigation is warranted. The primary influence on the orbit of the Earth is the Sun. While the Sun is far away, it is singular and this is very important. Most of the stars in space are not singular. From a great distance a star may look like a single object, but most stars are actually two stars revolving around each other. There are also some arrangements, which involve three stars. While this is spectacular to observe, it would be disaster for a planet such as Earth. Two or more stars revolving around each other do not have an overall, constant, stable, gravity field. The gravity field around a pair of stars is changing as quickly as the stars move around. The portion of binaries and triples in the universe has been suggested as being between sixty and ninety percent. (24) This quantity gets even higher when stars, which have one or more very large dark neighbors, are included. As a result of the ongoing effort to find a star, which has an Earth-like planet nearby, several stars have been discovered which have a regular repeatable deviation or wobble in their otherwise straight-forward movement. Analysis of these movements enables astronomers to determine what types of objects are causing them. In almost all of these cases it has been determined that a very large object, similar to or even bigger than Jupiter, is orbiting the subject star and orbiting close enough to cause the star to move slightly out of a straight line as it travels. In some cases, several large orbiting objects are computed to be near these wobbling stars. (25) This increases the percentage of stars, which are not singular, to an even higher range. Consequently, one could even ask whether there are any other stars in the sky besides our own Sun, which are singular. This would, of course be impossible to tell. It is certain though that the percentage of non-singular stars is very high and that none of these stars would be a satisfactory host for a life-bearing planet such as Earth. An Earth-like planet could never achieve a stable, nearly-circular orbit near any of these multiple-mass stellar assemblies. They are, therefore, not appropriate candidates as life-supporting sources of heat to support life on an Earth-like planet.

In order to have a repeatable, circular orbit a constant, steady, predictable, gravity field is more than important, it is absolutely necessary. It has also been observed that the other planets in our solar system also have stable orbits. Jupiter is very large. It therefore has a strong gravity field, which has an influence on all of the other planets, including Earth. However, if Jupiter were closer to the Earth, the gravity of Jupiter would disturb the orbit of the Earth. As it is, Jupiter is very far away and so does not cause the Earth to be thrown out of its precise life-enabling orbit. On the other hand, the gravity of Jupiter is understood to have a stabilizing influence on the orbit of the Earth. (26) The other planets may also have a stabilizing influence. If they do, by disturbing any of these other orbits, the orbit of Earth would also be disturbed. Kepler, one of the astronomical geniuses of the seventeenth century, studied all of the planetary orbits and thought that there was some overall harmonic relationship among them. The orbital stability of the Earth is so precise that it would not be surprising if Kepler's intuition turned out to be valid and that all of the other planets were involved in the maintenance of the stability of the Earth's orbit.

The Earth orbits around the Sun and in so doing describes a pathway, which might be thought of as being the rim of a flat plate. This imaginary plate is referred to as a plane. When the orbits of other planets are discussed, the respective planes of their orbits are also mentioned. All of the major planets of the solar system have orbital planes, which line up. All of the planets are travelling in almost exactly the same plane. Since all of the planets may be influencing the orbits of all the other planets, if one of them somehow deviated from the common plane, the stability of the orbits of every other planet would be affected. For example, if the plane of the orbit of Jupiter shifted until it was ninety degrees away from its present position, all of the other planets would be tugged slightly back and forth from their present stable orbits. Such ongoing perturbations would be detrimental to the Earth and its ability to sustain life.

The Moon is another major factor in the maintenance of a stable Earth orbit. In fact, the Earth and the Moon may be thought of as a double-planet system rather than two separate, nearby objects in space. (27) The size of the Moon is an appreciable portion of the size of the Earth. When the size of other moons in the solar system is respectively compared to the size of their host planets, they are found to be quite small. Other moons as large as our Moon do exist but they are comparatively much smaller than their associated planets.

The Moon, when compared to the Earth, is quite large and consequently, it has a major influence on the Earth in several ways. The most obvious influence is the tides, which the Moon causes on the Earth. The tides indicate, even to a relatively casual observer, that the Moon is pulling quite hard on the Earth. (The actual force of attraction between the Moon and the Earth is 2.23 x 10 (16) tons. (38)) While the tides obviously affect the oceans, they also affect the continents. It has been estimated that the continents also rise up with a tidal effect but instead of rising up several feet, they only rise a few inches. The gravitational pull of the Moon has some effect on the ongoing life-enabling activities on the surface of the Earth and it is possible that it has an effect on the activity in the interior of the Earth as well.

As the Moon orbits around the Earth it pulls on the Earth. The result is that the center of the Moon's orbit is not at the center of the Earth as it would be with the much smaller artificial satellites such as the orbiting astronomical observatory. The center of the Moon's orbit is offset from the center of the Earth and the Earth is slightly offset from this same center in exactly the opposite direction. The situation is similar to a person who is holding a heavy ball by a rope and swinging it around. As he swings around, the ball tends to pull him toward it. In order to swing it without falling over, he must lean back away from it slightly. In so doing, his center of mass is no longer the same as the center of rotation but is slightly offset. If the ball was heavier or he tried to swing it around faster, he would have to lean back even more to keep from falling over. As long as he continues to swing the ball around, the three items, - ball, rope and man - may be thought of as a system where the two main parts are both orbiting around a common center. This center is not at the center of either of their masses but at an intermediate location. The situation might be understood even better by visualizing a system where the ball was as heavy as the man. In such a case, in order to swing the ball around, the man would have to lean back until he was as far from the center of orbit as the ball. Therefore, while it may be stated for certain that the Moon is orbiting around the Earth, it is actually orbiting around a point slightly offset from the center of the Earth. At the same time and in order to accommodate the continual pull of the Moon, the Earth is also orbiting around this same point.

While a person swinging a ball is similar in appearance to the Moon swinging around the Earth, it is also similar from a stability point of view. The man-rope-ball system has greater stability than the man would have by himself (assuming for a moment that this man is more like a wooden statue which cannot really deal with any upsetting forces.) Similarly, the Earth-gravity-Moon system has greater stability than the Earth would have by itself. Orbital stability-upsetting transients such as a passing asteroid or comet will have less effect on the Earth accompanied by the Moon because the gravity coupling with the Moon makes these two objects act more like one large object than two separate isolated objects. It follows that any Earth-like planets in far-away settings will be more prone to wandering from a precision orbit than the Earth and will consequently suffer the effects of such orbital variations. One result of orbital variation is temperature variation and temperature cannot vary more than a few percentage points before it will be outside of its life-enabling range. It is a basic fact of science that our Earth is barely within the correct position-range in the first place. As ideal as the Earth seems to be, if it wandered a little either way, it would be outside of the life-enabling temperature range. If this should happen to the Earth, the result would be a barren lifeless place and not a teaming-with-life place like it is now.

In the case of the Earth and the Moon, where these two objects are continually tugging on each other, the result of their mutual pull is a very stable orbit for the Earth. Without this stabilizing effect, the Earth would still orbit around the Sun but would tend to deviate from a tight, repeatable stable orbit.

The orbit of the Earth is a "window of life" and it is a very narrow window. Any excursion outside of this window would be disaster for life on the Earth. Three main factors contribute to the stability of the Earth's orbit. A stable well-behaving singular Sun is a minimum necessity. Unfortunately, there is a serious possibility that the mass of the Sun is increasing. This is because numerous objects, including comets and asteroids go into the Sun on a regular basis. An example of this type of activity is the giant comet Howard-Kooman-Mickels1979 that went into the Sun in 1979. The head of this comet was about the same size as the Earth. (49) Comets are continually going into the Sun as well as other material simply because the gravitational attraction of the Sun is so over-powering. If the mass of the Sun is increasing it means that the Earth will gradually spiral into it and the current 'Goldilock's' orbit is only temporary and will not be maintained. Having a Jupiter at just the right distance is the second major factor. The third factor is the existence of a large nearby moon. If any of these three factors were either missing or modified, the orbit of the Earth around the Sun would be less predictable and any such development would be disastrous for life on Earth.

3.9 Moon Distance and Earth Rotation

While the Moon is an important factor helping to stabilize the Earth's orbit around the Sun and an important factor in that particular window of life, the distance from the Earth to the Moon is independently an important factor in the viability of life on the Earth and therefore may also be considered as a window of life. We may recall that the Earth-Moon system orbits the Sun following a very stable orbit. The stability of this orbit – which stability is absolutely essential for life on Earth – is thought to be enhanced by the Earth and Moon acting as a coupled unit. If the Earth did not have the Moon, its orbit would not be quite as predictable as it is with the Moon involved. While some would consider this aspect of the Earth-Moon relationship speculative, the changing distance of the Moon from the Earth, and the implications this involves, is far from speculative.

The Moon is primarily responsible for the tides. It is the distance from the Earth to the Moon that determines the magnitude of the tides. If the Moon was farther away the tides would be lower, and this would not affect us very much. However, if the distance was less, it would certainly affect us. For example, if the distance was one-half of the present distance, the tides would be four times as high. It would therefore be impossible to live in many coastal areas under such conditions or along any inland waterway that was connected to the sea. If the Moon were only one-quarter as far away, one can only imagine the difficulties involved. In such a case, the tides would be at least sixteen times as high. Also land tides would be increased from the present few inches to several feet resulting in earthquakes and the opening of life-destroying fissures in the ground. A circumstance such as this would resemble the current situation on Io, a moon in close proximity to Jupiter. 'The gas giant (i.e. Jupiter) produces huge tides on the moon that cause massive pressures, melting its rocks and driving the movement of its large 800 km (497 mi) mantle.' (46) With respect to the tides, the distance to the Moon may be thought of as a one-sided window. Further away is not of immediate concern but if the Moon was closer, the viability of life on Earth would be lower.

The distance from the Earth to the Moon is not constant but is changing slowly. Fortunately, the Moon is drifting away from the Earth, not being pulled closer to it. This, in turn, is resulting in lower tides. While this process is so gradual as to be unnoticeable, the distance is increasing continuously. (28) In order for this to happen, the Moon must be pulled into a higher orbit. Just as it requires energy to lift a Satellite into orbit, a great deal of energy must be exerted on the Moon to raise it to a higher orbit. But it is happening because the Earth is transferring energy to the Moon by way of the tides. This requires energy which is coupled or transferred to the Moon by the way the tides operate. The Moon pulls up the ocean when it is overhead but since the Earth is rotating, and since it takes time for the force of the Moon to actually raise the tide, the location of the resulting maximum tide is slightly ahead of a straight line between the Earth and the Moon. (Please refer to the diagram, 3.9 Tidal Effect of the Moon) The tidal bulge is thereby leading the Moon and pulling it forward in its orbit. The result of this pulling-forward activity continually raises the Moon into a higher orbit. As the Moon is being pulled higher, the rotation of the Earth is slowing down. As the Moon gets further away, and the rotation of the Earth gets slower, our days will get longer and the tides will get lower. The final result will be an Earth which rotates at exactly the same speed that the Moon orbits. When this happens, the Moon will remain directly overhead of some particular location on Earth and will remain locked in that location. Only one-half of the population of the world will then be able to see the Moon and there will not be any lunar-caused tides on Earth at all. (There would still be small tides caused by the Sun but they would not have any influence on the position of the Moon.)

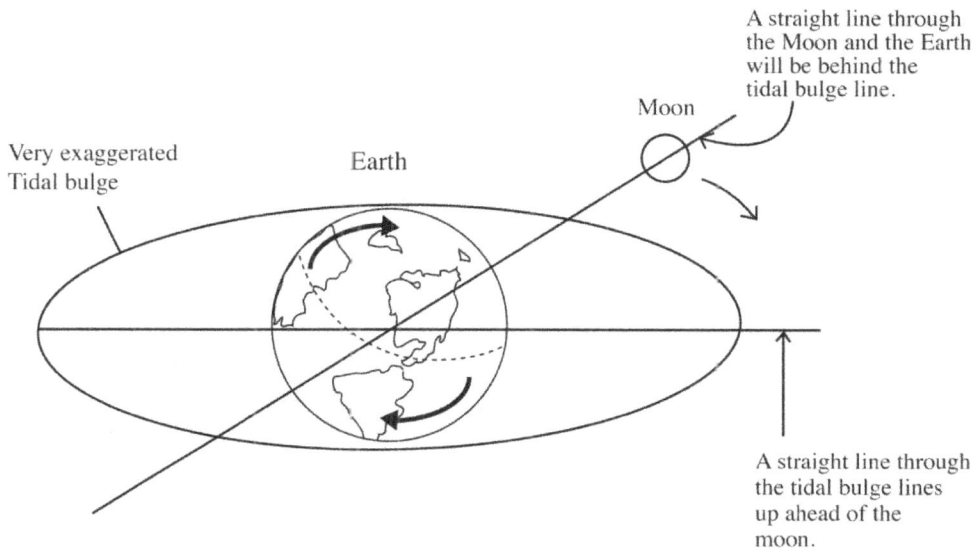

A straight line through the Moon and the Earth will be behind the tidal bulge line.

Moon

Very exaggerated Tidal bulge

Earth

A straight line through the tidal bulge lines up ahead of the moon.

3.9 Tidal Effect of the Moon

a) Moon pulls up the tidal bulge
b) Earth rotates before tidal bulge reaches maximum height
c) Tidal bulge is always ahead of the Moon
d) Tidal bulge pulls Moon forward and hence higher

It may initially appear that the two results of tidal action – a higher orbiting Moon and a reduced rotation rate for the Earth - would be interesting but not of any real concern. This may be a valid short-term conclusion but from a long-term perspective the situation is much more serious. The energy, which the Earth receives from the Sun, is the primary factor which determines the temperature of the Earth's surface. It is most convenient that the Earth keeps rotating to spread out the Sun's energy resulting in reasonably even temperatures between day and night. However, as the Earth's rotation slows, greater extremes of temperature between day and night will result. If a day on Earth was 48 hours instead of 24, it would get much hotter during the day and much colder at night. Mercury already has this problem. A day on Mercury lasts 88 times as long as one of our days. This, as well as the proximity of Mercury to the Sun, causes the surface temperature to increase so much that lead runs like water and rocks become almost red-hot. If a day on Earth was twice as long, daytime temperatures could easily exceed body temperatures and night-time temperatures would be much lower. Consequently the viability of life on Earth would be reduced.

The Moon will continue to recede from the Earth and the rotation of the Earth will continue to decrease until the two are matched. At that time the Moon will be further away than it is now. If all of the rotational energy of the Earth was transferred to the Moon and was used to raise the Moon into a higher orbit, the final result will be a day-night cycle which will be more than 1100 hours (i.e. 46, 24 hour days) long. (35) This will not be acceptable for life on Earth. The slower rotation will produce a daylight period which is approximately three weeks long. The Earth will therefore overheat during the day. Then when night comes, it will get too cold and everything will freeze solid. Life on Earth will not be viable at that time and it will become less and less viable as the Moon approaches its maximum distance.

The current separation rate is about 3.8 centimeters per year. (1) Using this information and simply extrapolating backwards in time, the Earth and the Moon would have been touching about ten billion years ago. However, the actual time back to contact would have been much shorter. The Moon is very slowly receding from the Earth. This is due to the tidal effect of the Moon on the Earth. The Moon raises tides in the Earth's oceans. The bulge this creates, exerts a gravitational force back on the Moon,(because of the Earth's rotation) pulling it ahead and into a higher and higher orbit. The rate at which the Moon is receding is now measurable and it can be shown that the Moon would have been in contact with the Earth only two billion years ago, less than half the accepted age of the Moon. This ignores tidal energy loss which would reduce the time to one billion years. (12) With the Moon close to the Earth, the crust of the Earth would have been pulled up into an excessive tidal bulge. The travelling bulge in the crust (i.e. travelling because the Earth is rotating) would have fractured the crust into numerous pieces. However, it is not fractured enough to indicate that anything like this has happened. This raises a serious question about how long the entire system has been in operation and casts a doubt on the idea that the Earth is 4.5 billion years old.(Please refer to 'Fairytales for Adults' for more discussion on this matter.)

At the mid-latitudes, the tidal bulge travels around the Earth at more than 700 miles per hour. Each day always brings two tidal bulges. Therefore within a few hundred million years ago, (with the Moon much closer to the Earth) there would have been two tides per day – many miles high – racing around the Earthat thousands of miles per hour. How would any continent have survived such repeated onslaughts?

95

Due to the onrushing oceans and the tidal fracturing of the crust, no continent - not even one made of solid granite – would have been able to survive. Consequently, continents would never have formed. Any formation would have been comparable to island formation in the Amazon River. In the Amazon, where the current is about five miles per hour - depending on rainfall patterns upstream – islands form and erode all the time. What would happen in the Amazon if the current was 500 miles per hour? The conclusion of these lines of reasoning is that the Earth-Moon system is not very old.

Lunar distance presents two different types of windows – a circumstances window and a temporal window. The circumstances window has two different types of sides. A closer Moon would develop higher tides. A more distant Moon would result in greater temperature extremes between night and day. Both of these situations would make life on Earth less viable. Since a Moon which was much closer to the Earth (i.e. a fraction of the present distance) would have made life on Earth impossible, there has obviously been a limited amount of time for life on Earth to have been possible. The maximum time for the historical side of this temporal window is much less than the time back to Earth-Moon contact. Since this time could not have been more than a few hundred million years at most, there really has not been enough time for life on Earth to have evolved. Then, when we look into the future to the other side of this temporal window, we see only another few hundred million years until diurnal temperature variations will also make life untenable. Therefore the total amount of time available for this life-enabling temporal window on Earth to be open is only a few hundred million years in the past plus a few hundred more in the future. While this is somewhat disconcerting, the reality of the situation is much more terrifying because neither the Earth's rotational speed nor it's axial tilt will not be maintained for nearly that long.

3.10   Earth Axial Tilt

As the Earth orbits around the Sun, it defines a plane, which may be called the orbital plane. The Earth also rotates about its own axis as it circles the Sun but the plane defined by the equator does not line up with the orbital plane. This means that the Earth has seasons. Spring comes every year to northern Canada and summer even comes every year to the high Arctic. The difference between the orbital plane and the equatorial plane has been measured and is 23.4 degrees. This is the ideal tilt. If it were greater, seasonal variations in climate would be more severe. If it were less, smaller areas of the Earth would get spring and summer. If there wasn't any tilt, there wouldn't be any seasons whatsoever so spring would never come in northern Canada and the ice up north would never melt. Permafrost would settle deeper, the equator would over-heat and less of the Earth would be habitable.

The Moon orbits around the Earth but not exactly in the same plane as the Earth orbits the Sun. If it were in exactly the same plane, there would be an eclipse of the Sun and an eclipse of the Moon every month. As it is, eclipses come less frequently. The offset of these two planes is a little more than 5 degrees. (36)

The seasons of the Earth are more important than eclipses but the influence of the Moon on the axial tilt of the Earth is much more important than either one. The Moon functions as a gyroscope to stabilize the Earth's axial tilt. The Earth is the only inner planet that has a large enough satellite to achieve this axis tilt (Murray, 1993). Without the Moon's effect, the Earth, like the other inner solar system planets including

Mercury, Venus, and Mars, would tilt as much as 85 degrees off vertical (with the plane perpendicular to the Earth's orbit around the Sun) eliminating the seasons altogether. Laskar and Robutel, 1993). (37) In other words, a change in axial tilt would result in a change in temperature patterns around the entire Earth. This is something that really should not be changed as evidenced by the problems resulting from the current seemingly very minor changes that are happening. The commentator was so concerned about any changes at all that he also offered the following; 'Variations as small as 1.3 degrees might trigger an ice age.' (37) While we understand that ice ages are not triggered, the actual concern being expressed is that the Earth could be seriously chilled if the axial tilt only changed a small amount. Further it was offered that the Moon applies a slight torque which shifts the axis when the planet spins. Mars has two very small moons which cannot stabilize the axial tilt at all and chaotic tilt from 0 degrees to 60 degrees will happen. (37)

While these commentators are very knowledgeable and their conclusions are recognized as correct, they did not really explain how the Moon's gravity actually does stabilize the axial tilt of the Earth. The Earth has two bulges. One of these is the equatorial bulge which develops because the Earth is spinning. As it spins (because it is a fluid body) the equator bulges out. The amount of this bulge is quite substantial and makes sea level at the equator about fourteen miles further from the center of the Earth than sea level is at the North Pole. The second bulge is the tidal bulge. This bulge pulls up almost directly towards the main pulling body which is the Moon. The tidal bulge is therefore adjustable. It forms (almost) in line between the center of the Earth and the Moon. Therefore, if the axial tilt was greater, the tidal bulge would peak further north. The pull of the Moon on the tidal bulge is towards increasing the angle of inclination. If there wasn't any counter-force involved, the Earth would tip right over until the North Pole was facing the Sun. The counter-force to prevent this from happening is the gravitational pull of the Moon on the equatorial bulge. If the Earth tipped over a little more, the equatorial bulge would be further away from the plane of the Earth's orbit (the Ecliptic) and the gravity pull of the Moon would generate a greater restoring torque to offset the torque caused by the tidal bulge. These two torques or forces equalize when the Earth's axial tilt is twenty-three and one-half degrees (at the Earth's current spin rate). This happens to be the ideal tilt. Either more or less would make the Earth less habitable.

If all of the Earth's rotational energy went into raising the Moon into a higher orbit, the loss of long-term stability would be a reasonable expectation but in reality most of the Earth's rotational energy is lost to friction and this is happening at a rapid rate! In this case the energy is lost as the tidal action moves both the bulk of the Earth and the water in the ocean. This is resulting in the Earth's rotation slowing down. Ever since atomic clocks were introduced in 1972 the length of a day on Earth has been measured. The day is getting a little longer all of the time. The equatorial bulge will gradually shrink and by and by it will not be able to offset the tidal bulge.This will happen well within one million years and will occur when the equatorial bulge is no longer able to offset the tidal bulge. Consequently, the Earth has possibly only a few hundred thousand years of stability left! A similar period of stability would apply to the past meaning that the Earth could not possibly have been habitable for more than a few hundreds of thousands of years in the past and more probably only a fraction of such a time period.

3.11   The Decelerating Earth

Just when it appeared that things couldn't get worse, they did. All of the above discussion is quite valid if we assume that all of the rotational energy of the Earth is applied to lifting the Moon into a higher orbit. Since the rotational energy of the Earth can be reasonably approximated and the energy needed to lift the Moon into a higher orbit can also be approximated, a valid calculation can be carried out to determine how much higher the Moon would be raised as the Earth's rotational energy is transferred to the orbital energy of the Moon. However, this straightforward approach ignores friction. Whenever something moves there is always friction involved. In the case of the Earth and the Moon friction develops as the water in the ocean is sloshed around due to the tides. As the tidal water moves past islands, around continents, up estuaries and into bays there is a loss of energy due to friction. Besides the water in the ocean the bulk of the Earth will also be involved as the crust is slightly lifted up and then settles back down as the tidal bulge travels through. All of this means that a portion of the rotational energy of the Earth will not be available to lift the Moon so the Moon will not get nearly as high as it would if there wasn't any energy loss. However, the Earth will slow down simply because energy is lost and it will not matter how it is lost. As a result of these actions the rotational speed of the Earth should be slowing down at the present time and if it is slowing down enough to be measured, confirmation of the whole idea would be obtained. Coincidentally, the length of the day has been measured and it is gradually increasing. Currently the Earth slows down at roughly 2 milliseconds per day. After 500 days the difference between the Earth rotation time and the atomic time would be 1 second. Instead of allowing this to happen, a leap second is inserted in the Atomic Clock to bring the two times together. (48) While this does not really sound like very much time, since there are only 86,400 seconds in a day, a constant rate of slowdown at the present rate, would only require about 1,000,000 years for the Earth to slow down to one-half of its present speed.

A problem will develop because as the Earth rotates slower and slower because the centrifugal force generating the equatorial bulge will diminish. However, the counter-balancing force caused by the tidal bulge will not diminish because the Moon will not be much further away than it is now. With a reduction in the equatorial bulge, there will not be any restoring force to keep the Earth's axial tilt from dramatically increasing. The Earth will just tip over. The habitability of the Earth will 'tip over' with it and sooner than we would like to contemplate, the Earth will not be habitable at all.

The centrifugal force involved with the generation of the equatorial bulge involves a square law where the speed factor is multiplied twice. This means that the rotational speed only has to drop to about three-quarters of the present speed before the equatorial bulge would be reduced to only a little more than one-half of its present size and would not be able to offset the tidal bulge at all! Therefore, if it took 1,000,000 years for the rotational speed to drop to one-half (as current measurements indicate), the Earth could be starting to tip over within a fraction of that much time and possibly within a few hundred thousand years making the Earth increasingly less habitable well within the 1,000,000 year time frame! And just to make matters a little worse, an Earth that was rotating slower would have a more exaggerated tidal bulge helping the tip-over process a little more. Further, the stabilizing capability of the equatorial bulge goes through a maximum point as it's near side (i.e. nearest to the Moon) gets further from the Earth's orbital plane (i.e. as the Earth tips over a little more) but then reduces again as the far side of the bulge comes up and gets closer to the Moon. When this point is reached, tip-over will accelerate quickly. None of these realities bode well for the idea that the Earth has been here for 4.5 billion years because using the same information and calculating backwards, within a similar amount of time, (i.e. a few hundreds of thousands of years) there would also have been a loss of habitability. (i.e. the Earth would then have been sitting up straight.) To summarize, the hypothesized extreme age for the Earth does not correlate with the current observations that the rotation of the Earth is slowing down rapidly and that the days are getting longer. The extreme-age idea should therefore be set aside.

3.12   The Magnetic Field

The Earth is a magnet. Therefore it has a magnetic field. The term field implies that there is some influence remote from the object, which is causing the field. The Earth also has gravity. The force of gravity is similar to a magnet because there is an effect far away from the Earth itself. While nobody has been able to explain this remote effect, the magnitude of it can be calculated. It is understood that the force of gravity between the Earth and the Moon keeps the Moon in orbit around the Earth. Similarly, there is a force of gravity between the Earth and the Sun which is recognized as keeping the Earth in orbit around the Sun. We have Isaac Newton and Albert Einstein to thank for the mathematics involved but we do not have anybody to thank yet for explaining just how the force of gravity works. The theories which have been advanced so far are very tenuous but this is not important compared to the fact that the magnitude of the forces can be calculated.

As the force of gravity acts remote from the Earth and can be accurately calculated, the magnetic field of the Earth also operates remote from the Earth and its magnitude can also be calculated. While gravity is credited with holding everything in place on the surface of the Earth, the magnetic field of the Earth gets very little attention. It is handy for operating compasses. If a person is away from well-known areas, knowing which way is north will be helpful in determining which way to go. A compass and the magnetic field may therefore be helpful for our survival away from home but there is another way in which the magnetic field is helpful all the time.

The magnetic field of the Earth is quite strong. This field therefore extends well out into space and will have an effect on all incoming particles, which may be influenced by a field of this nature. Unfortunately the Earth is the target for a lot of particles from space including very small particles, which are called cosmic rays. These particles are called cosmic because they come from the cosmos, which is just another word for space. It may seem like a more exotic term but space is space and these particles come to the Earth from space at very high speed. Also they are very small. They are atoms but they are not complete atoms and this is the reason that the magnetic field is so important.

Since these cosmic rays are very small and move very fast they do, in many cases, come right through the atmosphere and go right into the Earth. They could go right through a person on the way. Since these particles are atoms of various size, they may interact with the human body at the atomic level. They may crash right through the surface layers of the body and may damage some part of the complicated structure which causes the cells in the body to divide. If this happens the cells will not divide properly. Tumors may result, which is a very unwelcome development.

The magnetic field of the Earth has an influence on many of these unwanted particles from space and actually traps them high above the Earth. In this way most of the particles which are rushing through space do not actually come to the surface of the Earth at all. The magnetic field has prevented them from getting here and in this way it is acting like a giant shield. It is definitely safer to live on the Earth when this shield is operating. It is also to our benefit that it is transparent so we can still see the stars and sunlight can get to the surface of the Earth.

These little particles, which would have come to Earth, become trapped high above the atmosphere in a region, which is called the Van Allen Radiation Belt. It is the magnetic field of the Earth, which causes this radiation belt to exist. If the magnetic field disappeared, the Van Allen Radiation Belt would also disappear. In such a case, the Earth would be fully exposed to all incoming cosmic radiation and the harmful effects that this would bring.

The magnetic field may therefore be considered as a "window of life" because as long as it exists, it will be much safer to live on Earth. On the other end of the scale, a stronger magnetic field might offer even more protection but just how much magnetism a human being can stand is not well known. It is safe to say that it is a lot more than we have at the present time, but there would, of course, be an upper limit. This, however, does not really concern us. We definitely benefit from the shield generated by the magnetic field and therefore the magnetic field is properly referred to as a "window of life".

Unfortunately, the magnetic field of the Earth is dying off. While this issue is currently being hotly debated, long term measurements indicate that the overall strength of the field is falling off. This is not good news, in particular since, at the current rate of falloff, it will be reduced to half of its present value in about 1400 years. (2) Further, by 4000 A.D. the Earth's magnetic field may have disappeared altogether. (44) The magnetic field of the Earth is therefore both a circumstances window and a temporal window and as a temporal window, it is rapidly shifting towards closure.

3.13   Solid Fertile Ground

Solid, fertile ground is a "window of life". Solid ground is ground we can stand on. Solid ground will support highways and tall buildings. Bridges may be built on solid ground and wells, mines, subways and pipelines may be placed in solid ground. On the other hand, none of these things can be easily installed in swamps, marshland or quicksand. Peat bogs present another major challenge for either road or railway builders. As an example, northern Manitoba has large areas of peat. This makes it difficult to install a high-speed railway from Hudson Bay to central North America which project will be of major interest when Hudson Bay becomes free of ice for several months every summer. Another example is central Brazil. The mighty Amazon River continually places and replaces both islands and shorelines as its currents rise and fall with headwater rainfall. This means that bridge building is impossible because any foundation that was placed today would be gone tomorrow. Solid ground is needed for all types of structures and fertile ground is needed to grow food. If there wasn't any solid fertile ground, or only very little of it, life on Earth would be drastically reduced or eliminated altogether.

Solid ground must be predictably solid. That is why, if something happens to make ground fluid instead of solid, it will be so unexpected that it will be almost unbelievable. However, certain types of ground may become fluid if too much water is introduced. As the water soaks in, the ground may become more and more like a fluid. If such ground had been solid for a long time, there may have been houses built on it as well as roads. People would have driven on these roads and lived in these houses day after day without thinking about the solidness of the ground at all. Why would they? If the car is parked in the driveway as usual and the television set is on as usual, why would anyone worry about the solidness of the ground? If this ground then became fluid it would be so unexpected that it would be unbelievable.

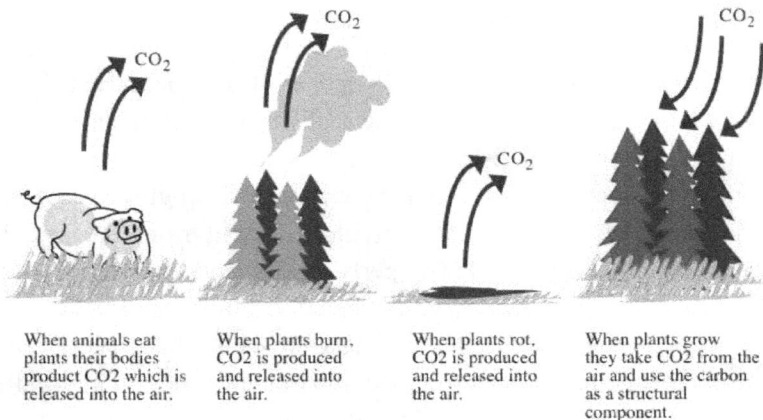

| When animals eat plants their bodies product CO2 which is released into the air. | When plants burn, CO2 is produced and released into the air. | When plants rot, CO2 is produced and released into the air. | When plants grow they take CO2 from the air and use the carbon as a structural component. |

3.13 The Carbon Cycle
Carbon is continually cycled through plants and back into the air.
Additionally, it cycles through plants, then animals, and then back into the air.

There is a location in Quebec where there were numerous houses, driveways and roads, which had been there for some time. It was therefore thought that they were solid and reliable as they always had been. However, one night a strange thing happened. A lady, who was a resident of one of the houses, heard the sound of water running under her house. Then she looked outside. Immediately next door, where the neighbors' house had been at sundown, there was an open space. Something terrible was happening. She alerted her husband. This event was so unthinkable that it was also unbelievable so he did not believe it. She left her house. At daybreak it was revealed that a great section of land near a deep river valley had become so saturated with water that it had flowed like water. The previously solid ground had become fluid due to the water and so it flowed like water and became a great mudflow. Several houses and other man-made structures as well as numerous trees were carried away.(3) Mudflows, such as this, provide examples of solid ground becoming fluid and hence of no further use for living, working or growing food.

An earthquake may also cause apparently solid ground to become fluid. During an earthquake the ground may shake back and forth. If the back and forth motion is great enough, buildings may be dislodged and roads may be damaged. In this case everything which is being shaken will be of no use until the shaking ceases and repairs are made.

An earthquake may also propagate along the surface with a back and forth motion, which is combined with an up and down motion. This will produce a formation of ground, which looks very similar to a series of water waves. Whereas a water wave transmits energy along the surface of the water, an earthquake can transmit energy along the surface of the ground in the form of ground waves. Certain types of ground are excellent for this type of transmission. When this type of wave passes, everything on the ground is disturbed just the same as floating objects would be with the passing of a water wave. As long as the ground keeps moving, life is not viable because nothing can be accomplished.

Neither can solid ground be used when a tsunami comes ashore. As a wave of this nature sweeps inland, trees and buildings and anything else, which can move, will be swept along with the wave. Even cars, which do not float very well, may be swept along. There is no point in contemplating use of this land under these circumstances. Survival is of primary importance and even this may not be possible.

Solid fertile ground may also become over-run with floodwater. In central North America, the land is very flat. It therefore only takes a small change in water level to flood a very wide area. As long as the land is flooded, it is really of no use. The floodwater may also be moving, which only makes matters worse. Moving floodwater not only makes the land temporarily unusable, it may actually sweep away fertile soil so the area can never again be used for crop production. The great Mississippi river is continually transporting soil. Usually this soil is taken away very slowly but sometimes during periods of heavy rainfall, it may be taken away very quickly. Much of the land, which is removed, may be deposited far away but even then it may not be of much use again for years. As an example, the Mississippi River delta is continually being built up with soil brought down the river. However, the soil which is brought this year will not be solid enough to be useful until much more is brought down in succeeding years.

Swamps are an important part of our overall ecosystem. Even in swamps where a few people live, they depend on some of the ground being above water and reasonably solid. If the whole world was swamp, there is no way that most of the animals or people could survive. Even if food were available, it must be recovered. This presupposes some minimal level of transport and utensil and support items such as these are usually procured from outside of the swampy areas. If this were not possible, survival in a swamp would be restricted to very few people at most.

Solid fertile ground is a "window of life". If there wasn't any, this window would be reduced and may even be closed altogether.

3.14   Carbon Dioxide – A Life Necessity

Carbon dioxide is present in our atmosphere. It is known to be an essential part of the cycle of life. In nature, carbon dioxide is produced and the carbon portion is used and recycled. This sequence is referred to as the Carbon Cycle, (please refer to the diagram, 3.13 The Carbon Cycle) which recognizes, first of all, that carbon dioxide is produced by all types of animals. It is produced in the body inside each cell of the body. In these mysterious places, the oxygen, which came in through the lungs, is combined with the food, which came in through the mouth. Actually most of the food is first changed into sugar. Sugar is a molecule, which is an assembly of carbon atoms. The food contains these carbon atoms but they were in a form, which could not be directly used by the body. After the ingenious processes of the body convert the food into sugar, it is transported by the blood, to all of the cells in the body. By converting the food into sugar molecules, a large number of carbon atoms may be conveniently carried to each cell throughout the entire body. The cell is able to receive this sugar molecule, pick off a carbon (C) atom and then combine it with some oxygen (O), which also came by way of the blood. It is most convenient that this process produces heat because this is the heat, which keeps us warm. When we exercise, a great many carbon-oxygen combinations take place so the body warms up. Then, if too much heat is produced, sweating occurs to help regulate body temperature. Unfortunately the carbon dioxide is of no use to the body so it is expelled out through the lungs and into the atmosphere, where it will float around as part of the atmosphere until it comes to a plant.

The plant has a use for the carbon. Carbon is the structural component of plants. In some ways it is like bricks which are the structural components of buildings. Trees as well as grass and flowers are all constructed from carbon atoms. This may be confirmed by heating a plant in such a way that everything but the carbon is driven away. For example, if sticks of wood are heated in a place where there is not enough oxygen to enable them to burn, almost everything but the carbon will be driven away. When only the carbon is left, the material is called charcoal. Charcoal is almost completely carbon but it will still have the shape, which the wood had before the heating started. The heating process removed the water and other things, which could evaporate or float away in the air. When this wood was alive as a tree, its leaves took in the carbon from the air and the tree was able to use it to grow larger. This is a most ingenious process to say the least.

The plant incorporates the carbon into its structure. The plant could just as easily be a carrot or an apple or a potato. All plants get their carbon from the CO2 in the air and then use it as a structural building block. In this manner the carbon becomes trapped out of the carbon cycle and will not be available for anything else until it is somehow released.

There are three ways for the carbon to be released and get back into the air again. The plants, that is, the wood or the carrots or whatever plant we have, could be burned in a stove. Burning combines the oxygen, which is part of the air, with the carbon, which is in the plant. This combination produces carbon dioxide, which will become part of the atmosphere and float away to wherever the air currents take it.

The second way for the carbon in the plants to get back into the air is by rotting. When a plant rots, the carbon in the plant is being combined with the oxygen in the air. This process is usually very slow and as it proceeds, the remains of the plant gradually disappear. Rotting is facilitated by moisture and microbes. If wood is kept damp, it will rot much faster than when it is kept dry. Rotting returns most of the plant material to the atmosphere. If this were not the case, forests would keep building up until they were hundreds of feet high and great mounds would occur wherever a forest had existed for a long time.

As discussed above, the third way for the carbon to get back into the air is by the plants being eaten. When a plant is eaten, most of the material of the plant is used by the creature, which eats it, to keep warm. The heat to keep warm was produced when the carbon from the plant combined with oxygen from the air. This process proceeds in each cell of the body.

With each of these three ways, carbon is returned to the air as carbon dioxide.

When a person breathes in, oxygen is taken into the body, which enables the body processes to proceed and we continue to live. However, the carbon dioxide, which was previously breathed out, is now part of the atmosphere and some of it is breathed back in again. When the carbon dioxide is breathed out, it becomes diluted by the rest of the air. So when a person breathes in, the actual percentage of carbon dioxide which is breathed in, is very small. As long as this percentage stays small, nothing adverse will happen to the body. (The current amount of CO2 in the atmosphere is about 387 parts per million (PPM)) (29)

In certain situations, the amount of carbon dioxide in the air increases until it becomes an annoyance. For example, if a room is crowded, there will be a lot of people breathing out carbon dioxide, and the level may rise to where it can be noticed. Carbon dioxide is not considered a poison gas. If it were poison, we would be poisoning ourselves all the time. However carbon dioxide will act like a poison if the concentrations are too high but instead of poisoning us directly, the greater amounts of carbon dioxide will displace some of the oxygen, which we need. As more oxygen is displaced and less becomes available to breath, the room will become stuffy. At this point, a person may sense that it would really be good to get a breath of fresh air. This is not the danger level but it is a warning level. A room will be noticeably stuffy when the amount of carbon dioxide has risen to two or three percent of the total air in the room. The oxygen level will still be in the eighteen or nineteen percent range whereas it is usually in the twenty-one percent range. The actual change in the amount of oxygen is not very great, but because the amount of carbon dioxide has increased to these levels, the room will feel stuffy and the air will be much less desirable to breath. If the levels continued to increase, the air would become totally unfit to breath and life would not be possible. This is the reason that life is completely impossible on planets where the carbon dioxide levels are significant. From a "window of life" viewpoint, the "window of life" will close before the level of carbon dioxide has reached the 10% range. (with certain short-term exceptions. e.g. animals can tolerate higher levels for short periods.) Above this level there will be serious anoxia (deficiency of oxygen in tissues) resulting in gasping, blood pressure drop and heart failure. (30)

On the other hand, while high levels of carbon dioxide are not acceptable, not having any is also undesirable. People certainly do not need it to breath and survive but plants need it to grow. Plant growth is enhanced when the level of carbon dioxide is increased. On occasion, this is done intentionally to enable plants to grow faster and bigger. For example, exhausting a natural gas furnace into a greenhouse will raise the level of carbon dioxide a little and enhance the growth of the plants. Without plants, we could not survive. Even if we only ate meat, the animals, which we had eaten, would probably have eaten some grass. Therefore, some carbon dioxide is absolutely necessary. While there isn't any particular minimum amount, which is necessary, there must be some or plants would not be able to grow at all. Plants are necessary for the survival of all types of animals. It may seem ironic that as the level of carbon dioxide increases, plants will grow better but animal life will become more endangered.

Carbon dioxide may therefore be considered a "window of life". There has to be some for life to continue, but if there was too much, life would not be possible.

3.15   Carbon Dioxide – Temperature Regulation

There is another way in which carbon dioxide is a window of life. This second way relates to the temperature control of the Earth. Carbon dioxide, as part of the atmosphere, is very important in helping to control and regulate the temperature of the environment. Carbon dioxide is sometimes referred to as a greenhouse gas because it makes the whole world operate like a greenhouse. A greenhouse is a place where plants are grown. The notion of a greenhouse is to have an area where the temperature on a cool day will be higher than outside. The covering on the greenhouse keeps the greenhouse warm inside even when the weather outside is several degrees colder. It does this by allowing the Sun to shine in. Glass is transparent to the light from the Sun so it shines right in. When it does, the interior of the greenhouse warms up enhancing plant growth. The plants will not grow outside on a cool day so if we need these plants before the season warms up, a greenhouse will be very useful. The greenhouse works because the Sun shines in very easily. However when it shines in and the interior warms up, the heat, which has been produced when the Sun hit the plants and other objects, cannot easily get out. This is due to the characteristic of the glass covering. Visible light goes through the glass very easily but heat does not go through the glass easily. Therefore, the greenhouse will be warmed as long as the Sun shines.

Carbon dioxide in the air acts like the glass covering on the greenhouse. It lets the Sun shine through, but it will not allow heat from the surface of the Earth to pass back through. This understanding forms the basis for the present concern over the increasing levels of atmospheric carbon dioxide. As discussed before, more carbon dioxide means bigger plants, but if the levels get too high, it will be more difficult for people and animals to breath. However, long before the levels get high enough to cause breathing problems, the greenhouse-warming effect will be very noticeable and the Earth will have warmed up significantly.

The level of atmospheric carbon dioxide started to increase during the industrial revolution. At that time, large amounts of coal were burned to fuel the factories. Coal is carbon. When the coal was formed many years before, all of that carbon was trapped out of the carbon cycle and stored away in a great carbon store-house. As this carbon was burned in the industrial furnaces, carbon dioxide was formed and released into the atmosphere. The atmosphere is very large so it would take a lot of carbon dioxide to make a difference. However a lot of coal was burned. Within a few years it became evident that the carbon dioxide level in the air was increasing. Measurements were taken and for more than one hundred and fifty years until the present time, the levels of carbon dioxide in the air have been monitored. Plants, of course, can make use of these increased levels of carbon dioxide but they cannot use all of it. Consequently, the level of CO2 in the atmosphere keeps going up and this effect may be causing the entire Earth to warm up by the greenhouse effect. There is a lot of evidence that this is happening. In addition, if the air warms up, the oceans will warm up creating another problem.

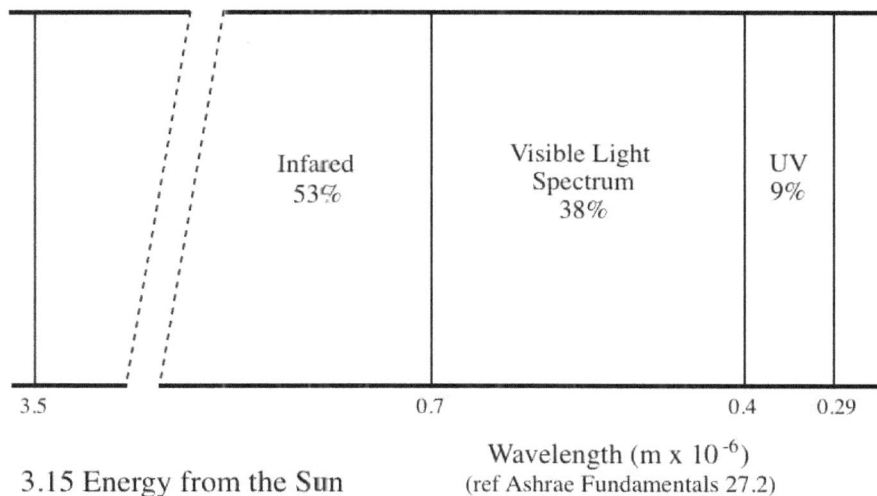

Infared 53%	Visible Light Spectrum 38%	UV 9%

3.5   0.7   0.4   0.29

Wavelength (m x 10^{-6})

3.15 Energy from the Sun   (ref Ashrae Fundamentals 27.2)

The oceans contain most of the carbon dioxide which is in the world, and it is held in the ocean water because most of the ocean water is cold. Of course the surface layer is quite warm in the tropics but the surface layer is not very thick and the ocean is very deep. Down below the surface the temperature drops until the maximum-density temperature of water is reached around 4 degrees C. Far from the tropics, even the surface is cold. The net result is that the average temperature of all ocean water is just above freezing and around 4 degrees C. (31) If the oceans could be warmed, the carbon dioxide, which is dissolved in the ocean water, would be released into the atmosphere because warm water cannot hold nearly as much. All of the carbon dioxide which would leave the water would be added to the carbon dioxide which was being produced by the ongoing burning of gas and oil. This may produce a compounding or runaway effect and the atmosphere could warm up even faster than it would directly from the burning processes. Then the curve, which represents the increasing temperature of the Earth, would be turning upward. This effect may be catastrophic because the temperature of the Earth would increase and the higher temperatures would result in still more releases of CO2 from the ocean creating a snowball effect with no real end in sight. This situation is of concern to the scientific community as well as others who are worried about the environment. It also causes concern with those who understand that there is a window of life involved with CO2 as well as with Earth's temperature and these windows may not be very wide. Life would not survive if either of these windows closed. As discussed above, some carbon dioxide in the air is necessary but too much may cause the Earth to overheat. Just how much temperature increase will be tolerable is not known. Neither is it known how much increase the carbon dioxide can cause. Of course, there are those who think that it will not cause any but since the level is going up quite steadily, any effects, which are involved, will become overwhelmingly apparent to everybody before long. Carbon dioxide can therefore be considered a "window of life" in two ways. With the first way, low levels inhibit plant growth and high levels inhibit breathing. With the second way, high levels will result in higher world-wide temperatures which will inhibit both plant and animal life.

3.16   Greenhouse Gases.

As discussed immediately above, CO2 is a greenhouse gas directly involved in regulating the temperature of the Earth. While there are several more gases which are properly called greenhouse gases, one of them is particularly important – basically just as important as CO2. Water vapor is a greenhouse gas. As the Sun warms the Earth, the heat that is produced radiates back up into the sky. Most of it would radiate right out into space if it wasn't for the greenhouse gases. The water vapor in the atmosphere reflects much of the heat right back to the Earth's surface and in this manner it helps to keep the temperature of the Earth in the habitable range. In the case of water vapor as with CO2 more moisture means higher temperatures. Within a fairly small range, this is OK but the problem with water vapor is that it produces a self-perpetuating viscous cycle. This means that the worse it gets, the worse it gets. This type of thing is also called a positive feedback loop where the results make the cause of the problem get worse. This can easily become a very serious problem for the Earth. By referring to the illustration; 7.3.1 Vapor Capacity of Air, the effect of increased temperature becomes plainly evident.

The reverse is also true. If the Earth should become chilled for any reason whatsoever the air would lose some of its ability to retain moisture. Therefore the greenhouse effect of the water vapor in the air would be diminished. This would cause more cooling. More cooling would result in less moisture in the air and so still more cooling would result. The end result would occur when the air could not really lose any more moisture so then the temperature of the Earth would settle out. Unfortunately such a temperature would be well below the comfort zone for the survival of animal life. The Earth could become uninhabitable from this effect alone. This raises the spectacle of how would the Earth ever be able to warm up if it was covered by ice and snow? It also eliminates the possibility of an ice age developing because the great amount of water which would be required to form the great ice fields would be trapped under a layer of ice in the ocean. This means that an ice age could never happen if the Earth became chilled!

Water vapor, as a greenhouse gas, is a Window of Life. A certain quantity is required to help keep the Earth in the comfort zone. If there is too much the Earth would overheat. If there isn't enough the Earth would chill. The self-perpetuating results of both overheating and chilling mean that this window is not very wide and could easily close making life on Earth impossible!

3.17 Ozone

A large part of the atmosphere of the Earth is oxygen. Most atmospheric oxygen occurs as a double atom because there are two oxygen atoms attached together. However, a small portion of atmospheric oxygen occurs as a triple atom where three atoms are attached together. It is still oxygen but when three atoms are attached together instead of just two, a molecule with a completely different characteristic is formed. This larger molecule of oxygen is called ozone and it absorbs the harmful ultraviolet portion of the incoming sunlight. (40) Much of the energy, which comes to the Earth from the Sun, is in the visible light range. (see diagram 3.15 Energy from the Sun) This is the part which our eyes can recognize. This is what enables us to see. The eyes of a person respond to energy which is in the spectrum of radiation, which is called the visible light spectrum. The Sun produces more radiation than just the visible part however, and the frequencies of these additional parts are both higher and lower than the visible light portion.

Height from Earth

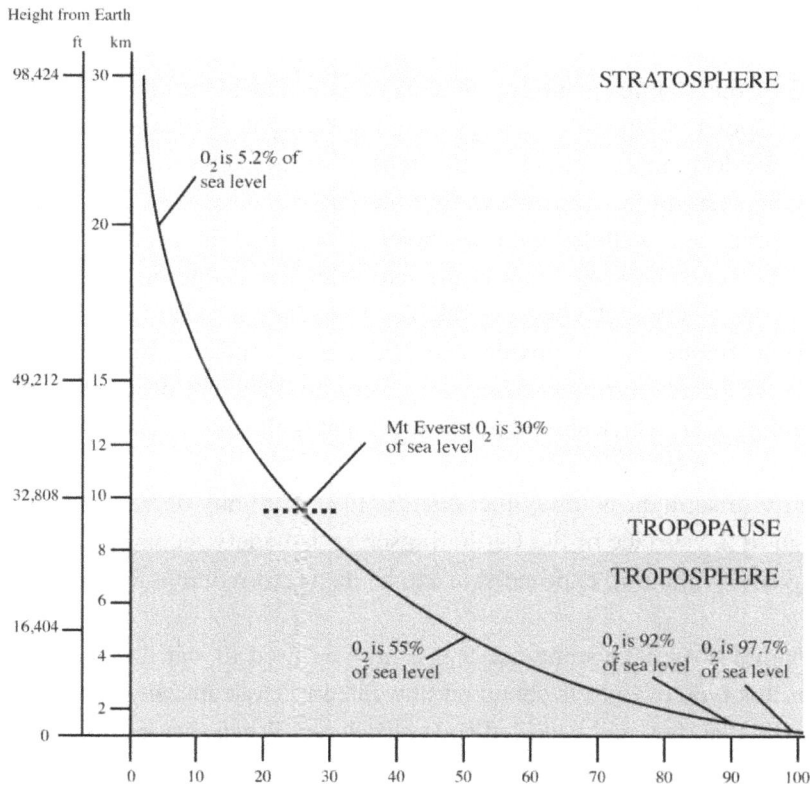

3.16 Atmospheric Pressure Compared to Sea Level

As elevation increases, the amount of oxygen, compared to sea level, decreases. On Mt Everest, there is only about 30% as much oxygen as there is at sea level. As the amount of CO_2 increases, it creates an effect similar to that of a higher elevation because the increased amount of CO_2 will decrease the amount of oxygen available. For example, from TABLE 2, if the amount of CO_2 were to increase until it was 5% of the atmosphere the amount of oxygen would decrease until it was only 11.6% of the atmosphere. From this perspective, it would be comparable to being 5km above the Earth (ie: 11 6% vs 21%).

The region below visible light is called infrared. This type of energy is heat energy. Our human eyes cannot see this energy but our skin can sense it and when it comes to us we feel warm. A warm stove produces infrared energy and this type of energy is very useful to us, but not for seeing. However this is not true for other creatures. For many different animals, especially those which go about at night, energy which is in the infrared region is very useful for seeing. The eyes of this type of animal have been designed to use energy in the infrared area just as human eyes use energy in the visible light area. In order for any creature to see anything, energy must come from the object, which is to be seen. When a deer moves through the forest at night, it does not bump into the trees. Neither does it stumble into the valleys. It simply sees where it is going because the eyes of the deer can see in the infrared region. Dogs, cats, foxes and wolves have this same ability. When they travel at night, it is not really dark to them. The only way in which people can make use of infrared energy for seeing is by using specially designed glasses or binoculars, which have been developed for this purpose. With these instruments, human eyes can receive a lot of information at night where otherwise it would just look dark.

Beyond the other end of the visible spectrum, the energy is called ultraviolet. This is because it has a higher frequency than violet. Ultraviolet energy is dangerous. It is dangerous no matter where it originates. Electric arc welders are an example of this type of danger. The arc, which is formed during the welding process, may be seen by the human eye, but the part which is seen is only a small part of the actual energy which is produced. Included with the violet color, which is seen, is a lot more energy in the ultraviolet range, which cannot be seen. This is the part which is dangerous and the reason why it is dangerous is because it has a very high energy level. The frequency in this region is much higher than in the infrared region. High frequency is equivalent to high energy and this high energy is dangerous to humans. The arc which is formed during welding has a very high content of ultraviolet energy. Eyes are easily damaged by welding arcs, which should not be looked at under any circumstances

About 9% of the energy in sunlight is ultraviolet energy. (32) With any decrease in ozone, more of this energy will get through to the surface of the Earth. Sunscreen is widely recommended as protection from ultraviolet solar energy and without it, exposed skin can be damaged within a few hours.

High energy of this nature does have some use and it may be used to sterilize things. Since the energy level is so high, when this type of light is shone on unwanted microscopic life forms, (i.e. bacteria) they will be damaged so badly that they will be killed and neutralized. When surgical instruments are placed in an ultraviolet light chamber, all bacteria on them will be killed making them safe to use again. Ultraviolet light is also used to treat sewage. Sewage is harmful, because it includes a lot of bacteria. As soon as these bacteria are killed, the water will once again be potable. When very strong ultraviolet light is shone onto a thin moving stream of sewage, all of the bacteria in the stream will be killed and the water will become safe to drink. (preferably after it has also been freed of particulate matter, odor and color)

Ultraviolet light is abundant throughout the universe. The Sun produces a great deal of it and so do all of the stars of the universe. The universe is therefore continually bathed in ultraviolet light which makes it difficult to understand how any life form could get a start unless it was well protected from all of this damaging energy. This ultraviolet light appears to a molecule just like a hammer does to a string of beads.

When the hammer comes down, the beads get smashed every time. They do not stand a chance. If ultraviolet light shines on a rock, any microscopic life forms, which are on that rock will be damaged and become lifeless.

All forms of life are very complicated. This complication includes numerous huge molecules such as proteins and DNA, which may include hundreds of atoms, all of which must be properly connected together. In some ways the situation may be compared to a chain. As long as all of the links in the chain are connected, the life form will be functional. If any link should become broken, we do not simply have two smaller life forms. Instead, we have useless material. Life may be compared to a very long chain. Even with the simplest forms of life, there are so many complicated parts involved that something as small as a bacterium could be compared to a chain which stretched all the way from the Earth to the Moon. If it should become broken anyplace along the way, the entire structure would be rendered useless.

A great amount of ultraviolet light comes to the Earth from the Sun. If it should get all the way through the atmosphere and shine on a person's skin, that skin will be damaged. Eyes are also susceptible and any amount is considered unhealthy. Also when there is an eclipse of the Sun, we should not look at the Sun because the ultraviolet light will damage our eyes. This situation would be even worse if the Sun were observed through binoculars because the binoculars gather and concentrate much more light and our eyes would consequently be damaged even more.

All of the damaging ultraviolet light from the Sun would get to the surface of the Earth and cause a great deal of harm except for ozone which is very high up in the atmosphere. Ozone is formed in the atmosphere and it has the beneficial ability to block ultraviolet light. This layer of ozone protects the Earth from receiving too much ultraviolet light.

Ozone protects the Earth from incoming ultraviolet light by absorbing it. As a result, this portion of the Sun's energy does not continue on to the surface of the Earth but is converted to mechanical activity among the ozone molecules. The ozone heats up and heats the stratosphere, (which is the layer of atmosphere including the ozone) until it becomes warmer than the troposphere below it. While the lower atmosphere (i.e. the troposphere) gets steadily colder as elevation increases, at the top of the troposphere, the tropopause, the temperature stops getting colder and starts to increase. In the next higher layer, the stratosphere, the air temperature steadily increases with elevation. In the lower regions of the stratosphere, the temperature may be minus 70 F. In the higher regions, it may be up around freezing because the ozone has absorbed the ultraviolet energy from the Sun and converted it to heat.

By heating the atmosphere in this manner, the ozone has provided a further benefit to life on Earth which we recognize as a barrier to rising storm clouds. When they reach the tropopause, storm clouds run into a 'no pass zone', which prevents them from rising further. This barrier puts a limitation on the updraft power of a storm. If there wasn't an ozone layer, there would be no limit to how high a storm cloud could rise, enabling the associated updrafts to be increased dramatically. As a consequence, hail would be much larger. With the greater updrafts, the swirl (or massive rotation) which commonly accompanies large updrafts would be greater and any tornadoes, which spun off as a result, would be a lot more damaging.

Unfortunately, ozone may be destroyed. If certain other atoms are present, (e.g. chlorine) the extra atom, which turned atmospheric oxygen into ozone in the first place, is knocked off and only atmospheric oxygen remains. We, of course, have good use for this oxygen but it does not block out the harmful ultraviolet light. The ozone depletion process is already happening and has been recognized by the scientific community as 'holes in the ozone layer'. Also because we are repeatedly encouraged to use skin protection due to apparently increasing levels of UVA and UVB rays (i.e. ultraviolet light) it is suspected that the ozone is really being depleted everywhere.

Ozone is a "window of life". It is needed to block out the harmful ultraviolet light from the Sun. It also generates a 'no pass zone' for storm clouds and thereby limits how high they can rise and their resultant energy levels. Without this limitation, hailstones would be much larger, the winds accompanying a storm would be much higher, the total rain produced would be greater and any spin-off tornadoes would have more energy and would therefore be much more devastating.

3.18   Oxygen

Oxygen is an important component of our atmosphere and makes up about twenty-one percent of it. Nitrogen constitutes a greater portion at seventy-eight percent. While oxygen easily reacts with a great variety of other elements, nitrogen hardly reacts at all and thereby provides a stabilizing influence. It is most fortuitous that our atmosphere formed with this particular combination because if the atmosphere consisted of more than twenty-one percent oxygen, it would be easier for things to burn. Forest fires would flare up more readily and anything else which could burn would do so more easily. This type of thing would become more and more exaggerated as the percentage of oxygen increased. During the early days of the American space program, a great tragedy developed during which three astronauts lost their lives while sitting in a space capsule which was still on the ground. (4) Somehow a fire started and because the atmosphere of space capsules at that time was almost one hundred percent oxygen, the fire flared up quickly and could not be put out in time to save their lives. Subsequent to that tragedy, the air in all space capsules was modified to be closer to the atmosphere of the Earth. While any increase in oxygen would enable fires to burn more easily, a little more would not result in dramatic changes to survivability for most of us. If, for example, the amount was twenty-three percent instead of 21% (0.21 ATA or 21% of one Atmosphere Absolute), life would still carry on quite readily for most people. Unfortunately, a significant percentage of test candidates showed toxic effects at 0.22 ATA after only nine days. (.22 instead of .21 is a very small increase) Thankfully however, most people can tolerate a 50% increase in oxygen pressure for indefinitely long times without the onset of pain. However, there is an upper limit. As the oxygen pressure increases beyond a 100% increase over normal, the time until the onset of pain is reduced to days. (5,6)

A similar situation develops in the other direction. If the percentage of oxygen is reduced due to an increase in CO2, trouble shows up right away. Carbon dioxide is not a poison gas. If it were a poison, we would have exterminated ourselves long ago as we produced it in our bodies. All cells of an animal body take in oxygen and carbon and produce carbon dioxide. The carbon dioxide leaves the body but we are not poisoned on its way out. It leaves in fact because the amount in the bloodstream is greater than the amount in the air in the lungs. The greater pressure in the body forces the carbon dioxide out into the air in the lungs which is then expelled out of the body with the next breath. It is readily possible however, if a room is crowded with a lot of people and the air in the room is not being circulated, that the amount of carbon dioxide will build up. Unfortunately, it does not have to build up very far before it will be noticeable and start to cause trouble. At first the room will feel stuffy. As the percentage of carbon dioxide increases, the room will become increasingly stuffy until breathing becomes quite difficult. At 2% (instead of the normal 0.04% i.e. 50 times normal) CO_2 a room will be stuffy. Instead of the regular proportions of 78% N_2 and 20.6% O_2, there will now be closer to 76.5% N_2 and 20.3% O_2 and 2% CO_2. The carbon dioxide has displaced some oxygen and lowered its percentage of the total. While the amount of oxygen which has been displaced is quite small, and the result is not overly serious, it will certainly be noticeable. This disagreeable trend will continue as the oxygen level is reduced. By the time 10% carbon dioxide has been reached, this air would be considered unusable. The relationship of the gases in the room would then be about 71% N_2, 18.9% O_2, and 10% CO_2. While there is still a lot of oxygen, the compounding effect of the CO_2 is making it very difficult for animal life to continue.

Carbon dioxide and oxygen are closely related. While it is quite appropriate to discuss the relative levels of carbon dioxide and oxygen as above, this would only be a local consideration and could not happen outside of a closed room. The reason for this is because one unit of carbon dioxide includes two units of oxygen. This means that if the burning of coal and oil were allowed to continue, until the level of carbon dioxide in the atmosphere had increased to 10%, the level of oxygen would have decreased by 20%. These relationships are shown in the following tables.

TABLE 1	Closed Room Example			
Atmospheric Gas	Nitrogen	Oxygen	Carbon Dioxide	Other
Normal amounts	78%	21%	trace	1%
Stuffy room	76.6%	20.3%	2%	1%
Unusable room	70.4%	18.7%	10%	0.9%

TABLE 2	Open Air Example				Equivalent elevation above sea level for O2
Atmospheric Gas	Nitrogen	Oxygen	CO2	Other	
Normal Amounts	78%	21%	Trace	1%	0 ft
Low CO2	78.4%	20.1%	0.5%	1%	1100 ft
Higher CO2	78.78%	19.19%	1.01%	1.01%	2640 ft
Approx. 5% CO2	82.1%	11.6%	5.26%	1.05%	14,800 ft
Approx. 10% CO2	86.7%	1.11%	11.11%	1.11%	73,000 ft

The relationships shown in Table 2, are also shown in the diagram, 3.16 Atmospheric Pressure Compared to Sea level.

The reason that both nitrogen and the trace gases become a larger part of the total is because every carbon dioxide molecule that is produced uses two oxygen molecules. With more CO_2 in the air there would actually be fewer molecules of air. The atmosphere would be heavier though because the CO_2 molecule is heavy and we are adding the carbon to the total weight of the air. Sea level pressure would be higher with more CO_2 in the air. (Please refer to the appendix for this calculation.) An even greater problem would be that the heavier CO_2 would tend to sink to the surface of the Earth. Breathing problems would therefore be increased more than the actual amount of CO_2 would indicate. The bottom line is that we really do not want an atmosphere with significant amounts of CO_2 in it. As we can see from Table 2, the related problem is the exaggerated reductions in oxygen compared to the increases in CO_2. While the increase in CO_2 would be very troublesome, the reduction in O_2 would be even worse.

While the tolerable limits of oxygen seem to vary somewhat depending on the nature of the study being conducted, and the physiology of the people being studied, it is clear that oxygen is a "window of life" and a narrow one at that. A change either way of a relatively small amount and this particular "window of life" will become narrower. The carbon dioxide window is closely interconnected. The carbon dioxide window would also close as its level increases and this would force the oxygen window into closure.

3.19 Wind

Wind is a window of life. Wind is atmosphere, which is moving. When wind passes over a body of water, it picks up moisture from the water and may carry it for thousands of miles. In this way land, which is far from water receives moisture and can be used for food production. If there wasn't any wind, the moisture-laden air over the water would just stay there. Then it would not rain on the land and without rain the ground would dry out and crack open. As a result it would not produce any crops so there would not be any land-produced food to eat

Wind is useful for transportation. Wind-powered boats have sailed completely around the world. This still happens. In fact, up until around 1850 wind-powered boats were the primary means of long-distance water travel. The entire Earth has been explored by sailors using wind-powered boats and invaders certainly made use of wind during their military maneuvers, even during the time of the ancient Egyptians and Greeks. (7) Since that ancient time numerous explorers and military groups have used the power of the wind to travel all over the world.

Wind is also useful for pumping water. Large areas of water-covered land have been drained using wind-powered water pumps. Wind-powered water pumps have also been used to pump water from wells. This type of water pump is still used in many places around the world right to the present time. Wind is very useful but if the wind gets too strong, it becomes dangerous.

In windy mountainous areas, the average annual wind speed determines how high trees will grow. As soon as the average annual wind speed exceeds a certain level, trees will not grow very high but will remain stunted. This type of observation has occasionally been made to help determine where to place wind-powered generators. (e.g. the Smith-Putnam) If the trees are small in a particular area, where the other factors which would favor tree growth are present, excessive average annual wind speed may be the cause of the stunted growth. If the wind should become more ferocious during stormy periods, these stunted trees will have a better chance of survival. Conversely, if the wind speed in a particular area has allowed the trees to grow normally, they would be in danger of destruction if a wind storm comes along.

People, of course, will also have a difficult time in a wind storm. Not only is there a danger from flying objects but people can be lifted right off the ground and blown away. Without shelter during severe wind storms, life would be completely impossible.

Wind is therefore a "window of life". A certain amount of wind is needed but there is a definite upper limit to how much can be tolerated.

3.20   Mutation Load

A mutation is an alteration (from the way the plant or animal is normally) in the genetic material of a plant or animal. Unfortunately, a mutation in a plant which seems desirable to people may not really be an advantage for the plant. Such a plant will usually become less fit to survive if it is left once again in the wild. (16)

An even greater concern however, relates to mutations among humans. Prior to the advent of modern medicine, a mutant, being less fit to survive, probably would not survive and the genetic modification would thereby terminate. However, as medical technology improved, many mutants may now be managed into adulthood and pass their mutated genes to the next generation. The genetic defect may thereby be preserved and show up in the children. Survival may be desirable for the particular person involved but it is far from desirable for the population as a whole. As a result, the number of diseases in the human population of the world, currently attributed to genetic factors is greater than 5000. (15) (It must be noted that these mutations are not being referred to as survival advantages. They are always referred to as diseases.) The number of mutation-related diseases is growing and is referred to as the mutation load. The mutation load is getting heavier and will continue to get heavier as more medical breakthroughs are achieved. Medical advances can never cure a genetic defect. They only identify procedures to enable the mutant to live a reasonably normal life. Unfortunately, this includes having children and passing the defect to the next generation. So the mutation load continues to grow. It is a significant "window of life" and it is shifting towards closure. It is in fact conceivable that the entire human population could become mutant and that the number of related diseases could number many thousands. At what point life would become totally untenable is hard to say but the accumulation of genetic defects in the general population is not good news. While that may be serious enough, accumulating them into any single person would be even less desirable. With the ever-increasing number of mutations being preserved and passed down to succeeding generations, this is bound to happen. This, unfortunately, is the direction we are heading. With this development, the "window of life" might not completely close but it will certainly become much narrower.

3.21 Vapor Canopy

As discussed in considerable detail in a later chapter, a water vapor layer or vapor canopy may have previously existed above our present atmosphere. This would have protected the Earth from a portion of incoming cosmic radiation and would also have provided a moderating effect on world climate. The Earth would have been universally warm without extremes of hot and cold. A layer of vapor above the oxygen-nitrogen layer would have been quite stable and would have readily remained in place as noctilucent clouds do at the present time. (33) However, if a major asteroid impact occurred, the water vapor layer would have been destroyed. As it collapsed, it would have partially cleansed the air of the debris from the impact. However, a post-impact winter would have settled in and the Earth would have spiraled down into a deep freeze because of the loss of the greenhouse effect of water vapor. Neither plants nor animals could have survived both the impact and the subsequent freeze up. Since a canopy does not exist at the present time and the Earth does experience extremes of hot and cold, the vapor canopy may therefore have been recognized as a "window of life". When it existed, the climate around the Earth was moderate and very favorable for the existence of life. Since its removal, the climates of Earth have been more extreme. The more important factor however, relates to the fact that since it does not currently exist, it would not be available to possibly abbreviate another post-impact winter which would certainly develop if the Earth is ever again struck by an asteroid. The immediate destruction caused by such an asteroid would be followed by a refrigerated Earth. Life on the Earth would therefore terminate. While the existence of this vapor canopy is only hypothesized, the fact that we do not have one means that the next significant asteroid impact will terminate life on Earth and the window of life on Earth will close.

3.22 Solar Orbit

While the Earth orbits the Sun, the Sun also has an orbit and it is around the center of the Milky Way Galaxy. Fortunately, the Sun's orbit is very specific. If it had any other orbit, the ability of life to survive on the Earth would be reduced. The Sun is in a special location in the galaxy at the co-rotation radius. Only here does a star's orbital speed match that of the spiral arms - otherwise the Sun would cross the arms too often and be exposed to supernovae. The Sun will not go too near the inner galaxy where supernovae, (i.e. extremely energetic star explosions), are common. It also orbits virtually parallel to the galactic plane whereas crossing that plane would be disruptive. (45)

3.23 Other Windows

The actual number of factors or windows of life, which must be well within a fairly narrow range for life to survive and thrive, is by no means exhausted by the above examples. As a further example ocean plankton (microscopic plants that grow in the ocean) being at the very bottom of the food chain are therefore necessary for every other link in the food chain including fish and the humans that eat the fish. Also ocean plankton consume a great amount of carbon, thus removing it from the atmosphere and helping to reduce CO_2 with its associated warming effects. (42) Reducing plankton production would therefore have a negative effect on the "window of life" and disintegrating plastic is doing just that.

A very large amount of plastic already contaminates the ocean and when it breaks down to the molecular level, ocean plankton take it up and thereby introduce it into the food chain. Could the "window of life" on Earth be closed by disintegrated plastic?

While the above list includes several of the more obvious parameters, which are absolutely necessary for our survival, a complete list would include several thousand factors. The human body is incredibly complex as any health care professional will testify. Other so-called lesser life forms here on the Earth, are hardly less complicated. These complicated inner workings of any and all forms of life each require that a complete set of favorable circumstances be in place in order for survival to be possible. Unfortunately, at the present time, several factors in our environment are shifting away from their respective optimal zones.

3.24   Elsewhere in the Universe

In spite of so many favorable factors, the "window of life" is still very narrow and tenuous on Earth. Circumstances are even worse elsewhere. The Moon has extreme temperatures, is treeless, lacks an atmosphere and has no significant supply of water. The surface of Venus is 900F, its surface pressure is 90 times higher than Earth (9) and the clouds only allow about 2.5% of the light from the Sun to get through to the carbon dioxide atmosphere. Far from being a sister planet with heavenly features, it more closely resembles the other destination intended for departed souls. Mars is so cold, airless and waterless it is unthinkable as a habitation for people. All pictures from the surface of Mars show terrain which we could only describe as an endless desert. (41) There is no atmosphere to provide protection from either ultraviolet energy, cosmic radiation or micrometeorites. Mercury is extremely hot, airless and waterless. When the Sun comes up on Mercury, it stays up for 88 days (34) The extremely long day together with proximity to the Sun results in surface temperatures hot enough to melt lead. The Sun actually comes up, moves westward, stops, moves back to the east, stops and moves westward again. As a result the surface temperature can rise to 750 F. (10) If the surface were clay, it would be almost incandescent. Jupiter is a great ball of gas, which swirls by so fast that even granite would be eroded. Saturn may be a great ball of ice and Uranus and Neptune are dark and so cold that their surfaces are more than 300 F below freezing. (11) There is no planet or moon in the solar system which has even a small fraction of the factors needed to enable life. The rest of the universe may not be any better. While several hundred stars have been identified which have planets, virtually all of these planets are so massive that their gravity would toss a small planet like Earth right out of the vicinity. In fact, having a large planet is similar to having a double star. A star must be singular like our Sun to provide stable gravity and energy environments. Otherwise there is no hope whatsoever that a planet similar to the Earth could have a stable, predictable, circular orbit and a constant energy input (both of which are absolutely necessary for constant temperature regulation). Most stars are multiple and some are systems of three large objects. Further, many apparently singular stars have a nearby object like a brown dwarf which is an object with characteristics between a planet and a star. In order for a life-supporting planet to exist, having a singular star as the host is the first necessity. There must also be an appropriately-sized moon at just the right distance to enhance orbital stability as well as axial tilt stability. However, even if an Earth-like moon-stabilized planet were found in such a system, it must be then be endowed with at least all of the "windows of life" listed in the foregoing sections (as well as a great many more) before there would be any remote possibility that complex life forms could exist there. The probability of this happening is so low that it only exists mathematically and for practical purposes is zero. (It has been argued that the probability of all of the necessary and favorable factors co-existing is indeed low but since it happened here, why wouldn't it have happened elsewhere? So it isn't the low probability alone that is of interest but the low probability that a very specific set of conditions will occur. Life cannot exist with just any set of conditions (having a low probability of existing because every remote planet also has a low probability of existing) but only with the very specific set of the necessary conditions. It is the probability that the very specific set of the necessary conditions happening is virtually zero that is of interest.) This leads immediately to the conclusion that 'life simply didn't happen elsewhere'. This makes Earth look more and more attractive and more and more like our only chance. (The question of life on other planets will be discussed further in the chapter entitled 'Predictions for the Future' as well as in 'The Impossibility of Extraterrestrial Life.')

3.25   Summary

Life on the Earth is dependent on numerous factors. Many of these factors have a range within which life may survive and thrive. If the life-enabling limits of any of these factors are exceeded, life will not thrive but will be extinguished. Collectively, these factors may be compared to a many-layered window where all of the various layers of the window must be lined up to allow the light to get through. Every window properly aligned, is the necessary condition for life to exist. If there is a significant shift one way or the other with any one of them, the light will not get through and life will not exist. A shift in any one window would be serious enough but if there were a simultaneous shift in several of the windows, there would be no hope at all that life could carry on. Life is very fragile. A very elaborate support system absolutely must be in place or lifelessness will prevail. The factors, which have been discussed above are only a few of the total number, which are actually required to maintain life. However, they do illustrate that the complexities of living things are simultaneously dependent on numerous parameters not only being present but being respectively within fairly narrow ranges. All is well as long as all of the windows are aligned. But all life on Earth will be affected if even one window shifts too far to either side.

While all the windows discussed above are circumstances windows, several of them are also temporal windows. In this second way, they are circumstances that are changing with time. They do not remain constant. The times involved are in several cases, quite large but in other cases they are not large. The size is, of course, related to one's world view in the first place. If extremely large amounts of time are integral to one's thinking, the largest temporal windows which have been discussed will not seem very large at all. In particular the saltiness buildup in the oceans and the retreat of the Moon from the Earth, will only involve a few hundreds of millions of years and this may not seem like a huge amount of time. To those whose world view involves much shorter times, the reduction of magnetic field strength and the increase in atmospheric CO_2 will involve comprehensible times, while ocean saltiness buildup and lunar departure are occurring too slow to be of much interest.

Unfortunately, at the present time, several of the windows of life are shifting. The fear is that the shift is away from optimum. Of course, when a shift like this occurs, there is no way to tell how far it will go or where it will end. The current warming trend is an example of this type of uncertainty. The Earth really does appear to be warming up. A certain amount of increase may be accepted without alarm but how far will the temperature shift? This is, of course, not known. The fear is that the shift will take us out of the habitable range and if this occurs, there will be no human life left on the Earth and possibly very little of any other type of life. Also the ozone layer is being depleted. Will the depletion activity terminate or will it continue? Depletion is not in our favor and it will be in our best interest if it does not continue. Possibly the most alarming shift is the increasing length a day. This will result in the Earth's axis of inclination increasing out of control and this will make the Earth uninhabitable within a time frame of 1,000,000 years and possibly much less. Many windows collectively form the "Window of Life", which will only exist as long as every individual one of them is in its particular life-enabling range. Otherwise life on Earth terminates and it appears that this will happen well within another 100,000 years!

4. Chaos and Bad News

4.1 Introduction

At one time in the past, all of the "windows of life" were lined up. The air temperature was well within the comfort zone. It was in a range which allowed human activities to proceed easily. At forty below on any temperature scale, very little human activity can proceed. Neither can it proceed when the temperature is above body temperature. However if a person can proceed with their activities either without a heavy coat or without sweating profusely in the heat we can state that we are operating in the human comfort zone. In fact, it appears from the evidence which is available, that once the temperature all around the entire Earth was in the comfort zone. Spitsbergen Island which is north of Norway in the high Arctic, is above the 75th parallel of latitude but coal has been found on Spitsbergen. (5) This is certainly evidence of a warm climate. Also high in the Canadian arctic on Axel Heiberg Island, the remains of trees have been found. (6) If they did not grow there, they at least had to drift there. Since both of these islands are currently ice-locked much of the year and only experience a very brief respite from the cold during the summer, it is totally inconceivable that any of these warm weather indicators could happen now. Therefore we may conclude that when the coal-forming plants were growing and the trees were growing, it must have been a lot warmer in the arctic than it is now.

Coal is material which was once living plants. The known amount of coal in the world includes enough material to equal fifty times the amount of material that may presently be found in the biosphere of the entire Earth. (7) The coal deposits therefore indicate that plant growth was once exceedingly abundant. Since abundant plant growth requires plenty of carbon dioxide, the atmospheric level of carbon dioxide must have been considerably higher than it is at the present time. Also vast areas of plants would require similar areas of solid fertile ground as well as an adequate supply of moisture. This moisture, however, could not have been in the excessive range or the plants would have washed away. It was within the range or window which enabled the plants to thrive.

Abundant plant growth means that there was plenty of food available for all types of creatures as well as for any humans that were present.

Such favorable conditions for both plant and animal life do not exist at the present time. In order to upset these ancient conditions around the whole world would have required an event of incredible magnitude and world-wide in extent. Such an event would have virtually closed most of the "Windows of Life" and they would not have reopened again until the period of chaos and uncertainty had passed. And they obviously did not reopen to the same extent as before. Neither plant nor animal life is abundant in the high arctic these days and life is an increasingly difficult struggle in tropical areas as well.

The event which occurred was the near-simultaneous impact of several large asteroids with the Earth. (10) Unthinkable chaos accompanied these impacts and the climate of the entire Earth was changed forever. After the chaos had subsided and the "Windows of Life" had reopened, life around the Earth became re-established and carries on to the present day. However living conditions on Earth are not ideal. The

119

current shifts in climate, together with long-term changes in some of the other parameters necessary for life, indicate that the Windows of Life are drifting towards closure. While drifting is better than the abrupt events of the past, the final result will be no less dramatic.

4.2 Asteroid Avoidance

An asteroid is a large rock drifting through space. There are a large number of asteroids drifting through space (at least one billion, (1)) but most of them orbit the Sun between Mars and Jupiter. Occasionally a small one comes to Earth as a burning streak and lands so hard that it makes a hole in the ground. These objects are called meteorites and a typical one would be as large as several footballs. By comparison, the asteroids which so seriously upset living conditions on the Earth long ago would have been measured in miles instead of inches and their destructive energy would have been far beyond everyday experience. After all, mountain ranges are not formed every day. The formation of a mountain range requires an unthinkable amount of energy but it is well within the amount of energy available from a large asteroid if one should come crashing to Earth.

During the last few years there have been several incidents of asteroids passing very close to the Earth. 'In 1936 Adonis (about 1.27 miles dia. or 2 km (75)) appeared and missed us by less than 12 days.' (72) In 1937, at a time when telescope technology was much less advanced than it is now, an asteroid named Hermes passed the Earth about 1.8 times as far away as the Moon. This asteroid is dumbbell shaped like two potatoes stuck together and is about one km. long. It came again in 1942 and this time was slightly closer at 1.6 times the distance to the Moon. (30) In 1989 there was a six hour miss by one which was 1 km. in diameter. (2) Improved detection methods are probably responsible for the rash of incidents during the following years. In 1991 an asteroid named 1991 BA passed the Earth on Jan. 18 at only 100,000 miles. This is less than one-half of the distance to the Moon. (3) In 1994 an asteroid named 1994XM1, which had the mass of a large house and was moving at 112,000 miles per hour, was within 14 hours of hitting the Earth before it was discovered. (4)

While all of these incidents are of interest, at least the asteroid in each case was detected before it actually passed the Earth. This apparently was not the case in 2002 when an asteroid was not detected until it was already past. It was not detected until June 17, 2002 but it had apparently passed within 120,000 km. (75,000 miles) of the Earth three days earlier. 'The asteroid was not detected until three days after it sped past Earth on June 14. …The asteroid, provisionally named 2002 MN, was travelling at more than 37,000 kilometers per hour (23,000 miles per hour) when it was spotted, Stokes said in a phone interview from Lexington, Mass., where he is associate head of the aerospace division of MIT Lincoln Laboratory. Light in weight but with a diameter of between 50 and 120 meters, (165 and 400 feet) 2002 MN was big enough to have caused the kind of devastation wreaked in Siberia in 1908, when an asteroid that exploded above Tunguska flattened nearly 2,000 square kilometers (736 square miles) of forest.' (70) A similar situation occurred again in 2010. 'This flyby (referring to asteroid 2005 YU55) will be the closest by any near-Earth asteroid with an absolute magnitude this bright since 1976 and until 2028. … having said that nobody saw 2010 XC15 during its close flyby within 0.5 lunar distance in 1976. (which) … wasn't discovered until late in 2010.' (78)

All of these encounters were serious enough but at least they appear to have been one time encounters. Asteroid Toro (3 km dia. or 1.87 miles 'Sometimes called "Earth's second satellite" (i.e. second to the Moon) due to orbital resonance, 1685 Toro … orbits the Sun in an 8:5 resonance with Earth and a 13:5 resonance with Venus.' (77)), on the other hand is a repeat visitor. This asteroid 'travels around the Sun eight times in five terrestrial years, meanwhile crossing Earth's orbital plane 10 times.' (76) With such a regular pattern, predicting a possible encounter should be much easier.

If an asteroid should strike the Earth, it would come with very little warning or possibly no warning at all. 'Right now, the most probable amount of warning we'll have for an asteroid impact is zero, because we don't know where most of them are, says Robert Jedicke, 46, a University of Hawaii astronomer.' (60) Even if an asteroid were as big as a house, it might not be detected until it was quite close. This was apparently the case with 1991 BA. Even though it was 30 feet in diameter (22), it was only noticed as it sped past. (25) Since not very many people are continuously looking for asteroids, unless one was quite large, and had previously been tracked, detection might not occur until just before a possible encounter. This was certainly the case with 1994 XM1. By the time it was detected and its speed was measured, it was within 12 hours of hitting the Earth. As mentioned, the astronomers had measured its speed but when they computed its trajectory, they were not certain whether it would hit Russia or Canada. (4) 'I asked them how much warning we would receive if a similar hunk of rock was coming toward us today. "Oh probably none," said Anderson breezily. "It wouldn't be visible to the naked eye until it warmed up, and that wouldn't happen until it hit the atmosphere, which would be about one second before it hit the Earth. You're talking about something moving many tens of times faster than the fastest bullet. Unless it had been seen by someone with a telescope, and that's by no means a certainty, it would take us completely by surprise."' (27)

A similar scenario involves comets. Comets usually become visible because the Sun illuminates their tails. However, if they travel close to the plane of the Earth's orbit or if they approach from the far side of the Sun, they may not be noticed. 'If their orbit is the same plane as ours they lie in the line of site of the Sun. This makes it very hard to see them. …Several examples have been discovered during total eclipses. …The comet Howard-Koomen-Michels 1979X1 was one of these. The cameras on the US Air Force Satellite recorded the comet closing on the Sun at 150 miles per second.' (73) 'Within a matter of seconds the comet, larger than the Earth, (i.e. the head alone was larger than the Earth) was destroyed.' (74) It is disturbing enough that this comet wasn't noticed until it was closer to the Sun than the Earth. It is even more disturbing when its size is recognized. An object that large would cause havoc and catastrophe even if it missed the Earth by tens of thousands of miles. The tidal waves alone would wash inland for hundreds of miles and the associated mayhem would involve the entire world!

This type of uncertainty is in stark contrast to the impression which is occasionally made by the popular media. For example in the same publication, the following statement was made. 'Amazingly, astronomers have calculated the future orbits for all of these, including asteroids 1999 AN10 and 1998 OX4, the only two known to have any chance of striking Earth in the next 50 years.' (4) There are two problems with this type of misleading statement. The first relates to the fact that new discoveries are made every year.

The second relates to the very real problem of making a precise orbital prediction. Such uncertainties are well recognized by experts in the field. '"Looking statistically at the asteroid population, maybe 50 times a year a 100-meter-class asteroid passes within a lunar distance of Earth," Stokes said, "But only a handful of such asteroids that have passed within the Moon's orbit have been spotted by asteroid search programs."' (70)

We have the asteroid surveillance team at Kitt Peak Observatory in the southern United States to thank for many asteroid discoveries. Even then, in several cases, potential threats to the Earth were only noticed a few hours before closest approach. But more pertinently, the program results in identifying twenty new asteroids in the vicinity of the Earth every year. (4) How then do we 'know' that there are only two that have any chance of hitting the Earth in the 'next 50 years'? The fact is we do not know any such thing. Further, with such uncertainties in mind, one can only speculate on how many approach and leave without ever being detected. After all, a telescope can only point in one direction at a time and the smallest of these objects are very difficult to see unless they are quite close.

Secondly, it is simply not possible to make a precise prediction of an asteroid's pathway. Compared to the planets, asteroids are not very large. Therefore their movements, which are the result of gravity in the first place, will be continually modified by gravity. In particular it is understood that the gravity of Earth is not exactly the same from different directions causing any approaching asteroid to deviate slightly this way and that. Hence the uncertainty in predicting the impact site for 1994 XM1. After all, both Russia and Canada are very large countries and this particular asteroid was not very big. Then, just to confirm these prediction difficulties, it missed the Earth altogether. This was certainly good news at the time but it further underscores the difficulty involved in making accurate predictions.

Avoidance is a two-part problem. The first, as mentioned, is the difficulty in accurately predicting an impact site. The second relates to the extent of damage that would result from an impact. From a damage perspective, it is conceivable that an object which was only a few feet in diameter could be avoided, but how could an object several miles in diameter be avoided? The catastrophic devastation from such an object would be so extensive that even the most organized community emergency plans would not be adequate. In fact, the whole community or even an entire state or province could become engulfed in the chaos and completely disappear.

We have observed that small incoming objects slow down when they enter the Earth's atmosphere. When the space shuttle first enters the upper atmosphere, it is going nearly 20,000 miles per hour. Very soon it slows down until it can land like an airplane. However, an object as large as the White House will not slow down and the Earth's atmosphere will have no effect on its speed at all. A large object will impact the surface of the Earth at practically the same speed with which it entered the upper atmosphere and it would pass through the entire atmosphere in less than one second. When a small object strikes the Earth, damage is restricted to a few feet around the impact site. (1) This would not be the case with a large object. Devastating results from such an impact would be felt even a thousand miles away.

The uncertainty concerning the possible sudden arrival of a comet is also recognized by astronomers. 'The threat of cosmic scale catastrophes remains very immediate. Indeed, civilization as we know it could be changed or even doomed within the lifetimes of our grandchildren by the collision of the Earth with a comet. Comets, which are composed of both rock and ice and which can appear from the depths of space at any time, are also capable of striking the Earth. The consequences of such events in the past history of our planet are only now becoming apparent. We can only imagine the damage they could inflict in the future. (32)

4.3 Lightning Similarity

If an asteroid approached our planet, the first contact it will have with the Earth is with the atmosphere. The effects of an asteroid passing through the atmosphere may be compared to the effects of lightning striking the Earth. When lightning strikes the Earth, the atmosphere is disturbed for the full length of the lightning bolt. A lightning bolt may originate in a cloud 15,000 feet above the Earth. As the lightning, which is actually an electrical current, proceeds between a cloud and the Earth, there are therefore 15,000 feet of atmosphere, which are disturbed. This partially explains why thunder, which is the sound that lightning makes, is so loud. The other reason why the thunder makes such a loud noise is the high speed of the electrical discharge. (The lightning that we see is actually an electrical discharge between the Earth and a cloud. It is as if someone charged a monster battery and then shorted it out.) As the lightning strikes, the electrical current of the strike heats the air. This heating effect is very rapid and the resulting local expansion of the air is very rapid. Both within and near the lightning bolt the air expands. (9) A pressure wave is thereby formed and it propagates outward away from the area of the lightning. The lightning is over very quickly but the pressure wave will continue to expand outward for a great distance. In fact it will expand outward until it eventually becomes undetectable. When lightning strikes, thunder, which is the sound we experience from this pressure wave, may be easily detected for several miles in all directions. On some occasions, the rumble of thunder may be heard for twenty of thirty miles from where the lightning actually struck. This is the result of an atmospheric disturbance, which was possibly 15,000 feet high and a few inches in diameter. What would happen if the disturbance involved the entire height of the atmosphere and was several miles in diameter?

4.4 The Tunguska Event

In 1908, a very significant atmospheric event occurred over central eastern Russia. A large object entered the atmosphere and exploded. An impact site was never found. However, a ground-zero location was found. Starting at this location, trees were flattened outward for more than thirty miles in every direction. It was also reported that a man was knocked unconscious while he was standing more than sixty km. away. (8) Since the object never actually touched the ground, the devastating effect was attributed to the pressure wave which the exploding object generated in the atmosphere. This wave propagated outward and was so strong that when it reached the Earth several thousand feet below, the forest was leveled. The only known witnesses to this event were on a railroad several miles away. It is fortuitous that the region was sparsely populated because being a nearby, unprotected witness would probably have resulted in death. If the trees were flattened, what chance would any witness have had as demonstrated by the report of a human casualty? It was speculated by the investigating astronomers that the damage, while significant, may have been caused by an object as small as 100 feet in diameter! (26)

4.5 Asteroid Shock Wave

When an asteroid enters the atmosphere, the atmosphere is pushed out of the way. It will be pushed out of the way as fast as the object is traveling. The reason this effect is so dangerous is due to the following two factors: (i) the speed of the asteroid, and (ii) the size of the asteroid. There are two ways to begin to understand this type of event. The first is by comparison with more familiar scaled down events and the second is by observing, if possible, a similar event.

4.5.1 Bullet Comparison

An asteroid passing through the atmosphere may be compared to a bullet passing through the atmosphere. After a bullet is fired, an atmospheric shock wave accompanies the bullet the length of the distance travelled by the bullet. This shock wave is a region of high-pressure air which is formed as the bullet displaces the air and that air, in turn, displaces more air. The shape of the shock wave is similar to the shape of the bullet, but it is larger than the bullet like the wave that a boat generates larger than the boat itself. As the bullet passes any particular location, air is rapidly forced out of its way and pressure builds up around the bullet pathway. Then, after the bullet passes, the air will move to reoccupy the space vacated by the bullet. Filling in the space takes longer than the time to move the air out of the way because the positive displacement force of the bullet is greater than the reoccupation pressure of the displaced air. The result is that the space behind the bullet is almost void of air for a short period of time. However, air which is directly in the path of the bullet simply must get out of the way and it must do so at a speed determined by the speed of the bullet. Air just outside the path of the bullet, will also be affected just as soon as the high-pressure shock wave has had time to spread out. The shock wave will spread out as a sound wave, which may radiate outward from the bullet pathway for a considerable distance.

4.5.1 Bullet and Shockwave

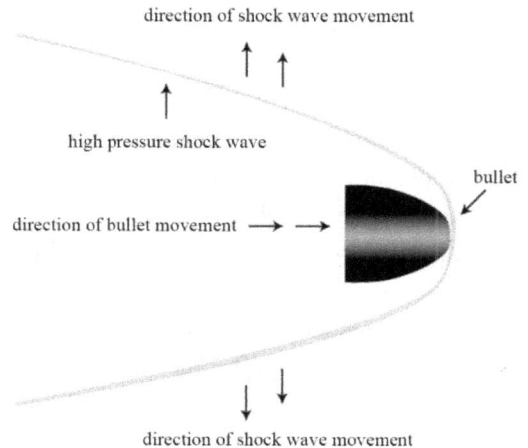

direction of shock wave movement

high pressure shock wave

direction of bullet movement → →

bullet

direction of shock wave movement

The shock wave formed by a bullet will form around the bullet as long as the bullet is in motion. It will spread outward from the path of the bullet because the bullet travels so rapidly it causes the air around it to move out of the way rapidly. This in turn forces more air out of the path of the displaced air, and so on, and this effect keeps occurring after the bullet itself has long since passed. Because the energy exerted on the air is so high, the resulting energy or 'shockwave' emanating from the path of the bullet also has high energy. This is difficult to see in the air, but when a bullet penetrates a solid object, such as an animal, it causes a lot of damage, which can easily be seen. Most of the damage is caused from the shock wave of the bullet and not by the bullet directly. When a bullet is fired into an animal, it makes a small hole roughly its own size. However, the shock-wave effect carries on into the animal with the result that the bullet opening is enlarged. As with the displaced air, which pushed outwards on the surrounding air, displaced material of the animal will move out of the way of the bullet so forcefully that it will push surrounding material out further. When a large animal is hit in a soft place, this larger opening will form and then close up again. However, if a high powered rifle is used on a small animal, the outward pushing effect of the displaced material of the animal's own body, (i.e. the material shock wave), will open the animal so drastically that it may not close again and in fact the animal may be blown to pieces.

While the expanding material shock wave may be devastating to a small animal and the atmospheric shock wave from a bullet may be damaging to nearby eardrums, the actual pressure involved in either case is not very great and will be reduced to inconsequential by the time it has traveled outward from the bullet pathway by only a few inches.

Shock waves of this nature will also be generated when an object such as an asteroid comes to Earth from outer space. There are two main differences between the damage resulting from the shock wave of an incoming asteroid and that of a speeding bullet. The first is obviously the size of the object. The asteroid is so much larger that meaningful comparisons of size are impossible. The second is the speed. The asteroid would be moving at more than ten times as fast as a bullet and possibly fifty times as fast. The shock wave or pressure wave effect will accordingly be much greater.

4.5.2   Shock Wave from a Golf-Ball Sized Object

While an incoming object the size of a golf ball would never be referred to as an asteroid, it will be useful as a starting point for understanding the nature of an asteroid-induced atmospheric shock wave. If a golf-ball sized object speeded into the atmosphere, the air directly in front of it would be pushed out of the way. A shock wave would be formed which would be very similar to the shock wave from a large bullet or a small artillery shell. The speed is of course much faster so a more intensive shock wave would develop. Since small objects like this will slow down quite readily, the shock wave would only be present in the upper atmosphere for as long as the object retained any significant speed. The air is very thin in the upper regions. Consequently, the magnitude of any pressure buildup would not be great. Never-the-less the pressure effect would exist and it would propagate harmlessly away through the upper atmosphere.

4.5.3   Shock Wave from a Basketball-Sized Object

An incoming object the size of a basketball would speed into the atmosphere, heat up and slow down but because of the size, it might not completely burn up. Some portion could reach the ground. Impact speed would be similar to the speed which any object would have if it fell from an airplane and would likely be in the range of 100 to 300 miles per hour (often referred to as terminal speed). At this speed, there would be very little effect on the atmosphere and any pressure buildup around the object's pathway would be insignificant. However until the speed of the object dropped from the unrestricted speed of open space to

125

terminal speed, there would be pressure buildup around the object's pathway and an outward-propagating pressure wave. Most of the pressure effect would still be high up in the atmosphere and there would be nothing significant at ground level. Compared to the golf ball example, the pressure effect, which would develop would be much greater because the object would retain significant speed until it was much lower down. Also because it is much larger, more air would be displaced.

The air directly in front of it would move out of the way. Outward travel distance for air which was right in the center would be the radius of the object. If it was ten inches in diameter, air would have to move out five inches to allow the object to pass. In effect, a cylinder of air the diameter of the object would be forced out and form a narrow circular tube of high-pressure air at the edge of the object's pathway. If all of the displaced air was compressed in this manner into a thin walled cylinder of air, for example, one-half of an inch thick, the pressure could be determined. In this example, a solid cylinder of air, which was eleven inches in diameter, is being forced into a thin narrow-walled cylinder eleven inches in diameter but only one-half of an inch thick. This is a cross-section area ratio of 6 to 1, which is not very great. Similarly, if the air became compressed into a narrow-walled cylinder which was 1/4 of an inch thick, the pressure increase would be in the 10 to 1 range. A shock wave of this magnitude would be noticeable and measurable but would not cause any damage unless fragile unprotected tissue (like eardrums) was very close.

As such an object moves down through the atmosphere, air would be continually sliding out of its way. Any pressure buildup directly in front of it would be similar to the increase at the perimeter of the pathway. It would simply be easier for air to move a short distance sideways than to keep piling up in front. Since the air would be stationary immediately prior to object's arrival, it would basically be jostled off to the side to allow the object to pass.

4.5.4   Jupiter Impact

The scientific community expects that if a large object should enter our atmosphere, a major disturbance would result. Accordingly, when a comet broke up in the vicinity of Jupiter and was in an orbit which would cause it to impact Jupiter, the scientific community became very excited. Telescopes were trained on Comet Shoemaker -Levy 9 during July of 1994, when it broke into several pieces as it neared the giant planet. The Sun illuminated these objects well enough for observers   on the Earth to see them all clearly. Observation was accompanied by great anticipation that a major astronomical event was about to take place. There was no disappointment. The impact sites were just out of site around the planet but fortunately, it was rotating in such a way that within a few hours, these sites came into view. Then it was observed that each comet portion had blasted an opening many miles in diameter, in the cloud cover of Jupiter. Estimates of the energy required to make such openings were made and the largest comet fragment produced an impact explosion that had 'hundreds of millions of times the force of the Hiroshima atomic bomb (one of the two atomic bombs dropped during WWII). It created a mushroom cloud that rose several thousand miles above Jupiter's cloud tops.'(28)

Each fragment of Comet Shoemaker-Levy 9 entered the Jovian atmosphere and generated a massive shock wave. Comparison with a bullet shock wave is valid as long as an appropriate scaling factor is included. However, while a bullet shock wave may be significant for several inches, the comet shock waves were significant for hundreds of miles. To see an excellent photo of these shockwave effects go to the NASA website and search for Jupiter.

4.5.5 Primary Shock Wave Formation

If an asteroid approached the Earth and entered our atmosphere, a region of very high pressure would be formed near its surface. As with the bullet example discussed above, the air in front of it would be pushed off to the edge to allow it to pass. If the asteroid were ten miles in diameter, the air at the center of the asteroid pathway would have to travel outward for five miles. Just outside the edge of the pathway, this displaced air would temporarily accumulate. It would be compressed into a relatively narrow-walled funnel-shaped region of very high pressure. The horizontal extent of this funnel-shaped shock wave would depend on the diameter of the asteroid (and its shape). The asteroid may be moving more than ten times the speed of a bullet, and the air would be forced out of the way at a similar speed. Therefore, there will be two main results from this activity. First, a region of extremely high pressure will be formed around the perimeter of the asteroid pathway. Secondly, (as with the bullet example) this high-pressure formation will have been given an outward push. So the net result of the passing of an asteroid down through the atmosphere would be the formation of an expanding cylinder of extremely high pressure air which was initially, and very temporarily, slightly larger than the diameter of the asteroid but expanding outwards at high velocity. This cylindrical pressure wave would continue to expand outward indefinitely. It would be damaging for thousands of miles and easily heard much farther away. In some respects, the more remote effect would be similar to the explosion of Krakatoa in 1883 in the Sundi Strait (which separates the islands of Sumatra and Java, both of which are now part of Indonesia). When Krakatoa exploded, it was reported that the sound was heard in Japan, which is several thousand miles away. (It was also reported that sometime later, the water sloshed in the English Channel as a result of the great waves generated by the blast effect on the water. (19))

4.5.6 Basic Physics of a Shock wave

As a pressure wave propagates outward from an asteroid pathway, the local atmosphere becomes compressed. Then just as quickly, the pressure will drop. While the pressure wave moves very quickly and will continue to move outward for a great distance, the air at any particular location hardly moves at all. It will be compressed momentarily and then return back to normal pressure. While it is compressed, the molecules of the air are crowded together very tightly. When any gas is compressed, its temperature will increase. Therefore when the atmosphere is temporarily compressed as the expanding shock-wave passes, the temperature of the air temporarily rises to correspondingly high levels and then drops back down as the high pressure region travels away. There are three detrimental aspects of the shock wave. Both the temperature increase and the pressure increase have a local but passing effect but the pressure increase also has a moving blast effect.

4.5.5 Primary Shock Wave Formation

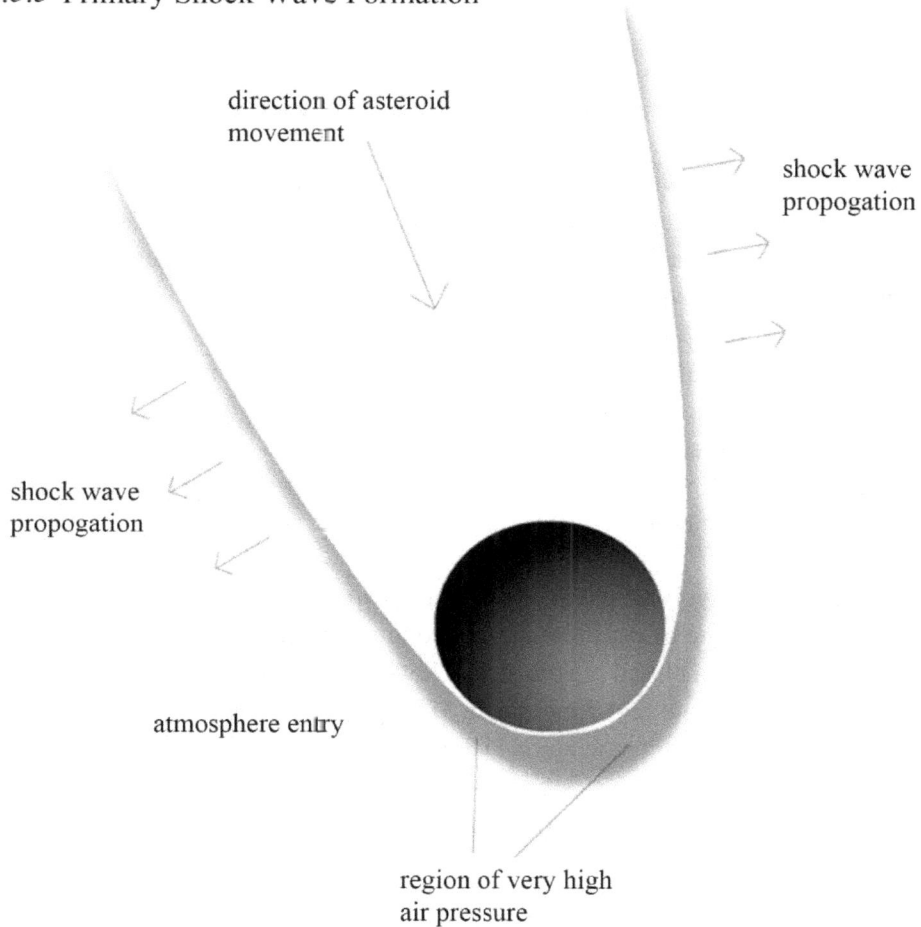

direction of asteroid movement

shock wave propogation

shock wave propogation

atmosphere entry

region of very high air pressure

In the case of an incoming asteroid, the high-speed shock-wave would have a blast effect similar to the blast from a chemical explosion or nuclear bomb, or to the effect of a supersonic airplane going through the sound barrier, only greater. It would be comparable to an expanding plate of steel or a speeding freight train and simply flatten everything in its way. As with the Tunguska Event in central Siberia in 1908, trees were flattened outward from ground zero over an area about the size of Delaware. This blast was attributed to the explosion, in the atmosphere, of a small comet or asteroid, possibly 30 meters across. (24) The shockwave from a large asteroid would have this type of effect on an entire continent. On the opposite side of the world, the shock wave, which would now be reduced to a sound wave, would be heard. In fact, the sound of the Krakatoa explosion was recorded by instruments all around the world indicating that the sound wave circled the world at least three times. (19) An asteroid shockwave would similarly circle the world many times.

The local effect of the pressure increase would catastrophically damage any soft tissue or any other material or object, which was either compressible or easily distorted. Bone, for example, is hard to distort but muscle and brain tissue is easily distorted. Any part of an animal body, which was compressible would compress. Any part, which included air passages, would collapse into a volume much smaller than the original size. This compression effect would occur even in places that were protected from the direct blast but which were connected to it by an atmospheric link. Therefore if an animal happened to be hidden, it still might not be safe because the high-pressure effect could propagate into hidden places.

There was a case during WWII, where a small group of soldiers were observed to be playing cards while seated on the ground in the battlefield area. Closer inspection revealed that they were all dead but still sitting as if frozen in place. It was concluded that the high pressure from a nearby explosion had caused their demise without actually moving their bodies.

The third effect of the shockwave would be locally elevated temperature. When air or any other gas is compressed, its temperature will go up and the more the pressure is increased the more the temperature will be increased. Therefore everything in the path of an expanding asteroid shockwave would experience a pulse of extremely high temperature. As soon as the shockwave passed, the temperature would drop back down again but the devastation would remain.

In nature, whenever a gas is involved, the three characteristics of a gas, volume, pressure and temperature, are always interrelated. In some cases, pressure will follow temperature. In the boiler of a steam engine, for example, heat is applied to a closed pressure vessel, which contains water. As the temperature goes up, some of the water will evaporate and form steam. The pressure of the steam (and the entire pressure vessel) will be dependent on the temperature increase which is provided by the flame in the firebox. When more fuel is added and the flame is built up, the pressure in the water container will increase. In this manner the pressure of the steam, is driven by and determined by the temperature achieved by the fire.

In other cases, temperature will follow pressure. For example, if a gas is compressed, its temperature will increase. The air, which exits from a large industrial air compressor, will be hot. It may be so hot that it is dangerous and some of the heat may need to be removed before the compressed air comes near personnel in the factory. In this case, the temperature of the air has been determined by the pressure of the air.

Similarly, when the pressure of the atmosphere is temporarily increased by a passing pressure wave, the temperature of the air will increase with the pressure and be determined by the magnitude of the pressure. It follows therefore, that as the pressure wave from an incoming asteroid passes any particular location, the temperature of the air will increase to a very high level. Fires will result. Anything, which can burst into flames, will ignite. In particular, grass, leaves and any wooden objects will start to burn. The scorching effect will be spread outwards as far as the pressure of the shock wave was able to raise the air temperature beyond the ignition point of any material in the way. Therefore, one result of a large asteroid passing through the atmosphere could be an entire continent catching on fire. While the pressure and temperature will drop back down very quickly, the fires would, of course, continue.

As the primary shock wave moves outward from ground zero, its peak pressure drops because it is forming an expanding cylinder which has a continually increasing circumference. The pressure drop will follow a very familiar pattern, which is variously known as a half-life curve or a time constant curve. In the mathematical world this type of relationship is called an exponential and it occurs repeatedly in nature. Suppose that the pressure of the shock wave was 10,000 tons per square inch when the pressure wave was formed at the perimeter of the asteroid pathway. If this particular asteroid had a radius of five miles and a perimeter of 31.4 miles, when the pressure wave had traveled out for another five miles, the total effective perimeter, or length, of the pressure wave would be 62.8 miles. Since the affected region is now twice as long, the pressure would be one-half as great and would have dropped to 5,000 tons per square inch. Similarly, when the radius of the expanding pressure wave had increased to twenty miles, and the perimeter or length of the wave was 125.6 miles (4 times the initial size), the pressure would have dropped to 2500 tons per square inch (one-quarter of the initial value). While these values are chosen for illustrative purposes only, they are realistic in two ways. The pressures of an asteroid shock wave would be unthinkably high and would not drop into more familiar realms until the pressure wave had traveled outward for hundreds or even thousands of miles. While any such pressure would be singularly devastating, the accompanying temperature would far exceed the combustion temperature of many burnable materials. Since the temperature would follow the pressure, the temperature of the expanding high pressure atmospheric shock wave would not drop below the ignition point of most combustible materials until the pressure wave was a great distance from ground zero.

These three results may be predicted from the well-known physics of gas behavior. However there is more than air in the air. There is also water in the air.

4.5.7   Primary Shock Wave Effect

There would be three different types of shock wave generated as the result of a large object speeding into the atmosphere. The first would be the basic perimeter or primary shock wave as discussed above. The second would be a spherical or secondary shock wave, which would be expected to start propagating at either earth or water contact. The third effect would be due to the shearing effect the passing object would have on the perturbed air column.

The primary shock wave forms as air is shoved off to the side to allow the incoming object to pass. With small objects the size of a bullet, an artillery shell, or a basketball, virtually all of the air in the object's pathway will be pushed off to the side and form an expanding cylindrical pressure wave. While such shock waves may be seriously annoying to anyone within a few diameters of the passing object, the pressure will drop very quickly as the pressure wave expands and will soon be reduced to low and damage-free status. This however will not be the case with a large object.

As a large object speeds into the atmosphere, a significant part of the air in its pathway will slide off to the side. As this happens, the primary shock wave or pressure wave is formed. While the incoming object would be moving very fast, the air has been standing still. Right near the perimeter of the object's pathway, air will remain almost stationary and only move over a little to make room. If the passing object is assumed to be basically round, it may be visualized that near the perimeter, the angle, which the surface of the asteroid forms with its own trajectory, is very shallow. Further from the perimeter, at locations a little further around in front of the asteroid, the attack angle becomes greater. Here it will be more difficult for air to slide along the moving surface and so some air will simply compress and ride with the asteroid. The compressed region then effectively forms a shallower slope so air a little further from the surface of the asteroid can more easily slide past and contribute to the buildup at the perimeter. This situation is shown in the diagram, 'Primary Shock Wave Formation' 4.5.5.

As air builds up in front of a moving asteroid, it would effectively be forming a more streamlined surface. This would allow some of the air, which was more directly in front, to slide off to the side and contribute to the massive pressure increase at the perimeter of the asteroid pathway. Suppose that all of the air, except a cylinder of it in the center with one-half of the asteroid's radius, was pushed aside and compressed into the primary shock wave. It would be impossible to know how tightly this air would compress and just how high the pressure would go. However, if a cylinder of air with an outer diameter similar to the asteroid and an inner diameter about one-half of the diameter of the asteroid, were to compress into a narrow-walled cylinder only a few inches thick, the pressure would be incredibly high. It would be expected to be in the tens of thousands of pounds per square inch. Air with such a high pressure would have the damaging effect of a moving wall of steel. Neither man-made objects nor any form of plant or animal life would be able to withstand the impact of such a shock wave as it propagated outward from the asteroid's pathway. Not only would the damage capability be comparable to steel but the speed of such a formation would be commensurate with the asteroid speed. Even though it would only be air, the speed and the pressure would ensure that in its pathway, nothing could remain standing.

4.6   Wooly Mammoth

Numerous carcasses of the wooly mammoth have been found frozen in permafrost across both Alaska and Siberia. Tons of tusks have been recovered and shipped to ivory carvers who would not have any use for them unless they were fresh. The flesh of some of these beasts was also fresh in appearance and has been used as fox bait or as food for sled dogs. It has even been eaten by people all of which indicates that it must have been frozen soon after the animals died or it would have simply wasted away through decay or scavenging. The point of greater interest however, is that these animals died very suddenly. Well preserved food fragments were found in the mouth and between the teeth of the mammal. This could only mean that the animal met with a sudden death and did not even have time to swallow its last meal! Further indication of sudden death was found in blood which was collected in great masses due to hemorrhage. Also bits of frozen blood were found which, when heated, turned into dirty, dark red spots indicating that the oxygen content of the blood had not been fully extracted. This condition frequently indicates a sudden death. (48) 'Microscopic examination of the skin showed red blood corpuscles, which was a proof...of a sudden death.' (49)

The animals not only died suddenly but they died by the millions. 'Such was the enormous quantity of mammoth's remains that it seemed...that the island was actually composed of the bones and tusks of elephants...The soil of these desolate islands is absolutely packed full of the bones of elephants and rhinoceroses in astonishing numbers….the islands are strewn with tusks cast up by the billows….herds of mammoth perished…the bones of elephants, rhinoceroses, buffaloes, and horses are found in this icy wilderness in numbers that defy all calculations…the soil...is full of the bones of elephants, rhinoceroses, and buffaloes…heaped up with skeletons of pachyderms, bison, etc.' (49)

One is reminded of the great buffalo herd that quite recently roamed the grasslands of central North America 'in countless millions'. (50)

The scenario revealed by the evidence is that the death of millions of animals was sudden and simultaneous. It would have been similar to the entire North American buffalo herd being killed at one time. From the mouth contents and the stomach contents of the Wooly Mammoth it is obvious that they were all grazing normally across miles and miles of grazing lands when in an instant they died! Subsequently they froze and they froze quickly enough to be preserved for future discovery in an apparently "fresh state". The evidence of hemorrhage gives the clue to the manner of death. As noted above, "the blood was collected in great masses due to hemorrhage". These animals all died when they were struck by a shock-wave. Soft animal tissue such as brain matter and blood vessels could not have withstood the trauma of the shock-wave which would have been produced by an incoming asteroid.

Following the explosion at Tunguska in central Siberia, trees were flattened for thirty miles. The high pressure produced by that explosion was like a moving wall of steel and knocked all the trees down. If a wooly mammoth had been there, it too would have been knocked down. It would not have survived. The expanding pressure wave would have not only hit the animal like a giant club but it would have raised the pressure of the atmosphere temporarily to a very high level. This would have destroyed certain parts of the interior of the animal resulting in hemorrhaging. Broken blood vessels and broken off tusks (i.e. the tusks are commonly found separately) mean sudden violent death. Millions of creatures died. The shockwave was devastating for hundreds of miles and all of the grazing animals for hundreds of miles were destroyed almost simultaneously.

The country of Russia has four known large impact craters. The objects which produced these craters would each have produced an expanding high-pressure wave which would have killed any animal which it contacted. The Popigai Crater is recognized as one of the largest in the world at 100 km diameter. It is located in north-central Siberia and would have produced a shock-wave which would have been devastating across most of Siberia. While commenting about what would result from an object about 5 miles in diameter one author has declared; 'the energy released was greater than that of 100 million megatons of TNT, 100 teratons, more than 10,000 times greater than the total U.S. and Soviet nuclear arsenals.' (51) A significant fraction of this energy would have been in the shock-wave and no animal could have withstood that much trauma. The Kara Crater is smaller than the Popigai crater at 65 km diameter but its shock-wave would have had a similar effect and it too is located in the northern part of Russia not far from the Arctic Ocean. The other two notable craters are located further south. The Puchezh-Katunki Crater at 80 km is less than 100 km. ne of Moscow while the Kamensk Crater at 25 km diameter is located near the Black Sea. (52)

While currently not technically part of Russia, Kazakstan boosts the largest impact site in Asia and possibly in the world. The Ishim Crater in the Teniz Basin has a diameter between 350 and 720 km (220 and 450 miles). (53) It is almost straight south of the Kara Crater. The shock-waves from all of these impacts would have overlapped so there would not have been anyplace in the northern half of central Asia that would have been free of life-terminating shock-waves. The entire herd including mammoths, rhinoceroses, buffalo and horses were wiped out by asteroid-produced shock-waves leaving their carcasses to freeze solid in the ensuing post-impact winter.

4.7   The Blinding Flash

As an asteroid charges through the atmosphere, much of the air in its pathway would be shoved off to the side and form the massive shock wave described above. However, some of the air in front of the speeding asteroid would not be able move to either side, but would instead, become trapped in front, creating a high pressure cone of air that would precede the asteroid on its path to Earth. This trapped air is trapped because the asteroid would be moving so fast that there simply wouldn't be time for it to move anywhere. (It is expected that any large asteroids or comets on impact pathways with the Earth would be moving so fast that they would drop down through the entire atmosphere in less than one second. Since we are talking here about objects which may be several miles in diameter, it may be appreciated that there is no way that any air which was directly in front could move that far aside in the time available.) As more and more air is added to this cone of trapped air, its pressure would keep increasing right up until the asteroid reached the surface of the Earth. (In effect, a cylinder of air, some portion of the asteroid's diameter and the entire height of the atmosphere would be trapped and compressed into a very small portion of its original volume.) The air pressure would continue to escalate and because of this, the air temperature would also continue to escalate. Compressed air grows hot, and the temperature below a speeding asteroid would rise to 60,000 Kelvin, (i.e. ten times the surface temperature of the Sun.). (33) When the air became hot enough, it would start to glow - red at first and then white as it became hotter. This result would probably occur just as the asteroid was approaching Earth contact, but it has nothing to do with touching the Earth. It would simply be the result of the high speed of the object. When the visible temperature was reached, the energy of the compressed air would radiate outward from the asteroid as a pulse of high intensity light, which would appear as a blinding flash. (12) The intensity of this flash would be too much for any naked eye to withstand, immediately blinding anyone close enough to witness it. The blinding flash of light would be the first type of energy to have an effect on the surrounding area and since it would be moving at the speed of light, it would cause its devastating results instantaneously.

The blinding flash effect would be similar to the blinding flash, which develops when an atomic bomb is exploded. The first element of destruction propagating from an exploding atomic bomb is a blinding flash. Anyone caught looking at this flash could be blinded. The light would be damaging directly but since light is a form of electromagnetic energy other forms of electromagnetic energy would also be produced including heat. Therefore a flash of light would impact everything around as well as a flash of heat. Any combustible surface would burst immediately into flame. Dry grass and wood as well as painted surfaces would become instantly ablaze. Also some portion of this electromagnetic energy would be outside of the visible range but would be detectable by electronic equipment including antennas, wires and circuits.

133

Some of this equipment would instantly overheat, even if it was buried out of site and not directly exposed to the radiant flash. This type of damage is always anticipated if an atomic bomb should explode nearby and great precautions are always taken to protect military equipment from such destruction. This is what happens when an atomic bomb explodes and this is what would happen if a speeding asteroid came down through the atmosphere to the Earth's surface.

The blinding flash would occur just as the asteroid reached the ground. This flash would be a pulse of light and consequently traveling outward from ground zero at the speed of light. (12) A pulse of heat would accompany the pulse of light. Anything which had not been ignited from the blinding flash would ignite from the heat pulse which means that fire damage from an incoming asteroid is a double-headed certainty.

4.8   The Secondary Shock Wave

As discussed, some of the air in front of the asteroid would pile up into a region of very high pressure. This air would be effectively trapped in front of the speeding asteroid with no place to go but to move along in front of the asteroid until it reached the ground. Then just as the asteroid touches the ground, it would explode outward forming a secondary pressure wave which would radiate away from ground zero in an expanding spherical pattern. This pressure wave would be another similarity to an atomic bomb detonation. When an atomic bomb explodes, one of the results is a spherical expanding pressure wave. In the case of an asteroid, this secondary pressure wave would follow quickly behind the primary pressure wave, which would also be expanding and already causing its particular type of chaos. The area surrounding the impact site would therefore experience two pressure waves, each of which would be devastating for several reasons including their extremely high pressure and high speed of propagation. The blinding flash and heat pulse would precede both of them.

4.9   Shear-Effect Pressure Wave

At the interface where some air is sliding past the asteroid and some is trapped and moving with the asteroid, a shear surface is formed. (The term shear is used whenever some material is sliding past other material. Even if nothing is moving, shear stress might be involved if one material is being stressed toward sliding past some other material.) Shear is actually a very common type of stress. When the tire of a car locks up and slides along the pavement, the interface between the tire and the pavement may be properly called a shear plane. This is the place where the stress is occurring and where the movement is occurring. In this example, shear stress would exist even if the tire did not move but was being tugged anyway. If the brakes were locked and the tire could not rotate, but someone was still trying to pull the car, the tire-pavement interface would be a shear plane and there would be shear stress involved as evidenced by the partial destruction of either the tire or the pavement or both. Sometimes a system must carry this type of stress without moving. Shear stress is simply the type of stress involved when one object is sliding, or is being stressed to slide, past another object.

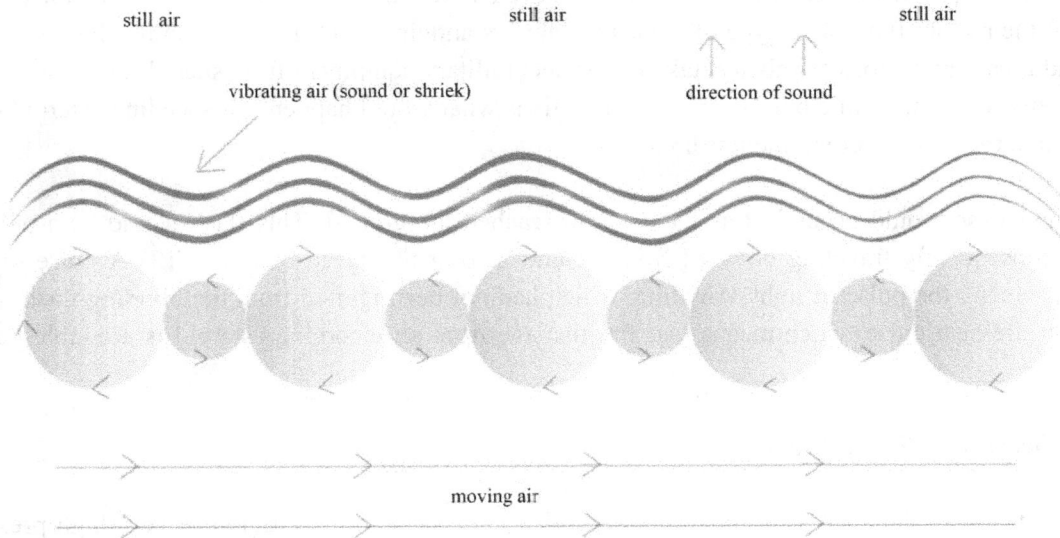

still air   still air   still air

vibrating air (sound or shriek)   direction of sound

moving air

4.9 The Shriek from a Jet Engine

When a mass of air moves past still air, circular air bearings or eddies form at the interface and proceed at about half of the moving air speed. The lateral movement of the bearing assembly puts a vibration into the still adjacent air in which the vibration creates a sound or "shriek". The pitch of the shriek is higher for air moving at a greater speed.

There is shear stress in the exhaust of a jet plane where the high-speed exhaust gases are moving past surrounding still air. Along this interface eddies develop. In this case they result in the unbearable screech, which formerly accompanied all jet engines. These eddies as well as the unbearable screech develop along the shear plane where the moving exhaust is shearing or sliding past motionless surrounding air. Air will not just slide over other air. Near the interface, the air on one side will be moving and on the other side the air will be still. In a very thin region in between, the air will form little rolls or eddies, which will roll along the interface. Down at the molecular level, if the molecules in the moving portion just slid past the molecules in the stationary portion, everything would be fine and there would not be any noise from a jet plane.

The interface region involved, however, is much more extensive than the distance between two molecules or even quite a few molecules. If all of the molecules of air were sitting quietly, it would be possible to visualize one molecule sitting still and the next one moving past. Molecules, however, do not sit still. They are always bouncing randomly into each other whether they are going very far or not. (i.e. the distance to a molecule one-quarter of an inch away or just to the next molecule.) They just keep moving. In fact they do a perpetual dance and they keep bouncing off of one another. The speed of the dance is determined by their temperature. The hotter they get the faster they bounce. This perpetual bouncing means that the air is a little more 'together' than simply being a group of molecules. There is always interaction among these molecules without any particular pattern or order. In this manner, molecules, in one region, effectively transmit their presence to other nearby regions. This is a positive development in

135

nature and accounts for the wind not all piling up against a wall that is in its way. Some portion of the wind will hit the wall but most of it will be warned out of the way. Similarly when the wind blows over a mountain, the effect of the mountain's presence will extend upward from the mountain peaks and this resulting disturbance in the air flow may be detected for a thousand feet of more above.

At the moving-air, still-air interface, little rollers of air are formed. These rollers roll along the interface. In fact, the entire interface plane is composed of tiny rollers moving along. The air, which is sliding away, is actually being moved away on a bed of tiny rollers. In the exhaust of a jet plane, these rollers are quite small and they move along at approximately one-half of the speed that the exhaust gas is moving. The rollers interface the stationary region with the moving region by developing a series of tiny air roller bearings. As these rollers move, they disturb the air, which is stationary causing it to ride up and down the roller pattern. The air is thereby vibrated and produces sound. Unfortunately, the sound from a jet engine exhaust is very unpleasant to hear resulting in a great deal of effort being extended to reduce it as much as possible. Reduction is achieved by introducing intermediate regions of moving air, which are moving at speeds in between the high-speed exhaust and the surrounding still air. In this manner the shear region is diluted and the shear effect is spread out. Tiny rollers no longer form but are replaced by larger rollers, which result in sound which is less objectionable.

With a speeding asteroid, there is a shear plane where moving air is sliding past still air. As with an airplane engine, very small eddies will develop but because of the much greater speed of an asteroid, these eddies will be much smaller. This will result in sound with a much higher pitch which could easily extend right into the ultrasonic region where only certain animals can hear it. Of course with an asteroid, no reduction in relative speed will be possible and the extremely tiny rollers, which will form along the interface, will produce a most unpleasant (if hearable) high-pitched scream. In comparison to the other effects of the speeding asteroid, this unpleasant sound will not be very important but it will certainly exist.

4.10   Cloud Formation

Clear air usually includes a certain amount of water. This water is not visible as long as it is in gaseous form. As a gas, water molecules are not connected together and as a gas, water will behave like other gases. If a sample of air which includes water molecules is cooled, a temperature will be reached at which the water molecules will loosely stick to each other. This is the dewpoint temperature and the water molecules will become visible as a fog or cloud. Usually fog is referred to if this occurs near the ground but if it happens high above the ground, the visible formation is called a cloud.

It is quite common in mountainous areas to observe a cloud on top of a mountain. The cloud may appear to be standing still but at the same time, the air may be moving over the mountain at considerable speed. If a person stood on the mountainside at the base of this cloud, he/she would observe the cloud to form as the air rushed up the mountain slope. Above the elevation at which it formed, the person would be in the cloud. Below that elevation, he would be under the cloud. It could happen that the person's head was in the cloud and his feet were beneath the cloud but the transition level is usually not quite that sharp. The cloud forms when the air cools (due to expansion) to a critical temperature. Then on the other side of the mountain, the cloud will dissipate as the air warms again (due to compression) on the way down.

Clouds may also form on a warm summer day if a certain area of ground becomes a little warmer than surrounding areas. This could occur over a freshly plowed field. The warmer air over the dark area of ground will rise until it reaches an elevation where the temperature has dropped to the critical dewpoint level. At this elevation, the water in the cloud will become visible and a cumulus cloud will be observed. This type of development may be of interest to both birds and sailplane pilots because the presence of the cloud indicates that a thermal (rising column of air) is present, which may give either of them a welcome lift to a higher elevation.

In a cloud, the water molecules, which were formerly invisible because they were disassociated from each other, have come into contact with each other and formed very small droplets of water (about 1/2500 inches in diameter (11)). This mixture of air and water droplets is no longer simply a gas but a mixture of gas and water. The water is in liquid form but the individual quantities of water are very small. However, the characteristic of such a mixture will be different from a multiple gas mixture. The manner in which the liquid water portion heats up will be as liquid water whereas the water vapor portion will heat up like a gas. In fact it is twice as easy to heat up water vapor (specific heat = 0.48) as it is to heat up liquid water (specific heat = 1.0). Usually, this would be of no interest to anyone. However if air which includes some water vapor should become exposed to a high-pressure shock wave, the matter will be of interest because the liquid water portion will not instantly heat up and then just as quickly cool down as a gas would. Instead the liquid portion will heat up and stay hot and then most of it will fall to the Earth as hot rain.

Water droplets which become entrained in moving air may be transported for great distances before being released as rain which might occur when the temperature of the air drops. Then the water molecules are not being jostled around nearly as much so they stick together easier. If enough of them cling together, a water droplet will be formed which will be too heavy to remain floating in the air so it will drop as rain. As a shock wave passes, something similar happens. In this case the water molecules are forced together by the pressure and then heated by the hot air. Subsequently, they will fall as hot rain.

4.11   Thunder Storm Shock Wave

During a storm, the relative humidity may be very high. Conditions may be just right for the gaseous water in the air, to start condensing. Then, with just a slight decrease in temperature or a slight increase in humidity, water droplets will form and it will begin to rain.

Sometimes during a thunderstorm, when the conditions are nearly right to enable rain to form, lightning strikes and thunder will sound. Almost right away it may start to rain. Sometimes it will already be raining when the thunder sounds, and then the rain will intensify. These results occur, in part, because the thunder, which is a pressure wave, caused all of the molecules in the air to be temporarily forced together. As the thunder pressure wave passes, the air is squeezed and the moisture will be effectively squeezed out of the air. This is comparable to ringing out a soggy towel. When it is just ready to drip, squeezing will produce a small stream of water.

4.12   Atomic Bomb Shock Wave

While a passing thunder clap may cause it to rain, there is another man-made atmospheric event which illustrates the situation even better. When an atomic bomb explodes, an atmospheric shock wave is produced. This shock wave propagates outward from the explosion location in a three-dimensional pattern the position of which can usually be seen as an expanding spherical ball of fog. As the pressure wave passes any particular location, fog will form and then immediately dissipate. The shock wave location appears as fog because the high pressure of the travelling shock-wave compresses the air. If there were enough water molecules in the air, the atomic shock wave would squeeze the water out of the air and cause it to fall to the ground as rain.

4.13   Shock-Wave-Produced Rain

The atmospheric shock wave produced by an incoming asteroid, would have a similar effect to a nuclear detonation and it could squeeze water right out of the air whether it was about to rain or not. As the shock wave passed any particular location, rain would start to fall. The amount of rain would depend on how much moisture had been in the air. If the humidity had been high, a noticeable amount of rain would fall. If the humidity had been low with hardly any moisture in the air very little or no rain would fall.

As air is compressed, and the pressure goes up, the molecules of air will be crowded together. There will be more of them in the same space and so they will bounce against each other a lot more. While the air molecules are being forced together, any water molecules that happen to be present will also be forced together. However, water molecules cling together more readily than air molecules. Once they have been brought together, they might not separate again. The pressure wave will pass and the pressure of the air will drop but the water molecules might have formed liquid water where previously there was only gaseous water. Some of the water, which was previously in gaseous form, could now be in liquid form and if there was enough of it, some rain would fall.

If the increase in pressure occurs very quickly, (which it would with any atmospheric shock wave resulting from an asteroid), the temperature of the air would also increase very quickly. In fact, the two effects will occur simultaneously. In this case the pressure increase is the initiating factor and temperature increase will track it. As long as the various components of the air, which include oxygen, nitrogen and water, behave as a gas, any increase in pressure will result in an instantaneous increase in temperature. Both oxygen and nitrogen occur as a gas over a very wide range of temperature and pressure because the individual molecules do not have any particular affinity for each other. They do not tend to stick together. The water molecule is different, however, and it is arranged with both of the hydrogen atoms toward one side. This unbalanced arrangement makes the water molecule slightly positive on one side and slightly negative on the other side. For this reason water is referred to as a polar molecule and it is this feature that enables water molecules to cling to each other. As soon as this happens, even if only very small droplets are formed, it is not really a gas anymore but has become a liquid with different physical properties. As previously mentioned, the specific heat is twice as high for liquid water as it is for water vapor. It takes twice as much heat to increase its temperature. The advancing pressure wave has therefore produced a situation where, instead of having a gas with temperature increasing directly with pressure, there is a mixture of gas and liquid. As a result, it will take twice as much energy to increase the temperature of the liquid portion as it would if it was still a gas. Also because it is a liquid, temperature increase will be dependent on energy being transferred to the liquid droplets from the surrounding hot air. In other words, the temperature increase of the water droplets will be due to transfer of energy rather than compression of gas.

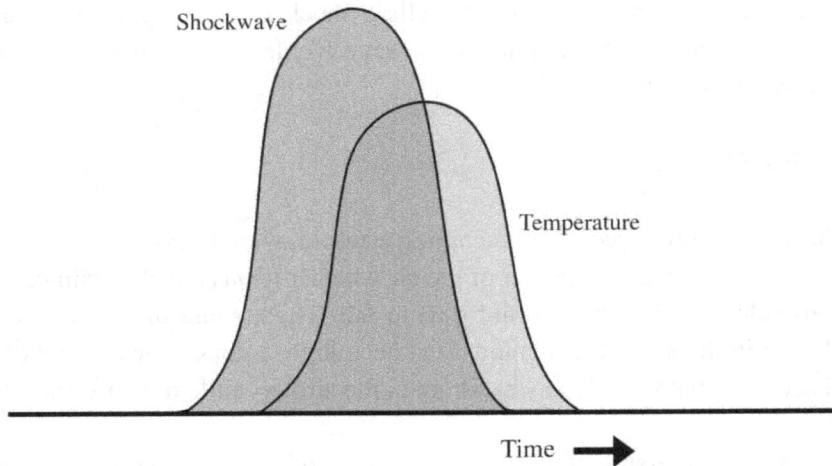

4.13 Shockwave - Temperature Lag

In nature when an initiating or driving parameter changes, dependent parameters will also change but they might not change instantaneously. In this case the temperature of the liquid droplets will continue to increase as long as the surrounding air temperature is higher. This will result in the increase in temperature of these droplets being slightly delayed. As the pressure impulse passes, the temperature of the droplets will not have increased nearly as much as the air and the droplets in the air will not have been able to track the very high-speed increase in atmospheric pressure. The temperature of the air will increase and cause the temperature of the water droplets in it to start to increase. However, before they get very warm, the pressure wave will have passed and the atmospheric pressure will have dropped. The temperature of the water in the air will continue to rise as long as the surrounding air temperature is still higher but then it will also start to fall. As a result, the temperature transient of the water in the air will be slightly out of phase or slightly delayed compared to the temperature transient of the air. The net result may be warm rain, which would cool, as it falls (by loss of heat rather than by a drop in pressure).

4.14 Asteroid-Produced Wind

Another effect of an asteroid coming down through the atmosphere would be wind. A major wind-storm would propagate outward from the impact site in all directions. (23) This wind would have similarities to two, more common atmospheric occurrences. Waterfalls, as they plummet downward, entrain air, which also plummets downward. When the downward moving air reaches the horizontal surface of the water at the base of the falls, the moving air turns outward away from the falls and travels horizontally. If people stand near the base of the falls, they will experience a wind blowing in their faces. Higher falls with more water flow generate greater wind. Fires have a similar affect. When an open fire burns, air is heated and rises upwards. Replacement air is drawn in and feeds the fire. If the fire is both vertically and horizontally large, large amounts of air will be drawn in. In a forest fire, because the fire involves tall trees and may extend over large areas, the replacement air will be moving very fast towards the fire and may even reach hurricane speeds. This type of situation is called a fire-storm and it may not be possible to even stand in such a wind because it will be moving too fast and will be picking up loose debris along the way. Just being in such an area at all would be hazardous.

An asteroid (or a comet) rushing down through the atmosphere would draw a large amount of air in behind itself. When this air reaches the surface of the Earth, it would move outward horizontally. It would be comparable to a waterfall because the air would be moving outward. It would be comparable to a firestorm because the air would be moving very fast. The wind-speed near the impact site could be several thousand miles per hour (20) In fact, 'Astronomers have calculated that a 10 km diameter (6.25 miles) comet. . .would produce wind-speeds of 2500 km/h at 2000 km from ground zero.' (10) Of course the speed would drop off exponentially as the distance from the impact site increased. At some particular distance from ground zero, the speed would drop to one-half of the original speed and then at a similar distance further, it would drop to one-half of one-half. In the meantime, anything which had not been flattened by the outward rushing pressure wave would be flattened by the wind, which followed.

4.15 Atmosphere Effect on the Asteroid

An asteroid as large as a house or larger would pass through the atmosphere of the Earth with very little effect on the asteroid itself. This is because a large asteroid is too big and too heavy for the atmosphere, which is very thin and light, to have any effect on it. If, for example, an asteroid entered the atmosphere at 25,000 miles per hour, and did not lose any speed on the way, it would reach the surface of the Earth in just a few seconds. Even a much smaller object the size of a basketball would come through the atmosphere in much less than a minute. For objects this small the atmosphere will have an effect called friction heating. Even though air is very thin and light, when an object moves through it at high speed, the air molecules will bang up against it and heat it up. When an object is quite small, its surface area is large compared to its volume. In such a case, the atmospheric heating effect will raise the object's temperature until it becomes red hot and readily visible. With a small object, there is a lot of surface area to impact the air compared with how much material there is to heat up. The energy, which goes into heating the object, will be subtracted from the energy of motion and the object will slow down. Exactly the same thing happens when the space shuttle returns to Earth. It heats up and slows down. (While the space shuttle is not very small compared to a basketball, it heats up in a similar manner because it is directed to enter the atmosphere at a very shallow angle. (In fact if it came in at an angle which was any shallower, it would just bounce off of the atmosphere and not come to Earth at all.) Most natural objects from space enter the atmosphere at much steeper angles and crash into the Earth within a few seconds.) In fact an object from space could heat up so much that it could completely burn up and leave very little material to reach the surface. When a meteor enters the atmosphere, it might heat up so much that it breaks up and becomes so completely burned up that nothing of any consequence will reach the surface of the Earth. For house-sized objects this will not be the case. If a stony object as large as a house entered the atmosphere, it would not heat up very much and would not slow down very much and so would reach the surface very quickly with very little visual effect. No matter what size an object was, there would be a friction-heating effect. However when an object is large, the energy of motion which is converted to heat, is a very small fraction of the total energy of motion of the object so that any reduction in speed will be insignificant. Further, the friction heating effect will not be great enough to raise the object's temperature to incandescence. Therefore, since large objects will not heat up, they will not be noticeable. Consequently, if you were not looking in exactly the right direction, you would not notice a large asteroid approaching the Earth. There would be no fiery streak to attract your attention. If an object were much larger than a house, (for example, one mile in diameter) the surface-heating effect would be completely negligible. While the atmospheric effect on a small object is very significant, and the effect of the small object on both the atmosphere and the surface of the Earth is very insignificant, the exact opposite is true for a large object. With a large object, the effect of the atmosphere on the object will be insignificant, and the effect of the object on the atmosphere and the Earth will be most significant.

After passing through the atmosphere where a cylindrical shock wave, a spherical shock wave, a loud screech, an electromagnetic pulse – part of which is a blinding flash - and a high speed windstorm are generated, an asteroid would reach the surface of the Earth and several more devastating worldwide events would follow immediately.

4.16   Water Impact

Water impact will be described first as it is easiest to visualize. There would, of course, be a splash but not an ordinary splash. As an asteroid enters the water, the water would displace to make room for it. The displacement speed would be similar to the speed of the incoming object, which, in the case of an incoming asteroid may be 25,000 miles per hour or more. However, it happens that 25,000 miles per hour is the escape speed from the Earth.

4.16.1   Escape Speed

All objects, which are not directly attached to the Earth, are held near the Earth by the force of gravity. The force of gravity is very well understood from a measurement point of view, but is not at all understood from a theoretical point of view. It is a force which acts between all objects in the universe in relation to their respective masses. Everything has measurable substance or mass - even air. This is fortuitous because we need our atmosphere to stay near the Earth and it is the force of gravity which keeps it here. If our atmosphere had no mass or substance, it would not stay near the Earth but would simply drift away. Without the force of gravity we would also drift away. However, even with the existence of the force of gravity, it is possible for things to leave the Earth. For example, a space shuttle can go into orbit around the Earth or it could leave the Earth and go far into space. In order for an object such as the space shuttle to leave the Earth it must be accelerated to a very high speed. This speed is the same for everything and it is 25,000 miles per hour. If an object, which is on the Earth or near the Earth, could be sped up until it was going 25,000 miles per hour, it would leave the Earth, travel out through space and never return to the Earth. (29) This applies to small objects or large objects or even to the air which surrounds the Earth. In fact, this is happening every day. Certain parts of our atmosphere are very light. Because of this lightness, these parts may be bounced around in the upper atmosphere so fast that they effectively bounce away from the Earth. This is happening with hydrogen in the upper atmosphere. Hydrogen is a very light gas. When hydrogen is released into our atmosphere and drifts upward to the upper regions of the atmosphere it will be bounced away from the Earth by the other heavier parts of the atmosphere including oxygen and nitrogen. This is a long-term problem for the Earth because water consists of both hydrogen and oxygen. If these two parts of water become disassociated in the upper atmosphere, the hydrogen will not remain but will leave the Earth altogether. Therefore, the Earth may be gradually losing its water supply.

In order for anything to escape from the Earth it simply needs to reach escape speed, which is 25,000 miles per hour. It is fortunate that both oxygen and nitrogen, the main components of our atmosphere, are a little too heavy to be easily accelerated to this speed or we would not have an atmosphere at all. While hydrogen can be accelerated to escape velocity by the action of the atmosphere, any object would escape if it could be sped up enough. Either the space shuttle or a 2008 Chevrolet could be sent out into space by speeding it up to 25,000 miles per hour. When we wish to send an object away from the Earth, we simply increase its speed. When the speed reaches escape speed, that object will leave the Earth. (Space probes being sent to other planets are accelerated to escape speed whereas objects intended to orbit the Earth are accelerated to a slightly lower speed.)

Escape speed is determined by the size of the object, from which one is escaping. Leaving the Moon is easier than leaving the Earth and the escape speed for leaving the Moon is much less than 25,000 miles per hour. The smaller an object in space is, the easier it is to leave it. This is a problem for space crews. It is very easy to leave the space shuttle because the force of gravity between the space shuttle and a crew member is very very small. Therefore very little speed is required to enable a crew member to drift away from the shuttle. In the case of the space shuttle, leaving is not at all desirable so a line is usually attached to prevent this from happening.

When an object hits the Earth at 25,000 miles per hour, some of the effects will be understood with respect to the escape-speed factor.

4.16.2 Asteroid Touchdown

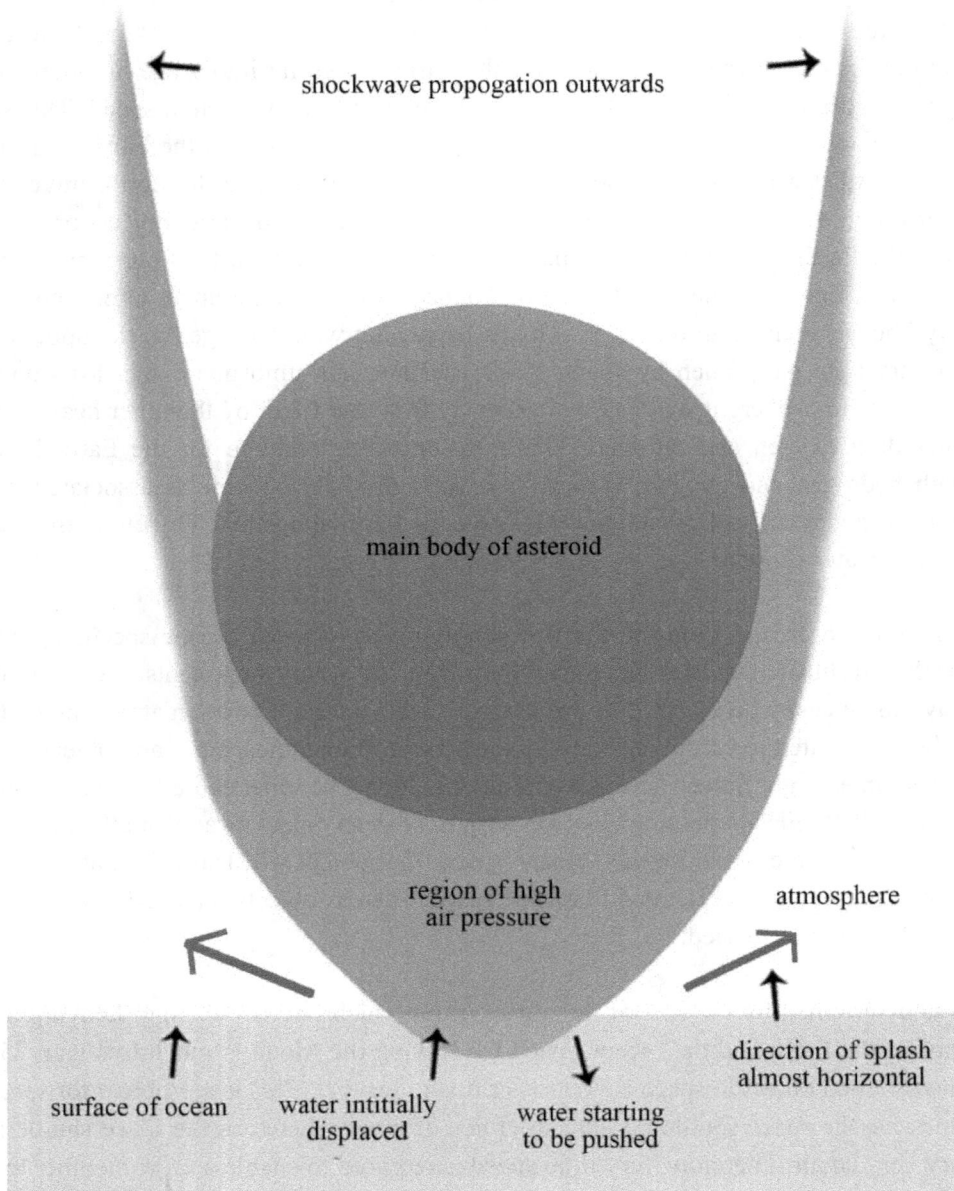

shockwave propogation outwards

main body of asteroid

region of high
air pressure

atmosphere

surface of ocean

water intitially
displaced

water starting
to be pushed

direction of splash
almost horizontal

4.16.2   Touchdown

As stated before, when an asteroid impacts on water at 25,000 miles per hour, water will be displaced. Some of it will be displaced at very high speed and this speed may actually exceed the escape speed of 25,000 miles per hour. This water would be hurled up into space. In such a case some of this water could be splashed all the way to the Moon. But even if that did not actually happen, it would be hurled so high that it could land almost anyplace. The splash from a large asteroid, landing on water, could actually cause it to rain on the entire Earth. Some water could leave the Earth, some could rain down on the other

143

side, and some could stay in orbit around the Earth for an indefinite period of time and fall as rain at a much later date.

4.16.3 Life in the Solar System

While there is no observed life throughout the solar system (due to multiple life-disabling factors including ultraviolet light from the Sun, material that erupts from the Sun and cosmic radiation from outer space, all of which continually permeate the entire system), there remains a possibility that some minute life forms or former life forms (i.e. fossils) might be found. There is an accompanying possibility that any such findings might have originated on Earth and were dispersed throughout the solar system following major asteroid impacts on Earth. Any object or substance on the surface of the Earth, which an impacting asteroid is able to accelerate to escape speed, will leave the Earth and might never come back. While this could include splash water, it could also include fragments of rock or other Earth material, which then opens the possibility that small life forms could be included. Could they survive the impact and could they survive the journey? Such questions remain unanswered. However, there is no question concerning the real possibility that material could have been propelled completely away from the Earth by any one of the numerous major impacts that have happened on the Earth.

In June of 1178, a 'fire on the moon' was reported by certain Catholic monks from southern England. More recent investigations suggest that an impact with the Moon had been observed at the Giordano Bruno crater site. The fire was probably incandescent gases similar to other lunar transient events observed over the years. (21) Of further interest was the report that three and one-half months later, the Earth experienced a major meteor shower. The Moon is much smaller than the Earth and consequently has a much lower escape speed. It is therefore easier for an object to leave it. But if debris could be ejected from the Moon by a relatively small asteroid, it would not be unreasonable to expect that debris could also be ejected from the Earth. (13)

4.16.4 The Primary Splash

As an asteroid enters the water, the first effect is a splash. This immediate effect will be called the primary splash which occurs because water, which is right in front of the asteroid, must get out of the way. Water cannot compress. The splash water will therefore squirt out between the asteroid and the remaining and as yet undisturbed water.

At initial contact with the ocean surface, the situation would be similar to a basketball being driven into a shallow spherically-shaped bowl, filled with water. As the ball enters the bowl, the water will squirt out between the ball and the bowl. If the ball could be driven into the bowl fast enough, the water could squirt right across the room or all the way down the block. In this comparison, the bowl represents the undisturbed water further down. This undisturbed water will not have time to move to provide the disturbed water with some place to go. Even though water has very low viscosity and is not particularly dense, any attempt to move it quickly will meet with resistance. When a canoe paddle is slapped on a water surface, a small amount of water will accelerate away as a splash. Then the paddle stops. Even though it may be brought down with all the force possible, it will not go very far into the water. Only a small amount of water will leave the area as a splash. All of the rest of the water will not be disturbed at all.

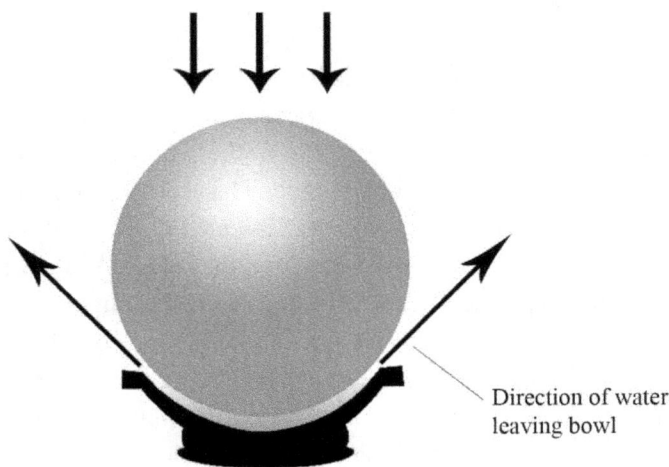

4.16.4
Water being forced from a bowl similar to the shape of a ball.

As the primary splash commences, water will be ejected almost horizontally. As the asteroid travels deeper, the splash water will be ejected closer to vertical. In all cases this water will be moving at speeds which compare to the speed of the asteroid. The more horizontally ejected water will travel through the atmosphere where the resistance of the air will be encountered. Therefore it will slow down. Most of this water will fall back to the Earth as rain. Some of the water, which is ejected closer to vertical, will meet with very little atmospheric resistance and therefore will not slow down very much. Some of it may reach escape speed and leave the Earth completely and to suggest that a portion of it could go all the way to the Moon would not be an exaggeration. Some of it might not have quite enough speed to escape but it might have enough to go into sub-orbit around the Earth and then return to the Earth as rain far from the impact site. Since it will be able to reach very high altitudes, it might not return to Earth right away. When it does, it could land almost anywhere. The splash effect of an incoming asteroid could therefore affect the whole world and if some water is ejected from Earth, it could also affect the Moon. While some splash water would return to the Earth as rain, some of it would land in the form of great life-destroying globules, which could be many feet in diameter. A major downpour of rain would be troublesome enough but globules would land with the effect of exploding bombs.

The several aspects of the splash will be the beginning of the second group of effects of the incoming asteroid. The first group included the shock-waves, the electromagnetic factors and the windstorm, which developed as the asteroid rushed through the atmosphere. The atmosphere will now have undergone two major types of disruption which will have repercussions worldwide. The splash from a large asteroid would involve many cubic miles of water. All of this water which does not either leave the Earth or go into orbit, will fall as rain and the splash from a single impact could cause it to rain over practically the whole Earth. Even though a large quantity of water will fall as rain, it will be spread out and the effect of the water will not be concentrated all in one place. Flooding and erosion could, of course, be expected in locations near the impact site but the major effect of the slash water will be on the atmosphere itself. This, together with the shock-waves and windstorm will upset the atmosphere and the nature of these effects will depend on the structure, which the atmosphere had before the asteroid arrived.

4.16.5   Transition Zone

Splash water is the water which leaves the surface of the ocean on asteroid impact and travels up through the atmosphere. The water which is ejected from the ocean closest to the impact site, will have the highest speed. Therefore it will be the water, which splashes the highest. As the horizontal distance from ground zero increases, the height of the splash will continually reduce until a region is reached where the water just wells up enough to barely leave the surface and then falls back immediately. The reason the height of the splash gets less and less with greater distance from the impact site, is because there is more and more water involved with less and less force to push on it. Right at the impact site, water is being forced out of the way of the asteroid at very high speed. Since water cannot be compressed, this impact-site water will push on the surrounding water and push it out horizontally in all directions. However, close to the impact site, the horizontal resistance will be very great in comparison to the vertical resistance. Some of the water, which started moving horizontally, will turn and move vertically like the water in the bowl when the basketball was dropped. As the distance from the impact site increases, the total horizontal pushing ability or horizontal impulse will be reduced. It will be reduced for two reasons. Some of the energy, which was originally available, will have been used up in displacing the splash water. The remainder, which is the ongoing cause of horizontal movement, is moving a continually increasing diameter of water. The horizontal pushing effort will therefore diminish. As it decreases, the height of the splash reduces. While the initial splash would go up for thousands and thousands of feet, this height will reduce until it is only a thousand feet. Further out it will only be a hundred feet and even further out it will barely be able to leave the surface and will almost immediately fall back.

4.16.6   The Primary Wave

Once the splash water had left the area, all of the remaining energy (which was imparted to the water during impact) would have been involved in wave formation. This would have resulted in a great circular accumulation of water around the impact site. It would not have had quite enough vertical momentum to get completely free of the surface but it would never-the-less have piled up like a mountain which might have reached a height which was a multiple of the asteroid diameter. It would not only have piled up but would have been moving horizontally at high speed. A monster wave would now have been formed and it would be moving outward from the impact site in all directions. Some of the splash water would have fallen back and rejoined this wave as it moved outward. (When a stone is dropped into a pond a similar phenomenon happens. In that case both the size of the stone and the height of the resulting wave will be measurable in inches whereas in the asteroid case the dimensions of both asteroid and wave might be in miles.)

A wave will propagate outward from the impact site just as it would when a stone is tossed into a pond and generates both a splash and a wave. Since an asteroid is larger than a stone, the accompanying splash and wave will be larger but the mass and the speed of the asteroid generate additional factors which compound the magnitude of the results beyond a simple scale-up in size. In other words it would be more than just a giant splash. As discussed before, the disturbed water that doesn't splash away will create a giant circular mountain wave which will have two characteristics of immediate interest.

The first one is the magnitude. When a stone is tossed into a pond, the water gets out of its way and all of the water which does not splash piles up as a circular wave. This wave will move outward because the stone gave it an outward push. Then, immediately, some water will rush back in to fill the opening where the stone entered the water. In the case of the asteroid, the outwardly-propagating water will have been given a phenomenal outward push. This water will move out of the way of the asteroid and because of the great magnitude of the force, it will continue to move outward for a much greater distance than just the size of the asteroid. The first effect of this incredible push will be a temporary opening in the water, which will be several times the diameter of the asteroid.

There was a relatively recent case of a shock wave effect on water, which is of interest to our present discussion. (14) During the WW1, there was a terrible explosion in the Halifax harbor. A ship, fully loaded with explosives, caught on fire. The fire caused the explosives to ignite and the ship exploded. It was reported that the explosion was so great that the floor of the harbor was laid bare. This means that the shock wave which was produced by the explosion pushed the water back with such force that the bottom of the harbor was exposed. (Of course the void quickly closed up again as the water rushed back into the opening.) With this in mind, It may be understood that the height of the primary wave could easily be several times the diameter of the asteroid because the hole, which the asteroid would blast in the ocean, would be much greater than the volume of the asteroid. As with the Halifax ship explosion, the asteroid would generate an opening in the ocean, which would reveal the ocean floor for many miles around. Some of the displaced water would form the splash but most of it would form the primary wave.

The primary wave could also be compared to the waves produced by the Krakatoa explosion which were about 130 feet high at the coastlines of the adjacent islands. These waves devastated hundreds of villages, killed about 35,000 people and carried large ships for miles inland. (18)

4.16.7 Ocean Floor Contact

As a result of asteroid contact with the ocean surface, both a splash and a primary wave would have formed. These effects would be propagating away from the impact site as the asteroid proceeds to crash into the ocean floor. Contact with the ocean floor would result in several more dangerous and far-reaching effects occurring. The primary splash would involve only water. The secondary splash would include solid material from the ocean floor.

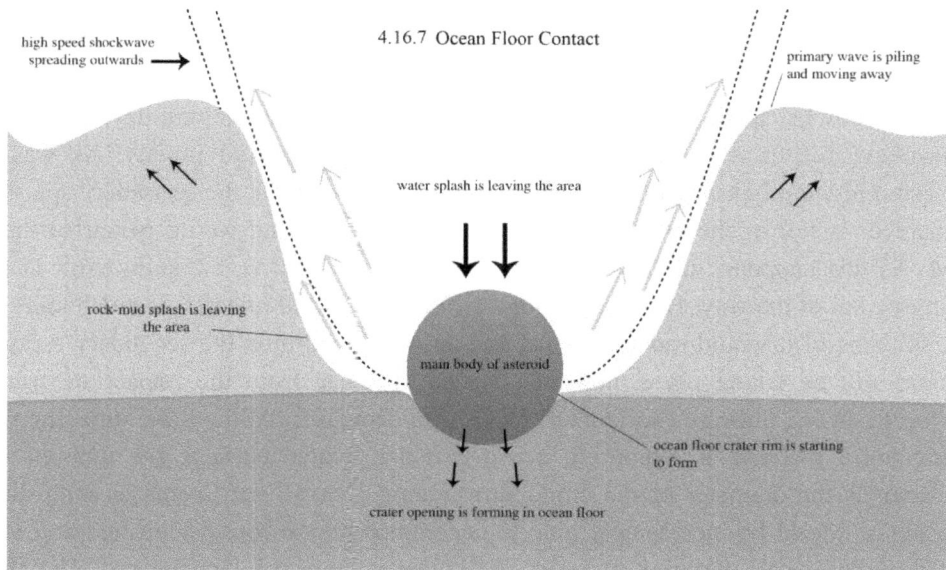

4.16.7 Ocean Floor Contact

high speed shockwave spreading outwards →

primary wave is piling and moving away

water splash is leaving the area

rock-mud splash is leaving the area

main body of asteroid

ocean floor crater rim is starting to form

crater opening is forming in ocean floor

4.16.8   The Secondary Splash

Just as the air and water before it was forced out of the way of the asteroid, when the asteroid contacts and forces its way into the ocean floor, the solid material which is directly in its path will also be forced out of the way. However, unlike the air and water, which were pliable, the ocean floor material will not yield easily. Now there will be a battle between the material which composes the asteroid, and the material which composes the ocean floor. If the ocean floor was composed of relatively soft unconsolidated material such as the sediment of the continental shelf, and the asteroid was composed of hard rock-like material (which is usually the case), the soft material would give way to the hard material. The effect, in such a case, would be comparable to asteroid contact with the ocean surface. With so much speed, mass and momentum involved, the softer material must give way. This would form a mud splash similar in some ways to the water splash. A mud splash would follow the water splash up through the atmosphere and out across the surrounding area. If, however, the ocean floor consisted of harder material, there would be more resistance to the asteroid's forward momentum. There would still be a secondary splash, but now it would include broken fragments of hard material from the ocean floor, as well as possibly broken material from the asteroid itself. A great mud-rock splash would form. This solid and semi-solid material would arc up into a fountain-like formation just like the water splash. It would then crash back to Earth over a wide surrounding area. No matter which type of material the ocean floor consists of, none of it would be expected to go into orbit because it would be heavier and consequently harder to move than the water above it. Never-the-less some of the solid material would be projected to great heights and would be spread over a very large area involving hundreds of square miles. As this solid and semi-solid material leaves the impact site, a crater is formed. All of the material which is removed to form the crater would be part of this secondary splash. In addition to the excavated crater material it is expected that some of the asteroid material might also become part of the great secondary splash. The water crashing back to Earth from the primary splash would be devastating enough to any life in the area but a great shower of mud and broken rock crashing down would mean certain death.

4.16.9   Crater Formation

A classical crater is bowl shaped, with a relatively flat bottom and sides sloping upward toward the rim. This shape develops due to what happens to solid material on asteroid impact. As was previously discussed, the asteroid would eject material from the ocean floor, causing the secondary splash. As this is happening, whatever is left of the asteroid is still moving forward, and would penetrate the Earth and become buried. As this happens, just as with the air and water before it, the solid earth in front of the asteroid must move out of the way. However, unlike air or water, solid material is much more resistant to being pushed, so some of it would move outward and upward, becoming the secondary splash. Outward movement would still be taking place, however. As the distance from the impact site (ground zero) increases, the force causing outward (horizontal) movement decreases. At the same time, the resistance of the surrounding solid material is becoming comparatively greater because the amount of material involved is greater as the diameter of the circle surrounding ground zero increases with distance. The surrounding material would be increasingly able to accommodate the forces with less and less material being blown away. The crater therefore becomes shallower as the rim is approached. The classic bowl-like shape would be expected for unconsolidated material. Softer material like water-soaked mud may also form a bowl but it would probably be shallower and the rim area may rise up and then slump back in. (If the water which formed the water splash, could have been removed from the ocean without moving the part, which pushed outward to form the water wave, a crater-shaped opening in the ocean surface would have been observed.) A crater in hard material would have another feature not expected with the softer examples. Just outside the rim of such a crater, the region may become elevated. The horizontal forces, which turned vertical inside the crater rim, also turned vertical to a lesser extent just outside the crater rim. In this area, the vertical effect would not be great enough to cause the rock to be blown away but since there would still be horizontal forces involved, these forces would have a vertical effect because deep below the surface, some displaced material had been forced sideways. Out as far as the crater rim the outward-moving material found a space by causing the original material to displace and splash up out of the way. As the sideways moving material moved beyond the rim of the crater, some of it would move in underneath the original material and so an elevated region around the crater would be formed. Since rock, unlike water, would not move back easily, it might just stay there in the new location. Further out from the center, there is less underground movement and less resulting vertical movement. Therefore as the distance from the impact site increases, the ground will slope down to the original undisturbed levels. The Manicouagan crater in Quebec has this type of formation. Immediately beyond the rim the land has been elevated and the vegetation is tundra-like. (34)

Another example of where this may have happened is adjacent to the Takla Makan Basin in China. While this area has never been formally identified as an asteroid impact site, it does meet all of the usual criteria. The nearly-circular Basin, which is about 1200 miles across, is almost completely surrounded by mountains. Immediately to the southeast is the Plateau of Tibet, which is a large area elevated to about 11,000 feet above sea level. The Himalayas lie concentric to the Basin, south of the Plateau. A large asteroid may have formed the Basin as it plunged into the Earth. Material that was pushed sideways to make room for the asteroid caused the surface to become elevated forming the Plateau. The shock and

149

lateral displacement of surface material caused soft recently-formed sedimentary layers to crumble forming the Himalayan Mountains. Since the Himalayas consist of sedimentary rock, which is a type of material laid by water, it is possible that the entire area was underwater when the asteroid hit. Further movement of water resulting from the impact may have altered some surface features or swept some of them completely away. In order to form a basin this large, a large object would have been required. The Earth would have convulsed from the impact. Some ground would have been blown away, some would have been elevated and some would have structurally failed in compression, shear or bending. A completely chaotic terrain, including the Deccan Plateau of southern India as well as numerous igneous provinces in the Indian Ocean, would have resulted and that is what remains. (35) (On the other hand, the arrival of India from the southern seas does not properly account for all of these features.)

Whether the elevated region rises up and slumps back or simply rises up and stays up is incidental to the causal effect, which it will have on the formation of the secondary wave.

4.16.10   The Secondary Wave

The primary wave formed when the asteroid formed an opening in the ocean and caused a great circular mountain of water to pile up around the temporary crater-like opening in the ocean. This great mountain of water would be caused to accumulate by the speed and force of the inrushing asteroid. It would also be given an outward push and would move away from the impact site at high speed. However, before it leaves the area, the asteroid would impact the ocean floor. A crater in the ocean floor would form and it would actually be the bottom of the hole, which was opened in the water. Both a water splash and a mud-rock splash would leave the area. As these effects develop, an elevated area surrounding the sea-floor crater would form. All of these things would happen very quickly and the ocean floor at the rim of the crater could surge up under the already-formed mountain of water, which has been referred to as the primary wave. Even without any disruption of the ocean floor, the primary water wave would have formed and moved away but if the ground swelled up under the primary wave, it will cause that mountain of water to rise up even higher. Also because the ground would now be sloping away from the impact site, the energy of the outward-moving wave would be increased, adding to its power and speed.

Even though a water wave might have great height and speed, any particular quantity of this water will not actually move very far from where it was prior to being disturbed. A water wave on the open ocean for example may have a trough to peak distance of ten feet and a forward speed of five miles per hour. However, any particular bit of this water will only move up and down as well as back and forth for a few feet. With a typical smooth surfaced wave train, each molecule of water will move in a nearly elliptical pathway and keep cycling back close to its starting place. Similarly with an asteroid-induced water-wave the water would initially blast out and pile up. The displaced water would not move away with the wave. Instead it would rush back into the temporary opening. In fact, it would rush back in so quickly that it would pile up in the center. It might pile up so vigorously that some of it shoots up into the air. Then as it comes back down another outgoing wave would be formed. This is the secondary wave and it would follow the primary wave. The water impact of an asteroid would therefore result in two great circular waves emanating from ground zero and travelling outward as far as their respective energies can take them.

4.17 Continent Inundation

4.17.1 France

The great waves produced by an impacting asteroid would more appropriately be described as mountains. If the asteroid had been five miles across, the primary wave might have been ten miles high. Waves this high – particularly when they have been given an enormous outward push – would easily travel right across an ocean. In particular the Chesapeake Bay crater has a diameter of about 50 miles. Material from the ocean floor was dislodged and blasted away over an approximately-circular area of the ocean floor about 50 miles in diameter. It is reasonable to conclude that the temporary opening that the asteroid created in the ocean also had a diameter of at least 50 miles. Some of the displaced water would have left the area as a splash. The rest would have mounded up into a giant outward moving circular mountain of water causing the primary wave to be on its way. If the inner half of the temporary, impact-created opening (in the water) left the area as splash water and the remainder piled up to form the primary wave, the ocean would suddenly have become several times deeper because the primary wave could easily have been ten or twelve miles high or even higher. It would have crossed the Atlantic Ocean within hours. When it came ashore in France it would not have stopped.

The famous French naturalist, Georges Cuvier, studied finds which he made in the gypsum formation of Montmartre in Paris and reported; 'It has frequently happened that lands which have been laid dry, have been again covered by the waters, in consequence either of their being engulfed in the abyss, or of the sea having merely risen over them…These repeated irruptions and retreats of the seas have neither all been slow nor gradual; on the contrary, most of the catastrophes which have occasioned them have been sudden…' (36) 'Numberless living beings have been the victims of these catastrophes; some, which inhabited the dry land have been swallowed up by inundations….the gypsum deposits of Paris contain over eight hundred species of shells, all of them marine….Much of France was once sea; …then it was land again…then it was sea again…recurrent upheavals…And as it was on the site of Paris, so it was in other parts of France and in other countries of Europe.' (36) '…in Burgundy: a hill…1430 feet high…remains of mammoth, reindeer, horse, and other animals….On the Mediterranean coast of France there are numerous clefts in the rocks crammed to overflowing with animal bones.' (39) 'In Prestwich's opinion the cause of the catastrophe was the sinking of the continent and its subsequent elevation, which was sudden…bringing chaos and destruction.' (40) Alternately, the giant waves produced by the impact of an asteroid would have been able to achieve these results without requiring the magical lowering and elevation of the entire continent. 'According to … Prestwich, the eruption of the sea was of a very violent nature; it spread to central France and the French Riviera, to Gibraltar, Corsica, and Sicily, and to the entire area that stretches to the lands of the ancient East.' (62)

4.17.2   England

In caves in England the most unusual assortments of bones were found. '...elephants, rhinoceroses, hippopotami, horses, deer, tigers, bears, wolves, hyenas, foxes, hares, rabbits, ravens, pigeons, larks, snipe, ducks,...hippopotamus, reindeer and mammoth... together...reindeer and grizzly bear...with hippopotami, lemming and reindeer with cave lion and hyena....hippopotami, bison and musk sheep together....found in gravel and clay (and given) the name of diluvium.' These unusual assemblages of animal remains suggested to the investigator that the evidence 'attesting the action of a universal deluge.' (36) 'In the neighborhood of Plymouth, clefts...are filled with...bones...mammoth, hippopotamus, rhinoceros...broken into innumerable fragments.' (38) Giant waves could have swept them all along, broke them into many pieces and then deposited them in the caves and fissures. It is otherwise difficult to explain how such completely different kinds of animals became buried together in the same locations.

'The Fens occupy an area of about two thousand square miles of Lincolnshire, Cambridge, and Norfolk counties, running east of Norfolk and around the wash, a gulf of the North Sea. ..."the forest trees all fell to the northwest. These fallen forests were mostly oak." Sometime after the hurricane that broke all the oaks came another calamity; the land "was now suddenly changed by an extensive invasion of the sea." ...the plain north of Cambridge was more than once invaded by the North Sea under circumstances that we would interpret as catastrophic.' (63)

'What agency of destruction could have accounted for...ten thousand square miles in extent (being) annihilated at once...The area ...comprises one half of Scotland, from Loch Ness to the land's northern extremity and beyond to the Orkney Islands...' (64)

4.17.3   Italy

'Of Monte Bolca, near Verona in Northern Italy, Buckland wrote: "The circumstances under which the fossil fishes are found at Monte Bolca seem to indicate that they perished suddenly...All of these fishes must have died suddenly...and have been speedily buried in the calcareous sediment.' (64)

4.17.4   Holland

'The sea did not slowly encroach ... it entered the land without much warning ... After ... man began ... to recapture ... land from the sea...discovered the bones of animals in vast and tangled masses, of extinct and living forms, generally ascribed to the Ice Age. So in the Dutch village of Tegelton, in a layer of sand, silt, clay and peat, ancient elm, ash, and grape were found with extinct fresh-water snails, with bones of elephants, mammoths, rhinoceroses, hippopotami, deer, horses and hyenas.' (63)

4.17.5 North America

4.17.5.1 Central

'In bogs ... in Michigan ... two whales were discovered. Bones of whale have been found 440 feet above sea level, north of Lake Ontario; a skeleton of another whale was discovered in Vermont, more than 500 feet above sea level; and still another one in the Montreal, Quebec area about 600 feet above sea level ... A species of Tertiary whale, Zeuglodon, left its bones in great numbers in Alabama and other Gulf States. The bones of these creatures covered the fields in such abundance that ... the farmers piled them up to make fences ... In Georgia marine deposits occur at altitudes of 160 feet and in northern Florida at altitudes of ... 240 feet ... Walrus is found in Georgian deposits.' (37) The only way for whales to become stranded and die miles inland and many feet above sea level is for them to have been carried there by giant waves and left to die as the water retreated. The Chiczulub Crater in Mexico was apparently formed by an asteroid (possibly) 15 miles across which would have produced a very large wave in the Gulf of Mexico. As with the Chesapeake Bay situation, to suggest that a wave several miles high was formed would not be exaggerating in the least. There would have been nothing to stop this wave from washing right over Florida, Louisiana and all the way into central North America via the Mississippi River drainage area. The central part of North America from the Gulf of Mexico to Hudson Bay is less than 1000 feet above sea level. (41) There would be nothing to stop a wave which was several miles high from going all the way to Hudson Bay. Once the headwater region of the Mississippi River is reached, it is downhill the rest of the way. Some of the water would have returned to the Gulf of Mexico, in which case the Mississippi would have carried a volume of water several times its present volume and an indication of the greater flow would be evident in the riverbed. Indeed the cross-section of the river does show that the river gorge was originally excavated to a much greater depth. (42) (Rapid melting at the end of the Ice Age would also account for greater historical flow. (43)) Subsequently, lesser flows have allowed it to fill in with silt.

4.17.5.2 Saskatchewan

In northern Saskatchewan, in the region around Snare Lake, there are a great many drumlins, all lined up from east north-east to west south-west. Drumlins are not symmetrical but indicate both a head and a tail. In this case the head is at the north-eastern end indicating that water flowed from this direction. (59) This region of Saskatchewan is several hundred miles from Hudson Bay which is connected to the Atlantic Ocean by Davis Straight. The land throughout the entire region is very flat and not very far above sea level. It therefore appears that a giant wave may have traveled from the Atlantic Ocean across Hudson Bay and continued for several hundred miles inland, forming the drumlins along the way.

The second feature in Saskatchewan which is evidence of massive water flow is the Cypress Hills. The Cypress Hills form a high, flat plateau about 300 meters above the next highest erosion surface and about 600 meters above the rivers to the north and south. Therefore about 600 meters of sedimentary rock has been eroded from the region around the Cypress Hills, because the substrate is generally horizontal. The gravel on both the Cypress Hills and Flaxville plateaus is well-rounded quartzite pebbles as well as

153

cobbles and boulders with abundant percussion marks. (Which means that they must have smashed into many other boulders on their way.) The current necessary to produce a sheet of gravel such as that observed capping the Cypress Hills would have been unimaginably large and difficult to imagine. (69)

4.17.6   Antarctica

In Antarctica, seals have been found far from shore. 'One of the unsolved mysteries of Antarctica is the mummified bodies of crab-eater seals found thirty miles inland and up to three thousand feet above sea level in ice-free areas.' (44) Also in numerous places across Antarctica there are 'labyrinths'. These are places where massive water flow has cut deep into the surface forming valleys in some cases several hundred meters deep. Only massive water flow can explain these features. (Please refer to Discover magazine July/August 2013)

The mummified seals and the labyrinths of Antarctica are evidence of giant wave activity. Asteroids crashing into the Earth would have produced waves which were capable of washing over Antarctica. In particular, the Burkle Crater impactor in the Indian Ocean could have provided some of these waves. Located about 900 miles south-east of Madagascar, in 12,500 feet of water, (45) the asteroid which caused the formation of this crater would, like all asteroids that impact upon water, have formed a temporary opening in the ocean. While the crater is only 18 miles across, it is located in very deep water. The entire depth of the ocean would have been forced aside as the asteroid plunged into it. The only place for deep displaced water to be manifested is at the surface. Water will not compress. Therefore a column of ocean water at least 18 miles in diameter and 12,500 feet deep was displaced and most it would have formed an outward-moving monster wave. Suggesting that it was 12,500 feet high would be conservative. It would more realistically have been several times this high. Antarctica is only about 3000 miles to the south. When this wave arrived it would have washed right over the continent, carrying hapless marine life with it and leaving displaced boulders and water-formed channels in the bedrock in its wake.

4.18   Rim Definition

What is a crater rim? The rim is primary evidence of asteroid impact. The region would have been totally disrupted out as far as the rim. The rim therefore, is the limit to which the impacting asteroid would have irreversibly dislodged planet material. Beyond the rim, disruption may also have occurred but the disruption forces were not great enough to totally remove material. Following impact, the areas beyond the rim would appear similar to the way they were prior to impact. The rim therefore is the defining feature enclosing an area from which the asteroid ejected material. Beyond the rim disruption may have been very serious but material would not have been totally and irreversibly dislodged.

4.19   The Earthquake

An incoming asteroid would smash the ground harder than it is possible to imagine. Part of this smashing would involve moving the ground horizontally. Whether initial contact was with water or solid ground, entry into either the ocean floor or an area of dry ground would certainly result in an earthquake. Locally, the ground would displace and not return to its original location. Far away from the impact site, the ground would displace and return repeatedly or quake. With a small earthquake the ground might shake for several seconds. With a large earthquake, the ground would shake for a few minutes, settle down and then shake again with after-quakes. With a small earthquake the ground would move enough to be detected. With a large asteroid-induced earthquake the ground would move so drastically that nothing could remain upright. Trees would be uprooted and topple. The back and forth movement of the ground would shake trees right out of the ground. Rivers would be displaced from their beds. Great cracks would open in the ground and may or may not close up again. If a large asteroid hit the Earth, the ground would shake for thousands of miles in every direction. As the tremors progress along the surface of the Earth, great cracks would open and close and anything in the wrong place could be swallowed entirely. It has been calculated that if an asteroid ten km. in diameter were to hit North America, the entire hemisphere would be destroyed including, of course all man-made structures. (15) Large buildings like skyscrapers would topple. The horizontal displacement of the ground would be so great that the lower stories might move a thousand feet out from under the upper stories. Therefore the upper stories would simply fall down. Mountains would form and crumble. Soil and vegetation would be catapulted in every direction. Great fissures would open in the Earth.

For both comparison and general discussion purposes, earthquakes are assigned a number on a scale. This number is intended to enable everyone who hears about the earthquake to have similar ideas about its destructiveness. The commonly used scale, the Richter scale, is named after its inventor, Charles Francis Richter. On the Richter scale an earthquake with a value of 2 is ten times as devastating as an earthquake with a value of 1. Every higher number is ten times as devastating as the number immediately below it. If an earthquake received a rating of 7, it would be considered ten times more destructive than one with a 6 rating. A rating of 8 is ten times as strong as a 7 rating. A rating of 9 is therefore 1000 times as strong as a 6 rating. Very few earthquakes in the history of the world have ever received a rating of 9. However, the Sudbury impactor in Northern Ontario may have generated an earthquake with a rating of 14. (71) Unfortunately, the Richter scale is not really appropriate for such extreme cases but the idea is appropriate. Earthquakes with such unthinkable magnitudes do not happen normally. However, if a large asteroid hit the Earth, the ratings, which could be properly assigned to the resulting earthquakes, could easily be of these higher magnitudes.

The Richter scale deals with the relative magnitude of the movement involved during an earthquake. While it does not directly address the energy involved, the energy of an earthquake is understood to increase about 31 times for every integer on the Richter scale. Therefore, an earthquake with a rating of 2 would be 31 times as energetic as one with a rating of 1. Similarly, one with a rating of 7 would be 31 times as energetic as one with a rating of 6. It follows that a rating of 8 is 961 times (i.e. 31 times 31) as energetic as a rating of 6. More simply put, earthquakes with Richter ratings in the upper integers like 7 or 8 or 9, are thousands of times more devastating than those with low integer ratings. (16)

If an earthquake with a rating of 4 occurred a few hundred miles away, a house might shake a little as if a heavily loaded dump truck drove by on the street. With a magnitude of 6, a few hundred miles from an epicenter, mud dwellings would crack and crumble. If a magnitude 7 earthquake occurred within a

155

densely populated area like a town or a city, properly constructed buildings, elevated highways and bridges would probably become structurally compromised and hence unusable. With a magnitude of 9, there are very few structures of any type that would not be damaged beyond use. (17)

The primary movement of the ground, as an asteroid crashes down through the Earth, is sideways. Asteroid is pushing on rock, which is pushing on more rock. Somewhere, something has to give. Rocks would shake and crack and crumble. Some regions would sink. Some regions would be elevated and appear as new plateaus complete with sharp vertical sides. As further shaking continues, some of these new structures may, in turn, be destroyed, but some would remain in place after everything has settled down.

4.20   Tsunami

If an asteroid should hit the Earth, it is absolutely certain that there would be an earthquake. If the earthquake displaced any land formation which was in direct contact with the ocean, there would also be a tsunami.

A tsunami is a water wave, which is caused by the movement of a land formation which is in direct contact with the water. Tsunamis can be caused by earthquakes, submarine landslides, asteroid impacts or volcanic eruptions. (61) The occurrence of an earthquake raises the possibility that a tsunami will result but it is not the general shaking of the ground that would cause the tsunami water wave. In order to have a tsunami, a region of land that is in direct contact with the water of the ocean must shift quite suddenly. 'Tsunamis are …caused…by the…deformation of the seafloor.' (61) The reason that the ground shifts is not important. When it shifts it pushes on the adjacent water and forces it to displace. This displacement will form a wave, which will then proceed to propagate outwards across the ocean.

With an incoming asteroid, the reason for the concern is the magnitude of the forces which are involved. The ground under the ocean (or along the shore) may shift very quickly and it may shift for some distance. The adjacent water will displace directly in relation to the magnitude and speed of this shift and it is this displacement that would generate a tsunami. Once formed a tsunami will race across the ocean as a series of low, fast waves about one meter high, traveling at speeds of 800 to 960 kph (500 to 600 mph). (61)

In 1755 at Lisbon, Portugal, devastation struck. 'Just before ten in the morning, the city was hit by a sudden sideways lurch now estimated at magnitude 9.0 and shaken ferociously for seven full minutes. The convulsive force was so great that water rushed out of the city's harbor and returned in a wave fifty feet high, adding to the destruction.' (55) 'The waves from this same disturbance traveled across the Atlantic and reached the West Indies in 9 ½ hours ' (56) Since the distance involved is about 3500 miles, the speed of this wave must have been almost 400 miles per hour.

The Morokweng Crater is located in South Africa and it is about 70 km (44 miles) across. When the asteroid which formed this crater hit the Earth, the entire continent of Africa would have shaken.

Tsunamis would have spread out from every coast of Africa. There is a second large impact crater also located in South Africa called Vredefort and it is 300 km. (187.5 miles) in diameter. (46) It is absolutely certain that when this crater was formed the entire continent was displaced and all of the water adjacent to every shoreline would have been displaced along with it. Giant tsunamis would have spread out from every coast of Africa, including the Mediterranean. Waves would have washed across southern Europe. Thirdly, there is another feature of African topography that appears like an impact site and it is even larger than Vredefort. The Congo Basin is nearly circular, more than 1000 miles across and surrounded by mountains. If this feature was formed by an asteroid, more than Africa would have been shaken. The entire world would have moved and tsunamis would have formed at numerous coastlines. Giant waves would have been propagated around the Earth in every direction.

A similar situation would have happened on the east coast of North America when the Chesapeake Bay crater was formed. In this case the crater is only 50 miles across (47) but it is partly in the water already. As the coastline was disturbed during the impact, a tsunami in the Atlantic Ocean would have resulted. Further, even though the west coast of North America is 3000 miles away, it would have been displaced which means that tsunamis would have been generated in the Pacific Ocean as well. The Chiczulub Crater on the Yucatan Peninsula in Mexico is also half in the water. In this case the impact would have displaced most of Mexico's coastline including the western side bordering the Pacific Ocean. A giant tsunami originating at this location would have proceeded out across the Pacific Ocean and would easily have travelled all the way to Japan and Russia.

In the Czech Republic there is a circular formation (i.e. the border) which appears as an impact site and which is described as being 200 miles in diameter (320 km). (54) It appears that an asteroid caused this formation, in which case tsunamis would have formed in the Black Sea, the Mediterranean Sea the Baltic Sea and the Atlantic Ocean.

A tsunami, which is proceeding outward from an under-water origin, might not be noticeable on the surface of the ocean. Even if it is moving at 500 miles per hour, it might not be noticed. There are two reasons why it could be inconspicuous. First, it may have a very long wavelength which is stretched out over a horizontal distance of several hundred miles. If this is the case, the slope of the wave-front would not be noticeable at all. In fact it would not even look like a wave. Secondly, the disturbance would involve a great depth of water. In this case the disturbance at the surface would only be a few inches high. No one would be concerned about a water wave which was only 18 inches high. In April 1946 there was an earthquake on the steep slopes of a trench off the island of Unimak in the Aleutian chain (a chain of islands running west from Alaska). In the open ocean the waves produced from the Aleutian quake were only about a foot or two high and would not be noticed by ships at sea. Their length, however, was enormous, with a distance of about 90 miles between succeeding crests. It took the waves less than five hours to reach the Hawaiian chain, 2300 miles distant, so they must have moved at an average speed of about 470 miles per hour. (57) Another commentator offers; At sea these mountains of water are virtually imperceptible.. A ship could sit completely unaware as a deadly tsunami goes past. (61)

The problem will only show up when this type of wave reaches shore. It would still be going 470 miles per hour and now all the energy, which was distributed down through the water for possibly a thousand feet or more, becomes concentrated in the shallow water near the shore. When a tsunami approaches a shore it builds up and breaks in a mountainous cascade of water. (61) It might rise up into a giant killer a hundred feet high and sweep inland, causing destruction of property and loss of life. When Krakatoa exploded in 1883, 'Waves as high as 130 feet washed over their coasts.. .The lighthouse fell in one piece, and all the houses of the town were swept away in one blow like a castle of cards. All was finished. There, where a few minutes ago lived the town of Telok Betong, was nothing but the open sea....two hundred ninety-five such towns were wholly or partially destroyed, and 36,419 people were either missing or killed, mostly by drowning.' (18)

The energy of such waves would not be dissipated until they had traveled a great distance inland. After causing enormous damage on their way in, they would retreat back to sea. The retreating currents would be almost as devastating as the original wave. In either case anything, which is not firmly attached to the ground would be swept along into a great accumulation of debris. This type of development has happened repeatedly in recent times and is expected to happen whenever an earthquake disturbs a shoreline. 'The most destructive tsunami of recent history was the one produced by the magnitude 8.5 Chile earthquake in May, 1960. One hundred eighty persons were left dead or missing in Northern Japan and Okinawa. Twenty others died in the Philippines. Coastal areas of the United States and New Zealand were damaged, and all towns along the coast of Chile between the 36th and 44th parallels were destroyed or heavily damaged.' (58) Such tsunamis are serious trouble for people living in low-lying coastal areas, but the earthquakes involved in all of these cases are very small compared to the earthquakes that would have resulted from the impact of any one of the asteroids mentioned above. In all of those cases, the surface wave would have been more than noticeable and wherever it came ashore it would have washed right over vast continental areas.

As with the primary water wave, there would be nothing to stop a large tsunami and it would become a world traveler. It could easily travel for a thousand miles or more and in fact it would not stop until it ran into a substantial landform. Even then the landform could be overrun before the energy of the wave was spent. 'If the impact occurred in the ocean, towering tidal waves would spread out over much of the planet and additional trillions of tons of water would be vaporized.' (31)

Water waves, which would result from an asteroid impact, would have a great deal of energy. This energy would be represented by the amplitude or size of the wave as well as its speed. While storm waves only a few meters high can travel on for hundreds of miles after the storm itself has died out, it follows that a super wave several miles high, would easily travel all the way across any ocean. Unless the shoreline was a solid wall of sufficient height, such a wave would not stop at the shoreline. It would just keep on going. As it rushed inland, everything would be washed away. If it had enough energy, it might wash right over a continent and still keep going. If, however, it was running uphill as it went inland, it would eventually come to a stop. Then it would retreat. While going back to sea, the retreating wave could wash overburden right off the continent. Any silt or other washable or movable material would then be deposited in the ocean next to the shore and form a continental shelf. New lakes may be formed inland.

The ocean would become very muddy. Great masses of floating material such as trees, could become stranded and eventually form either peat or coal, depending on how much they became covered by earth.

It is a curious fact that all of the oceans of the world are connected. A massive wave could propagate through these connection areas and keep on travelling around the world. Eventually it would dissipate but it is conceivable that it could circle the world several times before dying out. (19)

4.21 Sound-Wave

When an asteroid smashes into the Earth, it will make a great noise. This sound wave will travel both through the air and through the ground and propagate outward in all directions. Since the speed of a sound wave increases with the density of the material it is travelling through, and since ground has greater density than air, the sound wave would travel faster through the ground than it would through the air. As a result, two sound waves would be heard instead of one. If a witness was located, for example, a thousand miles from the impact site, the sound of the crash would be heard through the ground several minutes before it would be heard through the air. These sound waves must be distinguished from the atmospheric shock wave, which was generated as the asteroid passed down through the atmosphere. The sound waves will travel at the speeds, which their respective mediums enable, whereas the high-pressure atmospheric shock-wave will be travelling at a much higher speed. The sound wave, like the shock-wave before it, would be a compressed region of atmosphere. It would therefore have damage potential but not of the magnitude of the major shock wave.

4.22 Volcanoes

A large asteroid would penetrate deep into the crust of the Earth before it came to a stop. The effect of its arrival, however, would go even further and cracks would be formed in the crust of the Earth and propagate down as far as the molten interior region. These cracks could allow some of this molten material to come up to the surface of the Earth. A volcano would thereby be formed. Volcanoes have a great deal of variety in their characteristics. While some only ooze molten rock, others explosively throw up quantities of molten rock, which splashes down on the surrounding area. Some volcanoes erupt a great deal of water vapor. Still others throw up ash, which settles on everything and later turns into a type of rock called tuft.

4.23 Earth Wobble

The Earth has a wobble. 'The Earth wobbles ever so slightly, as though it were a spinning top gradually slowing down.' (65) There are actually several wobbles and at least one of them has been given a name, the Chandler Wobble, after Seth Carlo Chandler, who discovered it. With the Chandler wobble, the North Pole swings out 12 m (about 40 feet) from its center. (65) The axis of the Earth's wobbles like a top and traces a complete circle every 23,000 years. (66) This makes the Pole Star describe a small circle in the sky instead of simply remaining as the stationary location of the extended axis of the Earth.

The Earth has been repeatedly struck by some very large asteroids. The energy involved in a single asteroid strike is mind-boggling. 'No other natural event is as powerful, devastating, or potentially catastrophic as a major impact. Consider one capable of producing a 50 km (31 mi) wide crater....the energy expended is thousands of times greater than the simultaneous detonation ... of all the nuclear explosive devices manufactured to date.' (67) From the list of asteroid impacts given in the chapter entitled, Conclusions, it is clear that there have been numerous impacts producing craters much larger than 50 km. For example, the Sudbury Crater is variously given as being in the 200 to 250 km dia. range and it would have released more than one million times the energy of an earthquake of magnitude 9.0 on the Richter scale. (67)

The energy from a single impact is more than enough to form a mountain range as evidenced by the mountain rings which commonly define craters. For example, the Caloris Basin on Mercury is defined and surrounded by a ring of mountains. It may be appreciated that the ring of mountains only represents a portion of the total energy involved. Where did the rest of the energy go?

The energy of an incoming asteroid that was 10 miles in diameter (16km.) would be about $2.1 \times 10(31)$ ergs. This may be compared to the energy required ($10(29)$ ergs) to raise a mountain range.(68) This means that one large asteroid has enough energy to raise 200 mountain ranges. Since numerous asteroids have struck the Earth, hundreds of mountain ranges would be expected. Such enormous amounts of energy would also be capable of displacing the Earth from its orbit. (Please refer to the appendix for comparison energy levels.)

With all of these facts in mind, it is reasonable to suggest that the wobble of the Earth was caused by one or more of the major asteroid impacts that the Earth has experienced. Further, if the Earth had not been able to basically absorb the energy of these large impacts, it would be expected that the wobble of the Earth could easily have been much greater.

4.24   Summary

If an asteroid should come to Earth, there would be several serious and far-reaching effects. No matter where it contacted the Earth, it must first pass down through the atmosphere. Since it would be traveling at high speed, atmospheric pressure waves or shock waves would be generated. The pressure in these waves will be extremely high as they form and they would be given an enormous outward shove as well. As they travel outward, everything would be flattened. The increase in pressure would result in an increase in temperature enabling the expanding pressure wave to ignite everything that was combustible. The ground would be immediately scorched and the fires would continue until they ran out of fuel. Further out, the shock wave would be reduced to a loud sound and finally, after it had circled the Earth several times, it would peter out. As the asteroid compressed air in front of itself, a point would be reached where this air became so hot that a blinding flash would develop. While the light from this flash occurs at frequencies in the visible portion of the electromagnetic spectrum, there would also be an enormous and damaging electromagnetic impulse in the invisible portion, which would be able to destroy sensitive electrical circuits for miles around. Also included in this initial group of damaging effects would be a windstorm, which would have a rushing, outward-blasting effect and blow away anything that wasn't already blasted away by the initial pressure waves. Next, contact with the ocean surface will generate a great water splash. The early portion of the splash will travel upwards so fast that some of it might leave the Earth altogether. The rest would fall both near and far away as rain and great globules of water, which would land with the effect of exploding bombs. The water, which does not splash, would be pushed up into a mountainous killer wave. This too would be given a great outward push and so would leave the impact area at high speed. The asteroid would crash into the ocean floor and cause a mud/rock splash. Unlike the water splash, material from the mud/rock splash would probably not go into orbit but it would still be spread over a large area. The return of the mud/rock material to the surface of the Earth might bury some areas under many feet of debris. A large asteroid would continue deeper into the ground and cause the region around ground zero to swell up under the retreating water wave. This would raise the wave even higher increasing its energy. If the asteroid impacted on land instead of water, there would not be a water splash but there would still be water waves. The ground would shake on impact and the shaking and displacement would generate tsunamis. It would require a shift in some under-water structure to get the tsunami started but such shifts could happen miles away from the impact site. Even if a large asteroid landed in the middle of a continent, the severity of the impact would cause displacement of distant coastlines enough to start tsunamis. Sound waves would also be generated. One would travel through the air and another through the ground. Volcanoes would be formed. If the asteroid crashed deep enough into the ground, it would fracture the crust of the Earth all the way down to the molten interior region. If magma came up through some of these cracks as far as the surface, a volcano or an igneous province would have been formed. Destruction from an incoming asteroid would be widespread and it would take several different forms. Any one of these forms would be devastating but all of them taken together means that nothing could survive for thousands of miles around. Even with this terrible prognosis, the trouble is not over. While an additional aspect of the impact will cause trouble to show up on the other side of the world, an overwhelming sustained torrential downpour could initiate at the impact location and spread outwards in all directions. This possibility will be discussed in the next chapter to be followed by a discussion of the effects that an asteroid impact would have on the far side of the world.

From the time that an asteroid initially entered the upper atmosphere many miles above the Earth, until all of these effects would be propagating away from the impact site, less than one second would have passed. Of course, more time would be required for each effect to propagate and it would possibly be hours before some of the devastations reached the far side of the world. Ongoing trauma would continue for months.

5.   The Vapor Canopy Hypothesis

There is a possibility that, prior to any initial impact of an asteroid upon the Earth, the atmosphere of the Earth was structured differently than it is now. If there had originally been a region in the upper atmosphere, which, instead of containing atmospheric oxygen and nitrogen, contained mostly water molecules, (Please refer to the diagram, 'The Water Vapor Layer' 5.1.) a hot torrential downpour would have resulted from the expanding pressure wave generated by an incoming asteroid. Consequently, such a region of atmosphere would have been totally destroyed. This hypothetical atmospheric structure has been referred to as a vapor canopy because it would have been like a blanket or canopy enclosing the entire Earth and it would have had a significant moderating effect on the climate of the whole world. Would it have been possible to have had a region of atmosphere like this and if so, what characteristics would it have had?

5.1   Basic Physics

In order to speculate that a vapor canopy could have existed above the entire present atmosphere of the Earth, first, the basic physics, which would be necessary for this to be possible, must be recognized.

While it will probably never be possible to identify just how much water there might have been in this hypothesized upper layer of the atmosphere, the problem can be approached by speculating on the amount and then identifying the accompanying physical factors associated with such speculation. Suppose, for example, that there was a layer of water vapor above the present atmosphere and that it was equivalent in weight to about one-third of our present atmosphere. The present atmosphere weighs 15 pounds for every square inch of the Earth's surface. This is referred to as atmospheric pressure and what it means is that for every square inch of the surface of the Earth, there are 15 pounds of air pressing down. If we could put this small column of air on a scales, the scales would read 15 pounds. Then, if we could somehow add more air to the Earth, it too would be pressing down and the little column of air would have become heavier. From this we can speculate that if (instead of having more air on top of our present air) there had been a layer of water molecules on top of the atmosphere instead, we can see that our little column of air would have been a little heavier than it is now. Another way of saying this is if the Earth's atmosphere had more air in it, or if there was some other gas floating on top of it, it would be heavier. Atmospheric pressure would be higher. As suggested, if the additional amount was about one-third of the present atmosphere, there would be another 5 pounds so the total weight of the square inch of atmosphere would have been 20 pounds and atmospheric pressure would have been 20 pounds per square inch.

If such a layer of water vapor had existed and it was then squeezed completely out of existence, a layer of liquid water about ten feet deep would have rained down on the entire Earth. As rain, locally, this represents a storm, which is comparable to a very large hurricane. A hurricane has been observed to produce nearly four feet of rain. (18) Therefore ten feet of rain would be similar to the rainfall from two and one-half such hurricanes. A hurricane however, is a local event covering at most a few hundred square miles. The atmosphere, on the other hand, covers the entire Earth so the comparison would be to several hurricanes raining down on the entire Earth at the same time. If a monster hurricane produced ten feet of rain, serious local flooding would result. However, the flood water would soon run into rivers and lakes and find its way to the ocean. On the other hand, if ten feet of rain fell on the entire world, including the oceans, this much more extensive flood would alter the topography of the whole Earth.

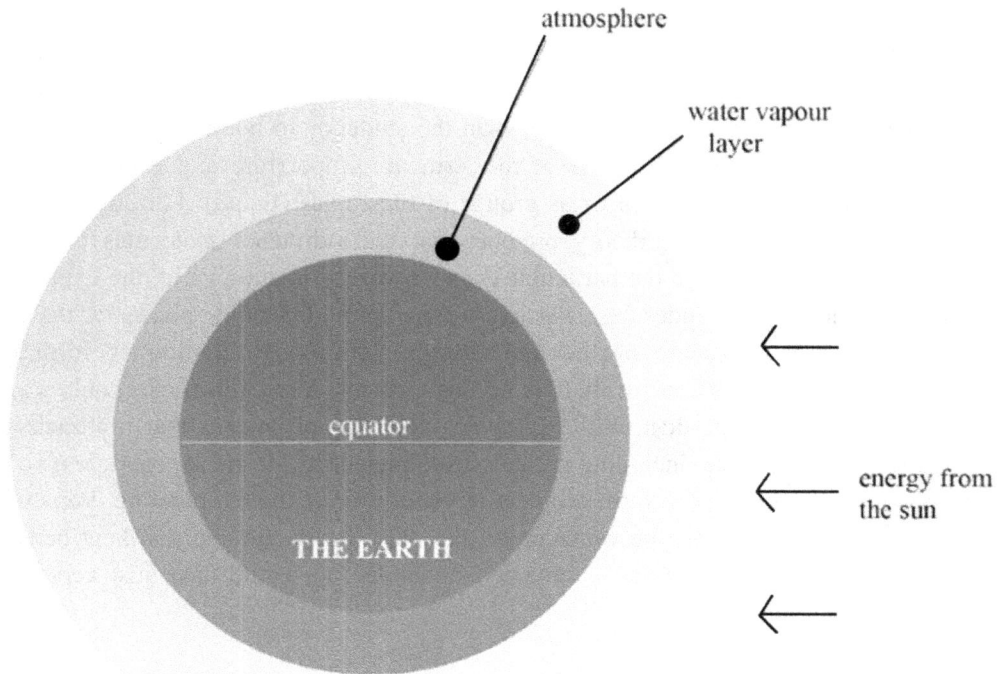

5.1 The Water Vapour Layer

5.2 Temperature Considerations

In order to have a water vapor layer above our present atmosphere, it must remain warmer than a very particular temperature which is referred to as the critical temperature. This temperature is called critical because if it was any lower the hypothesized water vapor layer would condense into rain. Above this critical temperature, the water would exist as a gas. Below the critical temperature, the water would exist as a liquid. Since a liquid cannot remain floating in the air, but would simply fall to the Earth, the entire idea would fall with it. Below this so-called critical temperature, any water above the atmosphere would not have been able to remain as vapor (or gas) but would have changed to a liquid. This is the way nature behaves in the present atmosphere. If, after a warm summer day when a lot of moisture has evaporated into the air, evening comes and it cools down, dew forms on everything. This happens because it becomes too cool to enable the water vapor in the air to stay as an invisible vapor, so it changes into water, which condenses on all of the cool surfaces. This is also similar to the way moisture forms on the surface of a glass of cool water. In the vicinity of the glass, the air is cool and it is too cool to enable the water in this region of air to remain as an invisible vapor so it appears on the glass as a liquid. If the glass was subsequently warmed up, the water on it would change back to a vapor and disappear back into the air. Therefore, in order to hypothesize that a layer of water vapor existed above our present atmosphere, it must be recognized that it had to have been above a certain very particular critical temperature.

If it was too cool, the individual water molecules would not have enough energy (speed) to bounce into each other fast enough to bounce away again. Water molecules have a certain amount of attraction toward each other. When this attractive force is greater than the tendency to bounce away, the water molecules will start to stick together. They are now at the 'critical' temperature and exist partly as individual molecules (vapor) and partly coalesced or as groups of molecules (liquid). For any particular pressure, there will be a temperature where theory predicts that the individual molecules of water would be expected to start coalescing. While the particular theory involved is quite valid, the expected results will only be obtained if the vapor includes dust. Particles of dust provide landing places for the atoms of water and make it easier for them to come together and form droplets. When tiny droplets form, fog is visible. The situation is dramatically different if there is no dust present. Also, in this case, only a very particular type of dust, called hydroscopic dust, will do. (Hydroscopic simply means that it attracts water.) Some types of dust are not hydroscopic including the dust from meteorites. If the appropriate type of dust is not present, it becomes very difficult to form droplets of water even if the temperature drops well below the expected critical level. As long as the water molecules stayed separate and just kept banging into each other, the water vapor layer would have remained invisible and would have just kept floating on the oxygen-nitrogen layer.

5.3   Atmospheric Temperature Chart

If water vapor is to remain in vapor form, this is equivalent to saying that it must be above its boiling point. However, due to the above-mentioned anomaly, it may actually be able to cool down much farther than expected and still remain as vapor. The following chart gives several examples of the theoretically expected, fog-forming temperatures as well as the lower anomalous temperatures for eleven examples which involve different amounts of water.

Example Number	Portion of atmosphere	Liquid water equivalent	Pressure psi	atm	Expected temp. to form fog		Anomalous temp to form fog	
1.	10%	3.3 ft.	1.47	0.1	115 F	46 C	59 F	15 C
2.	20%	6.6 ft.	2.94	0.2	141 F	61 C	79 F	26 C
3.	25%	8.25 ft	3.67	0.25	149 F	65 C	86 F	30 C
4.	30%	9.9 ft.	4.41	0.3	157 F	69 C	91 F	33 C
5.	40%	13.2 ft.	5.88	0.4	169 F	76 C	101 F	38 C
6.	50%	16.5 ft.	7.35	0.5	179 F	82 C	108 F	42 C
7.	60%	19.8 ft.	8.82	0.6	187 F	86 C	115 F	46 C

Example Number	Portion of atmosphere	Liquid water equivalent	Pressure psi	atm	Expected temp. to form fog		Anomalous temp to form fog	
8.	70%	23.1 ft.	10.28	0.7	194 F	90 C	120 F	49 C
9.	80%	36.4 ft.	11.76	0.8	201 F	94 C	125 F	52 C
10.	90%	29.7 ft.	13.23	0.9	204 F	96 C	130 F	54 C
11.	100%	33.0 ft.	14.7	1.0	212 F	100 C	134 F	57 C

In this chart, example number 6 indicates the minimum temperature which the base of a water vapor layer above the atmosphere must have in order to remain as invisible vapor and not turn into fog. (If the amount of water involved was 50% as heavy as the present atmosphere.) In this example, the equivalent weight of one-half of our atmosphere would have been floating on top of our present atmosphere in the form of water vapor. If it all had precipitated out, 16.5 feet of liquid water would have fallen down onto the entire Earth. With this extra material above our present atmosphere, the atmosphere would be heavier and atmospheric pressure would be 1.5 times 15 pounds per square inch or 22.5 pounds per square inch. The theoretical minimum temperature to keep this much water in vapor form is about 179 degrees F. However, if there wasn't any hydroscopic dust mixed in with the water molecules, the temperature might have to drop down as low as 108 degrees F or even lower before any fog would form. Similarly, example number 2 shows that 79 F might be needed to form fog if there was about 6.6 feet of water in the vapor layer. This temperature is within the comfort range for people meaning that such a circumstance would not present any threat to human survival. If the vapor layer contained any amounts of water greater than this, the minimum temperature at the bottom of the water vapor layer would probably have been above the comfort zone. (This would not necessarily have been any threat to human survival either.) Of course for any elevation above the bottom of the vapor layer, the required minimum temperatures would be even lower and would continue to drop off as elevation increased. It does mean however that for any amount above 20%, the temperature at the bottom of the vapor layer would probably have been above the comfort zone for people on the surface of the Earth. Therefore, for all of these cases, having a water vapor layer on top of the atmosphere would have been like having a warm blanket around the entire Earth, isolating it from the bitter cold of space.

5.4 Atmospheric Pressure Profiles

Atmospheric pressure decreases as the height or distance above the surface of the ground is increased. (Please refer to the diagram; 5.4 Atmospheric Pressure Profile) The pressure drops fairly rapidly through the lower elevations and then the decrease rate reduces at higher elevations. The pressure variation follows the familiar half-life curve or time-constant curve. Mathematically, this type of relationship is called an exponential and it has a very particular mathematical formulation. Atmospheric pressure very closely follows this type of relationship. In the diagram, pressure variation is shown as curve number 1. Three other curves are also included to show how atmospheric pressure would be increased if there was a layer of water vapor above the present atmosphere. From these curves the height of the base of the postulated vapor layer may be determined and this information has been identified on the diagram. For example, curve number 2 (the 25% of present atmosphere added-on-top case) shows that the height of the base of the water vapor layer would be about 7.5 miles. (Note the checkmark on the diagram.) From the Atmospheric Temperature Chart included in 5.3 above, the theoretical temperature to ensure the water at the base of this particular vapor layer remains as a vapor would be 149 F. The anomalous or lower possible temperature would be 86 F. The other anomalous temperatures for the examples shown on the diagram are not very much above temperatures, which are commonly reached on the surface of the Earth.

5.4   Atmospheric Pressure Profile

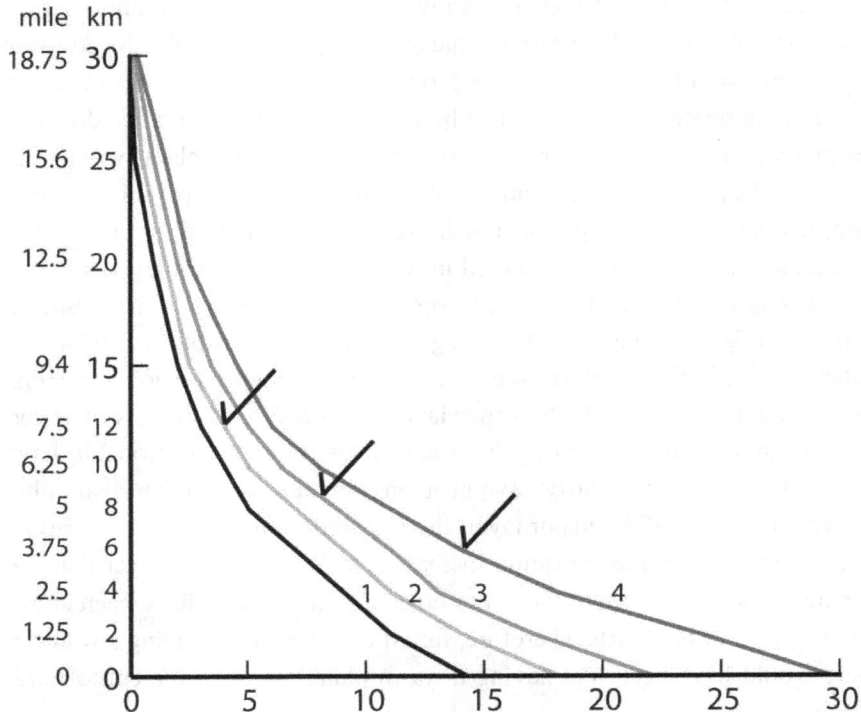

Line No. 1 Atmospheric pressure at the present time
Line No. 2 Atmospheric pressure with a 25% vapour layer addition
Line No. 3 Atmospheric pressure with a 50% vapour layer addition
Line No. 4 Atmospheric pressure with a 100% vapour layer addition

Arrows indicate the top of the atmosphere and the bottom of
the vapour layer. Please refer to the Appendix for further
discussion.

5.5   Thermal Stability of Atmosphere with a Canopy

Usually, air which is high up in our atmosphere is cooler than the air at the surface of the Earth. This is understood from the way in which all gasses behave as their pressures are varied. If air, which is near the surface of the Earth, is raised to a higher level, it will cool down. It cools because its pressure will drop as

167

the elevation increases. The atmosphere of the Earth always behaves like this and means that most of the time it is stable. However, occasionally these stable conditions are upset and storms result. If, for some reason, either the temperature at the surface of the Earth becomes hotter than usual or, the temperature high above the Earth becomes colder than usual, the atmosphere is recognized as being unstable and either thunderstorms or tornadoes are expected. This may be understood by considering the results of raising a quantity of air upwards from the surface of the Earth. As air is raised, it cools by expansion. It cools simply because as it is raised, it expands which directly results in cooling. Usually, as air rises up from the surface of the Earth, it will cool and reach a temperature which is very close to the temperature of the surrounding air at that height. If this does not happen and the air cools but is still hotter than the surrounding air, this rising air will rise up even further. It will, of course cool some more. Again, if the surrounding air is still colder, the rising air will now be rising at an increased rate and it will form an updraft, which will draw replacement air in at the bottom. If this rising air includes moisture, it might reach an elevation where the entrained moisture chills to its condensation point whereupon it will then fall out of the rising air column as rain. This, however, only compounds the problem because now the heat of condensation of the rain-forming moisture heats the resulting drier air so it rises even faster and a thunderstorm develops. This is also the way in which a hurricane develops, because a hurricane starts off like several thunderstorms linked together. Tornadoes often accompany thunderstorms and are usually formed at the perimeter of the rising air column. As the air rises, it also rotates. The rotation may cause a tornado to form. All of these stormy results begin with air which is too warm near the surface or too cold higher up.

On the other hand, if the air high above the Earth was warm instead of cold, the atmosphere would be very stable vertically and there would not be any storms. In this case, if a column of warm air started to rise, it would encounter air, which was still warmer and so it would just settle back down again. Therefore, if the atmosphere included an upper layer of vapor and the temperature at the base of this hypothesized vapor layer was at the anomalous temperatures or higher as discussed above, the atmosphere would be very stable. Any vertical air movement would be restricted to the lower regions and could only occur if the temperature at some particular elevation were a little cooler than a quantity of air would be if it had just been brought up from the surface of the Earth (to this same elevation). (Several examples of the temperature-elevation relationship are included and discussed later in this chapter.)

5.6   Physical Stability Factors

5.6.1   The Influence of Dust

If water molecules in the atmosphere are forced together, droplets will form and rain will result. This is not the usual procedure in nature however, where the coming together of the water molecules is greatly facilitated by dust. The dust in the air, (which in this case is understood to be hydroscopic dust), provides a location for the water molecules to start condensing or clinging together. This process is occasionally encouraged by manually seeding the clouds. If rain is badly needed, and a likely-looking cloud appears, an airplane may be dispatched to drop small dust-equivalent particles into the cloud to facilitate the water molecules to aggregate. If there is enough water in the cloud, rain could result.

The opposite situation also holds. If there is no dust in the air, it is very difficult to get a water droplet to form. It will not form easily and as long as the air remains above the dewpoint (or the theoretical condensation temperature discussed above), rain will not be expected. Even at temperatures much lower than the dewpoint (i.e. the anomalous temperature discussed above), rain still might not happen.

5.6.2   The Buoyancy of Water Vapor

When moisture is added to air, the air becomes lighter. At first this would appear to be incorrect. If two things are added together, the resulting mass should be as heavy as the two original materials added together. However, this is not the situation with gasses such as air. If you add one gas to another, there will still be the same number of molecules in any particular space as there was with either gas by itself. More space would certainly be required for these gases, but in any original volume of this space, the original number of molecules would still be present. Therefore, when water vapor is added to air, because a water molecule is lighter than either of the two types of air molecules, the mixture of air and water vapour will result in a lighter gas. Therefore, when moisture is added to dry air, the resulting damp air will be lighter than any similar volume of dry air and the damp air will rise up and float on top of the dry air. In the atmosphere, any air which includes moisture will be found on top of air which does not contain moisture.

A portion of atmosphere, which was entirely water vapor, would be lighter than either atmospheric oxygen (O2) or atmospheric nitrogen (N2) and so it would float on top of our present atmosphere with no tendency to sink into it. The demarcation between the two regions would be quite abrupt.

5.6.3   Noctilucent Clouds

Noctilucent clouds, or night-shining clouds, occur about 50 miles above the Earth. They are quite faint and may only be seen for a short period of time right after sundown. Later, when the Sun is even further below the horizon it will not shine on these clouds. Apparently, they were first observed shortly after the great explosion at Krakatoa in 1883 and have been up there ever since. (62) Krakatoa is recognized as the most powerful explosion that the world has ever witnessed and it did drive moisture up to extremely high altitudes. Moisture at such a high elevation cannot come down. If it should drift lower hypothetically, it would be warmed, boil and rise back up again. Exactly the same situation would exist with the hypothesized vapor canopy. The existence of noctilucent clouds therefore provides direct evidence that a vapor canopy layer would be quite stable in the upper atmosphere and would just remain there for years until something significant occurred to disturb it and bring it down. An incoming asteroid could provide such a disturbance.

5.7   Basic Criteria for a Canopy

There are two basic criteria, which must have been in place, if the atmosphere of the Earth at one time included a water vapor layer. First, the water vapor layer must have been above a particular minimum

169

temperature (i.e. the critical temperature) just to remain as water vapor. Secondly, the temperature at the surface of the Earth must have been in the comfort zone or animal life could not have existed and the "window of life" would already have closed.

The temperature of the Earth is determined by the characteristics of the atmosphere and by the incoming solar energy. The energy which approaches the Earth is virtually always the same but the way in which the atmosphere deals with this energy will determine the actual temperature which results at the surface of the Earth.

5.8   Solar Energy Distribution

While virtually all of the energy that heats the Earth comes from the Sun, the way in which it is either absorbed or reflected determines the temperature at the Earth's surface. Solar energy is divided into infra-red energy (53%) and visible light energy (38%). There is also a small amount of ultraviolet energy. (About 9% of the total) (65). Due to the presence of water vapor in the air, on the way through the atmosphere, some of the infrared energy is absorbed and some of the visible light energy is scattered. A little less heat energy (infrared is also called heat) is therefore available to heat the Earth and the Sun is not quite as bright as it is above the atmosphere. At the surface of the Earth, some energy is reflected and some is absorbed. The reflected portion of the visible light enables things to be seen. If no light reflects from an object, or if there isn't any light to even shine on it (i.e. after sundown) that object cannot be seen.

Reflected visible light energy is referred to as albedo. If the albedo is 1, everything is being reflected. If it is 0.5, only one-half of the visible light impinging on it is being reflected. Forests have an albedo of 0.1, oceans 0.2, deserts 0.45 and clouds possibly 0.5 depending on their dust content. This means that forests are absorbing 90% of the incoming visible light energy because their albedo is only 0.1 or 10%. (64) Similarly oceans absorb 80% because their albedo is only 0.2 or 20%. The average for the whole Earth from the perspective of space is 0.44 (63) because clouds are brighter than either forest or ocean and reflect more energy back to space (and there is always a certain amount of cloud). This means that 44% of the visible energy from the Sun will be reflected back to space. Also the atmosphere will absorb some of the infrared energy and some of the ultraviolet energy. The net result is that approximately 51% of incoming solar energy reaches the surface of the Earth. (Please refer to the diagram 5.8 (1) 'Solar Energy Distribution') Some of this energy will be used for plant growth but most of it will heat the Earth. Without a vapor canopy the temperature of those areas, which are directly under the Sun during the day (i.e. the tropics), commonly rises above the comfort zone while the polar regions are commonly below the comfort zone. Temperate regions are in the comfort zone much of the time. If there was a layer of water vapor above the atmosphere enclosing the entire Earth, this situation would be dramatically changed.

Water absorbs infrared energy and it scatters visible light. Absorption results in warming and scattering results in less light being available to see things. A layer of water vapor enclosing the atmosphere of the Earth would absorb virtually all of the incoming infrared energy. Of the visible light which was being scattered, some of it – possibly 50% - would be directed toward the Earth with the result being a brighter sky and a dimmer Sun. Suppose that all of the infrared energy was absorbed and that 50% of the visible light energy was scattered and that only the remaining half reached the surface. Of course, a water vapor layer would completely absorb all of the ultraviolet energy. With these assumptions in mind we have the following:

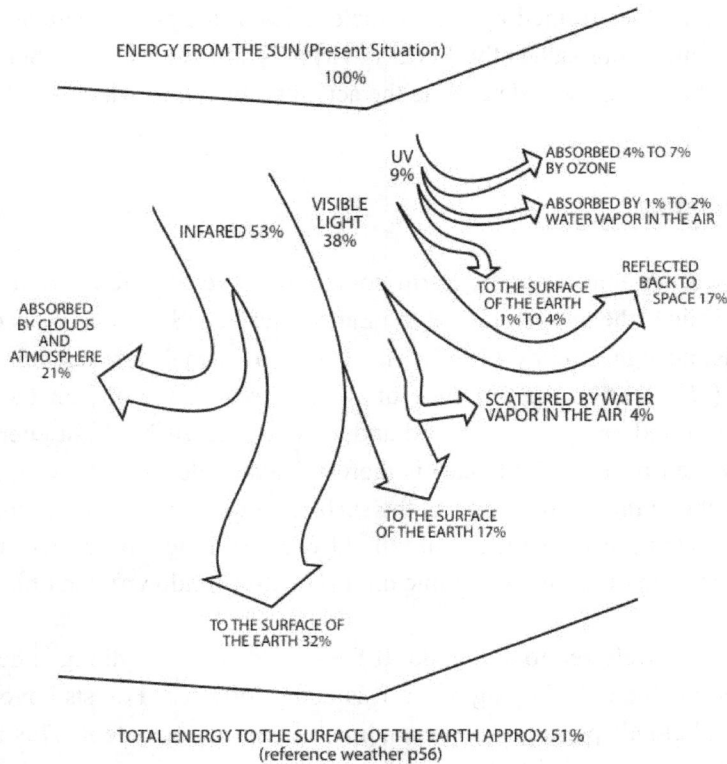

ENERGY FROM THE SUN (Present Situation)
100%

UV 9%

VISIBLE LIGHT 38%

INFARED 53%

ABSORBED 4% TO 7% BY OZONE

ABSORBED BY 1% TO 2% WATER VAPOR IN THE AIR

TO THE SURFACE OF THE EARTH 1% TO 4%

REFLECTED BACK TO SPACE 17%

ABSORBED BY CLOUDS AND ATMOSPHERE 21%

SCATTERED BY WATER VAPOR IN THE AIR 4%

TO THE SURFACE OF THE EARTH 17%

TO THE SURFACE OF THE EARTH 32%

TOTAL ENERGY TO THE SURFACE OF THE EARTH APPROX 51%
(reference weather p56)

5.8 (1) Solar Energy Distribution Without a Vapor Canopy

None of the quantities shown are either exact or consistent. They all shift slightly with location, time of year, and atmospheric content.

Solar energy available at Earth's surface with a water vapor layer above the atmosphere:

Energy =   100% - (53% (all the infrared energy) + 50% of the visible energy + 9% (all of the UV energy))

=   100% - (53% + 50% of 38% + 9%)

=   100% - (53% + 19% + 9%)

=   100% - 81%

=   19%

Therefore it may be concluded that with a water vapor layer surrounding the Earth, a little less than one-half as much solar energy would reach the Earth's surface as it does at the present time. (i.e. 19% compared to 51%) The primary difference is that in the vapor layer case the heating component (i.e. the infrared) would be above the surface in a medium, which can distribute the energy more efficiently. The

171

diagram, 5.8 (2) Solar Energy Distribution with a Vapor Canopy, shows these assumptions and relationships.

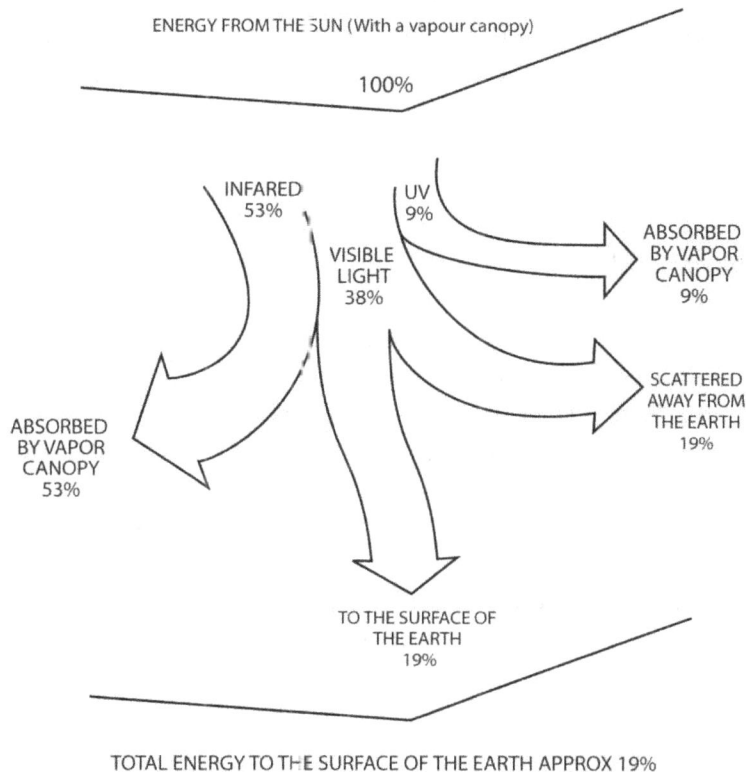

ENERGY FROM THE SUN (With a vapour canopy)

100%

INFARED
53%

UV
9%

VISIBLE
LIGHT
38%

ABSORBED
BY VAPOR
CANOPY
9%

SCATTERED
AWAY FROM
THE EARTH
19%

ABSORBED
BY VAPOR
CANOPY
53%

TO THE SURFACE OF
THE EARTH
19%

TOTAL ENERGY TO THE SURFACE OF THE EARTH APPROX 19%

5.8 (2) Solar Energy Distribution
With a Vapor Canopy

Regardless of the energy output from the sun, the quantities shown would have been
more consistent with the presence of a vapour canopy.

5.9   Present Energy Distribution

At the present time, when we do not have a vapor canopy above our atmosphere, the energy received from the Sun is not well distributed around the Earth. Fortunately, the oceans act as a temperature regulation mechanism and provide a thermal flywheel effect which helps to distribute the Sun's energy over the cycle of day and night as well as from season to season. However, the Earth is primarily dependent on winds and ocean currents to distribute the energy between the equator and the poles. Unfortunately, air has very little ability to retain heat with dry air having the least. It is the dry air of the Hadley Cell circulation systems that is currently the most widespread mechanism in operation transferring heat from the tropics to areas both north and south. The water in the great ocean currents has much greater ability to transfer heat but ocean currents only operate in a way that influences a few areas near their

172

pathways. Therefore present energy transfer mechanisms are not very efficient, resulting in significant temperature discrepancies between the equator and the poles. The Sun shines right straight down on the equator but it hardly shines at all on the North Pole. This situation is partially remedied by movement of both air and water away from the equator. All of the weather patterns of the world result from this movement, including storms which are concentrated, intensive, energy transfer mechanisms. The greater the energy discrepancy between heated zones and unheated zones, the more intensive both weather patterns and storms become.

5.10   Energy Transport

The energy, which would be received by the Earth from the Sun, would be the same with a water vapor canopy as it would be without it. However, the greater energy transfer ability of the water vapor in the vapor canopy, would result in solar energy being distributed over the entire Earth much more efficiently. The reason this would happen is because of the nature of water. Water is able to absorb and hold heat better than almost anything else. With a layer of water enclosing the Earth, the means to transfer energy and distribute it around the Earth would be much improved. Ocean water can and does transfer energy but ocean water is not nearly as free to move over the entire Earth as a layer of water in the upper atmosphere would have been.

As shown in the Atmosphere Pressure Profile diagram 5.4, most of the material in the atmosphere, whether it is vapor in the vapor canopy or air in the lower layer, is concentrated near the bottom. Since the bulk of the water vapor would be in the lower part of its layer, the incoming heat energy from the Sun would be mostly absorbed and concentrated in this region. At the same time the upper levels are more exposed to space and due to heat loss would be cooler. Solar energy would therefore warm the lower regions more than the upper regions. When water vapor is warmed, just like air or any other gas, its density will decrease. Further, if a region of vapor were warmed slightly more than a horizontally-adjacent region, there will also be a horizontal density differential. The amount of horizontal temperature differential, which the Sun can produce in one day, is shown on the diagram 5.10 (1) Vapor Canopy Temperature Increase. There are therefore two types of density gradient conductive to movement, vertical and horizontal.

Whenever the water vapor was warmed by the Sun, its density would have dropped causing it to rise upward while the surrounding cooler and denser regions would have moved horizontally to restore density uniformity. A circulation pattern would therefore have been set up, with the vapor at the upper elevations moving away from the directly heated region to be over the surrounding slightly cooler regions on either side. At the lowest level of the vapor canopy, horizontal movement would have been toward the hottest area directly under the Sun. In the next layer, slightly higher up, the warm vapor would be moving pole-ward. In this manner a global circulation pattern would have developed and become established in the water vapor canopy. Some flow will short circuit and not follow the maximum-distance route all the way. The pattern of circulation would continue as long as there was any difference in density to provide the driving force. Since the Earth is a better radiator of heat than the vapor layer would have been, the circulation pattern would have been augmented by heat loss from the surface of the Earth during night-time. With a vapor canopy, energy from the Sun would be distributed around the entire Earth by a very efficient energy transfer engine. Surface temperatures would therefore remain in the comfort zone with the equatorial regions being cooler than they are now. Polar areas would have been warmer.

As the Earth rotates, the movement would not cease. As long as there was a region of water vapor, which was warmer than adjacent regions, the movement would continue. As mentioned, some movement would be short-circuited and all of the moving vapor would not make it all the way to the extreme northern or southern regions. However the Earth is a sphere, not a flat plate, so as the vapor currents travel away from the area directly under the Sun, they would be converging and the flow would be accelerated. The heat transfer effect would thereby be improved.

5.10 (1) Vapour Canopy Temperature Increase
(in one day)

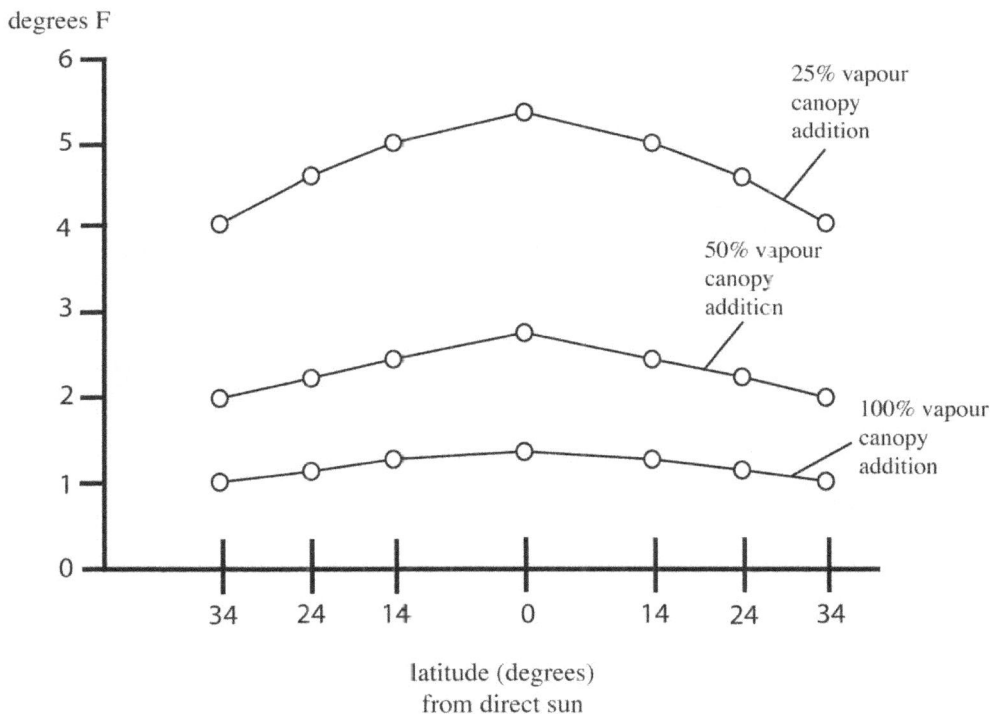

degrees F

25% vapour canopy addition

50% vapour canopy addition

100% vapour canopy addition

latitude (degrees)
from direct sun

Please refer to the appendix further discussion.

Curiously, the basic energy transfer mechanism, which our present atmosphere provides, operates in a similar fashion. The term 'basic' is appropriate because, the mechanism (i.e. solar energy) causing the air near the equator to rise and move towards the poles is continually happening. The Hadley Cell circulation and the Ferrel Cell circulation move energy away from the equator. (51) However, what would have been basic and readily recognized patterns of flow, are significantly modified by the existence of both land and water surfaces on the Earth (The fact that water freezes and turns into ice further disrupts any possibility of regular flow patterns.). If the surface of the Earth was either all water or all land, this would not be the case and patterns of circulation and energy transfer would be very similar in our present atmosphere as they would have been in a vapor canopy layer. Basic flow patterns would be affected by the rotation of the Earth in either case which effect is referred to as the Coriolis Effect. Atmospheric flow would not occur directly from equator to pole but would proceed along a curved pathway. Northward flow higher up would curve to the west and southward flow in the lowest layer would also curve to the west. In this manner the vapor over any particular area would be constantly changing and would not be simply travelling to the pole and back. With this arrangement, any particular 'parcel' of vapor returning to the equator, would be many miles to the west of where it left the equator as it rose and proceeded northward. (This flow pattern is shown in the diagram, 5.10 (2) Vapor Canopy Flow Pattern).

With an atmosphere, which included a vapor canopy, the basic cycle of energy gain and energy loss would be operating far above the surface of the Earth, which would be continually covered by this warm blanket and not exposed to the cold of space as it is now. The temperature over the entire Earth would consequently have been within a fairly narrow band. The blanketing effect relates partly to the ability of the vapor layer to reabsorb heat coming up from the surface of the Earth. This together with the more efficient energy transfer ability of water compared to air, would have resulted in the temperature around the Earth being within the comfort zone for humans everywhere.

As vapor within the vapor canopy rose, it would expand. At the reduced pressure higher up, the

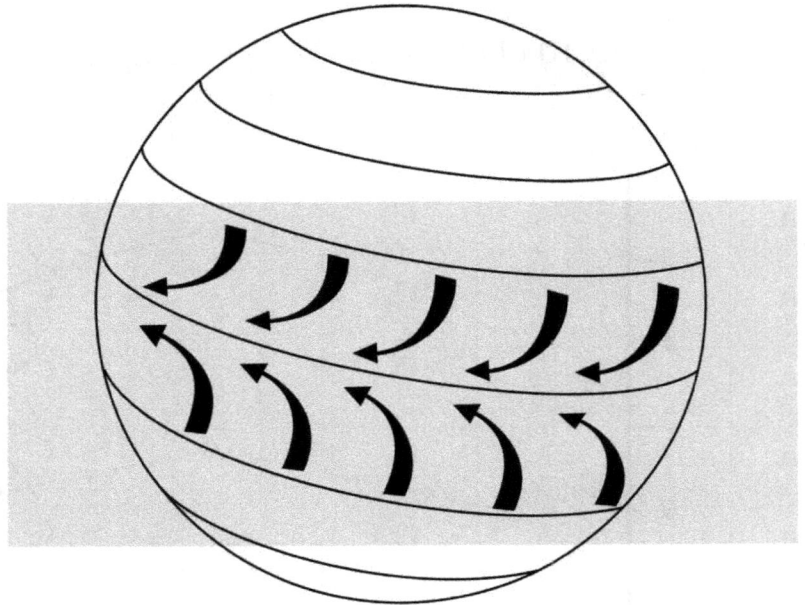

5.10 (2) Vapour Canopy Flow Pattern

In the lowest layer of the vapour canopy, flow would have been toward the equator while curving to the west. Directly under Sun, the vapour would have risen and travelled north westward toward the polar regions. This bending to the west is called the Coriolis Effect.

critical temperature (i.e. the temperature at which fog will form) is reduced. Consequently it would be even more difficult for the water molecules to condense and form fog. Therefore if the lowest vapor layer remained above critical temperature, all higher layers would also. These are the layers that would be involved in the transfer and distribution of heat around the Earth. The vapor layer would not be a good radiator of heat (i.e. it is not black) so the vapor moving toward the poles would not lose all of its heat. Then as it dropped down toward the tropopause and returned to the equator, it would absorb any heat being given off by the Earth. Consequently, the entire circulation pattern would operate within a fairly narrow band of temperature. Much higher up above the circulating vapor currents, heat would be lost to space. This would, of course be necessary to keep the Earth from overheating but it might also result in making the upper regions of vapor visible. It would become visible if it condensed into fog or formed ice crystals. Lower down the temperature leveling and blanketing effect of the lowest layers would have kept the entire Earth within a temperature zone that would have been quite comfortable for humans and animals alike. The net result of having a water vapor canopy is that a similar amount of solar energy would be involved in heating the Earth, as it is at present, but it would have been distributed over the entire Earth much more efficiently, thereby making the temperature around the world much more uniform.

It is true that the stratosphere of the present atmosphere does warm up as the ozone absorbs the ultraviolet energy from the Sun. It thereby might have provided a global heat transfer mechanism, except that the temperature within the present stratosphere increases with elevation. The rightward bulge in the temperature curve (as shown in the diagram, 'Atmospheric Temperature Profile' 5.11) indicates a no-pass-zone where there is no upward movement of air. Consequently, there is no upward movement of air in the stratosphere and no global air circulation pattern in the stratosphere. It therefore does not contribute to transferring solar energy from the tropics to the poles.

5.11   Temperature Profile of Atmosphere with Vapor Canopy

As mentioned, the temperature profile of an atmosphere with a water vapor canopy layer is shown in the diagram, 5.11 Atmospheric Temperature Profile. The leftmost line represents the present atmosphere. The other three, lines 2,3 and 4 each represent an atmosphere loaded with a water vapor layer. There is also a line in this diagram which shows the temperature that a parcel of air would have if it simply rose and cooled by expansion. This will be referred to as the reference line. It must be noted that it is the slope of the reference line and not its actual location on the chart that is important. This slope tells us how a parcel of air would behave at any particular altitude for any one of the four atmospheres represented. If, for example, the temperature curve of an actual atmosphere sloped more to the left than the reference line, it indicates that the air is colder the higher up we go and that any parcel of air that rose upwards in such an atmosphere would be able to keep on rising. The temperature of the surrounding air would not inhibit its upward movement but actually enable it because the rising air would still be warmer than the surrounding air. As mentioned previously, the reason that thunderstorms form is because it is colder than usual high up in the atmosphere. They form because upward movement of air is enabled by the colder upper temperatures. Any region, where the atmospheric temperature curve sweeps to the left as higher elevations are reached, indicates where upward atmospheric activity could occur. This is the type of temperature profile that would be expected within the vapor canopy (but not from the lower atmosphere into the vapor canopy). Therefore upward movement of vapor within the vapor layer would happen. As a result of upward movement, the temperature would drop due to expansion. Also some heat within the vapor would be lost to outer space. When this vertical activity was coupled with horizontal movement away from the hot equatorial regions, the thermal energy from the Sun would have been distributed much more evenly over the Earth than it is with our present atmosphere.

On the other hand, if the slope of the atmospheric temperature curve was a little steeper than the reference line, upward movement of any parcel of air would be inhibited. At any elevation, if the slope is steeper than the reference line, a no-pass zone would exist. In this diagram, any region of atmosphere, where the temperature curve slopes to the right on the way up, emphatically indicates a 'no pass zone'. Vertical activity cannot occur through such a region because the regions, which are higher up are warmer. Any rising air column would therefore be stopped. As shown, the lower layer of the vapor canopy would be quite warm and the temperature curve at the very least would be almost vertical and more expectedly curve to the right. Because of this stability feature, the oxygen-nitrogen region of the lower atmosphere would never mix with the water vapor canopy region above it. The shape of the temperature profile would vary with latitude, time of day and season but it would retain the basic no-pass zone feature because the

5.11 Atmospheric Temperature Profile
(Illustrative Only)

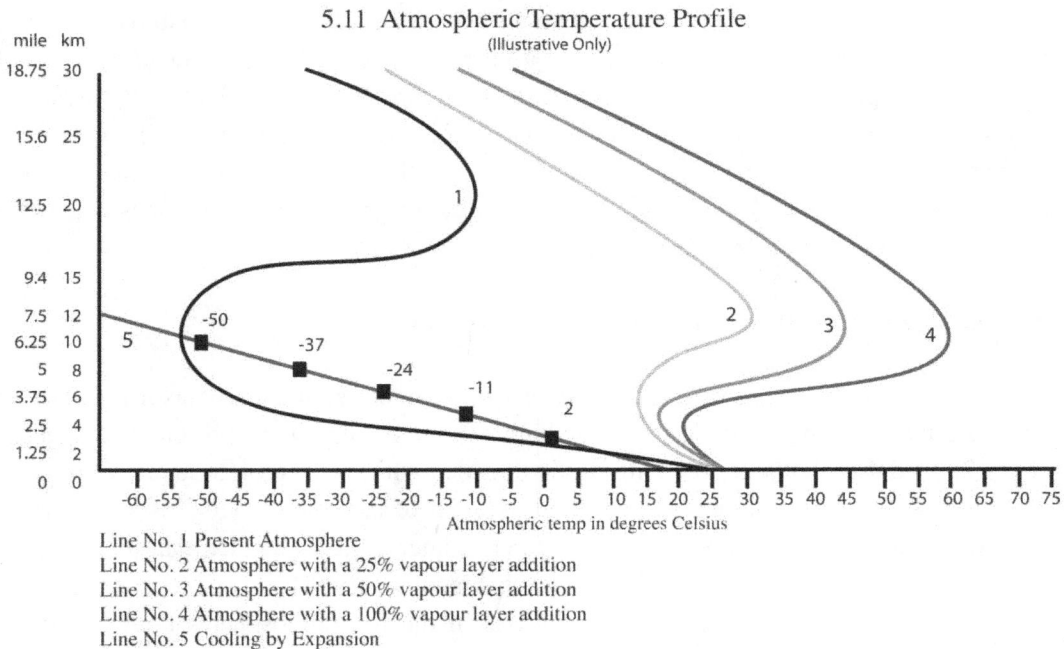

Line No. 1 Present Atmosphere
Line No. 2 Atmosphere with a 25% vapour layer addition
Line No. 3 Atmosphere with a 50% vapour layer addition
Line No. 4 Atmosphere with a 100% vapour layer addition
Line No. 5 Cooling by Expansion

water vapor in the canopy would preferentially absorb and retain incoming infrared energy from the Sun. In this manner the no-pass zone would be maintained and stable separation of the lower atmosphere and the upper vapor canopy would be guaranteed.

An atmosphere which included a water vapor canopy would have a definite 'no pass zone' at the junction of the oxygen-nitrogen lower layer and the water vapor upper layer. An atmosphere such as this would have a temperature profile very similar to our present atmosphere, which actually includes two 'no pass zones' – one in the stratosphere, (as shown - elevation between 5 and 15 miles) and one much higher up in the thermosphere, (not shown - which is of no consequence at the surface of the Earth).

5.12 Temperature Characteristic of Present Atmosphere

The present atmosphere of the Earth consists of four layers - so identified because of their particular temperature characteristics. These four layers are: troposphere, stratosphere, mesosphere and thermosphere.

The troposphere is the lowest layer and includes most of the clouds and all of the weather and other activity, which affects all of us on the surface of the Earth. In this layer the temperature normally drops as altitude is increased until an altitude of six or seven miles is reached. At this elevation, the temperature may be -70F. From basic physics it is understood that when pressure is reduced, temperature will drop. Therefore if a quantity of air is raised to a higher elevation, its temperature will drop as it expands. As mentioned above, storms develop in the troposphere because the surface of the Earth becomes warmer than the temperature required for stability or the temperature higher up becomes colder than the temperature required for stability. This stability relationship is shown in the diagram, 5.11 Atmospheric Temperature Profile, and is simply the temperature that any parcel of air would have if it was raised and cooled by expansion. (i.e. the reference line in the diagram) If stability is upset by either a colder than usual region above or a hotter than usual region at the Earth's surface, there will be upward vertical movement of air, updrafts will be formed and storms will result.

In the troposphere, thunder-clouds form and may rise to great heights. However, they do not usually rise higher than the top of the troposphere. This limitation is readily observed as thunder-clouds rise and reach an elevation where vertical movement stops. Further movement will only be horizontal and the top of the cloud will spread out in an anvil shape. At this elevation, the cloud has reached a 'no pass zone' or a ceiling, where further vertical movement is inhibited. This boundary, called the tropopause, is usually quite well defined and indicates the top of the troposphere. The layer above is called the stratosphere. There are two exceptions to this basic expectation. If a storm develops a stronger-than-usual updraft, vertical movement will continue into the stratosphere. However, even the most powerful storms will reach an altitude where the surrounding air is the same temperature as the rising air. The updraft will then peter out. For example, a storm may rise to 60,000 feet before the rising air cools by expansion and becomes as cool as the surrounding air at this elevation. Then the storm will not rise any higher but since more air is still coming up from below, the air at the top must move aside. The anvil top will then develop and may be readily seen from the ground. The cloud spreads out both ways showing that it has now become too cold to rise any farther.

While the temperature drops as elevation is gained up through the troposphere, the exact opposite happens in the stratosphere. As the elevation through the stratosphere increases, the temperature increases. The tropopause defines a basic boundary for vertical movement because at this level, the temperature no longer decreases with increasing elevation but starts to increase instead. Even if the temperature remained constant with increasing elevation, a boundary would still exist because any quantity of air, which started to rise into such a region, would cool due to being raised and would soon be surrounded by air, which was warmer. Therefore it would stop rising and just settle back down. A 'no pass zone' would still result. Further, if the air in the stratosphere continued to get cooler with elevation, but not as cool as a quantity of air would, if it was simply raised, there would still be a 'no pass zone'. As things presently exist, the temperature rises as we progress upward through the stratosphere, so the no-pass zone is quite extensive and therefore well guaranteed.

Ozone is produced and concentrated in the stratosphere. As the ozone absorbs the ultraviolet energy from the Sun, it is warmed. A major side benefit for those who dwell on the surface of the Earth, is that the harmful ultraviolet energy has been intercepted before it reached the surface and caused harm to numerous forms of life, including humans. Due to the absorption of the ultraviolet energy, the air in the stratosphere warms from around -70 F at the base of the layer to +40 F near the top. (66) As we proceed upward through this layer the temperature curve therefore slopes to the right, which clearly indicates a 'no pass zone'. Also it is so extensive that no air from the lower troposphere would ever be able to pass up through it.

The ultraviolet energy from the Sun is intercepted and absorbed by the ozone in the stratosphere and the air in the stratosphere is thereby warmed. At night, most of the stratosphere is in the dark and is not being heated. It will therefore cool down a certain amount before the Sun comes up again. However, it will only cool down a few degrees before the Earth rotates far enough to enable the Sun to rewarm it to the temperatures of the previous day. A similar situation would exist within a vapor canopy layer. It would cool down a few degrees at night before rewarming again the next day. A no-pass-zone would remain in both cases because the temperature curve will always slope to the right as elevation increased.

One of the great tragedies of losing the ozone layer would be the loss of the rightward bulge in the temperature curve. If this should happen, the lower atmosphere would no longer be isolated from the upper atmosphere and storm clouds, with their entrained updrafts, could rise upwards for many miles. In such a case the power of the updrafts would be greatly increased along with the destructiveness of the storms. Unfortunately, this appears to be happening as evidenced by the recent reports of storm clouds at 75,000 feet as well as by reports that the ozone is being destroyed.

Above the stratosphere is the mesosphere where the temperature drops to the neighborhood of -130 F before rising dramatically in the uppermost layer, the thermosphere, to over +2500 F. Temperatures at this level have no effect on the surface of the Earth because there are only a few widely separated air molecules at these great elevations.

5.13   Atmosphere Comparison

The thermal profile of the present atmosphere may be readily compared to the thermal profile of the postulated ancient atmosphere, which may have included a water vapor canopy. Both of these profiles may be considered along with the atmospheric pressure profile discussed earlier. If the ancient atmosphere included the equivalent of a 25% water vapor layer, the 'no pass zone' separating the vapor layer from the oxygen-nitrogen layer would occur at approximately 7.5 miles up. This, coincidentally, is approximately the level of the tropopause in the present atmosphere, which is also a 'no pass zone' for present weather systems. If the ancient atmosphere included a 50% water vapor layer, the demarcation level would have been lower at approximately 5 miles, which is lower than the present tropopause but not very much.

5.14 Evidence of a Vapor Canopy

5.14.1 Fossil Evidence

If the above speculation is correct and the atmosphere at one time included a moisture layer on top of an oxygen-nitrogen layer, atmospheric pressure at sea level would have been greater. The fossil record provides evidence that the atmosphere of the Earth was heavier than it is now and that atmospheric pressure at sea level was indeed greater.

Fossil dragonflies have been found which had wingspans which were more than two feet wide. (1) We certainly do not have any insect that large today. In fact it would be impossible because of the way that insects breathe. Insects, like all other creatures, need oxygen throughout their bodies but insects do not actually breathe. 'Insects don't have lungs; they rely on a network of tubes connected to openings called 'spiracles' that run down the side of the abdomen. Insects can pulse their abdomen and flap their wings to ventilate the spiracles, but as the insect gets bigger the proportion of the body that is taken up by spiracles gets rapidly out of hand.' (71) Air is admitted to the insect body through the skin as well as by the spiracles. Air is moved through the skin and along the tubes by atmospheric pressure rather than by the forced flow of air as in mammals which have lungs and a pumping system. Therefore, if atmospheric pressure was greater, the air could move further into the insect, thereby enabling a larger insect to develop. Current atmospheric pressure is not able to force air into insects very far, so they cannot grow to be very big. Apparently this was not the case in the distant past and insects could and did grow to be much larger. What would a dragonfly with a two-foot wingspan sound like?

There was another ancient creature, the pterodactyl, which provides evidence that atmospheric pressure was formerly greater. These were large creatures with very large wings, which appeared to be fully operational and intended for flying. Their bones were hollow and air-filled like the bones of birds. Analysis of their brain cavities indicated that they had massive (several times as large as a bird's) flocculi (the region of the brain that integrates signals from joints, muscles, skin and balance organs. (69)) and this may have been due to their large wings. (30) However it appears from an analysis of the size of these creatures compared to the size of their wings that they could not have become airborne at present atmospheric pressure. (2) Consequently someone decided that they should be depicted climbing up cliffs so they could glide back down. 'One of the most interesting pieces of evidence for the previous existence of a vapor canopy is the pteranodon, an enormous flying reptile with wing spans of up to seven meters (twenty-three feet), whose remains are found in the Cretaceous sediments. Experts long debated whether the creature would have had sufficient muscular strength for powered flight. They concluded that it must have lived on cliffs and sailed on the updrafts above the sea in search of fish. ... an even larger flying reptile, the pterosaur, with an estimated wing span of fifteen meters (fifty-one feet), was reported ... in 1975. ...there is no means by which it could have left the ground. However, if the atmospheric pressure was at that time twice as great ...the giant reptile could just have become air-borne.' (60)

More recent analysis of the fossils of these great creatures indicates that due to the existence of a small bone near the end of the wing, the lift that the wing could have generated would have been improved by 30%. (70) If this was the case, pterodactyls and pterosaurs could possibly have become airborne with present atmospheric pressure but certainly with a small percentage more.

If these creatures were indeed flyers, their large wings (which were otherwise appropriate for flying) would have been useful rather than just being in the way. Flying is much more desirable than cliff-climbing and much easier on the feet.

5.14.2   Carbon14 Evidence

The vapour canopy would have consisted primarily of water molecules and would have enclosed the atmosphere of the Earth completely. With such an arrangement, certain components of cosmic radiation would have been attenuated before reaching the nitrogen-oxygen layer. In particular, neutrons are readily absorbed by water and this is the reason that ordinary water is not generally used to control a nuclear reactor. If the neutrons in a reactor are absorbed they can no longer affect the reaction, which will therefore slow down too much. (37) Similarly, a layer of moisture above the oxygen-nitrogen layer would have absorbed most of the high speed incoming neutrons thereby shielding the nitrogen underneath. Very few neutrons would have penetrated down to the nitrogen layer to change nitrogen into carbon14. As a result, very little carbon14 would have been formed. This would also have caused a discrepancy between the production and decay rates of carbon14, (which there is), making coal appear older (i.e. it has a low C14 count) than it is in reality.

5.14.2.1   The Young Age of Coal

Coal is commonly thought of as being very ancient (i.e. tens of millions of years old) but if this were the case its carbon14 count would be virtually zero. (Since the half-life of carbon14 is only about 5730 years, (57) after 20 or 30 half-lives the carbon14 count would be negligible and hence not meaningful) As it is, carbon14 commonly indicates that coal has an age of 50,000 years. Unfortunately, this is not old enough to agree with the assumption that the carboniferous period was millions of years ago. An alternate explanation which also recognizes the evidence that human artifacts have been found in coal (24) would be that coal was formed within the last few thousand years and that the plants from which it was formed had not been exposed to very much cosmic radiation (i.e. due to the vapor canopy). The vapor canopy hypothesis therefore accommodates the human artifact evidence simultaneously with the low carbon14 count. The swamp theory for coal formation does not accommodate any of this evidence.

If coal has been formed quite recently, as the evidence discussed in the above section entitled 'The Swamp Theory of Coal Formation' shows, (34) the carbon14 count should indicate only a few thousand years instead of 50,000. (33) The low count may therefore be evidence of a vapor canopy having been present while the coal-forming plants were growing. (It is also possible that there was a higher level of CO2 at that time which may also have been partially responsible for the low C14 count.)

5.14.2.2 Production and Decay Rate Discrepancy

Carbon14 is being continually produced in the atmosphere by the action of incoming cosmic radiation on atmospheric nitrogen. (57) It is also, because it is radioactive, continually decaying and changing back to nitrogen. Both the production and decay rates have been determined and the decay rate is considerably less than the production rate. At face value this simply indicates that the carbon14 production process only started a few thousand years ago and that the decay rate has not had enough time to catch up to the production rate. It the carbon14 production process has been going on for a long time, (e.g. for 30 or 40 or 50 half-lives) the two rates would be equal. Since they are not equal, it is appropriate to ask why. It may be that the loss of the vapor canopy allowed the production level of carbon14 to be increased in which case the discrepancy is evidence for the canopy's existence in the first place. (There is a more detailed discussion of carbon14 in chapter 10.)

5.14.3 Evidence from a Universally Warm Earth

There is evidence that the Earth was once universally warm (as discussed in several other chapters as well as below) . Since a vapor canopy would have caused the Earth to have been universally warm, evidence of a universally warm Earth supports the vapor canopy hypothesis.

5.14.3.1 Evidence from the North

Included in this evidence are the trees of Axel Heiberg, Ellesmere, Ellef Ringnes and Amund Ringnes which are islands in the far north of Canada. All of these locations are far above the Arctic Circle and far north of the present tree-line. Countless numbers of trees (i.e. logs) are found lying across these islands. (21) Included among the logs are numerous tree stumps which appear in some cases, to have been undisturbed since they were placed. Neither are many of these trees fossilized (i.e. turned into stone) but appear simply dried out. They are not partially decayed either. The wood is both sawable and burnable.

The conclusion, that a warm climate once existed at Axel Heiberg Island, was clearly stated in a report in Science magazine. 'An ancient tropical paradise, complete with turtles and crocodile-like lizards … a hot steamy world … home to huge heat-loving lizards called champsosaurs … daytime temperatures hovered between 25C and 35C', captures the flavor of the report. 'Anywhere a champsosaur is found could not have been very cold.' (29) The bones of champsosaurs were recovered from what appeared to have been the remains of a fresh-water lake. It was the further conclusion of the authors that the warmth was caused by carbon dioxide in the atmosphere and that even the deep oceans were warm at 18C in the Arctic.

As well as trees, there are 'wood-bearing alluvial sediments containing plants of temperate climates'. In this regard the Canadian Arctic has similarities with the Russian Arctic 'where references have been made to the coastal bluffs of the island of New Siberia as the "Wooden Hills" '. (4) Fossil forests are found on the New Siberian Islands. 'In New Siberia (Island) on the declivities facing the south, lie hills 250 or 300 feet high, formed of driftwood……..Other hills of the same island, and of Kotelnoi, which lies further to the west.....(the trunks of trees) lie flung upon one another in the wildest disorder, ...wood hills...consist of carbonized trunks of trees, with impressions of leaves and fruit.......these desolate islands were covered with great forests, and bore a luxuriant vegetation.'(39)

Finding a wooly mammoth in the far north would not be surprising as they are thought of as being suitable for cool weather. However, finding elephants and rhinoceros would certainly be suggestive that the area was once much warmer. 'The New Siberian Islands....as well as the islands of Stolbovoi and Belkov...the soil of these desolate islands is absolutely packed full of the bones of elephants and rhinoceroses (and wooly mammoth) in astonishing numbers.' (54)

There are really only two options to explain the presence of trees and fossil animals in the frozen north. The trees were either carried to these locations or they grew in these locations. Either way the climate of the Earth must have been a great deal warmer at that time than it is now.

More evidence is available at Spitzbergen Island, which is an island north of Norway and about as far north as Axel Heiberg Island. Coal has been found at Spitzbergen. (5,38) The material which would later form coal would have required a warm climate to grow. Coal is understood to have been formed from the remains of plants. These plants had to have grown (and plant growth is only possible) when the appropriate heat, light and nutrients are available. Since the ice pack, the extreme weather and the ocean currents currently prohibit both ocean transport and in-situ growth, it may therefore be concluded that the plants, which formed the coal, grew in an appropriate warm location and the climate allowed them to be brought to this location. Alternately, the weather was somehow appropriate and they simply grew in this location in the first place. With either of these possibilities, as with the trees of Axel Heiberg Island, the Earth must have been much warmer than it is now.

Fossil palm trees have been found in Alaska. (8) This type of tree usually grows at a much warmer location. Since there are no palm trees either growing in Alaska or being carried to Alaska at the present time, these trees must have grown there some time in the past or they were carried there some time in the past. In order for either of these options to have happened, the climate must have been a lot warmer than it is now.

Even Antarctica provides evidence of having been warm as 'extensive coal measures are found'- at least seven seams of coal from three to seven feet thick. (27, 40) How this material could have washed in from elsewhere is most difficult to explain. The only reasonable conclusion is that the material grew in-situ and that the weather was warm enough to enable this growth. In addition to the coal, dinosaur fossils and tropical trees are found (72) while Spitzbergen, so far north of Norway can only claim dinosaur footprints to go along with the coal as evidence of a warm climate. (28)

While a universally warm Earth is commonly acknowledged, it is occasionally attributed to CO2. While CO2 is a heat retention gas, it is certainly not a heat transport mechanism and would not eliminate temperature disparities between the equator and the poles. While there is some evidence that the level of CO2 was higher during the time that the Earth was universally warm, (56) and that it would have contributed to the retention of the heat that got through the vapor layer, it could not have accounted for the equity of temperature that existed between the equator and the poles.

5.14.3.2   Submarine Canyons

Submarine canyons are evidence of a universally warm Earth because they are direct evidence that the ocean was warm. Submarine canyons always appear as extensions of large rivers right out into the ocean. Some of these formations are vast canyons that are hundreds of miles long and continue down for thousands of feet right to the bottom of the sea. Sea canyons are deep and winding V-shaped with their walls sloping down at a steep angle to a narrow floor.Their youthful appearance seems to relate them to the world of the Ice Age. (55) Indeed, they do relate to the Ice Age and were formed when the Ice Age ended. The cold melt-water from the great ice fields flowed down to the sea and because it was colder (i.e. denser or heavier) than the sea water, it flowed along the bottom of the sea and carved the canyons. If the ocean had been as cold as the melt-water, the submarine canyons would never have been formed. This, of course, assumes that the ocean water was fresh, or at least less salty than it is at the present time. However, since the density of water decreases by more than 4% as its temperature is increased from freezing to boiling, even if there had been some salt in it, the cold melt-water would still have been denser than an ocean that was several degrees warmer than it is now. This enabled the cold melt-water to flow straight to the bottom and form the submarine canyons on the way down.

5.14.3.3   Warm Earth Evidence Conclusion

A vapor canopy would have kept the Earth warm from pole to pole. However the effect of a universally warm Earth would have been an increase in ability of the air to hold water vapor, a greenhouse gas. This effect is a major concern for many scientists at the present time because there is an increase in the amount of $CO2$ (another greenhouse gas) in the atmosphere. The greenhouse effect of increased water vapor around the entire Earth would have been much more significant than the current concern over an increase in $CO2$. However, if there had been a vapor canopy in place, the amount of solar energy reaching the surface of the Earth would have been significantly reduced. This would have offset the effect of a greater concentration of water vapor due to a warmer world. The evidence that there was plant life and a warm climate in the polar regions provides direct evidence that a vapor canopy existed. Otherwise the Earth would have been overheated and inhospitable to life.

5.14.4   Gigantism Evidence

At one time the Earth was home to many very large creatures. From the ground, fossils, the evidence of extinct creatures, are continuously being recovered. While a range of size is commonly found, some of them were very large. The dinosaurs, for example, stood several times as high as man and some specimens reached lengths of nearly 100 feet. Tyrannosaurus Rex was about 47 feet long. Triceratops and Stegosaurus were in the range of 20 to 25 feet. (7) However, the dinosaurs were not the only large creatures. Baluchitherium was a very large horse-like mammal, 18 feet at the shoulder. (6) None of these particular extinct creatures have recognizable modern counterparts but there are many that do. These include: Rhinoceros (the size of a two story house); Raccoon (large and as dangerous as black bears); Guinea Pig (large as rhinos); Giant Teratorn (a flying bird with 25 ft. wingspan, 10 feet from beak to tail, standing 6 feet high when on the ground; (3)), Titanus (flightless bird 10 ft. high and approximately eight hundred pounds);Woolly Mammoth (about twice as massive as present day elephants and 15 feet tall); Crocodile (51 feet long); Sabre-toothed Tiger (9 feet long); Dragonflys (25 inch wingspan); Great Shaggy Bison (7 1/2 feet tall); Man (9 1/2 feet tall (32)); Pterosaur (35 foot wingspan (35)); Megatherium (18 feet long (62)); Beaver (large as black bears); American Lion (25% larger than present day African Lion); Jefferson's Ground Sloth (size of an ox); Short-faced Bears (tall as a moose); Camels (about one-fifth larger than present camels). (9) We must also include the Dinohyus Hollandi, a giant swine that stood six feet high. (41)

From Peru comes evidence of giant penguins. These creatures stood between four and five feet high and according to the report, have cousins which were even larger. It is also curious that they lived where the climate was warm whereas present-day penguins are more at home where the climate is cold. (31)

A discussion of gigantism would not be complete without mentioning snakes. The ancient giant snake, named Giant Titanoboa was about 45 feet long, too wide to go through a normal door and may have weighed more than a ton. This discovery in South America included 'turtles and giant crocodile-like dyrosaurs'. (60)

There were also giant, rhino-sized marsupials, Diprotodonts, but 3 yards long, 6 feet tall at the shoulder, and weighing 2 tons. (73)

There is no doubt whatsoever that many ancient creatures were very large, including humans. Since nothing happens without cause, what caused so many ancient creatures to be so large? The possible link between this gigantism and the hypothesized vapor canopy may be the Hypothalamus Disregulation Theory. (10) The basic premise of this theory is that having a higher level of oxygen available in the hypothalamus would positively affect both gigantism and longevity. Actually the Hypothalamus Disregulation Theory discusses the combined effect of greater CO_2 along with higher oxygen pressure. The CO_2 would dilate the blood vessels allowing more oxygen to reach the hypothalamus. There is, of course, a serious limit concerning how much more atmospheric pressure both animals and man can safely tolerate. Appropriate studies have apparently shown that up to almost two atmospheres may be acceptable. (11) (Higher pressures have been maintained in deep sea diving experiments where the higher pressure had a positive effect on healing. (16) Also high pressure oxygen has been used in the treatment of fractures wherein the healing rate in rats was accelerated by 25%. (26)) If the Hypothalamus Disregulation Theory is correct, slightly higher atmospheric pressure (possibly combined with a greater level of CO_2) would have resulted in larger animals. Of course, the required higher atmospheric pressure could have readily been provided by a vapor canopy. Therefore, if the link between gigantism and higher atmospheric pressure is valid, gigantism may be cited as evidence for a vapor canopy.

5.14.5   Plankton Evidence

Evidence of greater oxygen pressure has been obtained from an analysis of the shells of small extinct, ocean creatures. (17) From a study of oxygen isotopes retained within the structure of these little creatures, certain investigators have concluded that oxygen pressure was once 35% compared to the present 20%. From a preliminary viewpoint this represents a 75% increase above present oxygen levels and could have been achieved in several different ways. One way would be to have the present amount of nitrogen and simply have more oxygen in the atmosphere. In this case, with more oxygen, atmospheric pressure would have been higher. If, for example, there was 50% more oxygen, it might be represented as 30 units instead of 20. Since nitrogen is 80 out of 100 at present, the total with 10 more oxygen would be 30 plus 80 or 110. With this arrangement, the percent of oxygen has only increased to 30 out of 110 or to 27.3%. Similarly, if the oxygen level was doubled to 40 units, the percentage of oxygen would be 40 out of 120 or 33%. 35% oxygen would therefore represent more than twice as much as there is at present and would have resulted in an atmosphere which was approximately 20% heavier than the present one.

Alternatively, the vapor canopy hypothesis provides a consistent and valid explanation for an ancient oxygen level of 35%, which, as mentioned is twice as much as there is at the present time. If there had been a burden on top of the present atmosphere amounting to 75% of the present atmosphere, ancient oxygen levels would have been at the 35% level. If this burden had been provided by a vapor canopy, it would have weighed 75% as much as our present atmosphere and the postulated higher level of oxygen would be explained by the greater load on top of the atmosphere instead of by actually having more oxygen. There would not have been any more oxygen, only a higher pressure of oxygen (as well as the whole atmosphere). It is therefore submitted that the greater amount of oxygen which is thought to have existed in ancient time, is evidence of a water vapor layer above the present atmosphere, which layer was approximately 75% as heavy as our present atmosphere. The problem of explaining what became of the extra oxygen is thereby solved. With the vapor canopy hypothesis, there wasn't any extra oxygen at all but only greater oxygen pressure. This greater pressure was lost when the canopy collapsed.

5.14.6   Illumination in the High Arctic

There is a second way that the trees of Axel Heiberg Island provide evidence for a vapor canopy. The trees found on Axel Heiberg Island in the high Arctic of Canada are mummified. Numerous stumps are found positioned in the ground just like they would be if that is where they had grown. (12) The tree trunks are broken from the stumps. Since there is no sign of decay, the conclusion must be reached that the trees grew in this location and were catastrophically destroyed. The only other suggestion that has been occasionally put forward is that Axel Heiberg Island was further south when the trees grew. Subsequently it drifted north to its present location. (12) However drifting would have required many years during which time the trees would have rotted. They did not rot. They were simply knocked down, buried and frozen before rotting took place.

A catastrophe of this nature would be similar to the Tunguska Event in Siberia discussed earlier. In that case trees were not only broken from their stumps but carried many meters away. (13) With the Tunguska

Event, as with the trees of Axel Heiberg Island, it may be concluded that the trees grew right where the stumps are found today.

Many of the stumps and logs are found in a mummified state. The wood is still wood and has simply dried out and not become fossilized. There are also fossilized trees on Axel Heiberg. Either case implies that nearly as soon as the trees were knocked over, they were protected from rotting. The preservation of much of the associated material including nuts, bark and leaves, is so good that it is 'almost indistinguishable from the litter on the floor of a modern coniferous forest'. (14) From this evidence as well, it may be concluded that an extensive forest once grew where all of this forest material is found today.

However, trees cannot presently grow where the trees of Axel Heiberg are found. (12) Even allowing that the temperature may have been warm enough, the amount of light available this far north is considered inadequate for tree growth. It is completely dark for several months every winter and during the summer, the Sun is low in the southern sky. It would be physiologically impossible for any forest to exist in the northern regions at the present time because there simply is not enough light(regardless of temperature). As photosynthetic organisms, problems exist for plants in the Arctic. For instance, at Resolute on Cornwallis Island the long winter night begins on November 4, when the Sun does not rise, and ends February 5, when the Sun again rises. (12))

It is certainly true that no trees grow this far north today. The bitter cold could be sited as sufficient to prevent tree growth but the lack of light is also considered sufficient, even if the climate was much milder. It is reasonable to conclude that the trees of Axel Heiberg Island as well as the other islands of the high Canadian Arctic and Antarctica, present a great dilemma for many theories of the Earth because their existence cannot be explained with respect to current conditions on the Earth. However, there are three types of atmospheric phenomena which collectively point to a solution to this dilemma.

5.14.6.1 Tunguska Explosion

The Tunguska Event has been discussed above with respect to the physical effects the exploding object had on the countryside of central Siberia. Trees were certainly flattened but in addition to this immediate and localized result, a great deal of dust was stirred up and spread far and wide throughout the atmosphere. It spread across both Europe and Asia. Usually it would be expected that so much dust would block out the Sun, which may have been partially true, but in this case, it redistributed the Sun. A letter-writer to the London Times reported being able to read a book at night by the sunlight which was reflected from the dust in the atmosphere. (13) The night sky was lit up. London, England is located at approximately the 50th parallel of latitude north. For a location like this to be illuminated at night by reflected sunlight, the light must be transmitted around the Earth from wherever it is shining. At summer solstice, which occurred barely one week earlier than the date of the report, sunlight would have been continuously available at the Arctic Circle which is only 17 degrees of latitude north of London. To reach London, the light from the Sun would have to bend/refract/reflect around the Earth for 17 degrees which is equivalent to saying that a beam of sunlight would be bent 17 degrees from its original path. At summer solstice, when the Sun is sitting on the horizon at midnight at the Arctic Circle, it will be about 182 miles above the Earth at London. (Please refer to the appendix for this calculation.) There is no atmosphere at this altitude so the sunlight could not have been simply reflected to London. Therefore, it must have been bent and refracted around the Earth at a much lower altitude.

The Tunkuska explosion occurred on June 30, 1908. It would be expected that by that date all of the ice and snow from the previous winter would have melted. However the explosion stirred up the surface of the Earth with considerable vigor and the Tunguska River is nearby. In fact, the event is named after the river and its associated valley. It is therefore likely that some water was thrust into the upper atmosphere along with whatever dust and other debris was blown loose. In any event, whatever was projected high above the Earth was sufficient to carry sunlight from at least the Arctic Circle to lower England. (This discussion assumes that the letter writer was reporting from London. If this person was elsewhere, a different analysis is needed.)

5.14.6.2   Noctilucent Clouds

The second type of atmospheric phenomena suggestive of how sunlight could propagate beyond the horizon is the Noctilucent or night-glowing clouds. These very faint clouds extend for over one million square miles. (19) Their unusual feature is that they exist 80 km. (50 miles) above the Earth. (44) This means that they may be seen long after sunset because at this altitude the Sun can still be seen when it is 9 degrees of latitude below the horizon at ground level. These clouds are therefore reflecting sunlight 9 degrees past the sunset line.

The Noctilucent clouds appeared about two years after the volcanic explosion at Krakatoa in 1883. (43) It has been postulated that the massive energy from the explosion, (which was heard in Australia 3500 km. away and sent both atmospheric pressure waves and water waves completely around the world) projected dust and water vapor into the upper stratosphere. At this altitude, water vapor or ice particles are trapped and will not settle out. Lower down in the stratosphere, it is warmer because ozone traps ultraviolet solar energy and keeps the air relatively warm. Therefore if water in any form settled lower it would boil, (because atmospheric pressure is extremely low at this altitude) and ascend back up again. Similarly, over an equatorial desert, rain might not reach the ground. As it descends into warmer regions, it often evaporates (i.e. boils) and drifts away.

The other possible source of water vapor for the noctilucent clouds is incoming material from space. It is possible that small comets come into the atmosphere and because they would be mostly water instead of rock, they would breakup far above the surface of the Earth. A portion of their material could then remain trapped in the upper atmosphere and contribute to the noctilucent cloud effect. (42)

5.14.6.3   Atmospheric Ice Crystals

The third type of atmospheric condition which is suggestive of how sunlight could have reached Axel Heiberg Island involves ice crystals. Whenever an atmospheric cloud of ice crystals occurs when the Sun is low in the sky, secondary suns may be seen on either side of the Sun. These refractions of sunlight are called sundogs. There are usually two sundogs because the light is being bent horizontally as it passes through the layer of ice crystals. Therefore with Sundogs sunlight is being bent 22 degrees. This means that if the layer of ice crystals extended higher up (until it was 22 degrees higher than the Sun) a complete semicircle of sunlight would be seen instead of two elongated suns on either side. Also, if for some reason the Sun could not be seen directly due to an intervening obstacle, the sundogs would still be seen. In this effect, the ice is simply refracting the sunlight 22 degrees causing a solar image to appear 22 degrees away from a direct line between the Sun and an observer. Therefore if an observer was 22 degrees above the ice cloud, a single sundog would be expected. Similarly, if the Sun were below the horizon, but still shining into a cloud of ice crystals, a sundog could be formed and bend the light downward 22 degrees to an observer. Alternatively, when the Sun is high in the sky and a layer of ice crystals is in between, a complete halo will be seen around the Sun. It is simply characteristic of a cloud of ice crystals to refract light 22 degrees.

5.14.6.4   Atmospheric Refraction

Fourthly, at both sunrise and sunset, bending (i.e. refracting) of sunlight occurs. In these cases, the light is being refracted by the atmosphere to make the Sun appear red. This happens because red is the component of light which bends the least. The refraction, in this case, is very slight, but it makes the Sun appear above the horizon whereas it might actually be below it. In this manner, sunlight curves slightly around the Earth and extends slightly above the Arctic Circle at winter solstice.

5.14.6.5   Looming

There is a fifth type of light bending that also occurs in the atmosphere. The more familiar phenomenon involves air which becomes increasingly warmer toward the ground. In this case light from distant objects will bend upward and cause a mirage to form. The observer will see something much higher up than it is in reality. Alternately, if the air gets warmer with elevation (as it would with a vapor canopy) an inverted mirage might occur. This is called looming because an image of a distant object will be seen looming over the horizon. (15) In this case light is being bent downward enabling an object which is over the horizon to be seen.

5.14.6.6   Illumination Summary

There are several ways in which a vapor-canopy-loaded atmosphere could have caused Axel Heiberg Island to have been illuminated to a much higher level than it is now. Temperature inversions would have been expected in such an atmosphere because the vapor canopy may have stayed warmer at night than the surface of the Earth. Whenever this would have happened, looming may have developed and sunlight may have been bent around the Earth and illuminated areas which would otherwise have been dark. Secondly, having water in either vapor or ice form at very high altitudes, would have allowed the Sun to shine by reflection on areas far beyond the reach of its direct rays. This primary reach may have been extended by secondary and even higher order reflections until there was light far beyond any location where the Sun was on the horizon. Thirdly, ice crystals in the vapor canopy may have bent sunlight as in the sundog effect. As with reflection, there may well have been secondary bending and widespread scattering causing sunlight to reach and shine on land which would have otherwise been well into the shadow of the Earth.

In order to illuminate Axel Heiberg when the Sun was on the horizon further south at the Arctic Circle, the amount of bending required from any combination of the above possibilities, would have been about 15 degrees of latitude. This is further than either the noctilucent cloud example but similar to the Tunguska cloud effect. It is further than we would expect from looming. However, since the vapor canopy would have extended from about seven miles up to about fifty miles, a much greater amount of reflective and refractive material would have been involved. The possibility of propagating sunlight far beyond the horizon consequently comes much further into the realm of feasibility. Further, it may not have been necessary to provide an abundance of light all year to enable the trees of Axel Heiberg to grow. They may have done quite well with reduced light for several weeks every winter. However, with so much reflective material available, it is also possible that the areas in the far north were never really dark due to 'a visible water heaven, scintillating with light' (58). In any case, the trees of Axel Heiberg Island may be explained by the vapor canopy hypothesis and they therefore provide evidence to support the vapor canopy hypothesis

In the above discussion about illumination over London, England, 17 degrees of latitude are involved whereas only 15 degrees of latitude are involved in the Axel Heiberg situation. Both of these cases involve amounts of latitude which fall within the three definitions of twilight. These include Civil, Nautical and Astronomical which respectively begin when the Sun is 6, 12 and 18 degrees below the horizon. At the end of civil twilight, the brightest stars are visible and the horizon is clearly defined. At the end of Nautical Twilight the horizon is no longer visible. At the end of Astronomical twilight the indirect illumination from the Sun is less than the light from the stars. From these definitions we can conclude that with a little help from an upper atmosphere that was loaded with water vapor and ice crystals, light could conceivably have reached Axel Heiberg Island every day of the year. (please refer to Ice Bound, by Jerri Nielsen, page 147, published by Hyperion, New York for a brief discussion of this topic.)

5.14.7   Noctilucent Clouds Again

Noctilucent clouds not only illustrate that sunlight can be reflected far beyond the horizon where it would otherwise not normally shine, they also illustrate that a layer of vapor in the upper atmosphere can remain in place indefinitely. The noctilucent clouds have been in place for more than one hundred years. If they had been gradually deteriorating, they would probably be gone by now. Their stability provides evidence that even when mixed with air, vapor can remain in the upper atmosphere indefinitely.

5.14.8   Ancient Human Records

There are numerous accounts from ancient mythology that a water heaven existed above the atmosphere. '… numerous mythological accounts of the ancient Earth … tell of a visible water heaven, scintillating with light.' (58)

'In ancient Egypt, the heaven was regarded as an ocean parallel with that on Earth. The Sun god travelled in a barge through this ocean which surrounds the world. This watery heaven was the god Canopus.' (59) In Indian religious literature, 'the idea of an upper or heavenly sea is frequent'. The Greek word for heaven may mean 'there waters' and it was located above the upper air paralleling the Hebrew idea of waters above the expanse and not in it. In the Babylonian creation account, Tiamat was a water ocean half of which formed the sky. Indirect evidence for a vapor canopy comes from a Persian account when it refers to there being neither 'cold wind nor hot wind…enabling both mankind and animals to increase at an alarming rate'. This favorable state was followed by a flood and cold weather. From Polynesia reference is made to the sky being low followed by a situation where the sky retreated to its present position. In a Sumerian account, the waters above were maintained up there by a metal vault, possibly made of tin, as tin was thought of as 'metal of heaven'. (20)

'The ancient peoples of Mexico referred to a world age that came to its end when the sky collapsed and darkness enshrouded the world. (Could this have been the post-impact dust cloud?) The Chinese refer to the collapse of the sky which took place when the mountains fell. (An asteroid impact would have brought down the vapor canopy and it would also have formed and destroyed mountain ranges.) The tribes of Samoa in their legends refer to a catastrophe when "in the days of old the heavens fell down". The Lapps make offerings accompanied by prayer that the sky should not lose its support and fall down. The primitives of Africa ... tell about the collapse of the sky in the past. The tribes of Kanga and Loanga also have a tradition of the collapse of the sky which annihilated the human race.' (45)

The ancient Hebrew account is explicit and correlates the above reports. '…and God proceeded to make the expanse (where birds fly…) and caused the division of the waters which were down underneath the expanse from the waters which were up over the top of the expanse, and it came to be so.' (61)

The exact meaning of ancient accounts, such as these, can never be firmly established, but never-the-less, they are all very suggestive of a water layer above the atmosphere. If this is the case, they are supportive of the vapor canopy hypothesis.

5.14.9   Absence of Ancient Deserts

Many of the great deserts of the world have records made by humans that indicate these areas were once anything but desert. With respect to the Sahara we have: 'Drawings on rock of herds of cattle, made by early dwellers in this region, were discovered by Barth in 1850. Since then many more drawings have been found. The animals depicted no longer inhabit these regions, and many are generally extinct. It is asserted that the Sahara once had a large human population that lived in vast green forests and on fat pasture lands. Neolithic implements, vessels and weapons made of polished stone, were found close to the drawings. Such drawings and implements were found in the Eastern as well as the Western Sahara. Men lived in these "densely populated" (Flint) regions and cattle pastured where today enormous expanses of sand stretch for thousands of miles.' 'It appears that a large part of the region was occupied by an inland lake or marsh known to the ancients as Lake Triton.' (48) The Arabian deserts have similar information.

'In the southern part of the great Arabian desert, ancient ruins, almost entirely obliterated by time and the elements, and vestiges of cultivation are silent witnesses of the time when the land there was hospitable and fruitful; it was as copiously watered and luxuriously forested as India on the same latitude. Orchards covered Hadhramaut and Aden. It was a land of plenty, paradise on Earth, but following a sudden catastrophe, Arabia Felix turned to a barren land.' (49) Similar evidence comes from farther east. 'Like the Sahara and Arabian deserts, other deserts of the world disclose the fact that they were inhabited and cultivated sometime in the past. On the Tibetan plateau and in the Gobi Desert remains of early prosperous civilizations were found and occasional ruins surviving from those times when the great barren tracts were cultivated. (50)

The fact that deserts currently exist in all of these regions and that they occupy a band at these latitudes that extends all the way around the world is not coincidental. The deserts exist in these areas due to the present pattern of atmospheric circulation. In particular, deserts exist where the descending air column of the Hadley Cells comes to the surface of the Earth and feeds the trade winds.

'Intense heat reaches the tropics throughout the year and produces powerful convection currents. Warm air rises, creating a band of low pressure around the equator. The air that rises eventually meets the troposphere, where it can rise no farther. gradually cools and sinks back to Earth's surface about thirty degrees north and south latitudes. Some of the air from these latitudes, forced out by the sinking air, moves back to the low pressure at the equator: this airflow is known as the trade winds. The circulations that rise in the tropics, sink at 30 degrees and flow back to the equator are known as Hadley Cells.' (51) When the warm moist air at the tropics rises, it cools and loses its moisture load. 'As it rises, however, it cools – the temperature drops three degrees Celsius with every thousand meters of altitude – and cool air cannot hold nearly as much water as warm air, so the moisture comes out in the form of tropical downpours. High above the equatorial regions, therefore, there is a constantly replenished layer of chilled, recently-dried air – which is then pushed away, to both the north and the south, by more warm, moist air rising from below. This cold dry air comes back down.....When it hits the surface it is both hot and dry. This is what causes the world's deserts. The deserts are not randomly distributed around the planet. Most of them are arranged in two bands girdling the planet north and south of the equator, precisely at the latitude where the Hadley Cells bring their hot, dry air down to the surface.' (52)

The conclusion from all of this evidence is that the deserts of the world exist due to atmospheric circulation patterns. If these patterns did not exist, the deserts would not exist. It follows that if these areas were once luxurious forests and pasture lands, the Hadley Cells, which currently cause all of these deserts, did not exist. If they did not exist, the current global air circulation patterns obviously did not exist. A total lack of atmospheric circulation such as this may be explained by the existence of a vapor canopy. With the canopy in place, there would not have been the great weather-causing air circulations of the present time, but a stable, uniformly warm climate which supported "vast green forests, fat pasture lands, copiously watered and luxuriously forested... a land of plenty, paradise on Earth." (49)

5.14.10   The Existence of Life on Earth

The Earth is teeming with life. There is life everywhere; in the air, in the soil, in the water and all over the land. Included in the results of even one large asteroid impact would be massive worldwide destruction. Everything would be destroyed. Ocean waves would repeatedly overrun the continents and earthquakes would shake the ground until nothing could withstand it. As discussed elsewhere in this book as well as numerous other references, the devastation would be multi-faceted and many "windows of life" would close. These unthinkable events would be sufficient to terminate life on the Earth. Then, when these various catastrophes were finally settling down, a post-impact winter would be in progress. The atmosphere would be so full of dust and smoke that the Sun would not shine at all. A taste of this has been experienced in recent time when volcanic activity produced cloud cover that lowered the temperature of the entire world. 'Scientists estimate that Tambora ejected an incredible 35 cubic miles of rock, dust, and debris. This matter created a thick layer of volcanic dust in the atmosphere that girdled the globe and persisted for several years, screening out the sunlight that would normally heat the surface and causing the frigid summer and famines of 1816.' (46) An event such as this was devastating enough but it only involved one eruption. How would the Earth fair, if there were several thousand eruptions as well as several major asteroid impacts?

If plants are deprived of sunlight for several years and if the temperature remains below freezing for several years, plants will not survive. Without plants, animals will not survive. Without either plants or animals, humans will not survive.

The only way to enable survival would have been to abbreviate the post-impact winter and wash the dust and other particulate matter, out of the atmosphere within months not years. The atmosphere would have to be cleansed in order for the Sun to shine and life, after the major disruption of both impacts and volcanic eruptions, to become reestablished. However, impacts did happen and life really is here. Therefore, it can only be concluded that within a reasonable time, the atmosphere must have been cleansed of the dust and debris to enable life to carry on. The cleansing agent could have been, in part, the hypothesized vapor canopy. If this were the case, the existence of life on the Earth is evidence of a vapor canopy having existed. It would have been terminated by even a single major asteroid impact but as it collapsed, it would have partially cleaned the atmosphere and the Sun was allowed to shine much sooner than would otherwise have been possible. This, in turn, would have enabled numerous forms of life to regenerate and become reestablished within a short period of time. (Further comment on post-impact winters, is provided in a later chapter.)

While the absence of the Hadley Cell circulation pattern explains the absence of deserts, its development subsequent to loss of the vapor canopy also helps to explain the cleansing of the atmosphere. As mentioned, if the atmosphere had not been cleansed following the impact, a post-impact winter would have persisted reducing the possibility of survival. The atmosphere simply had to be cleansed within a reasonable period of time. As the vapor canopy collapsed, it brought down a lot of dust. As the canopy disappeared, the oxygen-nitrogen layer expanded and in fact became the whole atmosphere. The vapor canopy had also shielded the oxygen from ultra-violet light so that ozone had not yet been produced. (53)

There was therefore no increase of temperature up through the stratosphere and consequently no tropopause. With the collapse of the vapor canopy, the remaining atmosphere consisting of oxygen and nitrogen would have expanded and cooled by expansion which would have also prevented tropopause development. Rising warm moist air from the equator could therefore have moved unrestricted to very high elevations before heading north. Therefore, when the Hadley cell circulation pattern first developed it was not restricted to the troposphere as it is now. The air rose much higher before moving north. Consequently the rain that poured down washed dust from much higher altitudes than it could now. Also, when the air was able to continuously circulate to higher elevations, more dust became entrained in it and was brought down to lower elevations where it could be washed out. Therefore the development of Hadley cell circulation helped to clean the air over the tropics earlier than it was cleaned farther north. The post-impact winter in the tropics was quite brief and solar energy reached the surface of the Earth much sooner than it did in the mid-latitudes. Later, as ozone was produced, a tropopause developed and the Hadley circulation was restricted to lower altitudes. Its original absence explains the lack of ancient deserts and its appearance helps to explain why a post-impact winter didn't persist and terminate life on Earth altogether.

The development of Hadley Cell circulation (initially involving higher elevations), with its attendant band of deserts where there had not been any, therefore supports the vapor canopy hypothesis.

5.14.11 The He3 Dilemma

While carbon14 is being produced by cosmic-ray neutrons, a heavy isotope of hydrogen, tritium, is also produced from deuterium. 'Tritium is unstable and decays rapidly by beta decay to an isotope of helium, He3. But it turns out there is too much He3 in the atmosphere to be accounted for by this process operating at present rates over extended periods of time. A factor which would increase the amount of tritium in the past would have involved a warmer atmosphere whichincluded much more water vapor so the process (of generating tritium from deuterium) would have been operating at a much higher rate than at present. (67) Since the vapor canopy hypothesis provides an explanation for the present excess atmospheric He3, the He3 is evidence supporting the existence of the hypothesized ancient vapor canopy.

5.15 Objections to a Vapor Canopy

5.15.1 Greenhouse Gas Objection

In order for the Earth to maintain temperatures well within the habitable range the greenhouse gases must be within an appropriate range. If the Earth was warm from pole to pole there would have been much more water vapor in the atmosphere. This would suggest that due to the greenhouse gas effect, the average temperature around the world would have been too high for life to exist. However this would not have been the case with a vapor canopy in place because the incoming energy from the Sun would have been intercepted well above the surface of the Earth. A much smaller portion of the incoming energy would have reached the surface (refer to diagram 5.8 (2)) and the heating effect of the greenhouse gases near the surface would have been reduced from what it is at the present time. The corollary to this fact is that; 'if at one time the Earth had been warm from pole to polethere must have been a vapor canopy in place to prevent overheating'.

5.15.2 Heat of Condensation Objection

One of the basic objections to the idea of a vapor canopy is that when water condenses, it releases a lot of heat. This is a basic fact of nature and this principle is applicable any time that water changes from vapor form to liquid form. However, the conclusion that the heat of condensation released by a collapsing vapor canopy would cause the Earth to overheat, necessitates several assumptions which may not have been valid.

a. The entire canopy had to condense. However, it is more likely that it was already partially condensed and included ice crystals and small water droplets rather than 100% individual water molecules. The ancient accounts of a water canopy suggest that it was visible (20) and the fact that Axel Heiberg Island had enough light to enable tree growth also suggests that the canopy was visible. Visibility indicates that ice crystals and/or small droplets are present. It may therefore have been in a partially condensed state all the time so a reduced amount of heat would be produced at the time of its collapse.

b. The basic objection ignores the fact that if an over-riding canopy collapsed, the atmosphere beneath it would have expanded. Expansion always results in cooling. Apart from any other cooling mechanisms, the expansion cooling would have been significant. (25)

c. The total elapsed time to condense is also a consideration. If the entire mass condensed and fell in one hour, the rate of energy release would have been high. However, if the process took one or two months, the thermal energy released during collapse would have been spread out, resulting in a much lower rate of energy conversion and fewer temperature problems, if there were any.

d. The collapse of the hypothesized vapor canopy would have been partially the result of vertical movement of atmosphere initiated by the incoming asteroids directly as well as both volcanic action and water wave action. These activities would all have developed vertical air currents causing great masses of air to be driven much higher above the Earth than one would normally expect. However, anytime that a shower of asteroids comes to Earth, the situation is far from normal. The result of all vertical air movement is cooling due to both expansion as well as exposure to the cold of space.

A condensing vapor canopy may be compared to a thunderstorm. When rain falls from a cloud during a thunderstorm, the water vapor in the air does condense, and the accompanying heat of condensation is released. A thunderstorm may produce several inches of water in an hour and every drop of this condensing water will release heat into the air. In fact, the heat of condensation augments the updraft and is part of the reason that thunderstorms happen. Thunderstorms can produce a lot of rain. 'On Saturday, July 31, 1976…in Northern Colorado…at Glen Haven some 14 inches of rain fell in three and one-half hours. At the height of the storm the rate of precipitation held steady at five inches per hour for 30 to 45 minutes.' (47)

Another comparison, which involves the release of the heat of condensation, is a hurricane. It has been recorded that Tropical Storm Claudette brought 45 inches of rain to an area near Alvin, Texas in 1979. (18)

However, neither during a thunderstorm nor during a hurricane, does the heat of condensation, which is continually being released, become a problem for people on the surface of the Earth. In fact there doesn't seem to be any effect from this heat at all. It doesn't increase the temperature at the surface. It seems that the heat is dissipated somehow in the clouds, but in any event, there are never complaints about overheating the Earth.

If an atmosphere included a water vapor layer, which could produce several feet of water, and all of this water condensed during a catastrophe, it would be like the rain from several hurricanes coming ashore. If they arrived over a period of a month or more, there would certainly be a lot of rain to deal with but there wouldn't be any more heat of condensation to worry about than there would be with the hurricanes.

5.15.3   Potential Energy Objection

It has occasionally been argued that the potential energy of a collapsing water vapor canopy would overheat the Earth when the water fell and hit the Earth as rain. The potential energy of any object, which is elevated, is converted to kinetic energy when the object starts falling. Most of this second type of energy is dissipated on impact, usually by dislocating the impact surface. If a vapor canopy high above the Earth started to fall as rain, there would be a lot of rain and flooding would certainly result. However, when Tropical Storm Claudette dropped almost four feet of rain in only a matter of hours (18), nobody complained about the Earth overheating due to the impact of the falling rain. A collapsing canopy might have produced two or three times as much rain as Claudette but it would have been spread out over a much longer time period. Consequently any heating effect would have been even less noticeable. Therefore it may be concluded that water hitting the Earth from a collapsing canopy, would not have overheated the Earth.

5.15.4   Exposed Rock Objection

It is understood that when rock is freshly exposed to the air it will absorb carbon dioxide (CO_2). In light of this reality, one of the results of the impact of an asteroid will be a reduction of the amount of CO_2 in the air.

When an asteroid hits the Earth there are several ways by which rock will become exposed. First of all there is simply the impact. An asteroid – even a small one – will blast an opening in the ground exposing rock which was previously hidden. Rock will be broken and ejected from the site. Some of it will be thrown for great distances and in the case of a large impact like the Sudbury impact in Northern Ontario, these distances – being commensurate with the size of the crater – could easily be several hundred miles. Erratic boulders can be explained this way. When an asteroid hits the Earth, rocks would be flying for hundreds of miles. Since rocks from Canada are found all across the northern USA, it is readily seen that a great deal of exposed rock can result from an impact.

The second major way that an impact will result in exposed rock will occur at the antipode. Here the shockwave from a large impact (like the one in South Africa) would push up on the underside of the crust on the far side of the Earth and violently elevate it, breaking it into numerous fragments in the process. A great deal of rock can be freshly exposed in this manner. This appears to be what has happened on the Moon, on Mars and on Mercury. On Earth we note that the antipode for the large South African impact is the Laurentian Plateau of Canada and the Northern USA. This is an extensive (three million square miles) region of chaotic broken rock and deserving of some explanation. When this region was broken a lot of CO_2 would have been absorbed from the air.

It seems that prior to the impact of the asteroids, the entire Earth was warm. Both water vapor and CO_2 would have been in abundant supply. In fact, with our present understanding of greenhouse gases, the Earth would have been too warm for life to exist. Since the current average surface temperature of $+15C$ appears close to optimal, the presence of even more water vapor and CO_2 would have resulted in overheating. Instead of there being abundant life, there would not have been any life. However, this was not the case as life during that ancient time appears to have been abundant.

The explanation lies with the vapor canopy. With a vapor canopy in place, part of the incoming heat from the Sun would have been interrupted high above the Earth. The greenhouse gases lower down were therefore dealing with a much reduced energy input. Life on the Earth would therefore have been possible.

Then the asteroids came. A great deal of rock was broken and CO_2 startedto be absorbed. However the vapor canopy was destroyed. With the loss of the vapor canopy, in order for it to have been possible for life to carry on, some of the CO_2 in the air as well as much of the water vapor in the lower atmosphere had to be removed or the surface of the Earth would have overheated. A simple temperature decrease (due to the loss of the vapor canopy) would have removed some of the water vapor and the exposed rock would have removed some of the CO_2. More CO_2 would have been removed during the coming years as the ocean cooled and absorbed it. The destruction of the vapor canopy in conjunction with the absorption of CO_2 by freshly-exposed rock meant that, from the greenhouse gas perspective, life on the Earth could carry on.

5.16   Vapor Canopy Conclusion

If a water vapor layer had surrounded the entire Earth in ancient time, the temperature around the Earth would have remained within a fairly narrow range all year much as it does now in the Faroe Islands of the North Atlantic. While these islands would be considered to be in the 'north', because they lie between Scotland and Iceland, because they are directly in the path of the Gulf Stream, their temperature remains steady all year. 'The islands rise from the Atlantic in cliffs so sheer they could have been sliced with a razor. Because they are planted squarely in the path of the Gulf Stream, their weather is moderate, with an extraordinarily narrow range of temperature – from 37 to 51 degrees Fahrenheit, all year-round.' (68)

A water vapor layer surrounding the Earth would have been of great benefit. The key to its retention would have been atmospheric stability. As long as the vapor layer was dust free and not seriously disturbed, it would just float year after year providing protection for the Earth. If the atmosphere were seriously disturbed, however, the whole thing would have come crashing down. As well as dealing with this huge quantity of water, the Earth would lose its warm blanket and subsequently chill to some lower temperature range.

If a large asteroid had crashed down through this ancient water-covered atmosphere, there are at least three ways in which the protective water vapor layer could have been destroyed. As has been discussed earlier, first the expanding shock-wave would have caused the water to condense. If the water molecules remained stuck together after the shock-wave passed, a torrential downpour would have resulted. Secondly, the descent of the asteroid would have drawn atmosphere in behind itself. As this material moved horizontally into the region through which the asteroid had just passed, and then downward behind it, a donut-shaped pattern of air movement would have resulted. The associated upward movement would place air high up in the vapor layer. Immediately, the water portion of this new mixture of water and air would precipitate because air simply cannot hold very much moisture.(22) Condensation of the vapor into rain would release the heat of condensation. The upward movement of air would thereby be accelerated. More air would be drawn into the upper atmosphere from the lower atmosphere and more vapor would be mixed in with it. More precipitation would result. (Please refer to the diagram, 'Vapor Capacity of Air' 7.3.1, to see just how much vapor air can hold.) Such a region of instability might have expanded outward indefinitely until the whole Earth was involved. In this manner, the entire vapor layer might have been brought down by a single asteroid. The third difficulty would develop from any significant volcanic activity. Upward motion of effluent from a volcano would have driven air up into the water vapor layer and, as before, disaster would result. Stability within the atmosphere would have been the key to retention of the vapor layer and an incoming asteroid would have totally upset this stability.

An asteroid would have upset stability in several ways and the vapor canopy would not have survived. In addition to the numerous devastating effects discussed in the previous chapter, the Earth would also have to deal with rainfall several feet deep over its entire surface. The flooding and erosion from this much flood water would be difficult to visualize. Possible benefits would have included extinction of some of the forest fires (caused by the shock wave) as well as removal of volcanic dust but removal of the warm blanket from above the atmosphere would have caused the Earth to start cooling to some lower temperature range. Polar areas would have chilled first. But, due to the enormous heat retention of a universally warm ocean, its temperature would not drop instantly but would drift down over several hundred years. (23) If the volcanic dust and the smoke from the forest fires caused the air over the land to chill to below freezing before the oceans cooled substantially, an ice age would commence. (This is discussed in a subsequent chapter.) The cooling ocean would absorb more CO_2, which would reinforce the chilling effect. Atmospheric pressure would have been reduced by the removal of the vapor canopy and any affect this and the reduced CO_2 levels may have had on aging or gigantism would start to show up during the following years. Reduced CO2 levels would also result in less vigorous plant growth. All plants would, comparatively speaking, be stunted.

Further discussion of the vapor canopy concept has been included in the Appendix. Therein a few more of the complexities involved in modeling the atmosphere are reviewed. Also, there will be several other effects from an incoming asteroid and most of them will be even more devastating than the atmospheric effects just discussed.

6. More Bad News

6.1 Introduction

Asteroids of all sizes are very common in our solar system. Although the total number is not known, it has been suggested that there may be more than a billion altogether. (68) This large number would include the very small ones which may only be a few inches across and which would be called meteors if they should streak across the sky. Also included would be those which would only partially burn up in the atmosphere leaving some portion to make it all the way to the surface of the Earth. In this case they would be called meteorites and may occasionally cause damage to property. (On November 30, 1954, a four-kilogram meteorite crashed through the roof of a house in Alabama. (15))

While the very small asteroids are of interest if they burn up in the atmosphere or crash through a house, it is the large ones which seriously concern us. 'Calculations indicate that there are about one million whose diameters measure one kilometer or more.' (16) Further 'about 1000 asteroids are larger than 30 kilometers across, and of these, more than 200 asteroids are larger than 100 kilometers.' (16)

If an asteroid should impact the Earth, there will be several serious consequences. The relative magnitude of these consequences will depend on the size of the asteroid, but not directly. This is because the amount of material in an asteroid, and hence its weight or mass, is dependent directly on the volume rather than the diameter. If the diameter of a spherical object is doubled, its volume increases by a factor of eight. Therefore, the energy and hence the destructive power will have increased by a factor of eight. If an asteroid, which is one mile in diameter, crashed to the Earth, there would be catastrophic effects on the impact continent as well as various degrees of devastation world-wide. An asteroid which is two miles in diameter would cause eight times as much trouble but a greater portion of the devastation would occur on the far side of the Earth. Anything larger than a house, however, is very bad news for the Earth.

Some asteroids have an irregular shape and while they are usually rounded, shapes may be oblong with enlarged sections as well as surfaces which do not have any particular regularity of shape. The moons of Mars are examples of space objects, which have an irregular shape. Their roundness does imply a liquid phase and such objects appear like liquid blobs, which were cast into space to cool and which acquired whatever shape their particular (at that moment) viscosity and limited self-gravity could impose before they cooled to the solid state.

While the amount of damage that would be caused by an asteroid crashing to Earth, would be dependent on its size and energy, the type of damage will also depend on its size and energy. A very small one (i.e. a baseball size meteor) would heat up and burn out thereby interacting only with the atmosphere. One, as large as a living room chair, would also heat up and partially burn up in the atmosphere and anything that was left would fall to the ground and make a small crater. The energy of this larger object has therefore been dissipated in two ways: heating itself and excavating a crater. One as large as a football stadium would lose very little energy coming through the atmosphere which means that most of its energy would be dissipated on and near the surface of the Earth. Similarly only a portion of the energy of an asteroid which was several miles in diameter would be dissipated on and near the surface. Virtually all of its energy would be dissipated deep in the Earth and far from the impact site, even to the far side of the world.

As discussed in a previous chapter, a large asteroid would produce an atmospheric shock wave, a water splash and a water wave. As a crater is blasted out of the surface of the Earth, a mud/rock splash would result. The ground around the crater would be uplifted. The uplift might appear as a sloping plateau with the highest part around the rim of the crater. It might also appear as a ring of mountains, which surround the crater and actually form the rim of the crater. The ground would tremble as a great earthquake is produced. The earthquake would result in one or more tsunamis forming. These tsunamis might originate many miles from where the asteroid landed and as with the water wave, they would travel for thousands of miles. Volcanoes would erupt. The great sound of the crash will start a sound wave, which will travel both through the air and through the ground. All of these effects will result in a great deal of catastrophe but will still only account for a small portion of the energy, which the asteroid had as it entered the atmosphere.

6.2 Asteroid Speed

While it is impossible to predict what the speed of the next Earth-impacting asteroid will be, the actual speeds of several asteroids have been measured and this information can be used as a expectation guide. For example, the speed of asteroid 1994 XM1 'had the mass of a large house and was moving at 108,000 kilometers per hour. (67,500 miles per hour)' (18) Asteroids are always expected to be moving at high speed. 'Most of these objects are hurtling at such high velocities – 11 to 100 kilometers per second (39,600 km./hr. to 360,000 km./hr. or 25,000 miles/hr. to 225,000 miles/hr.).' (19) Comets also move at very high speeds and are expected to be moving even faster than asteroids. 'Hurtling at velocities up to twice those of asteroids, a comet the size of Halley's – about the width of a small city – could hammer the Earth with such force…..' (43) The speed factor is of great interest because the speed is of considerable consequence in determining the energy. The energy causes the damage and being able to identify the speed would be helpful in estimating the energy and hence the potential damage. Further, the speed factor is of particular importance because the energy is dependent on the speed when the speed is multiplied by itself. This is called a square law. When something is multiplied by a certain factor twice, that factor will be of very great importance in the final result. In this case, the energy is the result of multiplying the mass of the asteroid by the speed and then multiplying by the speed again. The formula for energy therefore appears:

Asteroid energy = mass x speed x speed = mass x speed(2).

In any example like this, if the speed is doubled, the energy increases by a factor of four. The effect of speed on the amount of energy is therefore very influential.

Both asteroids and comets move at extremely high speed through space. If they should approach the Earth, they would, due to the increasing pull of Earth's gravity, speed up. Further, the influence of the Earth's atmosphere would have a completely negligible effect on the speed. Therefore at contact with the Earth, they would be still moving at or above the extreme speeds that they had in space.

6.3   Accounting for Asteroid Energy

A very small portion – possibly one or two percent - of the enormous energy of a large asteroid would be accounted for by atmospheric effects. Another few percent would be accounted for by the effects listed above. (e.g. earthquake, waves etc.) The rest of its energy, which will in fact be more than half of its energy, will be transferred into the interior of the Earth and do remote damage. The evidence and reasoning to support this contention follows.

6.3.1   Crater Expectation

The bowl-shaped Barringer Crater in Arizona is approximately 4000 ft. in diameter and 570 feet deep at the center. (20) By estimating the amount of excavated material and by making a reasonable assumption of the asteroid's speed at 35,000 miles per hour, (20) a calculation of asteroid diameter can be made and the Barringer Crater is thought to have been made by an object 50 feet in diameter (20). It would seem reasonable to expect that larger objects would form larger craters and to a certain extent this is correct but as the size of an asteroid increases into the large category, (i.e. several miles across) there usually isn't any crater at all. Instead only a circular elevated formation of jagged hills, uneven ground or an outward declining plane is observed. On Mars, for example, the Hellas Basin is surrounded by a circular formation of jagged hills with a diameter of approximately 1430 miles. (17) The area enclosed by these hills is flat and follows the general curvature of the planet, although it is lower than the areas outside of the ring of hills by about 4 km. (2.4 miles) (21). If this formation was made by a large asteroid and the same assumptions and calculations were made as with the Barringer case, a crater with a depth of 203 miles would be expected. Since the depth of the Hellas crater is a small fraction of one percent of the its diameter, and the height of the surrounding circular formation is so low compared to its diameter, only a very small portion of the (expected) energy of the Hellas asteroid may be accounted for by the visible evidence. For a large asteroid like Hellas, an enormous crater would be a reasonable expectation. However, where large asteroids are involved, there really aren't any craters at all.

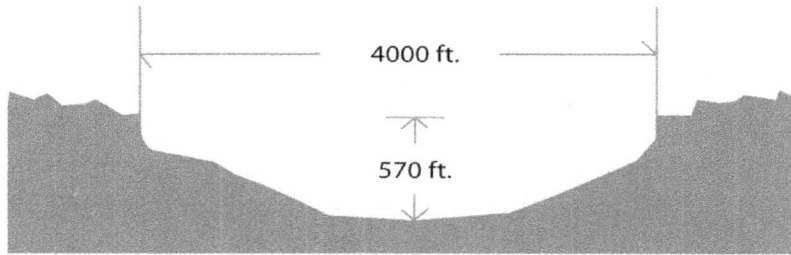

4000 ft.

570 ft.

6.3.1 Barringer Crater, Arizona

6.3.2 Hellas Crater on Mars

The Hellas Crater is 2.4 miles deep. However, depth in this case
is not significant in comparison to diameter when considering
other craters Hellas does not appear like a typical crater,
save the ring of relatively small mountains that surround it.

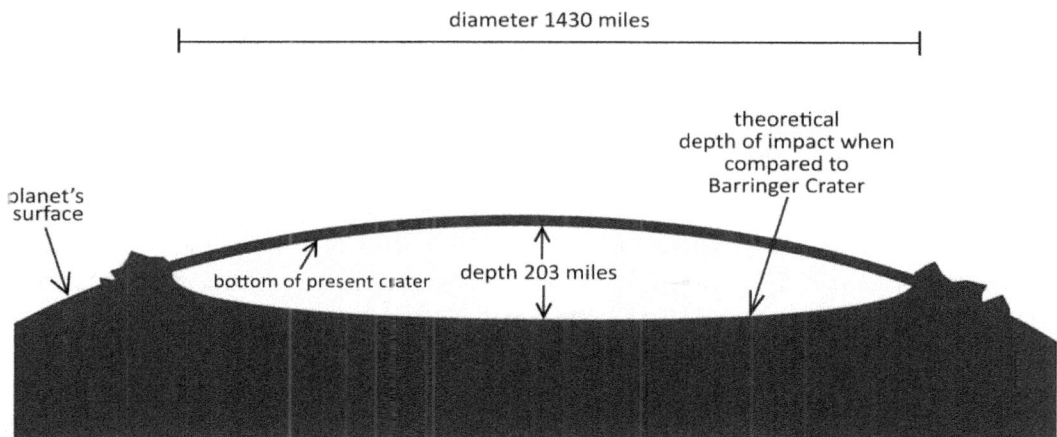

diameter 1430 miles

theoretical
depth of impact when
compared to
Barringer Crater

planet's
surface

bottom of present crater

depth 203 miles

6.3.2 Crater Flatness

The craters for all large asteroids are not bowl-shaped. Instead they are flat. That is, they are flat in the same sense that the Earth is flat. Over small areas it is reasonable to say that the Earth is flat. Over large areas, which cover a few square miles or more, we understand that the flatness is actually a curvature and is, in fact, the general curvature of the Earth. The surface of the Earth is curved because the shape of the Earth is spherical. The Earth is shaped like a ball so when large areas of the surface are being considered, it is understood that the shape of the surface is the shape of the surface of a sphere or ball. While this is not really of much interest when we are walking around the yard or driving into town, it is of considerable interest when we are travelling around the world. When great distances are being contemplated by ship or airline operators, the curvature of the Earth is always taken into consideration.

Examples of crater flatness (i.e. flat as opposed to dish or bowl-shaped) are numerous. The three great craters of Mars are all flat. This includes the largest one called Hellas, which is about 1430 miles in diameter. (17) Argyre is slightly smaller at 1100 miles and the diameter of Isidis is approximately 680 miles. (17) The great Sudbury Crater in Ontario has a major diameter of about 150 miles but the center does not have any particularly distinct geological features at all and there isn't any significant rim to consider. (23) The Manicouagan crater in Quebec is approximately 50 miles across and it is actually higher in the center than further out near the edge! (24) (Rim is the correct term but just to emphasize the point, the Manicouagan Crater does not really have any rim at all.)

6.3.3 Fluid Phase

The shape of objects in space is determined by their own gravity and viscosity. An object like the Earth has a spherical shape. The self-gravity of the Earth pulls everything towards the center. While this force of gravity is acting, the material of the Earth is fluid enough that it can assume a spherical shape. If the Earth had been formed from extremely rigid material, it would probably have had an irregular shape. A large portion of the surface of the Earth is water. It may be more readily appreciated that because water flows very readily, it will quickly settle into a shape, which is determined by the forces which are acting upon it. Similarly, if a portion of the surface of the Earth became molten, such a portion would respond like water and assume a shape, which would minimize its potential energy. This would be the case if an area of the crust of the Earth were pushed aside, exposing the hot molten interior. The molten rock from the interior would flow into the space where the missing crust had been, rise until it reached the approximate elevation of the surrounding region while assuming the shape of the general curvature of the Earth. After it cooled and hardened, this area would appear like a large flat plain. This, of course, assumes that nothing stirred it up as it cooled. For example, if solid material was floating in the molten region or if ongoing eruptions disturbed the surface, there would be some irregularity to the surface after it cooled. (But it would still generally follow the curvature of the Earth.)

The generally-flat central regions of all of the large craters on the Earth, the Moon, Mars and Mercury, indicates that after the asteroids struck and passed on into their respective interiors, these areas became fluid-filled prior to hardening into their present gravitationally-determined shapes. This could only have happened if the impacts exposed hot molten interiors, which were then allowed to rise up as far as gravity would allow and fill the recently blasted-out areas. These molten surfaces would then harden into shapes which were generally the spherical shapes of each respective planet or moon.

6.3.4   Impact Site Cross-Section

The cross-section of the Arizona Barringer Crater appears as shown in the diagram above. The depth, at the lowest area in the center, is approximately one-eighth of the crater diameter. (26) This arrangement does satisfy our intuition because a shallow crater such as this seems like a reasonable result for an asteroid which was 50 feet in diameter.

By extrapolation, it would be expected that a larger crater would form as the result of a larger impact. However, large bowl-shaped craters are never found. Instead only diameter indicators are found and the center regions are either not excavated at all or only excavated to a very small fraction of the crater diameter. As an example, the depth of the large Mars crater Hellas is much less than one percent of its diameter. (21) When a situation like this is drawn to scale, there does not seem to be any crater excavation at all. Similarly, the Manicouagan Crater in Quebec has a shallow depressed area inside the rim and a center region, which is not depressed. (27) In fact, it is elevated. The Sudbury Crater in Ontario is the most extreme example. In this case a crater rim cannot be discerned but, based on other geological evidence, is understood to be about 150 miles in diameter. (28)

6.3.5   Molten Interior Proximity

The cross-sections of both the Manicouagan Crater and the Sudbury Crater can be drawn showing the thickness of the Earth's crust and the proximity of the molten interior. The crust of the Earth is thought to be about eighteen miles thick. (22) At 150 miles the diameter of the Sudbury Crater is therefore almost ten times as great as the crust is thick. Similarly, since the diameter of the Manicouagan Crater is about sixty miles, it is about three times as great as the thickness of the crust. The Chesapeake Bay Crater is similar in size to the Manicouagan Crater and therefore it too is several times as large as the thickness of the Earth's crust. The Chesapeake Bay Crater has the distinction of being partly in the ocean where the crust is much thinner. (22) The Chicxulub Crater with a diameter of 112 miles is smaller than the Sudbury Crater and larger than the Chesapeake Bay Crater and is about six times as large as the distance through the crust to the molten interior of the Earth. In all of these cases, the distance through the crust to the molten interior is only a small fraction of the distance from one position on the rim of each crater respectively, across to a second position on the opposite side.

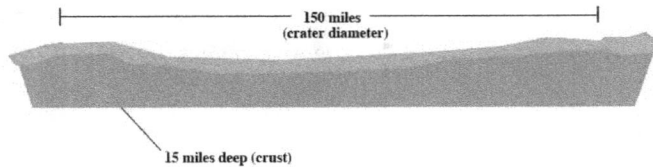

150 miles
(crater diameter)

15 miles deep (crust)

6.3.5 Sudbury, Ontario Crater

The Sudbury Crater in Northern Ontario has no surface features
which are significant with respect to its diameter. It is commonly
given as 150 miles in diameter. The crust of the Earth is thought to be
15-20 miles thick. so the crater is nearly 10x as large as the crust is thick.
It is therefore reasonable to expect that an asteroid which was able to form
a crater 150 miles in diameter would be able to punch through the crust
and travel far into the molten interior.

6.3.6   Absence of Dislocated Material

With the Barringer Crater, a lot of material has been removed in comparison to the diameter of the crater. With larger craters this is not the case. The amount of material, which has either been moved or removed, appears as either very little or non-existent in comparison to the crater diameter. As an example, the Hellas Crater on Mars is more than 1400 miles across but is only a little more than 2 miles deep. In comparison to the diameter, the excavated material is practically non-existent. Also the rim is comparable in height to the depth of the excavation. By simple comparison to the Arizona Crater, one would expect a rim several hundred miles high. Such a feature is not found. In every case where large craters are observed, any apparently dislocated material within the rim is always negligible when compared to the crater diameter. This is not to say that a large amount of material wasn't excavated when the asteroid struck. It certainly would have been. However, displacement energy calculations cannot proceed because there isn't any way that the volume of the possibly-excavated material can be calculated. It is clear that at least some material must have been dislocated out as far as the rim. Otherwise the region would not have been able to refill with molten rock and end up as a flat surface.

There appear to be two possibilities to explain the flatness of large craters. The asteroid punched through the solid crust into the molten interior and formed a temporary opening in the Earth that was actually bowl or crater-shaped. The material that was splashed away to allow the formation of this bowl, was ejected from the area completely and spread around the crater rim as well as further out for several crater diameters or more. Then the molten magma rose and refilled the opening before cooling into the general curvature of the Earth. Meanwhile the asteroid, or its remains, plunged further into the Earth and was swallowed by the molten interior.

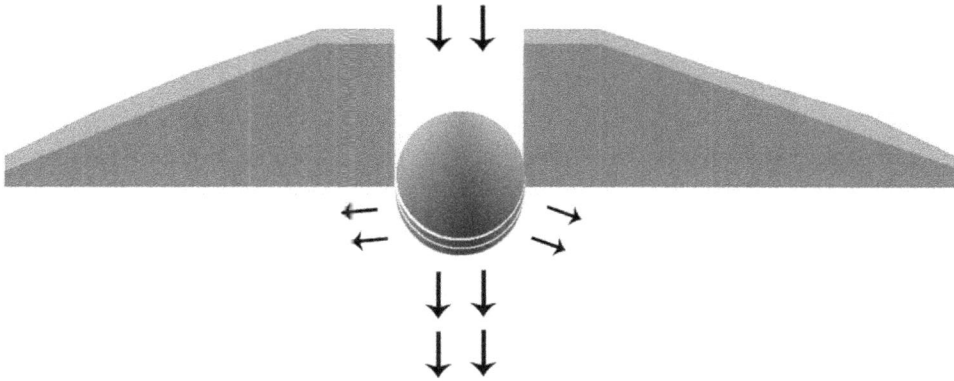

6.3.6 Asteroid Punch-Through

A large high-speed asteroid would be moving fast enough to punch right through the crust of the Earth. Neither the crust or molten interior materials would have time to move aside. Material in front of the asteroid's direction of movement would be included in an extremely high pressure zone which would extend ahead of the asteroid into the Earth's interior.

Alternately, the asteroid punched through the solid crust into the molten interior and completely disrupted and broke up the crust in the surrounding area out as far as the rim, but did not eject all of it from the area. The disruption however allowed the broken pieces of crust to settle back into the magma bending-stress free. In other words, pieces of broken crust floated on the magma and generally assumed the curvature of the Earth. Meanwhile, as above, the asteroid or its remains plunged further into the Earth and was swallowed by the molten interior.

6.3.7   Lunar Mascons

On the Moon, beneath several of the 'seas' or Maria (large, flat, slightly darker areas which have occasionally been referred to as forming a face) there are mass concentrations called 'mascons'. (29) These mascons are regions of high-density material - at least they have noticeably higher density than the surrounding lunar material. Also, the Maria indicate a recent (very few crater marks) fluid (virtually flat) state where liquid from the interior must have welled up and oozed over the lunar surface. The Maria on the surface and the mascons beneath the surface are suggestive of impacts by very large objects, which broke through the lunar surface and plunged into a liquid zone underneath. If this is indeed what happened on the Moon, it wouldn't be unreasonable to expect that something similar happened on Earth.

206

6.3.8 Asteroid Size - Crust Thickness Comparison

Large asteroids such as Manicouagan and Sudbury in Canada, Chesapeake Bay in the USA, Chicxulub in Mexico, Acraman in Australia, Vredefort in Africa, Popigai in Russia, and Tanitz in Kazakhstan, are thought to range in size from about 5 to 15 km. across. (32) The crust of the Earth is understood to be between 5 km. and 28 km. thick. (31) All of these large asteroids are therefore comparable in size to the thickness of the crust. This reality leads to the possibility that these objects punched right through the crust. Even an object, which had a diameter, which was only a (significant) fraction of crust thickness, would be expected to punch right through the crust and travel much further down into the Earth. Once the notion of punch-through is allowed, it may be further speculated, as has been done in the Mars case, (46) that these asteroids may have been much larger than has usually been suggested.

6.3.9 Energy Distribution Conclusion

The evidence is suggestive that the large asteroids which have struck the Earth hit the surface and punched through to the molten interior region, material from which refilled the openings before cooling and crusting over. The energy of one large asteroid would be sufficient to form a mountain range. (30) However the large impact sites do not indicate this much trauma. It must therefore be concluded that the energy of these large asteroids (e.g. those whose diameter was comparable to the thickness of the crust of the Earth) was dissipated elsewhere. Most of their energy must have been transferred into the molten interior and dissipated far from the impact site.

6.4 Shock Wave Formation

It is always difficult to move something – even air – in a hurry. As mentioned in an earlier chapter, a blinding flash is expected as a large asteroid approaches the surface of the Earth because the air in front of the asteroid cannot get out of the way and becomes super-compressed and consequently super-heated as the asteroid rushes Earthward. While some air near the edge of the pathway would slip off to the side, the air in the center of the asteroid's pathway would not have time to move sideways. It would move sideways if more time was available, but it cannot because there just wouldn't be enough time. Even as the pressure in this zone becomes exceedingly high, (which would normally cause air to move), it would not be able to move aside within the time available because the asteroid would simply be moving too fast. (i.e. It would come all the way through the atmosphere in about one second.)

Since air is not very viscous, it is easily moved. When people walk down the street they never worry about moving the air to the side to let themselves proceed. Similarly, water, while it has a very low viscosity compared to other fluids, is quite easy to move. However, if a person tries to walk when they are standing chest-deep in water, they will meet with considerable difficulty. The liquid rock or magma in the interior of the Earth is much more viscous than water. Consequently it is much harder to move than water. If the speed of an asteroid is so high, that even air which has very little viscosity doesn't have time to move aside, it may be safe to assume that anything more viscous than air wouldn't have time to move aside either.

Therefore, if an asteroid was too large to be slowed by contact with the Earth's atmosphere or even with the Earth itself, it would retain most of its speed as it smashed into the Earth and the material in front of it, like the air in the atmosphere, would not have time to escape sideways. As with the trapped air, the solid rock in front of a speeding asteroid would also be trapped which means that if it cannot move sideways, it would be forced to move ahead of the asteroid and be pushed down into the interior.

In the interior, the molten rock, like the air, would not have time to move sideways either. Consequently, it would continue to be pushed ahead of the asteroid. Only after the speed of the asteroid had decreased would there be time for the magma to move to the side. Most of the energy of such an asteroid would therefore be transferred to the material which was directly in front of it. Lateral material movement would only be possible as the asteroid slowed, but by then a massive pressure wave would have developed in front of the asteroid and it would be speeding ahead of it deep into the interior.

This line of reasoning is actually applicable to asteroids of all sizes – even those which form a conventional crater. All asteroids, on initial contact with the Earth, would be moving so fast that most of the material directly in their pathway would be pushed ahead. The difference is that a small asteroid would slow down very soon after contact. When its speed had dropped enough – possibly within a depth which was similar to its own diameter - lateral material movement would proceed and a crater would form. It follows that larger asteroids would be able to push farther into the Earth before losing their speed. Consequently more material would be pushed aside and larger craters would form. If the asteroid was above a certain size, it would be all the way through the crust before it slowed enough to allow lateral material movement. In this case, all the material that had been directly in its pathway would still be in front of it. A conventional crater would therefore not be found.

The more viscous the material was in front of such an asteroid, the less it would be able to move out of the way. If the very air in the atmosphere cannot move out of the way fast enough, neither would the molten rock in the interior of the Earth. Without lateral movement of Earth material near the surface, there will be minimal surface disruption and hence minimal evidence that there actually had been an impact. In other words there wouldn't be any conventional-looking crater. The logical consequence of this line of reasoning is that, as the size of an asteroid increases, the destruction on the surface of the Earth would initially increase forming larger craters and then as the asteroid became larger still, the surface disruption would decrease until there was hardly any evidence of a crater at all. With the large, more-or-less craterless asteroids, most of their energy would have been transferred into the interior of the Earth and would have formed a massive shock wave. Thus the asteroid's energy would have been dissipated far from the impact site throughout the entire planet.

Whether the asteroid retained its basic shape and structural integrity is immaterial. The shock wave would develop because a great mass of material would have been thrust into the magma at high speed. The actual material, which would contribute to the shock wave would include:

a.   The portion of the asteroid, which didn't splash away from the impact site

b. Pieces of broken crust, which were pushed ahead of the asteroid, as it smashed through the crust.

c. The magma, which went along with the pieces of broken crust and the remainder of the asteroid.

All of this material would form an energy-loaded mass, which would, because of its rapid forward movement, pressurize the surrounding magma and produce an inwardly-propagating pressure wave.

6.5   The Crust Factor

6.5.1   What is a Crust

A crust is the thin solid layer, which encloses or covers a softer layer underneath. A layer of ice on a pond could be referred to as a crust. The ice is solid compared to the rest of the pond, which is liquid. Also, the crust can be very thin compared to the depth of the pond. A pond could be several feet deep whereas the crust of ice might only be a fraction of an inch thick.

An apple pie has a crust. Actually an apple pie will probably have both an upper crust and a lower crust. In between, the material is softer. The crust in this case helps to hold everything together.

A crust could form in a pail of paint. This is undesirable and the crust must be removed before the paint can be used.

A crust could also form in a vessel of molten metal. If lumps of solid lead are placed in a metal pan and heated, they will melt into a single quantity of liquid material. But if the pan is not kept hot enough, some of the lead at the surface will turn solid again and form a crust. This crust will appear jagged compared to the liquid lead, which will appear smooth. The crust is always the solid layer, which covers the liquid or molten material underneath.

6.5.2   The Crust of the Earth

From space, the Earth appears to be a sphere of solid material like a baseball. This material is mostly rock or soil (i.e. obviously neglecting the oceans) and is certainly solid enough to support large buildings, highways, bridges and hydro towers. There is even solid material at the bottom of the ocean. Mountains appear very solid and mountains exist on both dry land and under the ocean.

Mines are often dug into the Earth and sometimes a mine can extend underground for more than a mile. For example, the nickel mine at Sudbury is much more than one mile deep. (33) The rock at this depth is quite solid.

Compared to the size of the Earth, however, the region which we refer to as solid ground is not very extensive. Instead of a baseball, the Earth is really more like a water balloon. The diameter of the Earth is almost eight thousand miles but the solid region is only a few miles deep. As we descend deeper into the Earth, the temperature increases. Down in the Sudbury nickel mine the temperature is higher than a person can tolerate. Therefore, in order for people to work this deep in the Earth, the temperature must be lowered into the comfort range. The deeper we go into the Earth, the hotter it gets. When a region is reached where the rock is hot enough to glow red, the extent of the solid layer has almost been reached. Solid rock will begin to glow red when a temperature of about 700F has been reached. As the temperature increases higher than this, the rock will gradually get softer. Molten rock is expected to be found well within thirty miles of the surface in continental areas and within possibly three miles of the floor of the ocean. Deeper than this, the rock will be molten and while it is still rock and still dense, this is now a liquid region rather than a solid region.

The deepest borehole on Earth was made by a Russian group on the Kola Peninsula which is in northwest Russia. The drilling location was on Russia's Kola Peninsula, near the Finnish border and they hoped to drill to a depth of fifteen kilometers. The work proved harder than expected, but the Soviets were commendably persistent. They continued for nineteen years and by then they had drilled to a depth of 12,262 meters, or about 7.6 miles. The world down there was far warmer than anyone had expected, with a temperature at 10,000 meters of 180 degrees centigrade. (356F) (lead melts at 327.5C or 625.5F) which was much higher than expected. (34) (If the steel drill rods and bit holders become too hot they will become too soft and lose their structural value as drilling implements.)

When compared to the size of the Earth, the solid region is very thin and therefore properly referred to as a crust. It may be compared to the skin of an orange except that the crust is comparatively much thinner than the skin of an orange. It would be more comparable to the skin of a water balloon. The Earth, which appears to be so solid, is basically a huge red-hot ball of liquid rock covered with a very thin cool solid rock crust. The spherical shape is only possible because the Earth is mostly liquid rather than solid and the self-gravity of the Earth causes a spherical shape to be maintained.

If the crust of the Earth had been a thousand miles thick instead of less than twenty-five, a spherical shape might not have been realized. In such a case, there would have been a serious problem with atmospheric distribution. It is a problem even now when our mountains only extend upward from sea level for a few miles. At the top of Mount Everest, which is between five and six miles high, there is not enough oxygen for anyone to breathe, so nobody can go up there without taking some with them. What if mountains extended upward for a hundred miles or a thousand miles? No one could explore mountains, which were that high, without having the full protection of a space suite and an adequate supply of oxygen. (The volcano, Olympus Mons, of Mars, extends upward for 22 miles and is more than 300 miles in diameter at the base. (35) If this structure was on Earth and if a person was protected by a space suite, he/she could hike right up into space.) As it is, the crust of the Earth is relatively thin, the Earth has retained a spherical shape and oxygen is distributed reasonably evenly over most of the Earth's surface.

6.6 Impact Comparisons

If a large asteroid hit the Earth, it would crash right through the crust and travel deep into the molten interior. While a situation like this has never been witnessed, (and it never will be, because witnessing would mean certain death for the witness) the situation can, never-the-less be reasonably understood by consideration of the physical relationships which are involved. In order to assist this understanding process, other types of impacts, which can be observed, will be considered.

With all of these comparisons as well as with any real scenario involving real asteroids, the underlying scientific principle involves the conservation of energy. The energy of an incoming asteroid – no matter its size – will be conserved and be transferred to and show up as movement of Earth material. If the energy of all of the Earth material put in motion by an asteroid could be added up, it would exactly equal the energy that the asteroid had as it approached the Earth.

6.6.1   Rock and Pond

When a rock is tossed into a pond, a circular wave will form around the impact site and proceed to move outward in all directions. When a large rock is tossed into a pond, a larger wave will form because the large rock has more energy to transfer to the water than the small rock. Further, if the rock were driven into the pond as hard as possible, a still larger wave would form because the faster rock has more energy to transfer to the water than the slower rock.

A similar result is expected if a rock is tossed into a pond, which is covered by a thin crust of ice. Some of the energy of the rock will be dissipated in breaking the ice. The rest will disturb the water, which is under the ice. Of course if the rock is too small to break the ice, it may simply stay on top of it or become partially embedded in it.

If a rock was thrown very hard at the ice-covered pond, we can imagine it breaking through without any difficulty. In fact if the rock was thrown very hard, it might break through ice, which was as thick as the diameter of the rock. Now suppose that the rock was loaded into a cannon and fired straight down at the pond. In this case, the rock would be able to break right through ice which was several times thicker than its own diameter. Similarly, if a rock was fired from a cannon straight down at a thin layer of rock, it would break right through and travel on into the region which was under the rock. Such a rock may also be able to break through a layer of rock which was as thick as itself. The high speed means that the moving rock has a lot of energy. As with the ice example, some of the energy of the moving rock would be dissipated as it broke through the thin layer of rock and the rest would disturb any material that was underneath.

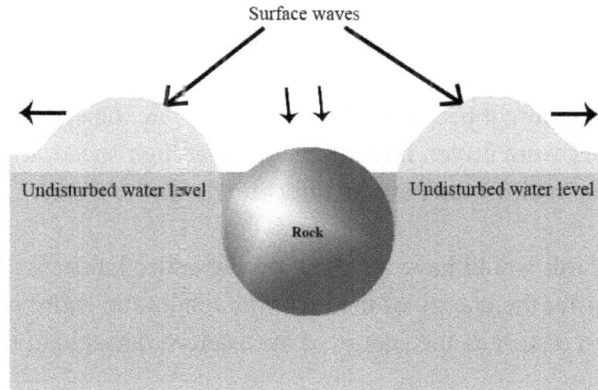

f.6.1 Rock-Pond Comparison
When a rock is dropped into a pond, the water is displaced, forming a circular
outward-moving wave. Most of the energy of the rock will be converted to the energy
of the wave.

From Einstein's famous energy equation; $E = mc(2)$, it is clear that the speed of light, c, is very influential on the outcome because it is multiplied twice. The equation for a moving rock is similar as, $E = mv(2)$. If it were possible to fire the rock at ten times cannon speed, ten times ten as much energy would be involved. While it is obvious that the energy of the rock increases as the speed increases, in fact, it increases much faster because the speed factor is multiplied twice. Therefore, if one rock is moving twice as fast as another rock, it will have 2 x 2 or 4 times as much energy. The energy of the moving rock also depends on the amount of material, which is in it (i.e. the mass, m). The energy of a speeding rock is actually its mass, times its speed, times its speed again. This is called a square law because we multiply by the speed twice.

With this basic understanding of physics, it can be appreciated, that a moving rock could be easily driven right through a layer of rock which is similar in thickness to itself, and still have a lot of energy left. After breaking through the postulated layer of rock the moving rock or its remains as well as broken pieces of the layer of rock itself might still be moving almost as fast as the rock was before any contact was made.

These ideas are scalable. If the moving rock was twice as big and the layer of rock was twice as thick, the same ideas apply. Even if the moving rock was a mile in diameter and the layer of rock was a mile thick, the moving rock would break right through because of the energy, which accompanies the high speed. If it were possible to fire a one-mile diameter rock into the Earth at ten times the speed of a bullet, (or ten times 1000 miles per hour) it will be readily appreciated that there would be no trouble breaking through a layer of crust, which was one mile or more thick. This would be the situation with an incoming asteroid. A large asteroid would be moving upwards of 30,000 miles per hour or thirty times bullet speed. (19) Further, if it was fifty miles in diameter, it can be appreciated that it would break right through the crust of the Earth, even if the crust was twenty miles thick. Then, after breaking through, it would still have a great deal of speed and it, or its broken remains as well as some broken crust, would continue on into the interior of the Earth until all of its energy was dissipated. Only then would it stop.

As with the rock that crashes through the ice of a pond, and yet retains a significant amount of its energy, this remaining energy now becomes of particular interest in this comparison because it will have an effect on the liquid material beneath. Since, in this case pond water is the liquid, the energy of the rock will have an effect on the pond water which will become a means to distribute the energy of the rock far beyond the impact location. If a large rock were driven into a large pond at high speed, it would not be unreasonable to expect that the energy of the rock would have an effect on the entire pond.

A large asteroid hitting the Earth would have an effect on the entire Earth. With this rock-pond example, it is the water which will transfer the energy of the rock throughout the entire pond. With an asteroid, it is the molten magma, which will distribute the energy of the asteroid throughout the entire Earth.

6.6.2   Ball and Diaphragm

The rock-pond comparison involved materials which are stiff and brittle (i.e. the rock and the ice). It is correct that both ice and rock have some elasticity, but compared to rubber, they are very stiff and brittle materials. It is not difficult to imagine that if a rock were driven at high speed into another rock, something would crack and disintegrate. Ice is even more likely to shatter. If a ball of ice were driven into a sheet of ice, which was floating on a pond, we would expect that something would break. In fact our expectations of shattering may cause us to overlook other effects. If the rock was replaced by a rubber ball and the crust of ice was replaced by a rubber sheet it will be easier to visualize some of the effects of a collision, which do not involve breakage.

Suppose we have a pond, which is covered by a one-inch thick sheet of rubber. Suppose also that a cannon is loaded with a rubber ball, which is one inch in diameter. When the ball is fired down at the sheet, it is possible that due to the high speed, it will go right through. It is certain, however, that the rubber sheet will be deflected. This deflection will produce a surface effect, which is of particular interest to us. We already understand that if the ball goes right through the sheet, there will be a sub-surface effect. But in this example the surface effect will be more apparent with the rubber than it would have been with the more rigid stone-on-ice example.

As the ball contacts the surface, the surface will start to deflect. As it deflects, a depression is formed where there once was a smooth flat surface. This depression indicates that the material which is under the rubber sheet (i.e. the water) has been displaced. As it was pushed out of the way, it had to go somewhere. In this case we imagine that the rubber sheet is strong enough to prevent a splash but still very flexible. Two main effects are expected from this type of impact. The energy of the ball (which has been forced to slow down by the sheet of rubber) will be transferred to the water in two different ways.

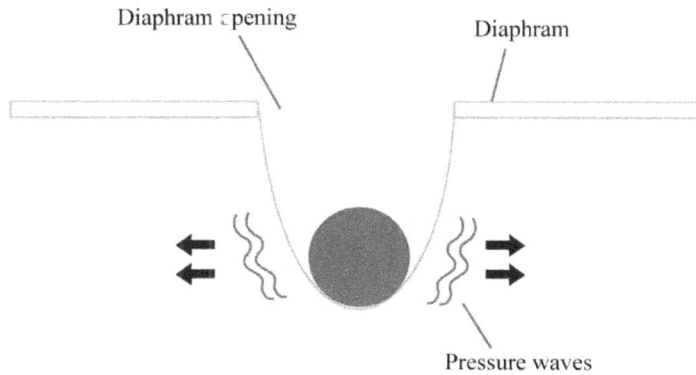

Diaphram opening   Diaphram

Pressure waves

6.6.2 Ball-Diaphram Comparison

First a surface wave would form. The magnitude of this wave would be dependent on the elasticity of the rubber sheet. If it was quite elastic, the wave would be quite high. In this case the wave would be easily formed as well as easily seen and it would be possible to see it propagate away from the impact site in all directions. The rubber sheet would go up and down just like the surface of the pond would have if a stone had been thrown into it. Some, perhaps most, of the energy of the ball would now been transferred to the water in the form of a wave and this wave would transfer this energy away from the impact site across the surface of the pond.

The wave is the most obvious form of energy propagation since it can be seen spreading out across the pond. A wave of this nature is a surface effect because its energy is contained in a region within a close proximity of the surface.

The second way the energy of the ball is transferred to the water develops as the ball travels deep into the water. In so doing it pushes some water ahead of itself. When most of the energy of the ball has been transferred to the water in this way, it will slow right down and almost stop. However in the meantime, a high-pressure region would have been built up in front of the ball and been given a significant push. The ball might stop, but the pressure wave ahead of the ball will not stop. While it is obvious that this pressure wave will travel onwards in the direction which the ball was traveling, it will actually propagate outwards in all directions. Unless it is interrupted by some object, it will continue to travel straight to the bottom of the pond as well as propagate outwards in a spherical pattern throughout the pond.

If the rubber membrane on the surface was very very flexible, most of the energy would be in the surface wave and very little in the pressure wave. If the stiffness of the membrane was increased, more of the incoming energy would appear in the pressure wave because the surface would not be able to deflect to receive any energy and thereby transfer it away. If the rubber membrane was really stiff, practically all of the energy of the ball would be transformed into the pressure wave.

Alternately, if the ball were driven into the pond through a hole in the rubber membrane, all of its energy would be transferred directly to the water in the pond instead of indirectly through the membrane. A surface wave would deflect the membrane from underneath and as above, a pressure wave would propagate throughout the pond. The deflection of the membrane by the surface wave would be similar to the previous case without the hole but would expectedly be of lower amplitude. A greater portion of the incoming energy would appear in the pressure wave and this energy would be dissipated throughout the entire pond. In this case, the membrane would not deflect due to the impact of the ball directly but rather would indicate the location of the traveling surface wave. Similarly, if an asteroid was large enough to punch through the crust of the Earth, the notion that a significant portion of its energy would be transformed into an interior pressure wave becomes more obvious.

The surface wave and the pressure wave are directly connected. The surface wave is the surface manifestation of the interior pressure wave. While the surface wave is travelling horizontally outwards from the impact site, the pressure wave is travelling outwards in all directions with an expanding spherical shape and it would be visible at the surface as the surface wave. This is actually all we can see of the interior pressure wave. If the membrane or crust, which covers the liquid region, was stiff compared to the energy of the pressure wave, a surface wave would not visibly form, but the pressure wave would still travel along the solid-liquid interface. In this case, wherever the surface crust was weak or flexible enough, it would deflect temporarily as the pressure wave passed. If the surface crust was too strong or too stiff, it would not deflect. However it would still be stressed. As a result, as the pressure wave passes any particular location, whether the crust was ice, rock or rubber, there might be some stress-related cracking or displacement. In the case of an asteroid crashing into the crust of the Earth, this effect could result in volcanic activity or vertical displacement of some portions of the crust. Escarpments could form or rift valleys could open. Near the impact site of a large asteroid, mountains could form.

6.6.3   Lead Bullet and Lead Block

Another example of impact effects may be observed when a bullet is fired at a block of lead. While the speed of a bullet is quite high, it is still less than 10% of the expected speed of an asteroid. However, this speed is sufficient to drive a bullet, which is only one-quarter of an inch in diameter, right through a block of lead, which is one inch thick. Therefore, wearing a suit of armor, which was made of lead one inch thick, would not be good protection from a speeding bullet.

When a quarter inch diameter bullet impacts a lead block, it will form an opening in the lead block, which is more than an inch in diameter. The material directly in the path of the bullet would simply have to get out of the way and it would be pushed aside at very high speed. It would be pushed aside with such force that it would not stop moving when it had formed a hole that was big enough to allow the bullet to pass, but would travel out even further forming a comparatively large crater-hole. Most of the displaced material would be blown aside so hard that it would splash away from the impact location altogether and it might even leave a slash-like appearance.

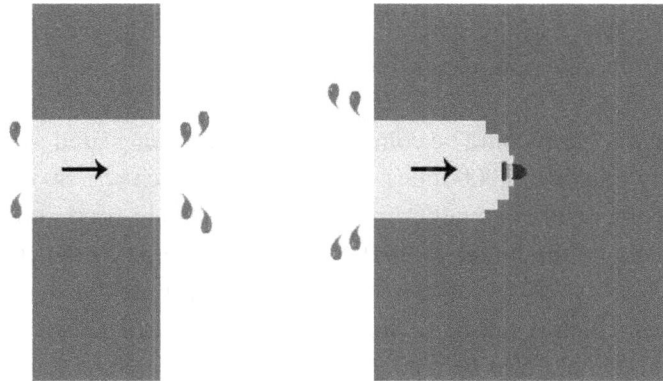

6.6.3 Lead-Lead Comparison

Steel covered lead artillery shells can be designed to go right through the solid steel of a tank. These are called armor-piercing shells and are used to destroy tanks and other steel-protected military ordinance. While these shells are only a few inches in diameter, they are able to blast right through steel which is several inches thick. An asteroid may be moving ten times faster than an armor-piercing shell, so expecting that a large asteroid will be able to punch right through the crust of the Earth is comparatively realistic.

6.6.4   The Battleship HMS Hood

During WWII, the British had a battleship, which was named the HMS Hood. This was one of their most powerful battle stations and when information was received that the Germans had put their newest and largest battleship to sea, the British responded by sending out their best ships including the Hood. The German battleship, the Bismarck, started firing at the Hood when it was still fifteen miles away, which at that time was a great distance to begin a battle. The battle however, was very short. One of the large shells from the Bismarck went right down the smokestack of the Hood and then exploded when it was deep inside the ship. This event was photographed by a person on a nearby ship. There was a brief period of lull and then every porthole on the Hood blew out, indicating that the interior had been blown into complete disarray.

Similarly, if a large asteroid should impact the Earth, it would disappear into the interior, there would be a period of lull and then the effects would be manifest on the surface of the Earth all the way across on the other side of the world.

6.6.5   Bullet and Lead Ball

Please refer to the appendix for the calculations for this section.

The impact of an asteroid with the Earth could be compared to a bullet being fired at a large lead ball. The diameter of the Earth is slightly less than 8,000 miles. If an asteroid, which was 80 miles in diameter, was being simulated, it would be comparable to a bullet from a 0.22 rifle being fired into a ball which was 22 inches in diameter. The diameter of the bullet in this case, is 1% as large as the target and is therefore comparable to an asteroid, which is 1% as large as the Earth, or 80 miles in diameter. An asteroid is expected to be made of material which is similar to the material of the Earth. (37) Since the bullet is made of lead, the ball in this example would also be made of lead. With these comparative sizes in mind, the amount of material which is in the two objects can be compared. Since the diameter of the bullet is 1% of the diameter of the ball, from the formula for the volume of a sphere, the ratio of the quantity of material in the two objects can be determined. The amount of material which is in a sphere is dependent on the radius of the sphere when it is multiplied by itself two times (i.e. a cube law). Therefore a ball with a diameter which is 1% as large as a second ball, will only have one millionth as much material in it. (i.e. $1/100 \times 1/100 \times 1/100 = 1/1,000,000$)

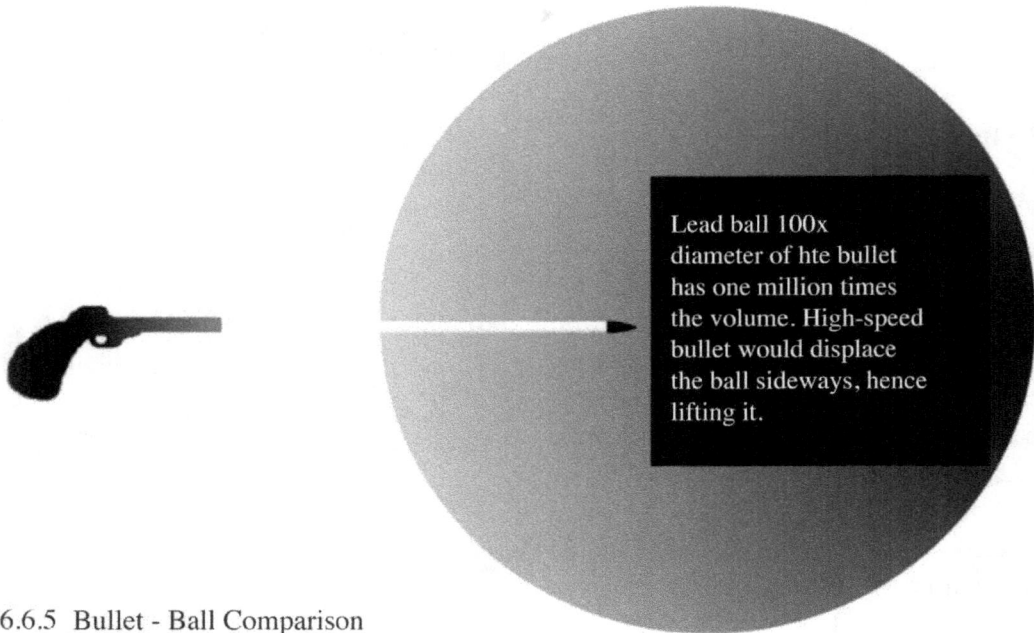

Lead ball 100x diameter of hte bullet has one million times the volume. High-speed bullet would displace the ball sideways, hence lifting it.

6.6.5 Bullet - Ball Comparison

If this large lead ball was suspended by a cable, the bullet could be fired at it and any motion could be observed. Prior to doing this, the motion which is expected could be calculated if the speed of the bullet was known. Of course the speed of a bullet would be known. However, compared to the speed which an asteroid would have, a bullet is hardly moving. Bullets usually move at speeds which are in the 1600 to

2000 feet per second range, which translates to a little more than 1000 miles per hour. Asteroids would be expected to move at least ten times as fast and more likely twenty or thirty times as fast. (19) If a calculation were done using a bullet speed of 10,000 miles per hour, it could be determined that the lead sphere would be pushed several feet sideways by the bullet. It is somewhat surprising that an object as small as a bullet is able to move a large round ball of lead, which is one hundred times as large in diameter. However, the bullet would be moving very fast and some motion of the ball would be expected. For this thought experiment, it will be assumed that the energy of the bullet was completely transferred to the ball. This may require that the bullet to be fired into an opening in the lead ball so when it contacts the ball, all of the material of the bullet will go into the ball and stay there.

If a similar comparison could be made between an asteroid and the Earth, the results should also be comparable. If the asteroid were only 1% as large in diameter as the Earth and only contained one millionth as much material but was travelling at 10,000 miles per hour when it hit the Earth, it would have enough energy to push the Earth right out of its orbit! Dr. Richard Muller in his book "The Nemesis, The Death Star" supported this conclusion when discussing the Chicxulub impactor, which would have been in the range of 0.06% of the Earth's diameter. He concluded that the entire Earth would have recoiled from the impact by several hundred feet. (76)

6.6.6   Bullet and Fluid Lead Ball

A second type of experiment can be visualized where the high-speed bullet is fired into a large lead ball but in this second case the large ball is fluid, encased in a thin solid crust. The Earth is a liquid sphere, which is enclosed by a thin solid crust so this second example would be a closer comparison to an asteroid impacting the Earth. In this case the crust, which would enclose the liquid ball, would be visualized as being only 1/4 of 1% of the diameter of the ball. This compares with the crust of the Earth, which is understood to be only about 20 miles thick (or about 1/4 of 1% of the diameter of the Earth). Therefore in this example, the bullet having a diameter of 0.22 inches would be four times as large as the thickness of the crust of this hypothetical crust-covered fluid ball.

If the high-speed bullet were fired at this crust-enclosed liquid sphere, it would punch right through the thin crust and travel well into the interior. The energy of the bullet would therefore be transferred into the interior of the sphere but instead of pushing the entire ball sideways, the energy would be transformed into a shock-wave, which would propagate outwards from the bullet towards the crust in all directions. The crust would become warped and dislocated. Any areas, which were too weak to stand the stress, would be fractured. Significant surface distortion would be expected. Would the bullet go right through and come out the other side? It might! If the ball were able to contain the bullet, distortion would account for the energy. Similarly, since the Earth has a crust, a large high-speed asteroid would be expected to go into the interior and its energy would be transformed into a shock wave which would distribute this energy throughout the entire Earth. The Earth may not explode, but the entire crust of the Earth would be catastrophically disrupted.

6.6.7   BB and Lead Ball

For a third thought experiment, we will consider a BB (instead of a bullet from a 22 rifle), being fired into a solid lead ball. A BB has approximately half of the diameter of a 0.22 bullet and so its volume would be approximately one-eighth as much as the bullet. As before, the materials of both the BB and the ball will be the same. With this information the ratio of masses of the two objects may be calculated. In this case the BB will be one-eighth of a millionth as heavy as the ball. Also, as before, the BB would be fired at the ball at 10,000 miles per hour. When the calculation for this experiment is done, the results show that the ball would move sideways several inches. Therefore by comparison, the Earth would be expected to move sideways if it were struck by an object, which compared in size as the BB compares to the ball. Since the BB is approximately one-half of one percent of the diameter of the ball, the asteroid would be half of one percent of the diameter of the Earth. In this case the hypothesized asteroid would be about forty miles in diameter which means that an asteroid this large would, by comparison, have enough energy to move the Earth out of its orbit.

6.6.8   BB and Liquid Lead Ball

The BB could also be fired at a liquid ball, which was enclosed in a thin solid crust. As above, assuming that the continental crust of the Earth is about twenty miles thick, the comparative thickness of the crust of our experimental ball would be about one-quarter of one percent of the diameter of the ball. Since the ball is 22 inches in diameter, therefore the crust would be about 0.05 inches thick or about 50 thousands of an inch thick. In this case the crust on the hypothesized liquid ball would be one half as thick as the diameter of the BB. If a BB should impact a crust of this nature at 10,000 miles per hour, it would certainly punch right through into the interior of the ball. Similarly it may be reasoned that if an asteroid, which was forty miles in diameter crashed into the Earth at 10,000 miles per hour, it would go right through a twenty-mile -thick crust into the interior. As the crust of the hypothesized ball would not be expected to slow the BB very much, neither will the crust of the Earth be able to slow a large asteroid very much.

6.6.9   Pin-head and Egg

The impact of a large asteroid crashing into the Earth can also be compared to a pin-head being driven into an egg. The shell of the egg is very thin and brittle and in this regard reasonably comparable to the crust of the Earth. Above a certain speed, which would not be very high, the pin-head would punch through the egg-shell and be absorbed into the interior of the egg. In this manner, the energy of the speeding pin-head would be distributed throughout the entire egg and not just accommodated by the shell. Of course, if speeds similar to asteroid speeds were used, the pin-head would be driven into the egg so fast that it would go right through. At lower speeds however, the energy of the pin-head would be absorbed by and distributed throughout the interior of the egg and it would have an influence on the entire egg shell. It would be very likely that the shell would crack and temporarily expand as the pressure wave welled up within the shell. If expansion occurred, the shell would no longer be able to cover the egg because the expanded structure would have a greater surface area. Cracks would therefore appear on the surface of the egg. A similar development would be expected on the Earth if a great pressure wave formed in the interior. Such an upwelling pressure wave would cause the Earth to temporarily expand. In such a case the present crust of the Earth would not be able to cover this expanded Earth, so the crust would fault or crack. Then, the area produced by the cracks plus the area of the original surface would account for the larger sphere.

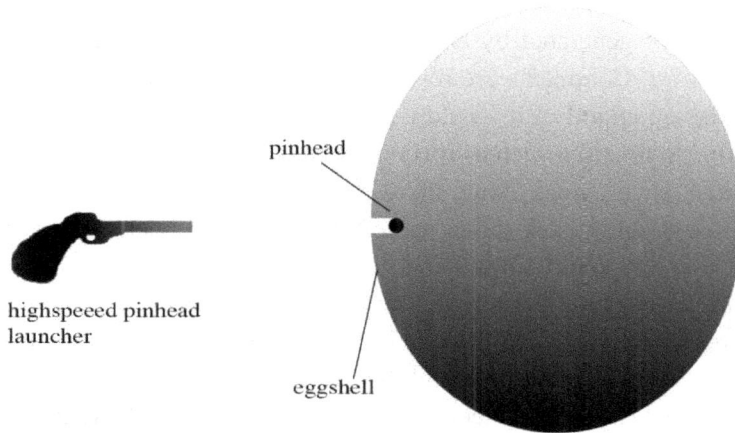

pinhead

highspeeed pinhead launcher

eggshell

6.6.9 Pinhead-Egg Comparison

The pinhead, at high speed, would break through the shell of the egg and continue into its interior.

A further comparison could be made by considering a pin-head sized ball made from eggshell material. In this case the projectile would have the same density and hardness of the eggshell. If it were similar in thickness to the eggshell, it may be visualized that if it were driven into the egg at high speed, it would punch right through and penetrate into the interior. Even though such a small object would not actually contact the entire inner surface of the shell, the entire inner surface would be affected by the miniature shock-wave, which would expand throughout the entire interior of the egg. Similarly, an asteroid impacting the Earth would generate a shock-wave in the interior of the Earth which would expand until it impacted the entire inner surface of the crust.

6.6.10   Hammer and Punch

The impact of an asteroid with the Earth can be compared to a giant hammer driving a punch into the Earth. Suppose there was a hammer 40 miles in diameter and 135 miles long and suppose that it was moving at 1000 miles per hour. Would it be capable of driving a punch, which was 20 miles in diameter, right through the crust of the Earth? Since we understand that the crust is between 3 and 20 miles thick, (31) it would be expected that this hypothesized hammer would be able to drive the punch right through the crust and into the interior. The energy of an asteroid, which was 8 miles in diameter and moving at

25,000 miles per hour, would be comparatively similar to the energy of such a hypothetical hammer. (See appendix for this calculation.) It might reasonably be concluded therefore, that an asteroid this large would punch right through the crust of the Earth and travel well into the interior.

6.6.11   Ice and Water

As the great pressure wave generated by an impacting asteroid, comes up under a continent, it would stress the continent upward. Continents are reasonably stiff and the first thing, which we expect, is that the continent would rise up slightly. It would rise up as the pressure wave comes up and then it would settle back down again. A situation such as this could be compared to an iceberg, which was floating on an ocean. When a large wave comes along, the iceberg would float up on the wave. Then as the wave passed, the iceberg would settle back down again. However if the same wave approached and passed under a sheet of ice, this ice would not respond the same way. It would rise up with the wave but the entire ice sheet would not respond as a single block of ice. The sheet of ice would fracture as the wave moved under it. The reason that it would fracture is because the rising wave would cause bending stress in the sheet of ice. The ice would bend a little and then it would break and float on the changing surface of the water in smaller pieces. If the thin sheet of ice were incredibly stiff, when the wave worked its way under the near edge, the entire ice sheet would be picked up. Then, as the wave traveled along under the ice, the near edge would rise higher and higher until the wave reached the middle of the sheet. At this location the ice would balance on the wave and then topple over the other way as the wave passed the center region. Finally, when the wave left the area, the ice sheet would settle back down. Iceberg behavior is something like this. It behaves as a unit because it is very stiff. It will not bend or break but simply rise up with any passing wave and then settle down again.

A pressure wave, which was welling up under the crust of the Earth, would be similar to an ocean wave, which was welling up under pack ice. (Pack ice is an area of floating chunks of ice packed tightly together – usually found in the high arctic regions.) As long as the ocean remains calm, ice will form in a continuous sheet. Then, it would stay in one piece until a force was exerted on it, which was too much for the ice to withstand. The ice would then crack and form smaller pieces. If a wave traveled by, these smaller pieces would bob up and down and more or less follow the movement of the wave.

A further comparison with pack ice recognizes that when pack ice is compressed due to wind, it may fail in compression and crumble into pressure ridges. Depending on its specific strength, the same thing could happen to a slab of rock that was subject to compressive stress.

When an enormous asteroid-generated pressure wave is travelling back and forth under the crust of the Earth, the crust might be able to ride the wave but if the stress was too great the crust would crack and smaller sections of crust would be formed. The smaller pieces would be comparatively much stiffer (like the iceberg), and would be able to accommodate the pressure wave simply by riding it.

If an incoming asteroid had enough energy to actually displace the Earth from its orbit but if the Earth didn't displace and all of the energy of the asteroid was absorbed instead, the entire Earth would roil and contort and there would have been disastrous movement throughout the entire planet. When the energy of the asteroid had been transferred to and distributed throughout the interior, it would show up at the surface with the result that there will be major disruptions of the crust around the entire world just like there would be if a giant ocean wave rolled into a sea of pack ice.

6.6.12 Loaded Beam

Oceanic crust is much thinner than continental crust. An assembly which includes a region of oceanic crust between two areas of continental crust appears structurally as a thin region in the middle of two thicker regions. These two thicker regions would respond like the iceberg, which was too thick and strong to do anything more than simply rise up under the stress of the upwelling pressure wave. The thinner oceanic crust in between the two continents would be more responsive to the increasing pressure underneath but because the oceanic crust is attached to the continents on either side, it would respond as part of an assembled system. As this thinner oceanic crust becomes pressurized from below, it can be compared to the situation where a plank, beam or plate, which is supported at both ends, becomes loaded throughout its entire length.

When engineers are instructed on how a beam or plate will behave when it is loaded, a diagram is drawn of the beam showing it being supported at both ends. Then the load is shown being piled onto the beam. Then some calculations are done and they will show that the maximum deflection of the beam will occur at the middle. This is not a big surprise because common sense and every day experience say the same thing. The deflection will always be the greatest at the middle because the total effect of the load causes the most stress at the middle. This type of stress is called bending stress. As more and more load is placed on a beam, it will bend more and more. When the load becomes too great for the beam to carry, it will break. When it breaks it will break at the middle.

Similarly, if the floor of the Atlantic Ocean became structurally loaded by a massive pressure wave from underneath, it would be bent upwards. As soon as the loading exceeded the stress-carrying ability of the ocean floor, it would fault or crack and like a loaded beam, it would fault near the middle.

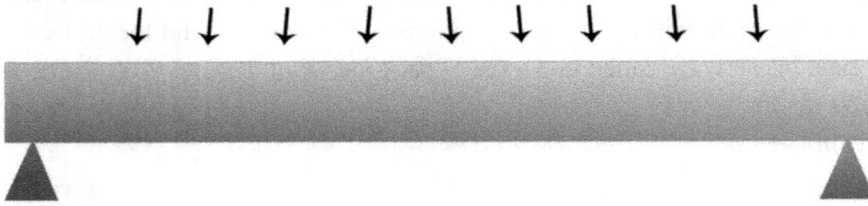

6.6.12 (1) Beam Comparison

The beam is supported at both ends and loaded along its entire length. When it bends,
it will bend mostly at the middle. When it breaks, it will break at the middle.

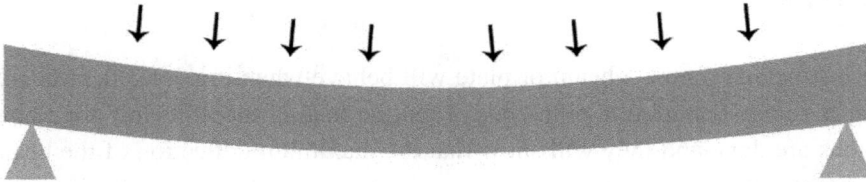

6.6.12 (2) Curved Beam or Plate

Either a beam or a plate could be curved before the load is applied.
The load would produce the most deflection at the middle as in the
straight beam case. If the curved beam is loaded to the breaking
point, it will also break in the middle.

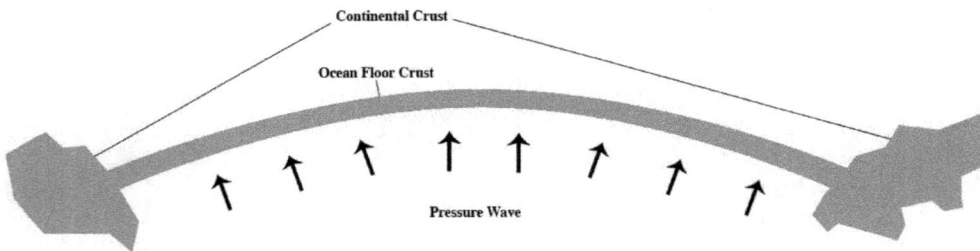

6.6.12 (3) Ocean Floor Pressure Loading

If the ocean floor were loaded by a pressure wave from the interior of the Earth,
it would be similar to placing a load on a curved plate. If the ocean floor cracked under
such a load, it would likely crack near the middle.

223

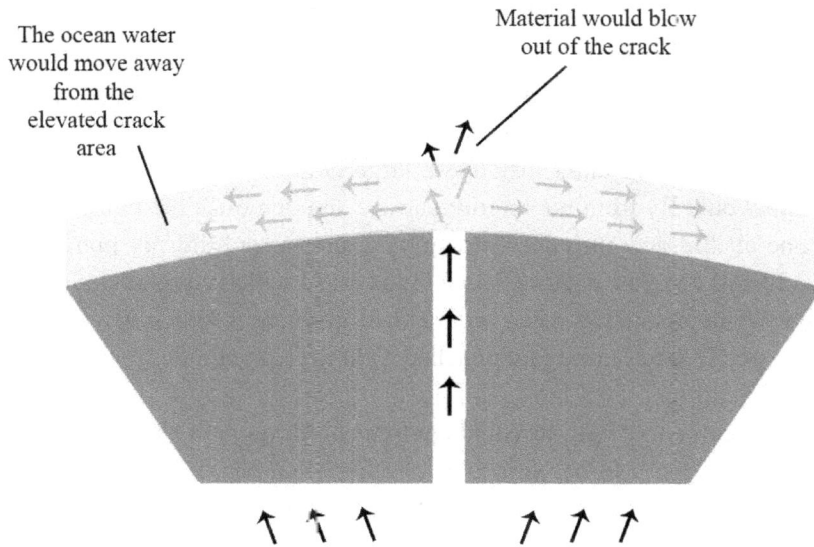

6.6.12(4) Ocean Floor Failure

6.6.13   Impact Comparisons Summary

All of these comparisons illustrate that the entire crust of the Earth would become stressed and distorted if an asteroid impacted the crust and dove deep into the interior. They also illustrate that large asteroids would be expected to punch right through the crust and expend most of their energy far from the actual impact site.

6.7   Impact Transmission

If a large asteroid should impact the Earth and go right through the crust into the molten interior regions, it would be expected to come to a halt someplace deep in the interior of the Earth. As it crashed into the interior, a shock-wave or pressure wave would be generated in the magma. (Soon the asteroid or its remains would reach the temperature of its surroundings, become molten and blend in with the other material inside the Earth.)

In   fact   a   shock-wave   would   be

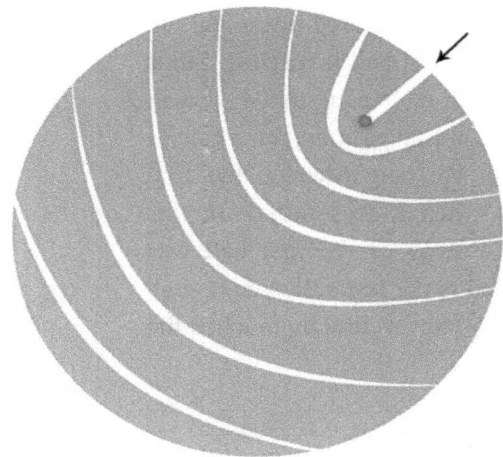

6.7 Asteroid Shock Wave

If an asteroid punched through the crust of the Earth, it would generate a shockwave which would travel throughout and cause enormous damage as it resurfaced on the opposite side. In this manner, the overwhelming energy of the asteroid would be distributed throughout the entire planet.

generated in the magma, even if the asteroid did not completely penetrate the crust and this type of situation can be compared to army weaponry. The objective of an anti-tank weapon is to penetrate the armor of the attacking tank and thereby take it out of service. This might be done by firing a projectile at the tank. If the projectile did not go all the way through the armor but instead stopped part-way through, the tank crew may still not be safe. A shock-wave would be generated in the armor, which would continue right through the armor and might actually dislodge a large chip of it (at considerable speed) from the interior surface. This chip would fly from the interior surface and demolish the interior as effectively as if the actual shell had gone all the way through. Similarly, if an asteroid did not punch right through the crust of the Earth, it could still generate a shock-wave in the crust, which could dislodge a large portion of crust from the underside. This dislodged piece would then generate a shock-wave in the interior very similar to what would happen if the asteroid had punched right through and did it directly.

This shock-wave would not come to a halt. It would continue on through the high-density interior right to the other side. A situation like this can be compared to other situations where the event-causing object comes to a halt, but the effect which it generates keeps right on going.

6.7.1 Suspended Balls Experiment

In the suspended-balls experiment, several solid balls (usually four or five) are suspended by strings so that they are lined up in a row touching each other. Then the ball at one end is lifted up and dropped so that it falls into the second ball. The second ball does not appear to move. Neither does the third ball. In fact, none of the balls in the line will appear to move but if they do move it will not be very much. But almost instantly, the last ball in the line shoots out almost as high as the first ball was raised. By dropping the first ball, the last ball has been moved. This is explained by a shock-wave being transmitted from ball to ball until the last ball is reached. Since the last ball is free to move, it will shoot out and up to a height which is similar to the height which the first ball had when it was raised. In this case the energy of the first ball was transmitted to the last ball even though the first ball never directly touched the last ball.

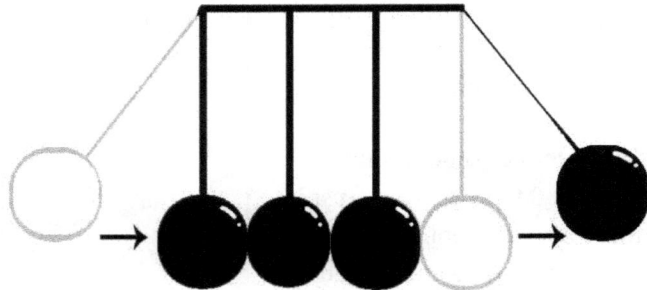

6.7.1 Suspended Balls Experiment

6.7.2 Billiard Balls

Billiard balls move on a flat table. The game is started by driving the cue ball into a group of balls, which are arranged at the far end of the table. When this is done, the cue ball will

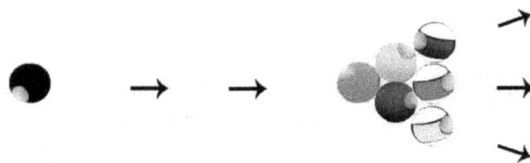

6.7.2 Billiard Balls

impact the first ball of the group and then slow down and stop moving. However the group of balls will spread all over the table. The cue ball will only have touched the first ball but the balls at the far end of the group will have moved away anyway and they will probably have moved away very quickly. In this case the energy of the cue ball has been transmitted to all of the other balls, including those, which were farthest away and which it did not directly hit. As with the suspended balls, the movement of the billiard balls is explained by a shock-wave. The shock-wave is generated by the cue ball and it travels from ball to ball. If any ball is free to move it will move. If it is hemmed in and cannot move, it transmits the shock-wave to the next ball, which will either move or continue to transmit the shock-wave until all of the balls have been involved. This type of situation is also comparable to an asteroid impacting the Earth. While the asteroid itself will not keep right on traveling, its energy will be transmitted, via a shock wave, into the interior of the Earth and have an effect far away from where the asteroid came to rest. When an asteroid crashes into the Earth, it will come to a stop, but the effect of its impact will continue much further. The billiard balls are all of similar density. Some portion of the impact energy of the cue ball will be transmitted to every ball in the group. What if the cue ball was travelling at ten thousand miles per hour? How many other billiard balls could then be affected?

6.7.3   Pile Drivers

A pile driver is a large machine, which drives steel pipes or steel beams into the ground. While the beam is held in place, the pile driver hammers on the upper end with the expectation that the bottom end will go into the ground. Now it seems pretty obvious that this should work but why should it work? We understand that, when one end of the beam is struck, the other end will move. When the beam is struck by the hammer of the pile driver, a high-speed shock-wave is generated and travels down the beam. It travels from molecule to molecule and from atom to atom all the way down the beam. The hammer stays behind. The beam is so 'together' that when it is struck at one end, the whole beam moves. This situation is comparable in some ways to asteroid impact because the interior of the Earth really is very dense - at least as dense as a steel pile. (36) In the pile driver case, the object, which caused the movement, will stop moving, but the results of the impact will be manifest at the other end of the pile and it will be driven into the ground.

6.7.3 Pile Driver

The pile driver impacts the top of the pile but the bottom end goes into the ground. The pile driver never actually touches the ground.

6.7.4 Hammer and Nail

When a hammer is used to drive a nail into wood, the wood is forced to yield to the penetrating nail. The hammer does not touch the wood. When the hammer hits the nail, the energy of the hammer is transmitted down the nail and disrupts the wood. It is completely possible for a hammer to drive a nail into a piece of wood for several inches without ever touching the wood at all. This would be similar to an asteroid driving deep into the molten interior of the Earth. Material, which was far ahead of the asteroid, would be affected, but the asteroid, would not necessarily have actually touched any of this remotely-affected material.

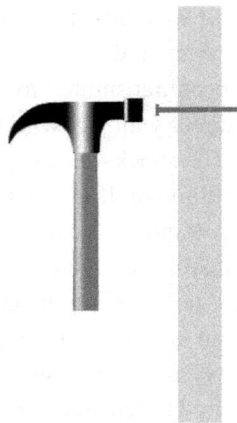

The hammer impacts the nail, making the nail penetrate the board's surface and exiting through the other side.
The hammer does not touch the board.

6.7.4 Hammer-Nail Comparison

6.7.5 Rock and Mud

When a rock is thrown into a pool of mud, it might bury itself depending on its speed and the density of the mud. If, instead, the rock was loaded into a high speed cannon, and then fired into the mud, how far would it go? Would it go five diameters into the mud? Would it go twenty diameters into the mud? The energy, which is associated with the high speed, would enable the rock to penetrate into the mud much further. This can be compared to an impacting asteroid because the densities of asteroids and the Earth are similar. (37) The material of the Earth is basically just hotter, less viscous and therefore easier to move if forced. If the incoming object were moving ten times as fast as a bullet, it would certainly be necessary for something to move. After it had finally stopped, the effect of its impact would be manifest far from the impact location. Even though the diameter of the Earth is almost eight thousand miles, this does not seem to be very far if a large asteroid which was, for example, eighty miles in diameter traveled ten or twenty diameters into the Earth before stopping. Then after it stopped, it would reasonably be expected that the effect of the impact would be manifest much deeper in the Earth and much further away.

6.7.6 The Planet Mars Case

If a large asteroid should impact the Earth, it would crash right through the crust and send a shock-wave through the interior. A similar situation appears to have happened on Mars. The planet Mars is smaller than the Earth (about 11% as massive as the Earth (38)) and it comes to within approximately fifty million miles of the Earth at its closest approach. While it always looked distorted from Earth-bound telescopes, it has been viewed many times by the Hubble Space Telescope as well as by small exploration satellites, which were sent to it. Many photographs have been taken. There are three very large impact sites on Mars. The largest has a diameter of about 1400 miles. The second largest has a diameter of more than

1000 miles and the smallest of the three has a diameter of almost 700 hundred miles. (39) These are very large craters. It is never possible to say how large the impact objects were for a lot of reasons but suggesting that they may have had diameters which were at least five percent of the respective size of the craters would be a very reasonable starting point. If this were the case, the largest of the three impact sites could have been caused by an asteroid which was fifty miles in diameter. While this is a very large asteroid, in the asteroid belt there are still more than 200 asteroids with diameters greater than 100 kilometers (62.5 miles). (69) While the size of these three craters is of interest, it is even more interesting that all of them may be observed from one side of the planet. Then, almost directly opposite the largest one, Hellas, on the far side of the planet, there are four huge volcanoes, a 6 1/2 mile high bulge and a very large and extensive trench formation. The largest volcano, Olympus Mons, rises 37 kilometers (121,360 feet, Mt Everest is 29,028 feet) above the surrounding plain and it has a diameter of 500 km. (312 miles). The opening at the top is more than 40 miles across. The trenches vary between 50 and 75 miles across and between 3 and 4 miles deep. They extend for hundreds of miles and are adjoined by additional trenches and chaotic terrain.

Nothing happens without cause so something caused these features to form. Since their formation was not witnessed, nobody can say for sure, but it is more than a little curious that these spectacular features are at the antipode (an antipode is the region directly opposite an impact site) of a region within a circle defined by the three large craters; Hellas, Argyle and Isidis. A reasonable explanation would be that these three impacting objects generated shock-waves, which traveled right through the planet to the surface on the other side (40) where they caused disruption. The volcanoes could have formed because the shock-wave forced molten material from the interior to come to the surface. The high pressure of the shock wave would have been relieved by the disruption. Since Mars is not very big, what would have happened if these impacting objects had been a little larger? Would they have punched right through Mars or would they have totally dislodged surface material to form an antipode crater?

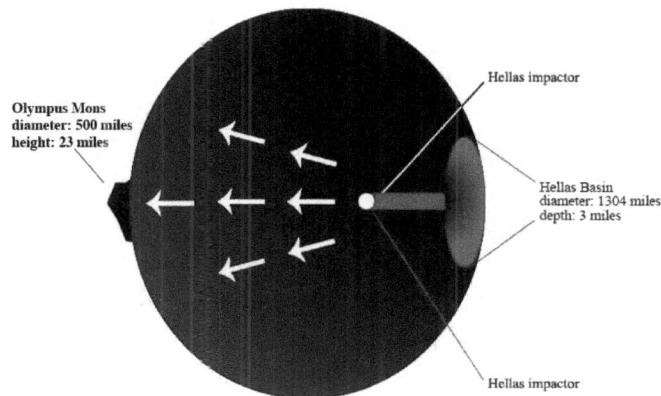

6.7.6 Planet Mars: Hellas Impact Diagram

Olympus Mons is almost directly opposite the giant Hellas Crater. Shock waves from the Hellas impactor could explain Olympus Mons, the nearby volcanoes and chaotic local terrain.

Mars is approximately 4200 miles in diameter. The largest crater might have been formed by an object which was 50 miles in diameter. The diameter of the planet would therefore only be 84 times as large as this object. Further, it could have traveled several diameters into the planet. It is most reasonable then to suggest that a shock-wave went all the way through to the other side of the planet. Whether the interior was molten or solid would not have hindered the shock-wave, which would have reached the other side in just a few minutes. The chaotic features on one side of Mars are most readily explained by shock waves originating from the three massive craters on the opposite side. This would make the chaotic features in their entirety the antipode of the three asteroids collectively.

6.7.7 The Moon

The near side of the Moon has several large flat areas called Maria (Latin for seas). Each of these areas is accompanied by a Mascon (mass concentration) deep below the surface. The situation is suggestive of large asteroids, having greater density than the material of the Moon, impacting the surface and causing molten material to ooze out and spread around. Shockwaves would have been generated by these impacts and would, without doubt, have propagated all the way through to the other side. If this were indeed the case, evidence of chaos should be present on the far side and it is present. The far side of the Moon has both a bulge and broken chaotic terrain called highlands. (please refer to National Geographic, July 2013) As the shockwaves from the impacts came up under the far side the surface was disturbed. The ground shook and was dislocated both horizontally and vertically. The highlands are the result. (see also 9.4)

6.7.8 Mercury

There is a very large impact site on Mercury called the Caloris Basin. Around on the other side there is an area referred to 'weird terrain' where the surface is all broken up. Since this area occurs at the antipode of the impact it is strongly suggestive that the two features are directly connected. The shockwave from the impact would have been the connection means and judging from the size of the impact it is no wonder that the antipode area is so broken up. (see also 9.6)

6.8 Interior Shock Wave Effects

As has been discussed, if a large asteroid punched through the crust of the Earth, a major shock-wave would be generated in the interior and it would travel right through the Earth to the other side. There are several possible effects from such a shock wave and they depend on the features and characteristics of the crust where it comes back to the surface. If the crust was very strong, it would only suffer uplift. If it was relatively thin, the pressure would exceed its stress limit and it would therefore fracture or fault.

6.8.1 Impact Continent Effects

When a large (e.g. one km. or more across) asteroid lands on a continent, the surrounding region would be involved in a different set of consequences than would a continent on the far side of the Earth. This is because the surrounding regions would experience the rapidly spreading pressure wave, whereas the remote continent would be affected by a pressure wave, which would be rushing up through the interior from the far side of the world and which would act on large areas simultaneously.

The impact continent is the continent, which experiences the direct impact. Included in the devastation-producing list of results would be; a blinding flash, atmospheric shock waves, the crater-generating mud/rock splash, the horizontal displacement of continental crust and an earthquake, all of which were discussed in the earlier chapter entitled; Chaos and Bad News. In addition to these effects, the impact continent would experience a high-pressure material shock wave which would spread out from the impact site in all directions. It would also spread horizontally through the crust from the impact location while it was simultaneously propagating outward through the magma under the crust. Wherever the crust was weak enough to allow it, a surface wave would form in the crust and partially relieve this extremely high pressure. The arrival of the pressure wave at any particular location would be the beginning of the earthquake at that location. The pressure wave would force the crust material to displace outward away from the impact site. Then, as it passes, the crust material would rebound back toward the impact site. It will continue to bounce back and forth for some extended period of time. This bouncing or shaking back and forth would be the earthquake. At the same time under the crust, the expanding shock wave would be concentrating the molten magma into a propagating high-pressure impulse. When the wave passed on, the pressure in the magma would drop back down. This under-crust change in pressure would work synergistically with the vibrating crust to cause extensive disruption and displacement throughout the crust around the entire world. Some regions would be particularly resistant to any displacement but some areas would be displaced. Displacement could be up or down or horizontal. Since the asteroid would be several miles across, the horizontal and vertical displacement near ground zero would also be several miles. Mountains would be formed. Further away, where all of the displacements and pressures are reduced, escarpments and rift valleys would form and portions of continents could rise or sink.

An example of submergence of land is provided by Velikovsky. 'Human artifacts and bones of land animals were dredged from the bottom of the North Sea … stumps of trees with their roots still in the ground were found … the area covered by the North sea was a place of human habitation … .resulted in a rather sudden sinking of this land.' (44) This opinion was shared by others trying to understand the confusing evidence. '… the cause of the catastrophe was the sudden sinking of the land and its subsequent elevation, which was sudden, and during which water from the heights broke upon lower levels, bringing chaos and destruction.' (48) Also, across portions of the North Atlantic, ocean floor material is similar to continental material instead of the expected basalts of most ocean floors. (45) Were these portions of the Earth's crust once much higher up above sea level? Why are they now so far below sea level?

Uplifted areas would include escarpments and plateaus. There could be areas where nothing but shaking occurred and then further away where the original crust was either weaker or thinner or where the pressure waves converged, there would be more vertical and horizontal displacement. Vertical displacement could include airborne boulders, as well as crustal faults. If the crust of the Earth became faulted, while the pressure in the molten magma was under-going great variation, volcanoes would form. All of these effects would happen on the impact continent.

6.8.2   The Antipodes

Just as a tsunami can travel for thousands of miles across an ocean and then cause devastation when it comes ashore, an asteroid-initiated shock-wave could travel right through the interior of the Earth and cause devastation when it surged up underneath the crust in far-away places. In some ways the Earth is a very large place. The diameter of the Earth is slightly less than 8,000 miles. If an asteroid were 80 miles in diameter, and it punched through the crust of the Earth, it would retain most of its speed and therefore most of its incoming energy. As it plowed into the molten interior, it would continue moving for at least several diameters of itself. It would be just too big to stop suddenly. It might not stop until it had traveled ten or more diameters into the molten interior, in which case, the asteroid would have moved into the interior of the Earth until it was ten percent of the way through. Someplace in the interior, it would eventually come to a stop. However the shock-wave, which it generated as it plowed in, would just keep right on moving. Since the asteroid would have been travelling at very high speed, the ongoing shock-wave would also be travelling at a high speed. The shock-wave would quickly travel right through the Earth and would be expected to reach the crust directly opposite (i.e. the antipode) in less than one hour. (47) Under these conditions, the Earth does not seem to be very large.

The speed of an asteroid is far beyond everyday experience and so is difficult to comprehend. In some ways, the situation would be similar to a tsunami travelling across an ocean. A tsunami can travel at five hundred miles per hour which is about one hundred times faster than the much-more-common wind-waves and it could easily cross the Pacific Ocean within a few hours. (50)

Initially, an asteroid-induced shock-wave is similar to a tsunami because it would propagate outwards from the impact site underneath the crust. As it got further from the impact site, the surface-referenced propagation pattern would gradually become more and more of an upwelling pattern. By the time the surface shock-wave had traveled about halfway around the Earth, the angle of impact which the interior shock wave would have with the underside of the Earth's crust would be about 45 degrees. While it is not yet coming straight up, it would now be moving vertically as much as it is moving horizontally under the surface. Therefore it would have both a horizontal effect and a vertical effect. These effects would be in the direction of both lifting the crust and pushing it away from the impact site. In this area the great shock-wave would be travelling along under the surface and stressing the crust both upward and forward at the same time. As this horizontally-travelling and vertically-pushing shock-wave moved along, the crust would be highly stressed.

Impact Site	Country	Location Co-ord.	Diameter	Antipode Co-ord.	General Location
Acraman	Australia	s 32 e 135	90 km.	n 32 w 55	N Atlantic
Chesapeake Bay	USA	n 37 w 76	90 km.	s 37 e 104	W Australia
Chicxulub	Mexico	n 21 w 89	170 km.	s 21 e 91	Indian Ocean
Congo	Africa	s 5 e 25	1300 km.	n 5 w 155	central Pacific
Czechoslovakia	Czech.	n 49 e 12	320 km.	s 49 w 168	New Zealand
Manicouagan	Canada	n 51 w 68	100 km.	s 51 e 112	S. Pacific Ocean
Popigai	Russia	n 71 e 111	100 km.	s 71 w 69	S. Shetland Is.
Sudbury	Canada	n 46 w 81	250 km.	s 46 e 99	S. Indian Ocean
Taklimakan	China	n 40 e 85	700 km.	s 40 w 95	S. Pacific Ocean
Tanitz Basin	Kazakhstan	n 52 e 68	350 km.	s 52 w 112	S. Pacific Ocean
Vredefort	Africa	s 27 e 27	300 km.	n 27 w 153	Hawaiian Islands

TABLE - Earth Antipode Table

Even continental crust is not very thick when compared to the forces, which are involved with an asteroid-induced shock-wave. If the crust was sufficiently strong, it would only tend to rise up similar to the tidal effect. If it were not strong enough to retain its shape as it became stressed, it would bend upward. If the stress was simply too great for the crust to withstand, it would crack. Since the stress level would be so high, cracking would be expected. Cracks would tend to open across the direction of travel. However, if the crust cracked because it was stressed, it could also displace. It could displace both vertically and horizontally. The displacement might not be uniform. If some regions displaced upward more than others, for example, plateaus and escarpments would be formed. If they displaced horizontally some regions would rise or sink and valleys or great rift regions would result. All of this activity would develop at the antipode of an impacting asteroid.

6.8.3   Crust Deflection

Unlike the gravity pull of the moon, which acts on a particular location for several hours, the moving pressure wave would only act on a particular location for a few minutes. However the deflection effect may be much more widespread than the region of high pressure. This type of situation may be compared to a transport truck, which is moving over a frozen lake.

In the far north, during the winter season, trucks, which are fully loaded, transport food and equipment over frozen lakes and rivers to settlements in the remote northern wilderness. These trucks move along prepared pathways, which are referred to as winter roads. Actually they are not really roads at all but simply pathways, which have been cleared across rivers and lakes to allow the trucks to travel. Travel does not commence in the fall until the ice is thick enough to support these heavily loaded trucks. Even then, though the ice might be strong enough to carry the trucks, it would not be nearly as stiff as a normal road. Wherever the truck is located, the ice would be depressed. Therefore, even though it might not break, (hopefully) it would be bent. The truck would actually be sitting in a depression. As it moves, the depression would move. If it moved quickly, the depression would move quickly. If it stopped the depression would stop and the truck would always be sitting right in the middle of it. The net result of this truck-ice-water system is that a travelling wave would be generated by the moving truck and this wave would move right across any body of water that the truck was moving across. The most obvious part of this wave would be the depressed part at the truck location. As the distance from the truck increases, the depression would be reduced. Wherever the truck sits still in one place, there would be a slight incline in all directions. As the truck started to move, it would start to climb this incline. However, it would never succeed because as it moved the depression would be continually regenerated and relocated, always with the truck at the center. The net result is that as the truck moves across a lake, a traveling wave would be generated in the ice. When the truck was gone, the shape of the ice surface would return to normal but not immediately. As the ice moved down with the approaching truck and then back up as the truck passed by, the ice over a much greater area than the truck would also have moved up and down in a wide-spread deflection pattern, which would not settle down for a considerable time after the truck had passed. The deflection, for example, one mile from the truck, would hardly be noticeable but it would be measurable and the ice, even this far from the pathway of the truck would continue to deflect after the truck had gone. Ongoing movement would occur because there is a relatively thin and flexible crust floating on a fluid body. As the truck moved, it transferred energy to the system of ice and water. Even an hour after the truck had passed, movement might still be detected.

A similar effect would develop when the massive pressure wave generated by an asteroid rolled along under continental crust. Both of these systems (i.e. the truck-water-ice system and the crust-magma system) have a thin crust floating on a fluid body. As the relatively thin ice is floating on water, the relatively thin crust is floating on molten magma. Therefore any motion which was set up in these crust-fluid systems would continue until internal damping mechanisms have dissipated the energy.

While the truck itself is not very big, the deflection pattern in the ice would extend outward for several hundred feet. Similarly, the asteroid-generated pressure wave which would be moving along under the crust of the Earth, might only be one mile in extent, but the deflection which it sets up could extend outwards for many miles - perhaps even for hundreds of miles.

When the surface effect of the internal pressure wave had traveled halfway around the Earth it, because it would be propagating through the Earth with a spreading spherical pattern, would have almost reached the far side as well. The direct distance from the impact site to a surface location which was halfway around the Earth is about three-quarters of the direct distance to the antipode at the far side. This half-way region would experience an upwelling, rolling pressure wave which would have an intense high pressure region, which was not very wide. The far side, (i.e. the antipode) however, would experience a massive pressure wave, which would well straight up from underneath and which would be spread out over a large area. This type of pressure wave would be available to lift an entire continent and an entire continent might indeed lift.

The primary difference between a continental crust and an oceanic crust is that the oceanic crust is much thinner and therefore much weaker. It is understood that oceanic crust is probably only about one-fifth as thick as continental crust. (22) If two different materials are of similar strength, but one is only one-fifth as thick as the other, the thinner one will only be one-fifth times one-fifth times one-fifth as stiff as the other. This means that it will be 125 times easier for it to bend or break. (i.e. 1/5 x 1/5 x 1/5)

The crust of the Earth is called the crust because it is so thin compared to the diameter of the Earth. Even if the crust were eighty miles thick, this thickness would only be 1% of the total distance from one side of the Earth to the other. The oceanic crust is far from being eighty miles thick, however, and might be only three miles thick. This is much thinner than the continental crust and because it is so much thinner, it would break much easier when similar stress was applied.

6.8.3 Truck-Ice-Water System

Ice deflection is greatest where the truck is located. As the truck moves, the region of maximum deflection moves with it. This creates a wave-like motion in the ice. Since the ice is floating on water, the ice will continue to deflect up and down after the truck has passed. Because the truck causes the ice to deflect, the truck is always going uphill.

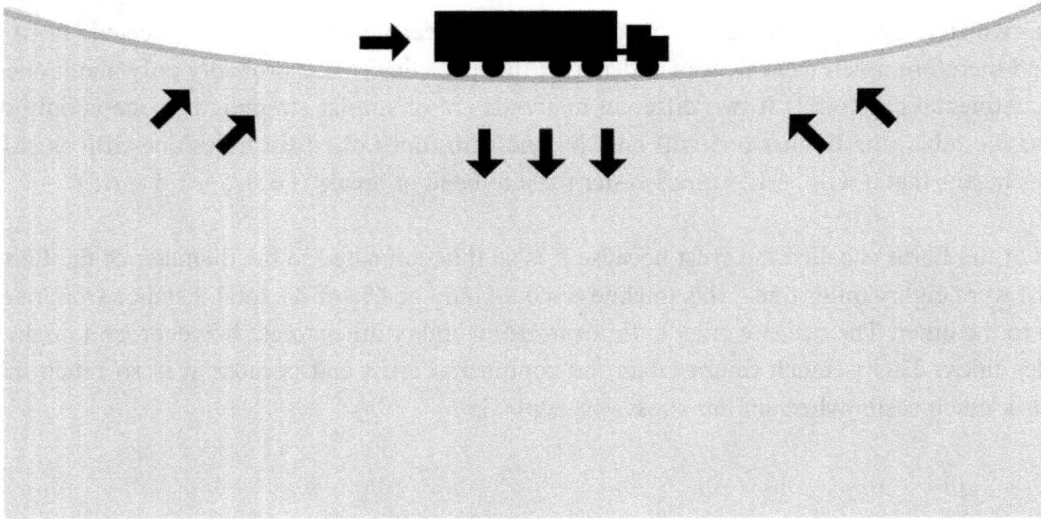

6.8.4   Sheet Flow Generation

If the oceanic crust became pressurized from underneath, it would mound up. The water above it would also mound up and then accelerate down the slopes on both sides. The entire depth of the ocean could become involved in a great flowing mass - sheet flow, which would be down the sides of the mound toward the continents. As this water moved horizontally, it would develop a wave-front because it would encounter water beyond the mound which would not have been accelerated. Behind the wave-front there would be a very large area of flowing water which would slide along on top of the stationary water further down. Water above the interface would be moving horizontally and water below it would be almost stationary. This could explain why certain seamounts (called guyots) have flat tops and appear to have had their tops sheared off. (51)

Marine volcanic islands which have been truncated by the waves... are called guyots. Guyots are flat-topped seamounts … which … were sheared off at sea level, possibly by rapid currents … There are thousands of guyots on the ocean floor, especially in the western Pacific Ocean.' (51) The tops of the guyots are between 600 and 2000 m below present sea level. Some of this variation could be due to subsistence and some could be due to the variation in the time of their formation with respect to variation in flow patterns. Recalling the Earth Antipode Table from section 6.8.2 above, it is clear that the Pacific Ocean is home to numerous antipodes, all of which would have caused upwelling of the ocean floor, massive sheet flow of ocean water and subsistence of the ocean floor in various places.

The upper layer of ocean water flowed horizontally and removed the unconsolidated upper material of these under-water formations. As concluded by H. Pettersson, former director of the Oceanographic Institute at Goteborg, Sweden; 'Tectonic catastrophes raised or lowered the ocean bottom hundreds and even thousands of feet, spreading huge 'tidal' waves which destroyed plant and animal life on the coastal plains.' (49) In agreement Velikovsky concludes; 'The ocean floor all around the globe bears witness that the oceans of the Earth were the scenes of repeated violent catastrophes….and tidal waves raced against continents.' (49)

This type of activity was very similar to present-day tidal bores. A tidal bore occurs when the incoming tide is moving so fast that it forms a wave-front on top of the stationary water of an estuary. Waves of this nature may be several feet high and extend all the way across the estuary. This phenomenon results because the water that is already in the estuary doesn't move back fast enough and the inrushing tide simply piles up and flows along on top. Eventually the entire depth of water may become involved in the flow but by then the speeding wave front will have traveled several miles up the estuary from the sea. An example of this type of water movement is provided by the tidal bore on the Qiantang River in China. During most of the month the bore advances up the river in a wave 8 to 11 feet high, moving at a speed of 12 to 13 knots with a leading edge which is a sloping cascade of bubbling foam. however during spring tides the advancing wave is 25 feet above the surface of the river. (79)

6.8.4 Sheet Flow Generation

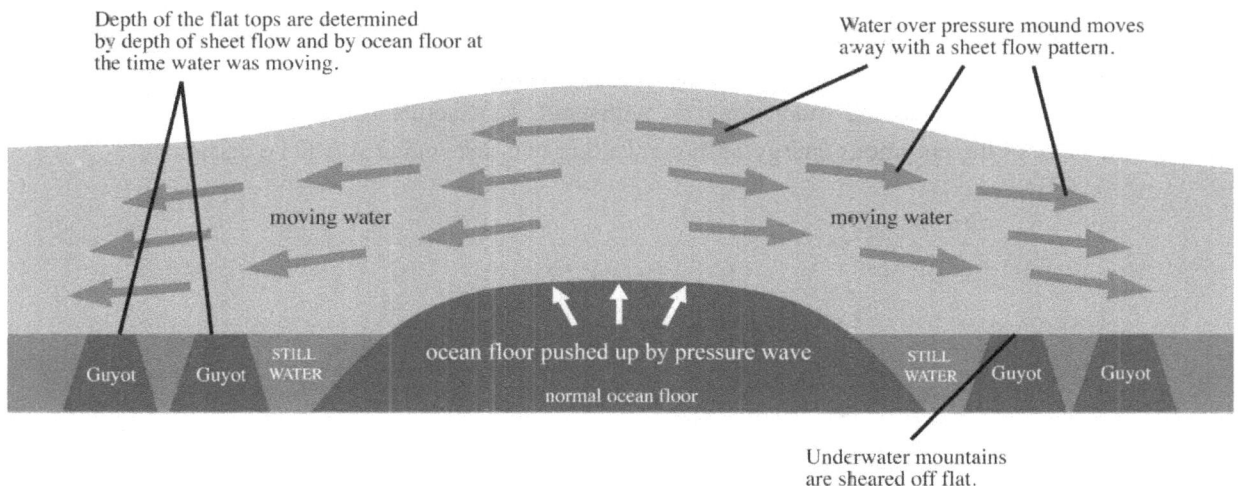

Depth of the flat tops are determined by depth of sheet flow and by ocean floor at the time water was moving.

Water over pressure mound moves away with a sheet flow pattern.

moving water

moving water

Guyot   Guyot   STILL WATER   ocean floor pushed up by pressure wave   STILL WATER   Guyot   Guyot

normal ocean floor

Underwater mountains are sheared off flat.

6.8.5 Fountains of the Great Deep

As discussed above under Impact Comparisons, oceanic crust which is being stressed by an interior pressure wave can be compared to a loaded beam. When a beam is loaded its deflection will always be the greatest at the middle because the effect of the load causes the most stress at the middle. This type of stress is called bending stress. As more load is placed on such a beam it will bend more. When the load becomes too great for the beam to carry, it will break. When it breaks, it will break at the middle.

An upwelling interior pressure wave (i.e. at the antipode of the Acraman asteroid that impacted Australia) stressed the Atlantic Ocean floor to the breaking point. It broke and the break occurred right near the middle. On either side of the break, the oceanic crust, (i.e. the floor of the ocean) rose up. (Please refer to 6.8.5 Fountains of Great Deep) The ocean water on top of this area also rose up and was thrust away from the fissure and flowed down the slopes on both sides. Great masses of ocean water were involved in this flow accompanied by monstrous waves which formed and propagated from the fissure towards the continents on either side. Hot molten magma was thrust up through the fissure by the pressure wave. This hot rising magma assisted in thrusting the water away but because it was hot, it caused water, which it directly contacted, to boil. This boiling was so violent that a great fountain of water was formed. Both water and steam were thrust high into the air. From either side it would have appeared that a vertical wall of hot gurgling, gushing water and steam had been thrust from the ocean depths by some great pump thousands of miles long. The pump, in this case, was the hot rising magma. There were two main factors involved in the formation of these fountains of the great deep. The upwelling magma and the boiling activity created a great fountain wall, which extended the entire length of the fissure (which was the entire length of the Atlantic Ocean).

By comparison, the Pacific Ocean floor crust fractured at numerous locations because it was not referenced to parallel shorelines as definitively as the Atlantic Ocean and because it is so extensive that it included the antipodes of several major asteroid impact sites. These include; Czechoslovakia, Tanitz in Kazakhstan, Popigai in Russia, Vredefort in South Africa and the possible; Takla Makan in China and Congo in Africa. In general, ocean floor crust would fracture wherever the stress produced by the upwelling pressure wave was too much for it to withstand. As fractures occurred, the stress would be relieved and some of the enormous energy of the upwelling pressure waves would be dissipated.

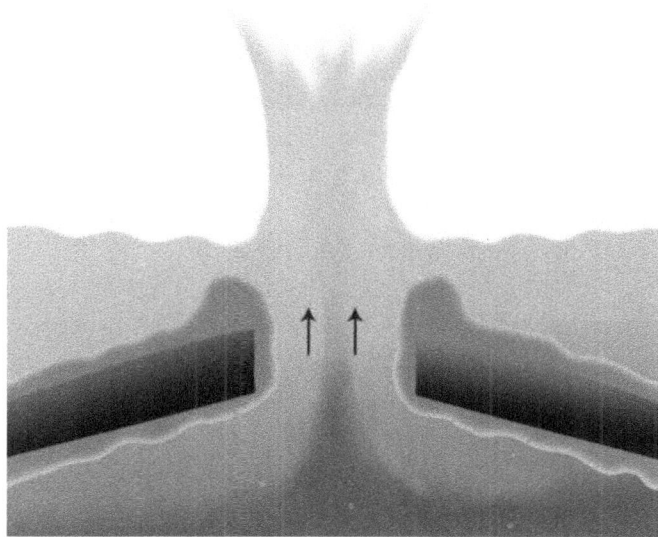

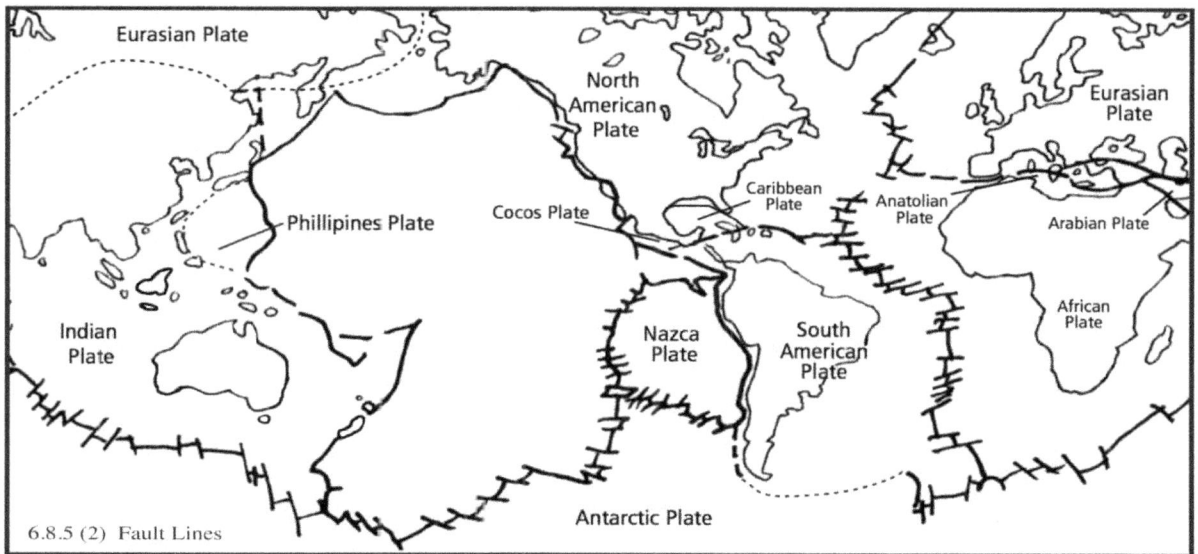

6.8.5 (2) Fault Lines

6.8.6   Lateral (Stretch) Cracks

If the ocean floor was forced up from underneath, it would actually become part of a sphere that had a surface area which was greater than it had before it rose up. Therefore, there would not be enough material to cover the surface of this larger sphere unless lateral cracks developed. The Atlantic Ocean

238

floor does include a lot of cracks, which radiate outward (at approximately ninety degrees from the mid-ocean rift) for many miles both ways. These cracks can be referred to as stretch cracks because as it was forced upwards, the ocean floor was stretched by an upwelling pressure wave. If the floors of the ocean were more or less flat, it would take an enormous force to fracture 40,000 miles of ocean rift and elevate it

o its present position. If this occurred, magma or lava would flow out and roll down both sides. Oceanographers have confirmed this, but one of the most striking proofs is that, as the ocean floors rose, the diameter of the Earth, at that point, would increase. Since circumference increases more than three times as rapidly as diameter, the edges all along the rift would be stretched and break into fractures perpendicular to the rift. These fractures are still a very prominent feature of the mid-ocean rifts, often extending from 100 to nearly 1000 miles on each side of the rift. (66)

Since an uplifted spherically-shaped ocean floor would have been part of a sphere which was larger than the original sphere, (i.e. the Earth prior to a large portion of crust being uplifted), cracks would open and the cracks would account for the greater surface area. The area of the open cracks plus the area of the uplifted ocean floor would be the area of the elevated surface. The cracks would extend as far back from the primary fissure as the ocean floor had been uplifted. The lateral fractures therefore define the extent of the uplift.

As the pressure from underneath built up, the lateral stretch cracks might even have opened first. In this case, the central or primary fissure would not be a straight or continuous line but would occur in straight-line segments, which would be randomly offset from each other. The mid-Atlantic rift occurs exactly like this. (41)

6.8.7 Continental Shelves

The direct ocean landing of a large asteroid would cause a water wave to spread outwards from the impact site. The pressure wave, which traveled all the way through the Earth, would also cause a great wave to form in the ocean when it came up on the far side of the Earth. Just as the impact wave would be capable of washing right over some continents, the wave generated on the far side (i.e. the antipode) would also be capable of washing right over the adjacent continents. These waters would rush onto the continents and cause extensive erosion. However, just like a tsunami, they represent a singular or transient event. They happen and then they are gone. The water would rush onto the continents and then rush off. The destruction would remain. An event such as this could cause an entire continent to be covered with moving water. Then just as quickly, the continent would reappear again as the water continued to the other ocean or returned back to the ocean from whence it came. In either case, the water would be carrying a very heavy sediment load. As long as the water was moving quickly, the sediment would stay in suspension but as soon as the water left the continents and reentered the ocean basins it would slow down and its sediment load would be dumped. Continental shelves would thereby be formed.

Water is always involved in the placement of sediments. However, flowing water can only transport a very particular type of very light and almost dissolvable material. Then, when the water slows down to a certain critical speed, or the water basically comes to a halt, its sediment load will settle out and fall to the bottom. In this manner deltas are formed at the mouths of rivers. Similarly, rivers fill up with sediment.

However water can also move larger grained material and it can even move boulders. As the speed of water increases, its ability to move larger objects improves. None of these larger objects will ever become entrained in the flow but will only be rolled along. It follows that the size of the moved object can be used to determine the power of the flow, which placed it. For example, sand and small pebbles can be moved by the waves near the shore of a lake, whereas boulders can be moved by a rushing mountain river.

If the ocean washed over a continent, innumerable amounts and types of sediment would result. Within a movement of water of this nature, there would be countless variations in the speed of movement as well as whether the movement was straight ahead, around or over something or adding to or removing material already placed. Even reversals and overlaps would occur. Similar variations of movement can be observed in any number of rivers.

All flow requires energy in order to be sustained. If energy is either dissipated or added (due perhaps, to local gravity-induced accelerations) the nature of the flow would vary and would not only vary from place to place but from time to time. Interference patterns would result from impediments as well as compounding flows.

As the great flowing masses of water dropped off the continents and re-entered the ocean basins, the entrained sediment dropped from the quieter water and formed the continental shelves. Sediment placement was also affected by the earthquakes, volcanoes, and the other geological upheavals, which were taking place including mountain building, escarpment and rift formation. A more chaotic scene cannot be imagined.

6.8.8   Igneous Provinces

Igneous rock is rock that forms when lava or magma cools. Province means district or territory. Therefore, igneous provinces are areas of the Earth's surface which consist of previously molten material (i.e. lava) which has oozed up onto the surface of the Earth, spread out and cooled. 'Marine geologists have discovered many new things about Earth's oceanic crust, with the advent of deep-sea drilling and the use of advanced, remotely operated vehicles. One of the most amazing discoveries is that Earth's surface (both oceanic and continental) is covered by massive basaltic lava flows … defined … as "Large Igneous Provinces" (LIPs).' (60) The technical description is given as '… voluminous emplacements of predominately mafic extrusive and intrusive rock whose origins lie in processes other than "normal" seafloor spreading. They include continental flood basalts and associated intrusive rocks, volcanic passive margins, oceanic plateaus, submarine ridges, ocean basin flood basalts, and seamount groups. LIPs represent major global events, whose genesis and evolution must have a direct link to mantle dynamics.' (60) In other words igneous provinces indicate that a major geological event has occurred but there is no explanation for it.

Some of these igneous provinces are very large and they are found all over the Earth. The group which is underwater north-east of Australia has a collective area larger than Europe. Also there is a strip immediately west of India which has an area equivalent to half of India. Similarly the igneous province south of Cuba is several times as large as Cuba. Their formation is therefore properly considered 'major volcanic activity' (60)

'Another variant is the fissure eruption. It does not fit the popular conception of a volcano at all. There is no dramatic mountain, no symmetrical cone. Fluid basaltic lavas ooze up from long, narrow fissures in the Earth and spread out over the landscape. The great lava plateaus of the world were created this way. One striking example is the Columbia River plateau, and adjacent Snake River Plain. Innumerable flows of lava spread out over large parts of what are now Oregon, Washington, and Idaho, covering an area of 200,000 square miles to thicknesses that layer after layer accumulated to as much as 3,000 feet. The volume of the Columbia River and Snake River Plateaus has been estimated to be 60,000 cubic miles. The Deccan Plateau of India, covering 200,000 square miles and the Parana Plateau of Brazil and Paraguay, 300,000 square miles are other famous examples.' (67)

It is also understood that immediately preceding breakthrough of the lava to form these features, the entire area would have been uplifted. It has been recognized that, preceding major volcanic activity, the uplift of the area above a LIP could be considerable; "Maximum uplift can reach 1000m in continental, and likely more in oceanic lithosphere and precedes initiation of major basaltic volcanism and would have been rapid. (60) Whenever the upwelling pressure wave (created by an asteroid impact on the far side of the world) fractured the crust, molten material would come to the surface and spread around. Igneous provinces might therefore indicate that an antipode is involved.

6.6.9   Magnetized Rock

'In all parts of the globe, rock formations are found with reversed polarization.' (61) Not only are rocks found with magnetic orientations which do not line up with the Earth's magnetic field, the variations in magnetic orientation vary both vertically and horizontally through the rock. (62) While it is known that when molten rock cools, it will acquire a magnetic component, the strength of this component is only about twice the strength of the magnetic field of the Earth; whereas the strength of the magnetism found in rocks with inverted polarity is much greater. 'The rocks with inverted polarity, however, are magnetically charged ten times and often up to a hundred times stronger than they could have been by terrestrial magnetism. "This is one of the most astonishing problems of paleomagnetism, and is not yet fully explained, though the facts are well attested." ' (63) The idea that the magnetic field of the Earth reversed numerous times and caused rocks to have various magnetic orientations, is therefore not substantiated.

Several studies have been carried out with ancient pottery to try to identify the magnetic field direction at the time the pottery was formed. 'The position of the ancient vases during firing is known. They were fired in a standing position as the flow of the glaze testifies. The magnetic inclination or the magnetic dip of the iron particles in the fired clay indicates which was the nearest magnetic pole, south or north. In

1896 Giueppe Folgheraiter began his careful studies….His conclusion was that in the eighth century the Earth's magnetic field was inverted in Italy and Greece. Italy and Greece were closer to the south than to the north magnetic pole. P. L. Mercanton…studying the pots…from Bavaria…came to the conclusion that about the tenth century before the present era the direction of the magnetic field differed only a little from its direction today yet his material was of an earlier date than…Folgheraiter.' (65) It is assumed in these studies that the pottery took up the magnetic field of the Earth as the clay hardened. If this was the case, knowing where and when the pottery was made would indicate the magnetic field of the Earth at that time. In other words, the pottery approach is not an independent dating system but depends on a series of assumptions concerning the date that the pottery was made. The further weakness of this approach is evident when it is realized that there is no way of knowing when the final magnetic state of the pottery was set. (Did it happen when the pottery was set in the oven or did it happen when the pottery was laying in the ground, in no particular arrangement, and subjected to passing shock waves?)

There is another explanation more in keeping with the extremely violent nature of the circumstances following a major asteroid impact. A mechanism for changing the direction of magnetized igneous rocks is magnetostriction which is caused by distortions of the rock due to shock or stress and has been demonstrated in the laboratory by simulating meteorite bombardments. Also, 'A large but transient magnetic field can be produced in a metal when fractured.' (64)

The collision of an asteroid with the Earth would cause three types of shock waves to propagate throughout the Earth. When the asteroid hits the Earth, shock waves spread out in all directions and their magnitude is extremely large. Secondly, the numerous impacts of debris thrown up by the impact would generate similar but lesser shock waves. Thirdly, when the ocean floor failed due to the bending stress produced by the upwelling pressure wave it failed with a sudden snap. A layer of rock several miles thick cracked and split apart due to overwhelming pressure from underneath. Great lengths of rock could have failed at exactly the same instant but more likely, failure initiated at one or more locations and propagated at high speed thereby forming the great faults currently found in the Earth's crust. When the crust actually broke it broke suddenly and this caused a major shock wave to travel outwards through the rock in both directions. Since there are numerous faults in the crust of the Earth, numerous shock waves would have been generated. With multiple impacts it is clear that a great many shock waves would have been traveling through the Earth in every conceivable direction. As with other types of pressure phenomena, whenever these shock waves crossed one another their intensity would add and the additive effect would have itself propagated as an interference pattern as long as the waves continued to travel. Many of these shock waves would have encountered the clay pots that were lying in the ground. It would therefore be reasonable to expect that their remanent magnetism would have been affected and might have had more to do with the shock waves than with any magnetic component acquired during their manufacture.

6.8.10   Erratic Boulders

Erratic boulders are boulders that occur unexpectedly. They are out of place and different from the basic types of rock found in their vicinity. If a mountain consists of granite and large granite rocks are found near its base, they would not be considered erratic. However, if rocks are found in a location remote from any similar types of rock from which they could have been broken, they would be called erratic.

Erratic boulders are quite common and have been found in numerous locations. 'The loose rocks lying on the Jura Mountains were torn from the Alps; in their mineral composition they differ from the rock formations of the Jura, showing their alpine origin. Rocks that differ from the formations on which they lie are called "erratic boulders". There are erratic boulders in many places of the world.' (1) Stones and boulders from Scandinavia and Finland are found in Britain, Germany and Poland. Stones from Ontario are found in Missouri. (2) Erratics are even found in coal. 'Erratic boulders are often found in coal, far from any lithologically similar stone. … waterworn quartzite fragment in the coal at Zaleski, Ohio … talcose slate in … coal … at Mineral ridge, Ohio … from Canadian Highlands … and a quartzite … in the coal, resembling a Huronian rock in Canada. Sedimentary erratic resembling local carboniferous rocks, are found in Belgium.' (3) Velikovsky's research led to the following; 'In the British Isles, on the shore and in the highlands, are enormous quantities of them, transported there from across the North Sea from the mountains of Norway. From Scandinavia boulders were also carried to Germany and spread over that country, in some places so thickly that it seems as though they had been brought there by masons to build cities. From Finland … to Poland and lifted onto the Carpathians. In North America erratic blocks, broken from the granite of Canada … were spread over Maine, New Hampshire, Vermont, Massachusetts, Connecticut, New York, New Jersey, Michigan, Wisconsin and Ohio.' (5) 'In innumerable places on the surface of the Earth, as well as on isolated islands of the Atlantic and Pacific and in Antarctica, lie rocks of foreign origin, brought from afar by some great force. Broken off from their parent mountain ridges and coastal bluffs, they were carried down dale and uphill and over land and sea.' (7)

These erratic boulders come in a great range of sizes. Some are quite small and some are 'upwards of 100 kilograms.' (4) 'The block near Conway, New Hampshire is 90 by 40 by 38 feet and weighs about 10,000 tons. Equally large is Mohegan Rock, which towers over the town of Montville in Connecticut. The great flat erratic in Warren County, Ohio weighs approximately 13,500 tons and covers three quarters of an acre. The Ototoks Erratic, thirty miles south of Calgary, Alberta, consists of two pieces of quartzite "delivered from at least 50 miles to the west," of a calculated weight of over 18,000 tons. Blocks of 250 to 300 feet in circumference, however are small when compared with a mass of chalk stone near Malmo in Southern Sweden, which is "three miles long, one thousand feet wide and from one hundred to three hundred feet in thickness…".' (5)

Two basic explanations have been offered to explain how Erratics arrived at their respective locations. One of them involves movement by ice and is the dominant explanation at the present time. Actually it is so pervasive that Erratics are often referred to as glacial Erratics, which incorporates the belief that glaciers were involved in their placement. 'Some erratic boulders, transported across the Baltic Sea, are believed to have been emplaced by floating ice. Erratics transported across Hudson Bay, are now believed to have been transported by icebergs or sea ice since they are found below the marine limit. Some far-traveled boulders in the northern Queen Elizabeth Islands could have been transported by ice rafting.' (6) The other explanation involves water and was the dominant theory until the ice placement idea became popular. Velikovsky reintroduced the water placement explanation quite recently by quoting the Swiss naturalist from the end of the eighteenth century, Horace Benedict de Saussure. 'The waters were carried toward these abysses with extreme violence, falling from the height they were before; they crossed deep valleys and dragged immense quantities of earth, sand, and debris of all kinds of rocks. This mass, shoved by the onrush of great waters, was left spread up the slopes where we still see many scattered fragments.' (8)

While Erratics are found all over the world, they are also found at elevations much higher than their supposed sources. 'These stone blocks lie on the Jura Mountains at an elevation of 2000 feet above Lake Geneva. Some of them are thousands of cubic feet in size, and Pierre a Martin is over 10,000 cubic feet. They must have been carried across the space now occupied by the lake and lifted to the height where they are found.' (8)

Moving water does develop very high forces and when it just keeps coming large boulders can and are moved. This can be witnessed occasionally during spring runoff in the mountains. Swollen streams move very large stones during these times so it is not a mystery that water can accomplish such a task. Also ice has been observed to move boulders. If the ice on a lake becomes jammed against the shore it might pick up a few stones. Subsequently the ice could float away and carry the stones with it. With such evidence in mind, it seems reasonable to accept that some erratic blocks were moved in similar ways. Such explanations seem stretched however, when the size, distance, elevation change and topography are simultaneously recognized. Moving a 10,000 pound stone by either ice or water does not really seem credible, especially when it must be placed at a much higher elevation than its apparent source. Both the force and energy required to place such monsters is very large and has not been explained by either an ice or water hypothesis.

On the other hand, an asteroid would have more than enough energy to carry out such a task. In fact, an asteroid with a diameter in the five to ten mile range would have enough energy to form a complete mountain range. Mountain formation under such circumstances would not be a docile affair. The trauma would include hurling rocks for miles into the air and for distances which were several times the diameter of any crater that was formed. In South Australia, the diameter of the Acraman crater is 90 km. This means – as it does with all craters – that material was dramatically disturbed out as far as the crater rim, which in this case is 90 km. The Sudbury Crater is also of interest. The diameter of this crater is 250 km. Material which was blasted away during the impact would have traveled far from the crater rim. It could have been propelled for several crater diameters before crashing back to Earth. In many cases, crater rims are actually mountain formations which would have been formed violently as the asteroid crashed into the Earth. One would expect a shower of rocks to accompany any such event. Either from the material ejected to form the crater or from the material which was violently forced up into a crater rim, Erratics could result. From such activity, Erratics should be found sprayed around the countryside, having been tossed for an unknown number of miles, and then rolling and skidding to their final resting places. 'In North America erratic blocks, broken from the granite of Canada and Labrador, were spread over Maine, New Hampshire, Vermont, Massachusetts, Connecticut, New York, New Jersey, Michigan, Wisconsin, and Ohio; they perch on top of ridges and lie on slopes and deep in the valleys. They lie on the coastal plain and on the White Mountains and the Berkshires, sometimes in an unbroken chain; in the Pocono Mountains they balance precariously on the edge of crests.' (14)

Since there have been numerous large impacts, there should be countless erratic boulders all over the Earth. Conversely, if one should find an erratic boulder, he should expect to be within a few hundred miles of a large impact site or much closer than this to a small impact site.

The second way that an asteroid could produce Erratics is by the remote effect. Whenever an asteroid crashes into a planet, it will generate a massive pressure wave which will not only propagate outward from the impact site creating several forms of trauma as it goes, but will also result in a pressure wave or impact impulse, which will travel all the way through the planet and come up under the far side. The location diametrically opposite an impact site is called an antipode and the pressure wave would come up at the antipode at a speed commensurate with the original speed of the asteroid. Rocks would fly up from the surface. Faults would form. Molten material would come to the surface.

The antipode table included in paragraph 6.8.2 above shows a list of antipodes for the largest known impacts. However, smaller asteroids would also have had an antipode and there are more than one hundred sites known where the impactor would have been at least several kilometers in diameter. (75) Erratic boulders should therefore be found all over the Earth and they are. (7)

The situation would be similar to Mars opposite the three largest craters. '…the pictures sent back by Mariner 6 and Mariner 7 showed that the Martian surface was not uniformly cratered but had large areas of chaotic terrain unlike anything ever seen on either the Earth or the Moon. (This comment clearly does not recognize the chaotic terrain of the Laurentian Plateau which has an area of about 3,000,000 square miles.) How can one explain how one entire hemisphere of the planet, shows scarcely any evidence of internal activity, whereas the first area to be investigated in detail on the opposite side of the planet has four enormous volcanoes? (Olympus Mons, Ascraeus Mons, Arsia Mons and Pavonis Mons) Immediately to the east of the volcanic province is a highly fractured area, and beyond that another extraordinary topographical feature was discovered, a series of huge canyons… These canyons, 80 to 120 km. wide and five to six and one-half kilometers deep, are much larger than anything found on Earth. Like the volcanoes, the great canyons of Mars suggest a fairly recent episode in the history of the planet characterized by large-scale events.' (9) 'Chaotic terrain exists just to the east of the Tharsus Bulge. It includes a maze of short, interconnected canyons, which has the name, Noctis Labyrinthus. This region has what appears to be a series of uplifted blocks or sections, and this also bespeaks of the impact of the Hellas Impact Asteroid on the opposite or anti-side.' (11) Pressure impulses generated by the objects which formed Hellas, Argyre and Isidis, would have come up under the opposite side and could account for the chaotic terrain as well as the giant trench formation, Valles Marineris. 'Called Valles Marineris, this 4,000 km-long network begins on the east side of the Tharsis Bulge and ends in an immense region of chaotic terrain between Chryse Planitia and Margaritifer Sinus.' (11)

The situation on Mercury is also of interest. The Caloris Basin is the largest feature on the surface of Mercury and is understood to be an impact site. Since it is 800 miles across and Mercury is only 3200 miles in diameter, the asteroid which created it should have caused havoc on the other side as well and this is indeed the case. 'The collision did more than just blast the Caloris Basin into existence. Some astronomers suspect that shockwaves from the Caloris hit converged on the other side of Mercury's globe, rearranging the surface into a blocky jumbled landscape (referred to as weird terrain).' (12) Other commentators offer similar explanations. 'Caloris, a behemoth 1,300 kilometers in diameter, is the largest of these craters….The impact was so violent that it disrupted the surface on the opposite side of Mercury: the antipode of Caloris shows many cracks and faults.' (13)

The presence of Erratics might therefore be suggestive of an antipode. Since the surface of the Earth is mostly water, antipodes of most of the known large impact sites are underwater. However, it is very reasonable to expect that there have also been major impacts directly in the ocean and these would have developed antipodes on land or at least include land which would be complete with broken terrain and erratic boulders.

It has even been suggested that rocks from Mars have come all the way to Earth. 'Some scientists have also studied Martian rocks found here on Earth. These rare samples of the Martian surface were blasted into space by the impact of an asteroid or comet and later fell to Earth as Meteorites.' (10) '…a meteorite, ALH84001, located in Antarctica in 1984, which came from Mars….' (25) If an asteroid impact on Mars could send material to Earth, it would be reasonable to expect that an asteroid impact on Earth could place erratic blocks – even big ones - almost anywhere on Earth! Therefore, wherever erratic blocks are found, one would expect to find either an impact site or an antipode within a few hundred miles. (For example, if the 200 mile-diameter feature bordering The Czech Republic is an impact site, it would help explain the erratic boulders found all over Europe.)

6.8.11   Coal Deposits

If huge waves should sweep over a continent, the soil would be eroded and swept away. Forests, grasslands and swamps would be dislodged and swept away as well and would form great rolling, tumbling, masses of logs, branches, leaves and whatever else had been growing. Where the surging mountainous waves and flows were deep enough, these masses would float, partially submerged. Whenever the water was not quite deep enough, they would be tumbled along over the newly-exposed bedrock. Several square miles of forest could be involved with a single accumulation. It could get snagged in some depression. When the forward motion of the wave slowed enough, these great entangled masses of plants would remain. The next wave might bury them. In some cases, repeated wave action and dumping of masses of vegetation would produce layers of carbon-rich material interspersed with sediment. This material would pack down under its own weight as well as the weight of any additional deposits, which were placed on top.

Coal is 100% carbon which has been pressed together and is believed to have been formed from plants because the outlines of plants are frequently found in coal. Also, plants are made from carbon as may be determined by removing all of the water and other volatiles from the plant. Heat appears to have been involved in coal formation. Heat would drive out the volatiles as it does during charcoal production. Heat is always available underground because the interior of the Earth is hot. If a great mass of plants were gathered up, pressed down, covered up and heated, a great carbon deposit or coal bed would be expected.

There are several different types of coal. Since coal is carbon, the different types of coal are explicable as originating from different types of plants. Peat is also formed from plant material. However, peat is always found at the surface. Consequently it is never packed down and never heated. Therefore in addition to originating from different types of plants, factors involved in variations in coal would include the amount of heat as well as the amount and type of cover. Exactly the same circumstances and source material would be expected to produce the same product but since the circumstances and source material varied considerably, coal should and does have a range of characteristics. Since all of the plants did not grow in the same location and all of the water flows were not the same everywhere, variations in deposition would also result.

6.8.12 Volcanoes Again

A major asteroid impact would generate three different types of volcano.

The impact site would behave like a volcano. After the asteroid had disappeared into the interior of the Earth, the opening in the Earth would fill in again. However, it wouldn't necessarily refill quietly. Instead a back-splash could develop. When an object like a stone is tossed into a pond, the water separates to let the stone enter. Then when the stone has passed, the water refills the opening. This activity results in a back-splash where the water rushes to the center of the opening, converges on itself and shoots up into the air. It may shoot up so violently that some material leaves the surface altogether. Both the rising up and the falling back result in concentric wave disturbances in the opening. Similar activity would result from an asteroid impact. As the molten rock refilled the opening, it would have converged on itself and shot up into the air. (This would be similar to a volcano shooting large globules of molten lava into the air.) This back-splash volcano could rise up for several kilometers before falling back and causing secondary back-splash activity as well as general chaos both in the opening as well as the surrounding area. The impact would have opened a passage to the hot molten interior and magma would refill the opening. While a small asteroid would make an impact crater which would remain, it would remain because the surrounding ground was solid material and was able to retain a shape. Conversely, a large asteroid would open a channel into the hot interior. When this channel refilled, it would not be able to retain any special shape because the refill material would be molten. When it refilled, the impact site would appear like a lake of red-hot rock with a few pieces of solid crust floating in it. This would not necessarily have been a belching volcano but since it was so hot, vapor would probably have risen from it. Since this semi-solid region would be fluid for some time, its shape would be determined by gravity. The net result would be that this impact-site crater would be almost flat and would follow the general curvature of the Earth. Just like the great craters, which are on the Moon, if one were able to stand in the middle of this crater, they probably would not be able to see the rim at all. It is quite possible that several areas within this large molten lake would spew effluent into the atmosphere but if they stopped spewing before the surface crusted over, the whole area would remain quite flat.

There is a second type of volcano that would have resulted from the earthquake. Firstly, the impact of the asteroid would have caused the Earth to quake. Consequently the crust of the Earth cracked. Any of these cracks, which went all the way through the crust, would have allowed molten material from the interior to come to the surface. Wherever this happened, a volcano would have been born. These volcanoes would have belched out several different materials including water, dust and ash as well as hot molten magma.

The third type of volcano would have resulted from the interior pressure waves, which propagated throughout the interior of the Earth. Ocean crust is thinner than continental crust and so it is much weaker. Therefore it is much less able to withstand stress. As an interior pressure wave welled up under the crust on the far side of the world, the crust became stressed and wherever it was too weak to withstand the stress, it faulted or cracked. In particular, the crust under the remote oceans (i.e. near antipodes) was stressed to the failure point in numerous places and material from the interior surged up through many of these faults and formed mountains on top of the ocean floor. Volcanic activity accompanied all of the faults formed this way and in fact most of the volcanoes around the world are concentrated near or right on top of faults. (55)

6.8.13   Iridium Layers

A massive cloud would have risen from the impact site. 'There were…the…ejection of trillions of tons of dust into the atmosphere…almost complete darkness (i.e. world-wide darkness) lasting several months.' (52) This massive cloud would have included some pulverized portion of the asteroid which would have included the element iridium. 'Iridium is much more common in meteorites, comets and asteroids than on the Earth's surface.' (53) Therefore the massive spreading black cloud would have included iridium.

The cloud originating from the impact site would have been augmented by all of the other volcanic activity, which was going on around the world until the world was completely covered by cloud. The cloud '… spread out over the entire globe.' (54) While this cloud remained airborne for several months or longer, it eventually settled back to the ground. As it did, a layer of dust formed on the ground and this dust included iridium.

Iridium has been found at more than eighty sites around the world. (53) It has also been claimed to provide evidence for more than one large asteroid impact.   'In a Geological Survey office in Ottawa …a …geochemist   ….studies evidence he has gathered from around the world related to earlier mass-extinction episodes …and …finds the same type of iridium boundary along with microtektites…created by an impact.' (55)

There are several ways that more than one iridium layer would have formed. First, if the dust cloud settled intermittently while other materials were being placed, iridium layers would have been interspersed with these other materials. Secondly, if more than one impact was involved and they occurred far enough apart in time to allow the placement of other material, there would be more than one iridium layer. Thirdly, the iridium-enriched dust might have been brought down by rain which would have poured down for awhile and then let up before pouring down again.

6.8.14 Mountain Formation

6.8.14.1 Mountain Material

Most of the continental mountains of the world are formed from sedimentary rock. Sedimentary rock is a type of rock which appears in layers and is understood to have been formed by water and perhaps even underwater. In fact, the most striking thing known about mountains is that they are mainly composed of sedimentary rocks, that is, of rocks which have been deposited originally in the seas. This includes the Andes, Caucasus, the Alpine heights, the Matterhorn and the Himalayas. (72)

6.8.14.2 Water Waves

Water waves consist of a series of crests and troughs where each crest is followed by a trough repeatedly. When the wind is very light, the crest to trough distance may only be half of an inch. Small ripples of this nature may be useful to a sailor who is attempting to maneuver his boat to his best advantage because they will tell him exactly which way the wind is blowing. When the wind gets stronger, waves can develop which have lengths of several feet from crest to crest. Where the ocean is concerned, waves thirty or forty feet high can be formed and they will also have correspondingly greater distances from crest to crest. There is abundant energy in these giant ocean waves. In fact there may be as much as one megawatt in one meter (3.3 feet) of wave-front. Since these energy levels are so high, numerous attempts have been made to capture some of it for industrial use.

An example of a water wave with even higher entrained energy is the tsunami. This is the type of wave that forms as the result of sudden material movement under the sea. The particular type of movement is quite secondary to the fact that some underwater material moved. It could be a landslide or it could be the lurching of a rock structure due to an earthquake. In this case, energy would have propagated from the source of the earthquake into surrounding rock, some of which was exposed to the ocean. As the rock lurched from the shock wave of the earthquake, it caused the water which was in direct contact with it, to move. Energy would thereby be transferred into the water and a pressure wave would propagate away through the water. Tsunamis usually travel at several hundreds of miles per hour. As they travel across the ocean they are not really noticeable because their energy has been distributed throughout the entire depth of the ocean. (This is in direct contrast to water waves formed by wind because all of their energy is contained in the top layer of water.) As a tsunami approaches shore, the situation changes dramatically. In the shallow water, the energy becomes concentrated and the water may rise up into a giant killer wave a hundred feet or more in height. As such waves wash over the shore and travel inland the energy which was originally generated by the earthquake, is propagated along the surface of the ground and converted to the movement of whatever material was in the way. Boats, houses and trees may be moved inland for thousands of feet. Then the movement peters out. It can therefore be said that the energy of the distant earthquake has been dissipated far from its source by the displacement of trees and other movable objects.

6.8.14.3 Earthquake Waves

The energy of an earthquake can also be propagated along the surface of the ground – right in the surface layer. In 1964 a very strong earthquake occurred in Alaska. The ground shifted repeatedly, buildings were damaged and the pavement came apart. Where the ground allowed it, the energy transfer mechanism had the appearance of a train of waves. 'The land became the ocean, the streets were rippling like waves.' (74) This was not an isolated occurrence. A series of earthquakes troubled the central USA during 1811 and 1812 and were called the New Madrid earthquakes because they were centered near New Madrid, Missouri. In this case waves were also produced. 'The Earth was observed to be rolling in waves of a few feet in height, with a visible depression between. (56) In order to form waves of this nature, the land had to have been pliable. If it had been rigid rock, these waves would not have formed.

6.8.14.4 Mountain Waves

Similarly, in the mountains, not only has the rock been folded and fractured but in many cases the various layers of sediment are seen to be bent into arches, with the bending occurring both up and down. When the rock is folded so a mound is formed, it is called an anticline. When it is folded so a trench is formed, it is called a syncline. Often these anticlines and synclines occur repeatedly across entire mountain ranges and each range (i.e. anticline or group of anticlines) can run for several hundred miles. The Rocky Mountains of North America, for example, consist of numerous parallel ranges which extend across both Canada and the USA.

In some cases, the anticlines seem to have bent too much and cracked open. In many cases the rock simply bent and remains that way. Even where the rock is fractured, it seems to have bent as far as it could and then fractured. Not only is it found bent, it is clear that bending was allowed because the rock was soft enough to allow it to happen. One commentator suggests that the rock was bendable because it was soft as wax. '… the limestones, folded as if made of wax, reveal themselves in great arches.' (72)

6.8.14.5 Comparing Earthquake Waves to Mountain Waves

The tops of the earthquake waves can be likened to anticlines. The valleys - the "visible depression between" – can be likened to synclines. Both earthquake waves and mountain anticline-syncline formations have often burst open leaving a v-shaped opening where the various layers may be traced on either side. 'Broadly speaking, the effect is the production of mountain chains consisting of strata (i.e. sedimentary layers) that have been strongly folded, fractured, and over-thrust in one or more dominating directions.' (73) Such activity would have been very similar to the fracturing and bursting open of the earthquake waves. 'By and large these swells (i.e. the earthquake waves) burst, throwing up large volumes of water, sand, … When these swells burst, large, wide and long fissures were felt.' (56)

Earthquake waves and mountain waves are very similar in appearance. The basic difference between them is magnitude. The New Madrid earthquakes waves were several feet high. For discussion purposes let their amplitude be considered as about five feet from crest to trough. Mountain waves, on the other hand, might have a crest-to-trough distance of ten thousand feet. In this example, the amplitude of the mountain waves is about two thousand times as high as the earthquake waves. (If an ocean wave fifty feet high was compared to a ripple half an inch high, the ocean wave would be about twelve hundred times as high as the ripple.)

The New Madrid earthquakes were given Richter magnitudes in the 7.5 to 7.8 range. (56) Since every integer on the Richter scale represents a magnitude increase of ten, the earthquake which formed the mountain waves - being two thousand times as high as the earthquake waves - would have a Richter magnitude of 7.5 to 7.8 plus 3.4 or about 11.0 to 11.5. (Richter magnitudes are not usually assigned in such a high range but the concept is applicable.) Each integer on the Richter scale also represents an energy increase. Whereas the magnitude increase is 10, the energy increase is about 31. (58) Therefore an amplitude increase of 10,000 represents an energy increase of 30x30x30x30 or 810,000. Therefore, roughly speaking, the earthquake energy involved in forming mountain waves 10,000 feet high would have been one million times as high as the energy level involved in the New Madrid earthquake. 'Shock waves from the impact (i.e. the impact of an asteroid) rattled the Earth with energy much greater than that of the largest earthquakes humans have experienced, probably a million times more energetic than the one that devastated San Francisco in 1906.' (71)   While, in general, such enormous earthquakes are not expected to happen, if a large asteroid impacted the Earth, earthquakes of such magnitudes would certainly develop.

The energy relationships of earthquakes and asteroids have been considered and values have been assigned. For example, according to one commentator, the energy of an earthquake could be 10(24) ergs and the energy to form a mountain range could be 10(29) ergs. (57) In this case, the mountain range required 10,000 times as much energy as the earthquake had available. Of course only a fraction of the energy of an asteroid would be transformed into an earthquake. If the asteroid had been 32 km. (20 miles) in diameter, its energy could have been 10(32) ergs or 1000 times greater than that required for the formation of the mountain range. (57) This would mean that only 0.1% of the energy of an asteroid 32 km. in diameter would be required to form a mountain range. In fact, an asteroid this large could supply enough energy to form ten mountain ranges and still only use 1% of its energy.

6.8.14.6   Asteroid Earthquake Epicenters

There would be two epicenters for earthquakes produced by an impacting asteroid. The most obvious would be ground zero where the asteroid touched down. A major shock wave would propagate outwards from ground zero because the ground would be so violently displaced.

An earthquake might also propagate outwards from the antipode (the location on the Earth directly opposite the impact site) if the crust of the Earth was fractured by the upwelling pressure wave. The rising pressure wave from the interior would cause the ocean floor to bulge upwards. When the stress level became too great, the ocean floor would crack and in the case of the Atlantic Ocean, it would be expected to crack near the middle. The vertical displacement following such a fault event, would elevate the ocean water and cause it to rush towards the shores. 'The ocean floor all around the globe bears witness that the oceans of the Earth were the scenes of repeated violent catastrophes when flows of lava and volcanic ash covered the precipitously rising or falling bedrock and tidal waves raced against continents.' (49)

While the ocean floor would displace vertically, it would also displace horizontally and there are three factors, which would contribute to this horizontal displacement. First of all, when the ocean floor cracked, it would separate. It would actually crack and separate with a shock and a loud noise and the accompanying horizontal movement would be a short, sharp, pressure transient, which would travel as a shock wave towards the shores. The magnitude of such a shock wave would have been much greater than anything that we could imagine. As it propagated across the ocean floor, the floor would have been momentarily stressed in compression. Wherever the crust was strong enough to accommodate this compressive stress, it simply transmitted it. (This would be similar to the way that the suspended balls and the soldier piles transmitted energy without really moving very much. Refer to 6.7 above) However, if it wasn't quite strong enough to withstand the shock wave as it passed, the ocean floor would have failed and buckled up into pressure ridges.

The opening formed by the crack allowed magma, which was being forced up by the upwelling pressure wave, to surge up through the opening. This would have caused additional horizontal movement. The third component of horizontal movement would have been due to the fact that the ocean floor on both sides of the crack would have been elevated. The ocean floor, in the elevated position was therefore being supported on a slope, which was declining towards the shore. The ocean floor would therefore tend to slide down this decline. The resulting horizontal displacement may have been as much as the vertical displacement which could have been quite significant. 'Tectonic catastrophes raised or lowered the ocean bottom … thousands of feet, spreading huge tidal waves …' (49) As the ocean floor was thrust upwards, it was also thrust towards the continents and the thrusting force would have been very high. With all of these unimaginably-high forces in active mode, something had to give. In this manner the antipode would have become the epicenter for an earthquake.

The rising pressure wave would have applied a horizontal thrust to the ocean floor, which would have slammed the ocean floor against the continents. This effect would be similar to the way in which the floating ice pack behaves in the arctic. As the wind blows across the ice, it imparts a horizontal force to the floating ice. The ice is thereby forced against other ice, which might be free to move, or might be jammed in and cannot move. The stress remains however whether anything moves or not. All of the ice, which is involved, might be able to withstand the stress. However there are numerous times when it cannot, and when this happens the ice will fail in compression and buckle. The locations where the ice has buckled are called pressure ridges. Everyone who has traveled over the arctic ice has encountered these pressure ridges and they present serious impediments to ice travel. Even though the pack ice might only be a few feet thick, pressure ridges can rise up for fifty of sixty feet or more into the air. They may extend for miles or for just a few yards. Some underwater formations appear very similar to the pressure ridges in pack ice.

When the ocean floor ruptured, either at an antipode or a ground zero, an earthquake-generating shockwave propagated horizontally towards the continents. When it reached the continents, pliable (i.e.

something like wax) layers of sediment (i.e. material that would become sedimentary rock when it hardened, instead of the hard material of the ocean floor) were encountered and the manifestation of the shockwave changed from a basically-invisible moving region of highly-compressed rock into giant ground waves which traveled inland through the surface layers. (This would have been similar to what happens when a tsunami comes ashore. The energy of the traveling tsunami pressure wave is not really noticeable at sea but when the shore is reached, monstrous waves develop and surge inland.) The surface layers of the continental crust were very recently deposited layers of malleable material and formed into giant waves which included anticlines and synclines. The magnitude of displacement was too excessive for some of these land waves so fracturing resulted. Sometimes material was thrust up and over on top of other material. Sometimes material was displaced several miles from its origin but primarily it was the shockwave that moved on for hundreds of miles and the material only dislocated locally.

When the ocean floor rebounded from the shock wave, it left the ocean-continent interface trenches almost void of material, which is the condition of many of them at the present time. (70) In some regions it smashed into the continent but the impulse was not great enough to cause very much displacement. In some areas, the folding and failing and buckling action continued inland for hundreds of miles. Neither would the sea floor be exempt from failure. The enormous shock wave resulting from the primary rupture of the ocean floor, compressed the ocean floor as it traveled outward toward the continents. Therefore if ridges formed parallel to this split, they would not be unexpected. (e.g. There are many ridges on the Atlantic Ocean floor that are parallel to the central ridge which is split. The split is referred to as a rift.)

While both ground zeros and their respective antipodes could have been epicenters for earthquakes, the energy available at a ground zero location would have been much higher and was therefore probably the main source of mountain-forming earthquake waves.

6.8.14.7   Summary Discussion

Many of the major continental mountain ranges of the Earth run parallel to ocean coastlines. This is certainly not a coincidence. Everything in nature is the way it is for a reason. The parallel arrangement suggests that some action of the ocean crust was involved. The Theory of Plate Tectonics attempts to explain the arrangement by the spreading action of ocean crust. However, continental crust is many times thicker and hence stronger than ocean crust. Over the many eons of time required by the Theory of Plate Tectonics, the continental sediments would have hardened making it very difficult for ocean crust to do anything with them.

The folding of the various layers indicates that the material was shaped after it was deposited but prior to consolidation and hardening. The always-has-been-solid ocean crust transmitted high-energy shock waves from asteroid impact sites into pliable continental crustal material. As with the suspended balls which basically only transmitted the shock wave to the last ball, the ocean crust transmitted the asteroid shockwave into the pliable continental layers of sediment which deformed up and down to relieve the stress that the shockwave created.

If the mountains had been formed simply by crushing action, domes would not be expected. To get the anticline-syncline formation required a type of energy which could form waves. Then, when these waves had passed, there wasn't any appropriate force available to return the material to a level arrangement so it simply remained in a broken, folded, over-thrust, chaotic state and has remained that way ever since.

As with water waves where any particular particle of water does not move very far, so the sediments did not move very far and any general shortening of the Earth's crust did not occur.

Also the height of either a wind-driven water wave or a tsunami will depend on the depth of the water at that particular location. Generally speaking, the shallower the water is, the higher the wave becomes. The heights of the mountain waves would also have varied with specific local circumstances and would have been responsive to any wave overlaps, reflections or interference activity.

As a tsunami approaches a shore, the characteristic of the energy transfer medium (i.e. the ocean water) changes (i.e. becomes shallower) so the way the energy is manifested also changes (i.e. the wave becomes bigger). Similarly, as the asteroid-generated giant earthquake waves passed from ocean crust into continental crust, the energy transfer medium (i.e. the ocean floor) changed (became the soft sedimentary layers of the continent), so the way the energy was manifested changed (i.e. giant waves formed).

After traveling inland for several hundred miles in many cases, the energy of the earthquake waves dissipated, just like the energy of tsunami waves dissipate after the waves have washed inland.

In this manner the continental mountain ranges were formed by the energy from a shower of asteroids, which impacted the Earth in numerous locations, but which on impact, were able to transfer part of their incredibly high energy into deformation of the Earth's crust all over the world.

6.8.15   The Pulsating Earth

When the pressure wave had traveled through the interior of the Earth, and come up under the far side, it would have dissipated some of its energy. However some of its energy would remain. As the pressure wave came up under the crust, the crust partially reflected it similar to the way that a vertical cliff reflects a water wave. Whenever the continental crust rose up, it soon settled back down again and the return of the continent back down would cause a type of reflected wave to start traveling back through the interior. The energy of the pressure wave would be partially dissipated by interacting with the crust and would be partly preserved and reflected. The oceanic crust would have a similar result. The ocean floor might crack and rise up but shortly thereafter, it would settle back down. As it settled down, a return pressure wave (albeit less intense) would be generated and it would start travelling back through the interior of the Earth. These reflected pressure waves would also traverse the interior of the Earth and once again come up under the crust. When they did, they would have had some disturbing effect on the crust and then they would start travelling back through the interior yet again. The Earth would now be pulsating. The period of these pulsations would be the time required for a primary pressure wave to go from one side of the Earth to the other through the interior. However, if there were multiple impacts, which were offset by several minutes or more from this basic period, the Earth would be pulsating with a random pattern. A different pressure wave could surge up to the surface just a few minutes after the last one. Some energy would be dissipated on every pass but as long as any energy remained, these great pressure waves would have continued to travel back and forth through the interior. The results at the surface were similar every time. Volcanoes would surge up again and the crust would shift again. However the effects would diminish with each pass through the interior until they were no longer noticed. The energy of a large asteroid is so incredibly high that pressure waves from a single impact might travel back and forth through the interior several times before dissipating. (Just as the atmospheric pressure waves from Krakatoa travelled around the Earth three times before dissipating.)

Even one major impact would cause the Earth to shudder and pulsate. The Earth might even resonate, which is a type of activity that occurs when an object is energized near one of its natural frequencies. In this case the Earth might resonate, shake, pulsate or shudder for a period, settle down and then do it again. This type of event would be similar to the impact on the Moon in 1178 AD. A group of men were observing the Moon on June 19, 1178 AD when flames and sparks were observed coming from the location of the present lunar crater Giordano Bruno. First they observed flame and sparks. Then there was a period of tranquility before the moon shuddered. Apparently there were several periods of shuddering before the lunar surface became black. The impacting object may only have been 2 miles in diameter but it formed the Giordano Bruno crater, which is about 13 miles across. This impact has been used to explain the excess libration of the Moon (rocking back and forth so we actually see more than one-half of the lunar surface) '... we can see about 59% of the Moon due to fluctuations in its orbit that see it 'wobble' from side to side and up and down, meaning we gain a slightly better view than just half.' (77)) as well as the major meteor shower on Earth 3 1/2 months later. (i.e. later than June 19, 1178 AD) (59) One can only wonder about the magnitude of the shuddering on the Moon if it was observed from Earth. It would have been completely untenable to have been on the lunar surface at that time.

When the analogy of the bullet, which was fired into a crust enclosed molten ball of lead is recalled, it will be remembered that the energy of the bullet was sufficient to push the molten ball sideways for some distance. If the ball was molten however, sideways displacement would have been accompanied by disruption of the entire surface. In fact the pressure wave, which would be generated in the interior of the molten ball, would have similar effects on the surface of the ball as an asteroid would have on the crust of the Earth. Major disruption would occur in both cases and continue until the interior pressure waves had dissipated all of their energy.

The surface of the Earth includes evidence of several very large impacts. All of these impacting objects were large enough to smash through the crust and generate interior pressure waves and all of these waves would have travelled back and forth through the interior. In such a case, interference patterns would have been generated. Pressure waves, like other types of waves may be added together. Water waves may be added together. If the waves, which are generated by a storm in one region of the ocean, make their way to another area where waves have been generated, these waves will cross each other and generate super waves. It occasionally happens that three wave trains interact. In this case, very large so-called killer waves are generated. These monsters are able to threaten even the largest ships. Waves can also be added together when a wave train reflects from a vertical shore. As the reflected waves travel back they will interact with the incoming waves to make larger waves. Small lakes with vertical cliffs along the shore are particularly susceptible to this type of wave.

If the pressure waves, which would have been travelling back and forth through the interior of the Earth, occasionally interfered or added up as they came up under the continents, their trouble-causing characteristics would have been increased accordingly. As with the interacting water waves, when pressure waves are added together, the effect of the addition disappears again as they pass each other and there is no reduction in the original power of either wave. As the pressure waves bounced back and forth through the interior of the Earth, they would have repeatedly interacted and separated before finally petering out when all of their energy has been dissipated. If several large asteroids impacted the Earth, the interior pressure waves could continue to reflect back and forth for days, weeks or even months. In this case, the types of damage, which was caused by the first impact, would continue and be repeated over and over again. A series of great water waves would repeatedly wash over the continents and the great fault down the middle of the Atlantic Ocean would repeatedly reopen. The Earth would be pulsating for months. The plates which formed when the crust of the Earth was fractured would be jostling and bobbing on the turbulent magma. Whenever they jostled apart slightly, volcanic activity would redevelop.

Nothing could have survived such a catastrophe. One commentator summarizes the situation; 'The picture that emerges … is a barren world with no trees, plants, animals or birds … All air-breathing, land-based animals had died and were becoming fossilized. The newly-formed stratosphere contained a thick blanket of volcanic dust from the extensive volcanic. It was a pitch-black depressing world. (78)

7. Post-Impact Developments

7.1 Post-Impact Winter

7.1.1 Introduction

A few years ago during the "Cold War" there was serious concern that the whole world could become entangled in a war which included atomic bombs. The devastation from any war which involved atomic bombs would be widespread and could leave the world in such a state that there really wouldn't be any winners. The immediate results of this type of war include the usual damage that might be expected in any major war such as broken bridges, cratered highways and demolished buildings. The nuclear factor adds radioactivity which is unseen and which continues its devastation during the succeeding years. Among the survivors there would be people with terrible burns, and loss of body functions including sight and mobility. Radiation sickness would be ongoing and mutants would show up among the children which were born during the next few years. Radioactivity near every bomb site would not die out for many years. Most of these effects would not be visible but they would be quite real and cause untold mayhem for a long time. There would also be a winter which would be quite visible and it too would last for a considerable time, - time in this case being measured in years.

7.1.2 Nuclear Winter

A nuclear winter is the period of cold following an all-out war during which nuclear bombs were dropped. Like any other winter, it would be cold because the cloud cover produced by the nuclear explosions would block sunlight from reaching the ground. Most of the sunlight would reflect from the cloud tops. 'Recent satellite measurements of the solar and infrared radiation balance have revealed that clouds do exert a major influence on the surface temperature. A regional increase in cloudiness causes cooler surface temperatures… The most surprising result from this new research is the large magnitude of surface cooling that results from increased cloudiness.' (14) With extensive cloud cover, the ground would not be warmed by the Sun and it would freeze. As with any winter north of the temperate zones, when the weather gets cold, the ground will freeze. Gardens cannot be planted. Roads cannot be built. Rivers and canals crust over so shipping comes to a halt. If it got cold enough, cars would not start unless they had been plugged in to an electrical source of power. All of these things are expected during any winter in a northern country. By taking precautions and planning ahead, winter conditions can be managed, even if they last for several months. The main difference with a nuclear winter is that it would happen quickly and spring would not come. It would remain winter right through the spring. There would be no letup in the summer and fall would be the same. It would be winter all year and it could continue for years.

There has been a great deal of interest in the nuclear winter that would result from a nuclear war and it has been compared to the effect of large prehistoric volcanic eruptions as well as to the hypothesized asteroid extinction of the dinosaurs. Mathematical climate models suggest that a nuclear winter would produce mid-latitude surface temperature drops in continental interiors, to well below freezing in a matter of days at any time of year. (23)

7.1.3 Cloud Production

When any bomb explodes, heat is produced. As the chemicals which make up the bomb react, heat is produced and produced very rapidly. The heat raises the temperature of everything around so if there is anything nearby that can ignite, it will catch fire. The smoke from these fires will ascend up into the air and form a great black trail high into the sky. If the bomb is strong enough, it will blow a bowl-shaped crater opening in the ground. The material which is blown out of the way to make this crater will be splashed around the area as well as being pulverized into dust and blown into the air. An atomic bomb is very powerful and when it explodes, an enormous cloud results. The pressure-wave from the explosion will form a crater in the ground but instead of making a relatively shallow depression a few feet deep, an enormous opening hundreds of feet across might be formed. As with a smaller bomb, some of the material from the crater will be splashed around the area. However a lot of it will be pulverized into dust and will rise into the air to augment the smoke. One atomic bomb can generate significant atmospheric loading from these two effects. The fear related to a war is that there could be any number of bombs exploded with associated amounts of material becoming airborne and contributing to the cloud cover. Great plumes of smoke would rise into the air for thousands of feet and then start drifting downwind. Wherever they drift, they would cast a shadow on the ground. Anyone standing in one of these shadows would be shielded from the Sun and immediately feel a chill. The smoke from different bombs would coalesce to produce a widespread cloud. All of the area under this cloud would be in shadow and the Sun's rays would be blocked from reaching them. A nuclear war could result in thousands of tons of smoke and dust being placed in the atmosphere and it would drift on for miles and miles. The entire sky could become overcast. The chilling effect would be very widespread and weather would be affected. It would simply cool down and stay much cooler than usual. If summer was expected or in progress it would suddenly be fall instead. While the fear factors anticipated from a nuclear war certainly include the devastation which would show up around the site of each explosion, and the radiation problems which would last for years, the additional concern is the chilling effect of the widespread clouds which would develop from the smoke and dust rising from ground zero. If a large quantity of nuclear bombs were involved, an extended winter would result. The effect of this winter would be very widespread. Since plants would not grow and everything would be continually frozen, people could not survive and so the devastation from the war would reach much farther than the site of the conflict and could, in fact, affect the entire world. With such a no-winners-at-all scenario the major powers seem to have backed away from seriously considering nuclear bombs in future conflicts and have made efforts to dismantle them.

7.1.4 Volcanic Winter

A volcano could produce a winter. Volcanoes come in various forms but most of them produce a great deal of smoke, dust and ash. The smoke might have similarities to the smoke from something burning but in many cases it will be water vapor mixed with dust. The solid matter, which we refer to as dust and ash, is mostly small disintegrated particles of rock. However, it doesn't really matter what the exact nature of the particles are as much as their effect, which is to cast a shadow on the ground. Some volcanoes produce an enormous amount of material with the result that the cloud they produce is very extensive. The effluent from volcanoes has occasionally been estimated and these estimates are sometimes given in cubic miles. It is hard to imagine how much dust and other sun-blocking material there is in a cubic mile of mountain.

A volcano named Tambora on the island of Sumbawa in Indonesia erupted in April, 1815. Two other major eruptions preceded this one. Soufriere on Saint Vincent in 1812 and Mayon in the Philippines in 1814 were also considered to be especially intense volcanic eruptions. Tambora was the most productive. It was estimated that Tambora ejected an incredible 35 cubic miles of rock, dust and debris into the air. Collectively, the material from these three eruptions created a thick layer of volcanic dust in the atmosphere 'that girdled the globe and lasted for years'. (1) The Sun could not shine on the ground and the Earth cooled down. The cooling effect was serious in faraway New England. A history of Madison County included the following report. 'There was frost every month. The crops were cut off, and the meager harvest of grain was nowhere near sufficient for the needs of the people. The whole of the newly settled interior of New York was also suffering from the same cause. The inhabitants saw famine approaching. …Every resource of sustenance was carefully husbanded; even forest berries and roots were preserved. The spring of 1817 developed the worst phases of want. In various sections of the county families were brought to the very verge of starvation.' (2)

New England is far away from the source of the problem, which actually occurred about half a world away, but it never-the-less was not immune. The effects of these eruptions were felt worldwide. 'Throughout the northern hemisphere, the weather was abnormally cold. In England it was almost as cold as the United States., and 1816 was a famine year there and in Germany and France. Actually the year 1816 was just one of a famous series of cold years from 1812 to 1817. Everywhere, temperatures were lower than usual, and in the United States the depression of summer temperatures was the lowest on record.' (2) Cold weather is often accompanied by snowstorms and there was no exception in the Tambora case. 'No account of major blizzards or extraordinary periods of cold is complete without at least passing reference to the year 1816, the famous "Year Without a Summer". In that year the problem was not so much a cold winter but an abnormally cold summer. It was preceded by a late and cold spring and followed by an early and cold fall, so the frigid summer of 1816 brought widespread hardship and misery. From June 6 to 9 frost occurred every night from Canada to Virginia. Ice was an inch thick on standing water in Vermont. Everywhere, people shivered, broke out their winter clothing, and watched helplessly as their gardens and crops blackened in the cold. Newly shorn sheep died, and millions of birds perished. A light snow in much of New England and Western and Northern New York State on June 6 was followed by a moderate to heavy snow in New England on June 7 and 8. In Vermont there were drifts 18 to 20 inches deep.' (2)

All of these miseries resulted from the eruption of just three volcanoes. Reports of some other eruptions follow a similar pattern. On June 6, 1912, Mount Katmai in Alaska exploded. This eruption is considered by some to have been the most violent of the twentieth century. It is only fortunate that it occurred in a relatively remote area. Otherwise the destruction would have been costlier. This volcano produced very serious darkness and in this case it lasted for several days. '…Mount Katmai ejected tremendous quantities of ash into the atmosphere.' A ship was in the harbor at Kodiak which is 100 miles from the mountain. 'At about 4 P.M. on the day of the eruption, Perry (the captain of the U.S. Coast Guard cutter Manning) noticed a peculiar black cloud approaching from the northwest. Soon course gray ash began falling, accompanied by thunder and lightning, and the sky turned dark. Five inches of ash had accumulated on the deck by the next morning, and by noon the ash fall resumed. Now everyone was becoming concerned. At 2 P.M. pitch darkness set in. The radio receiver gave forth only static. Through a sleepless night, the crew anxiously awaited dawn. It failed to appear; all sunlight was obliterated. The cloud of falling ash was so thick it was impossible to see a lantern at arm's length. …The decks had to be constantly shoveled and sprayed with water…On the afternoon of June 8, two days after the arrival of the cloud of night, the ash fall decreased, the sky turned reddish, and objects became dimly visible.' (3) At Kodiak, the total darkness lasted 60 hours. During the next few weeks, a reddish haze from the fine dust in the atmosphere was observed throughout the world.

While the explosion of Tambora was the most powerful up to that time, Krakatoa in 1883 has been considered the most powerful of all time, which is of course a very serious claim. The effects of this explosion were seen, heard and felt literally around the world. In May of that year, activity started with fountains of liquid lava and a giant column of steam. Apparently people came to the island just to get a closer view. On August 26, a series of sharp explosions occurred and they continued all night. Then during the morning of the next day, the mountain blew up. The sound traveled all the way to Rodriguez (an island east of Madagascar in the Indian Ocean) in four hours and sounded like the distant roar of heavy guns. Subsequently, the sound waves traveled all the way around the world three times. Monstrous waves were produced and swept many villages away. Of more immediate interest was the darkness which followed. 'The eruption, it is estimated, blew one cubic mile of material to a height of 17 miles, and the dust was carried completely around the Earth several times by the high-altitude currents. Soon areas within 15 degrees of the equator, then gradually the entire Earth, were witnessing brilliant glows in the skies after sunset and before sunrise because of the dust particles in the air.' (4) There was an accompanying decrease in solar radiation reaching the Earth. 'Measurements of the amount of solar radiation reaching the Earth's surface show a marked decrease (to less than 88% of the average) in 1884 and 1885, corresponding with Krakatoa's eruption. This supports the idea that the low global temperatures of the late 1880s and early 1890s were due to the dust from Krakatoa.' (4) The drop in temperature was not as dramatic as the drop associated with Tambora and the difference is attributed to the lower volume of material ejected into the atmosphere.

Darkness was also involved. 'The enormous load of ash sent hurtling into the air brought days of darkness. At a distance of 103 miles, the darkness lasted 22 hours, and at a distance of 50 miles, for 57 hours.' (5)

Other major eruptions involving large quantities of material could also be mentioned. 'When Bandai erupted in Japan in 1888, it cast up almost three billion tons of material and blew off one of its peaks. (6) Pelee, a volcano on Martinique in the West Indies, blew up at 7:50 on the morning of May 8, 1902. (7) The eruption in the Aegean Sea about 1400 BC has left the Islands of Santorin as the remains of the rim of the cauldron. Fourteen cubic miles of island disappeared in this event and it is thought to have caused the darkness reported in the Egyptian Papyrus Ipuwer which says that 'the land is without light.' Further, in 'the Papyrus Anastasi IV the years of misery are described, and it is said: "The Sun, it hath come to pass that it riseth not"' (8) That explosion might also be responsible for the darkness reported in the Book of Exodus and which accompanied the Children of Israel during their wanderings in the desert.

Mount St. Helens is a more recent example of the darkness which will result from volcanic dust. A giant column of smoke and ash rose to a height of sixty thousand feet in just a few minutes. The ash turned day into night and got into everything. (10)

The dust in the atmosphere not only causes darkness, which is unwelcome, it also results, because of the darkness, in cooling, which is even more unwelcome. Any surface that is isolated from the Sun will be immediately cooler than any nearby surface exposed to the Sun. The Sun simply must shine directly on the Earth in order for it to be warmed into a temperature range which is suitable for normal life activities to be carried on. Extensive cloud cover will result in extensive cooling. Persistent cloud cover will result in persistent cooling. Since there is only a narrow temperature range wherein normal life activities can be carried on, it is immediately apparent that the Earth and its residents cannot tolerate any significant upset in energy balance. The source of the Sun-blocking material is of secondary consequence to its effect. That effect will always be to cause darkness and cooling and neither of these developments is tolerable for life on Earth.

When the number of active volcanoes on the Earth is considered, it is a wonder that there isn't more trouble with temperature control. The number of occasionally active volcanoes is between four and five hundred and there are several hundred more that are considered extinct. Stirring up any significant percentage of these volcanoes at the same time would altogether likely be catastrophic for life on the Earth because the cloud cover produced would chill the Earth into an intolerable temperature range.

7.1.5   Asteroid Winter

Any object which can produce a high pressure wave near the surface of the Earth, would stir up a great cloud of dust and debris. As higher and higher energies are involved, more debris would be stirred up. The exploding object would not necessarily have to touch the ground to produce the dust but if the ground was contacted, the debris production would be worse. Any explosion near or at the ground would result in a massive cloud being formed. It is not surprising then, that atmospheric dust production, followed by darkness and cooling is expected to accompany all asteroid impacts. In fact this expectation is so high that every time major asteroid impacts are discussed, nuclear winters are declared to follow.

'The Yucatan Peninsula appears to have been stuck by an asteroid between ten and twenty kilometers in diameter (the Chicxulub impact crater)…exploding with the energy of 100 million megatons…ejection of trillions of tons of dust into the atmosphere…almost complete darkness lasting for several months, and the severe cooling which accompanied it.' (9)

Further comment on the darkness and its accompanying cold is offered by another author. 'The surviving creatures below (the vast layer of dust, smoke and debris from the impact) probably thought that night had come early, but it was night without a moon, without stars. The dinosaurs could not see their own claws in front of their faces. Morning would not come for several months. A few animals had avoided the initial destruction, and at first they seemed to manage surprisingly well. Most of the plant eaters could still find food, although the settling dust added a gritty texture to it all. Some of the carnivores were accustomed to hunting in the dark, although they had never experienced blackness such as this. But the ultimate source of all food, the Sun, had been effectively blocked out. Without sunlight, there was no photosynthesis, no creation of sugar and starch from carbon dioxide and water. Unseen by the animals, the plants were turning from green to yellow, and then to brown. Without sunlight, the temperature of the Earth began to drop. On much of the land the temperature soon fell below freezing. Only those fortunate few animals that had already begun to hibernate didn't notice. …Virtually every animal and plant died…It was a miracle that any of the higher life forms made it through…The reptilian dinosaurs, both on land and in the sea, had vanished forever.' (11) In this discussion the author is simply recognizing that whatever creatures were not killed by the impact, would die soon thereafter because of the darkness and cold. If there is nothing to eat, it only takes a few days to die. It is understood from comments such as this that the period of extreme darkness and cold was long enough to ensure that no animal would survive. In spite of this understanding some commentators like to hedge their bets and suggest that less than 100% would be killed. 'Hurtling at velocities up to twice that of the asteroids…a comet…could hammer the Earth with such force that most of the life forms on the planet would be annihilated. This is exactly what seems to have happened … when the dinosaurs were extinguished along with 70% of all other animal and plant species.' (13) Similarly, another commentator offers; The greatest of the mass extinctions occurred at the end of the Permian period when the vast majority of all species were exterminated. (16) However, justification is never offered to explain why a few would survive.

The Chesapeake Bay impact site is a relatively recent discovery. As the name indicates, it is located at Chesapeake Bay but its large size causes it to extend, not only across the bay but up onto the mainland as well as out into the ocean. It is a very large impact site at 85 km in diameter, which indicates that it was formed by a large object. Various estimates have been made of the impactor size and they fall in the range of 3 to 5 miles. A lot of material would have been stirred up and projected into the air from the impact. '… a large amount of water would have been vaporized, generating a very significant aerosol plume. Vaporized and fragmented rock and sediment would be entrained with the steam explosion to produce the plume. The Chesapeake Bay impact released 100 times the combined energy of all existing nuclear weapons and its environmental ramifications would have been worldwide, including a significant drop in temperature(similar to a nuclear winter) due to the dust.' (15)

In this example, one of the results of an asteroid impact is compared to a nuclear winter which presumably is referring to the results of exploding numerous atomic bombs in the atmosphere. In this case whatever effect the asteroid had is comparable to exploding all of the nuclear bombs in the entire world and then ordering up another 99 times as many and exploding those also. If all of the atomic bombs in the world were exploded in one day, a nuclear winter would certainly be expected. Actually it seems reasonable to expect it if only a fraction of the total arsenal was exploded. If 1000 atomic bombs were exploded, world-wide cloud cover would be a reasonable expectation. As the number of exploding bombs increases, there would expectedly be even more cloud cover. Supposedly the more cloud cover there was, the more difficult it would be for it to dissipate and let the Sun shine once again. The above comparison, however, is being made to 100 times the nuclear arsenal to emphasize that this impact was very serious and would have resulted in the annihilation of life on the Earth. More than 100% of animal life cannot be annihilated but if it could have been, in the case of the Chesapeake Bay impact, it certainly would have been.

Any animals not destroyed by the impact and its immediate results would be destroyed a short time later when their food supply ran out. How long would that take? If the food supply ran out in one month, most animal life would perish within the next month. Occasionally people go on hunger strikes or extended long-term fasts. In many of these cases, a person can continue to live for more than a month after their last meal. If this were also true for animals, we might expect a few to still be alive after a month. We would not likely expect any to remain alive after two months. With this line of reasoning, within three months of an impact, all animal life would have perished from starvation. Any further suggestion that some would still be alive only recognizes that mass extinction is really expected to follow an impact such as the Chesapeake Bay impact.

The notion of mass extinctions and nuclear winters seems to be no stranger to the popular press and its contributors. After discussing the Tunguska event, one commentator offers the following: 'Every 5,000 years, a 200-meter-wide rock strikes with the force of a 100 megaton bomb, which is capable of destroying a large city. Every 300,000 years, a one to two kilometer-wide asteroid crashes, initiating a short-duration global winter. Every 100 million years, a bigger asteroid hits, producing the world-wide calamities that have punctuated evolutionary history. There are 14 known mass extinctions in Earth's geological record. So far, only one – the Chicxulub impact – has been linked to an asteroid or comet.' (17)

Even much smaller asteroids are thought to have produced widespread cooling. The Ries crater of Southern Bavaria, is about 24 km. in diameter and is thought to have been formed by an asteroid about one mile in diameter. However, the effects are expected to have been world-wide. 'The blast wave would have destroyed all life to a radius of hundreds of miles, while the earthquake shock must have been felt around the world. Of the trillion tons of material blasted from the crater, perhaps a billion tons of fine dust remained suspended in the atmosphere, darkening the sky all around the globe.' (18)

An asteroid 1.5 km. in diameter would be expected to produce a crater in the 15 to 20 km. diameter range. There is a certain expectation that even an asteroid this small would have devastating results worldwide. '… wondering what newly-located 1997XF11 might do. He knew it was a near-Earth asteroid by its motion. Some early, simple calculations suggested that in the year 2028, the 1.5-kilometer-wide rock would approach precariously close to Earth. If it hit, millions would die from the impact and subsequent famine produced by a decade-long, planet-wide winter.' (76)

7.1.6   Combination Winter

The winter that would follow a major asteroid impact would actually be a combination of an asteroid winter and a volcanic winter. Following the impact, hundreds of volcanoes would erupt all over the Earth. The dust and ash from every volcano would augment the dust produced directly by the impact. The chill resulting from the asteroid dust cloud would be reinforced by the effluent from all of the volcanoes. As discussed previously, there were three types of volcano; igneous provinces, cone volcanoes and underwater rifts or ridges. Since all of these types of volcano are found world-wide, the dust developing from an asteroid impact would have been quickly augmented by volcanic clouds all around the world. Within a few hours after an impact, the entire world would have been in total darkness which would have been the beginning of a post-impact winter.

7.1.7   Repeated Winters

The Earth has experienced numerous asteroid impacts similar to or even greater than the Chesapeake Bay event as evidenced by the following. Tektites are small tear-drop shaped lumps of glass and are found over much of Central Europe. 'A dozen or so other tektite fields have been recognized around the world, each apparently corresponding to the melted debris from a very large impact.' (18) The tektite evidence has also been used by another group to identify major impacts and their accompanying mass extinctions. 'In the Geological Survey Office in Ottawa, … Wayne Goodfellow studies evidence he has gathered from around the world related to earlier mass extinction (i.e. earlier than the Chicxulub event in Mexico) episodes, 230, 365, and 445 million years ago. He finds … small glass-like spheres believed to be produced only at the high temperatures created by an impact. Statistically, a comet of this size should clobber the Earth every 100 million years.' (19) In other words, five major impacts are being recognized by these comments and more are expected to arrive on a regular basis.

From these particular sources, opinion varies from five major impacts, to "a dozen or so", to "14 known mass extinctions". A review of crater data generally supports these comments. The Earth Impact Database which is maintained by the University of New Brunswick in eastern Canada lists six craters which have diameters equal to or greater than Chesapeake Bay crater discussed above. These include the Acraman in Australia at 90 km., the Chicxulub in Mexico at 170 km., the Manicouigan in Canada at 100 km., the Popigai in Russia at 100 km., the Sudbury in Canada at 250 km., and the Vredefort in South Africa at 300 km. This database does not include the Ishim crater in Kazakhstan at more than 350 km. but does include six more with diameters exceeding 50 kilometers. Supposedly, a post-impact winter would have accompanied all seven of the large ones as well as, perhaps to a lesser degree, the other six as well. This line of reasoning implies that between seven and thirteen mass extinctions have happened on the Earth due to the combined effects of direct asteroid impacts followed by cold post-impact winters lasting months or possibly years.

7.2   The Transition

7.2.1   Introduction

A post-impact winter is expected to follow a large asteroid impact. However, a post-impact winter would not have been survivable. Life did survive as witnessed by the diversity of life on the Earth at the present time, even though many local eco-systems have been lost and many species have been lost. Therefore it may be concluded that a post-impact winter began but it did not continue long enough to totally terminate both plant and animal life on the Earth. The Great Ice Age happened instead. While a post-impact winter and an ice age do have several characteristics in common, the developing post-impact winter was truncated and transformed into the Great Ice Age instead because of the unusually warm ocean and by the timely demise of the massive dust cloud.

7.2.2   Dust Cloud Development

As discussed above, a post-impact winter is expected to immediately follow the impact of a large asteroid with the Earth. The direct cause of this winter would be atmospheric and surface cooling. This cooling, in turn, would be caused by the very large, dense, dust cloud that would form as one of the immediate effects of an impact. When an asteroid smashes into the Earth, some of the material of the asteroid and some Earth material would be pulverized. Surrounding material would be stirred up. The net result expected is the formation of a large spreading cloud, which will rise many thousands of feet into the air. 'The explosion produced by such a large impact would itself be damaging, but even more important would be the consequences of lofting 100 trillion tons of dust into the atmosphere. Scientists have calculated that this quantity of dust would have produced a pall over the whole globe with a duration of at least several months. During this period of global darkness, neither light nor heat was available to sustain life. With photosynthesis at a complete standstill and the Earth plunged into a deep cold, it is no wonder there was a great dying.' (32). From this understanding, the extent of the asteroid-produced cloud was the whole world. There would be no getting out from under such a cloud.

Secondary effects of the impact include volcanic activity which also would produce an abundance of cloud. The clouds produced by a single volcano can be even darker than a thunderstorm and have been reported so dark that a person could not even see his own hand when it was right in front of his face. (33) An experience like this would be similar to being in a mine when the lights are turned off. When this happens, it simply is not possible to see anything at all. Since the post-impact volcanic activity included the production of igneous provinces, cone volcanoes and underwater rifts – all of which would have produced cloud, it is easy to understand that, following a major impact, the entire world would have been darkened. It would not only have been virtually pitch black, the area covered would have involved the entire world. A more discouraging scene is hard to imagine.

7.2.3   Moisture Cloud Development

As discussed in the previous chapter entitled; The Vapor Canopy Hypothesis, there is abundant evidence that the Earth was once universally warm. This evidence includes the trees of Axel Heiberg Island, the coal at Spitzbergen Island and the coal and plant remains found on Antarctica. When referring to the ancient fauna of Axel Heiberg Island, which is an island close to the North Pole, it has been referred to as a tropical paradise. 'An ancient tropical paradise, complete with turtles and crocodile-like lizards ... a hot steamy world ... anywhere a champsosaur is found could not have been very cold.' (61) The authors further state that the deep ocean was warm at 18 Celsius. Others agree with this conclusion. Immediately prior to the Chicxulub impact, the Earth was warm. This was the Cretaceous Age of Chalk which is believed to have terminated when the Chicxulub impactor crashed into the Earth and killed all the dinosaurs. This world view is captured by the following comment which is quoted from Sir Author Holmes. 'The mean annual temperatures ... were 25C ... The general fall in temperature since the Chalk was deposited has been estimated from pollen and other plant remains as well as by the oxygen-isotope method.' (20)

While the ocean is recognized as having been warm, the effect of the post-impact volcanic activity would have warmed it even more. Hot lava oozed out of the Earth and spread around forming Igneous Provinces both above water and below water. Some of these provinces, as discussed in the previous chapter entitled, More Bad News, were similar in size to Europe. Also, the floor of the Atlantic Ocean was ruptured, hot lava welled up and formed the mid-Atlantic rift and its accompanying mountain ridges. Numerous other underwater mountain ridges also formed. Further, the ocean is home to hundreds of cone volcanoes, many of which are located close to the major faults in the Earth's crust. Much of this activity would have caused the ocean water to locally boil with the overall result being an ocean with an average temperature that was much warmer than it was prior to the impact. If it had been 18 to 25 Celsius prior to impact, it could only have become warmer than this as a result of the volcanic activity.

Warm water evaporates. The warmer it is, the more it evaporates. Therefore one further result of the impact of a large asteroid was to cause the entire ocean surface around the world to evaporate just as if the whole ocean had been placed on a stove with the burner turned on. The burner was in fact turned on when the hot molten lava from beneath the crust was forced up forming the underwater igneous provinces, the mountain ridges and the cone volcanoes all of which effectively heated the entire ocean. As it was heated, it evaporated at rates directly dependent on its higher temperature.

The prodigious evaporation caused the air over the ocean to become extremely humid. Almost immediately above the ocean surface moisture clouds would have formed in great banks covering thousands of square miles and extending for thousands of feet into the air. As these extensive cloud banks drifted away they were continually replenished by a seemingly endless supply of moisture from the overheated water. Any parcel of air that wasn't quite humid enough to form cloud right away would have done so as soon as the moisture-saturated air drifted over the cooler sun-deficient land.

An example of how the rapidly evaporating ocean may have appeared is provided by the Niagara River. The Niagara River flows from Lake Erie to Lake Ontario and partway along its journey it plummets over a precipice forming both Niagara Falls and the American Falls. The violence of the drop causes a great cloud of mist to continually form and rise several hundred feet into the air. Much of the surface of the ocean would have appeared very similar following the asteroid impact and would have continued this way as long as it was over-heated and evaporating at a high rate.

Another example involves the formation of the Island of Surtsey. This island formed in the Atlantic Ocean about seventy kilometers south of Iceland in 1963. As an example, it is much more representative of the situation following the impact because in this case both moisture and dust rose from the ocean surface as a dense breath-stifling cloud. The Icelandic National Tourist Office took numerous photographs of the island as it formed. When the ocean started to evaporate following the impact of the asteroids, much of the ocean surface would have appeared like it did during the formation of Surtsey. The fact that Surtsey formed right on the mid-Atlantic Rift makes it even more representative of the post-impact situation.

Storm clouds are primarily water vapor clouds. When water evaporates from a body of water, it will rise unseen into the air and simply drift away. However, as soon as its temperature drops to a critical level, the previously-invisible vapor becomes visible as a cloud. There are several ways in which this phenomenon can be observed. On a warm summer day, as the Sun shines on the ground, it will be heated and moisture will be released into the air. In particular if the Sun shines on a freshly-plowed field of dark earth, the moisture in the disturbed, uneven surface will be released into the air and rise upwards. As it rises it will cool because air expands as it rises and expansion results in cooling. At a few hundred feet up, a critical temperature will be reached and a cloud will form. This type of cloud is commonly called a cumulus cloud because of its pillowy appearance. Such clouds indicate an updraft and both birds and sailplane pilots make use of such information to locate an updraft and gain altitude. Another example can be observed during the winter when mist can be seen rising from a lake or pond. In this case, even though very little moisture is being released from the water, the air is cold enough to cause a mist to form. A third way the moisture-forming activity may be observed is in the mountains. As air rushes up the side of a mountain it will chill. If the critical temperature is reached before it gets to the top, a cloud will form. Often such clouds form at a very distinct elevation even allowing a person to have his head in the cloud and his feet under it. From a distance, it may appear that the cloud is stationary on the mountain top whereas in reality the air might be moving at considerable speed.

Following the impact, a great moisture cloud formed when the warmth of the ocean water caused the surface layer of water to evaporate into the air. The air was also warm over the ocean and so clouds might not have formed right away. However, the land had been chilled by the great dust cloud. Therefore as soon as the moisture-laden air drifted over the land, clouds would have formed in abundance. The water temperature was much higher than room temperature and massive quantities of moisture were released. (It was already at room temperature before the impacts so the volcanic activity just raised its temperature more.) Extensive cloud production and torrential rains resulted. This would have been very similar to the situation in the tropics at the present time where warm moist air rises and tropical cloudbursts result. As the Ice Age developed, the cloudbursts would have been virtually continuous as soon as the cooler land was beneath them. But, a little further north where the temperature was below freezing, snow would have fallen instead of rain and it would have fallen in abundance.

7.2.4   Two layers of Cloud

Immediately following the impact, the dust cloud formed. It rose to high altitudes and then spread out to cover the entire world. 'As the fireball reached the top of the atmosphere, it bobbed like a cork, floating on the cooler air beneath it, but it had nothing to hold it together and it began to spread out over the entire globe.' (54) Similarly, volcanic clouds will also rise to extreme altitudes as discussed previously with respect to Tambora and Krakatoa.

The moisture clouds on the other hand were not projected to extreme altitudes but formed and drifted along much lower down. Never-the-less these moisture clouds were also very effective at blocking light from the Sun from reaching the surface of the Earth. Occasionally when a thunderstorm passes overhead, the sky becomes dark. The clouds appear almost black. But clouds do not really have various shades from bright white to dark gray as they appear. Instead, the shade of a cloud is determined primarily by its thickness. The thicker it is, the darker it is because the greater thickness prevents more light from getting through. The darkest period of a thunderstorm occurs when the thickest part of the cloud is directly overhead. The reason that it becomes so dark is simply because these clouds have great vertical extent and sunlight cannot penetrate all the way through. For example, 'The moisture-laden air from the south began to mushroom upward until, within an hour, the tops of the thunderheads had reached an altitude of 60,000 feet in the evening sky, darkening the land before sunset.' (58)

Due to the impact it was already dark because of the dust cloud higher up so the addition of a moisture cloud only added to the totality of the darkness. Normally a moisture cloud appears with a degree of darkness dependent on how thick it is and how much light is shining on it. In this case there wasn't any light shining on these moisture clouds at all because they were in the shadow of the dust cloud higher up. Therefore they were dark all the way through with the result that no light from the Sun reached the Earth at all. It would be hard to imagine the utter despair that would have accompanied such a scene.

7.2.5   The Warm-Cold Anomaly

The consequences of the asteroid impact involved both heating and cooling. The volcanic activity heated the oceans until they were, on the average, considerably warmer than room temperature. The direct contact of molten magma with the water would actually have caused it to boil in many areas. The ocean would have become like the pools of water that form near hot springs. Some of these pools are warm enough to allow bathing but some are too warm to permit bathing as their temperature exceeds what a human body can stand. The volcanic activity would have raised the average temperature of the ocean into a range which was well above room temperature, resulting in prodigious evaporation.

On the other hand, since the double layer of cloud blocked the heat from the Sun as well as the light, there would have been cooling. In particular it would have been cool over inland areas because they were further from the influence of the warm ocean water. Since the heat from the Sun was cut off, the only heat that would have reached the inland areas would have been from the relatively-warm moisture-laden clouds that drifted in from the over-heated ocean.

The situation could therefore have been considered anomalous. While it was dark over both the water and the land, it was warm over the ocean and chilly over the land. The entire Earth had suffered the unthinkable trauma of an asteroid impact and was now in a state that was untenable for animal life. Further consequences of the impact would depend on the longevity of both of the two different types of cloud. As long as the dust cloud persisted, heat from the Sun would have been blocked. If the resulting chilling trend continued until the ocean cooled down, the Earth would have gone further and further into the deep freeze. This would have been the post-impact winter that so many commentators expect from a major impact. 'Though the impact occurred in one region of the world, its environmental ramifications would have been worldwide, including a drop in temperature similar to a nuclear winter due to ejected dust and aerosols.' (79) Such comments appropriately acknowledge the chill brought on by the cloud but do not recognize the heating affect that the asteroid impact had on the ocean. While the chilling conditions were suggestive of a developing post-impact winter, the unusually-warm ocean would not have permitted this to happen

The great dust cloud produced an immediate and significant chill over the land. This effect of cloud cover can be experienced even on a warm summer day. If direct sunlight is suddenly cut off by a thick bank of cloud, air temperatures will immediately dip and a sweater might be needed. Or simply standing in the shade of a cloud or a tree can be welcome relief on a hot summer day. Since the post-impact-winter-causing cloud would have cut sunlight off completely, temperatures would have plummeted and freezing would soon have followed. With these developments, a post-impact winter would have been underway within hours of a major asteroid impact. However, in order for it to have been sustained, the cloud cover must have remained intact not allowing any significant amount of sunlight to get through. If either layer of cloud started to break up, any discontinuities in the other layer would have allowed sunlight through, the Earth would have been warmed and perpetual winter conditions could not have been maintained.

A world-wide cold-causing dust cloud would have been a necessary condition for a post-impact winter. However in order for this winter to have developed properly, it would also have been necessary that the ocean not have been warm. An unusually warm ocean would have partially offset the chilling condition produced by the absence of sunlight and relatively warm air drifting in from the ocean would have tempered the chill caused by the sunlight-blocking clouds. The dust cloud would have been expected to persist for several years because the dust clouds produced from recent volcanoes have persisted for that long. 'Measurements of the amount of solar radiation reaching the Earth's surface show a marked decrease (to less than 88 percent of the average) in 1884 and 1885, corresponding to Krakatoa's eruption. This supports the idea that the low global temperatures of the late 1880's and early 1890's were due to the dust from Krakatoa.' (4) However if the ocean warmth also persisted this long, the offsetting factors of chill from the dust cloud and heat from the warm water would have prevented a post-impact winter from becoming established.

Significantly, the chill factor which would initiate a post-impact winter is one of the factors which would be necessary for the initiation of an ice age. It is therefore to be expected that these two phenomena are occasionally confused. 'We do not know how long this dust pall remained, or what effect it might have had on the world's ecology. If it were thick enough, it could have had long-term influence on the climate even triggering an ice age.' (31) While ice ages are not really triggered, this type of comment captures the expectation that both a post-impact winter and an ice age involve cold.

It was the combination of heat and cold that enabled the greenhouse effect to be retained and it was the combination of heat and cold that provided the necessary and sufficient conditions for the Great Ice Age to develop.

7.3   The Ice Age

If a large asteroid should impact the Earth, both moisture clouds and dust clouds would form and they would create a synergy, which would be detrimental to all forms of life. An abundance of moisture clouds would bring copious precipitation. An abundance of dust clouds would cause the Earth to cool. Excessive rain would result in flooding and erosion while cooling would shift the environment into the freezing zone and make survival more difficult or impossible. With excessive snowfall, all activity would grind to a halt. Excessive rain would do a lot of damage but it would drain away whereas snow would not drain away but would keep right on building up wherever it landed. When everything was buried in snow, life would not have been able to carry on.

7.3.1   Ice Age Criteria

There are two conditions that are absolutely necessary for an ice age to happen. The air temperature over the land where the snow is to accumulate into massive glaciers must be kept below freezing all year so that the snow that falls will not simply melt away. Secondly, there must be a prodigious amount of atmospheric moisture available to precipitate as snow to enable the massive glaciers of an ice age to accumulate. These two criteria are directly connected. In order to keep the land below freezing while the massive quantities of snow were accumulating, the two layers of cloud blocked sunlight from reaching the Earth for virtually twelve months of the year. In order to block sunlight in this way, the clouds must have been thick, extensive and continuous. By meeting these criteria, the moisture layer would also meet the criteria of being able to transport huge quantities of moisture at the required high rates. This in turn necessitates that the ocean remain above a certain critical temperature in order to be able to evaporate its water fast enough to keep the cloud continuity intact. As soon as ocean temperature dropped below a certain critical level, evaporation would have dropped off and cloud continuity would have been lost. The Sun would then have shone on the ground and warmed it allowing last winter's accumulation of snow to melt. The accumulation phase of the Ice Age would then have been over and melting would have proceeded.

Fortunately, ice ages are not triggered. The environment is not all wound up tight, in an extremely unstable state just waiting for some minor development to occur so an ice age can get underway. Things happen in nature because the necessary and appropriate physical conditions are in place so they can happen. Then, they not only can happen, they will happen.

Neither does an ice age happen because the Earth simply turns cold. It is a fact of nature that if the temperature drops too far, a crust of ice will form on water. This crust of ice will prevent any further escape of moisture into the air and so snowfall in the area will drop off. This happens very predictably in locations which are downwind of large lakes. Snowfall can be abundant up until the lakes freeze over but then it will drop off substantially. An example of this type of phenomena has been observed at Buffalo, New York. Being downwind of Lake Erie, Buffalo often experiences heavy snowfalls which have even amounted to more than three feet in one storm. 'Early in the season 40.5 inches of snow fell on Buffalo in a four-day storm ending December 2, 1976.' (82)

7.3.1   Vapor Capacity of Air

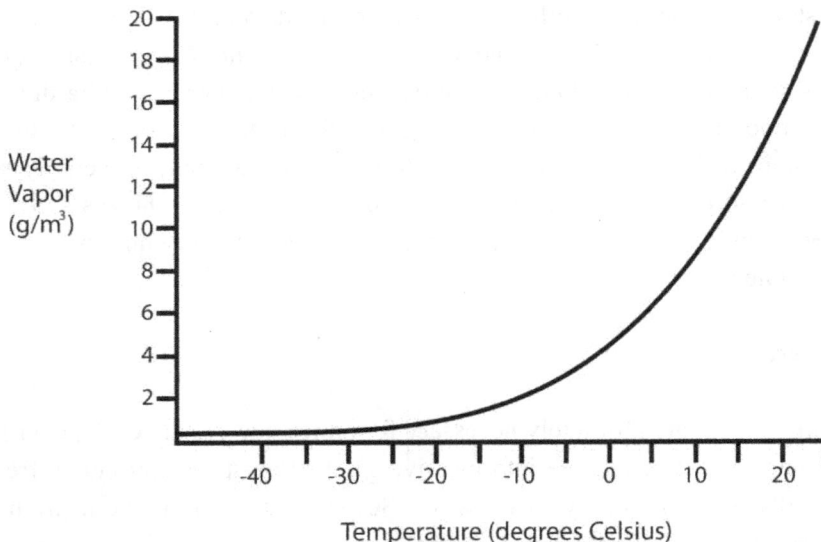

Temperature (degrees Celsius)

The second effect from extreme cold relates to the water vapor capacity of air. As air gets colder, it has less capacity to retain moisture. This is actually the reason that snow falls. It falls because the air chills and cannot retain the moisture. Very cold air retains very little moisture so if it should be chilled a little further to extract whatever remaining moisture it has, very little snow would result. Therefore if the Earth turned cold and the oceans froze or became covered with drifting ice, the ocean water would be unable to evaporate and the cold air would be unable to transport any significant amounts of moisture. In such a situation not only would the moisture source be turned off, but the moisture transport mechanism would also be turned off. Snowfalls under these circumstances would be minimal. An ice age would therefore not result.

Mathematical models of the Earth's orbit (used in the Milankovitch or Astronomical Theory) indicate that, over long periods of time, the Earth will drift a little closer to the Sun as well as a little farther away. The variation in solar energy received by the Earth during these excursions would be in the range of one or two percent. (40) A drop of two percent would certainly chill the Earth and an increase of two percent would heat it but when heat regulation and transfer mechanisms (including atmospheric circulation and ocean currents which operate throughout the world) are factored in, the actual temperature variation would be very small. (40) Such circumstances would not be sufficient to cause and sustain an ice age. For past orbital changes, both the seasonal and the mean annual temperatures fail to produce glacial advances of the magnitude that were required. (42) It was therefore concluded that the Milankovich Theory did not offer an appropriate explanation of the Ice Age. In fact, it is very unlikely that the Milankovich cycles can start an ice age because the cycles are just too weak. (88)

It must also be noted that the small changes anticipated by this theory would occur over extended periods of time involving thousands of years. (72) The evidence however, contradicts this assertion and indicates that the Ice Age ended abruptly and that melt-water formed massive flows with rivers being hundreds of feet deeper than they are at present. '... a larger stream ... must have flowed when the ice cover melted ... prodigious floods of almost unimaginable magnitude accompanied the melting of the ice cover.' (43) Of even greater consequence however, the Milankovitch Theory does not offer any hint at all to explain how more than 2000 feet of ocean water evaporated, or how it evaporated above the critical rate which was necessary to enable the glacial ice to actually accumulate.

As mentioned earlier, the further inhibition inherent in the 'cold Earth' (Milankovitch) approach relates to the greenhouse gases. Water vapor is a major greenhouse gas so if the entire world became chilled, the warming effect of the water vapor would be lost. This means that the Earth would become even colder and only stop getting colder when any further reduction in temperature did not result in any further reduction in water vapor in the atmosphere. The end result would be that the entire world would freeze solid and simply stay that way indefinitely.

A further aggravation with the whole approach relates to the reflectivity of snow. If the land should become chilled enough for even one inch of snow to cover it, the heat from the Sun would be reflected away. This would further establish the chill and an ice age would not result.

An ice age would be a period of time when snowfall continued for most or all of the year. There would be no intervening summer periods, which would be warm enough to melt all of the snow, which accumulated the previous winter. As long as a winter's accumulation of snow could melt again the next summer, an ice age would not be happening but only snowy winters. In order to have an ice age, more snow must fall than the amount which melts on an annual basis. This would result in a buildup of snow and an ongoing buildup, year after year, would be an ice age.

An example of a territorial ice age is found in Greenland. Currently, certain areas of Greenland are experiencing a buildup of ice. (41) It can therefore be concluded that these particular regions of Greenland are experiencing the build-up phase of an ice age. Meanwhile coastal regions are experiencing a decrease in glacier cover and can therefore be said to be in the terminal or end phase of an ice age. The great Ice Age of interest however, was not local or territorial but involved the entire world.

The importance of heat to effect the great Ice Age has been recognized by several commentators. 'If we take into account the area occupied by ice during the glacial epoch, much larger than the area of the present polar ice … The usual estimate of its thickness is between six and twelve thousand feet … the water must have come from the oceans … which must have been at least three hundred feet lower … an enormous quantity of heat was necessary …"Some eminent men have thought, and some still think, that the reduction of temperature, during the glacial epoch, was due to a temporary diminution of solar radiation; others have thought that, in its motion through space, our system might have traversed regions of low temperature, and that during its passage through these regions, the ancient glaciers were produced … many of them seem to have overlooked the fact that the enormous extension of glaciers in bygone ages demonstrates, just as rigidly, the operation of heat as well as the action of cold. Cold alone will not produce glaciers.'" (84)

There are two necessary criteria, involving both heating and cooling, which, occurring simultaneously, provide sufficiency, for a world-wide ice age. The heating criteria involves keeping the ocean water warm enough to evaporate the prodigious quantities of water required to build up the vast glaciers. Secondly, the water must be initially warm enough and kept warm enough to evaporate moisture at a rate which would maintain the moisture cloud cover sufficiently continuous to prevent the Sun from getting through and melting the ice. A certain minimum abundance of clouds are required to prevent the Sun from melting the snow. In other words, a certain minimum cloud cover was required to keep the ice-fields cold which means that evaporation had to proceed at a certain minimum rate or the ice would not have been able to accumulate. As soon as the evaporation rate fell below this critical rate, the dependent moisture clouds would have thinned out and allowed the Sun to melt the snow as fast as it accumulated. Alternately it could be stated that the evaporation rate had to be above a certain critical minimum to keep the ice-fields cold enough to cause snow to fall instead of rain. If rain fell instead of snow, the ice would have stopped accumulating. Then with a further slight decrease in ocean temperature (together with the resultant decrease in moisture production and cloud cover), melting would have increasingly exceeded accumulation and the Ice Age would have been over. It might seem contradictory that both heating (evaporation) and cooling (Sun-blocking) were necessary but without these two factors being present to the necessary degree, the Ice Age would never have happened.

Further, the absolute necessity of both factors being simultaneously present is understood from the effect of greenhouse gases. While it is intuitively clear that heating the Earth would not produce an ice age it is also clear from an understanding of the greenhouse gas effect that simply cooling the Earth would be a disaster because the entire Earth would lock up frozen solid. If the Earth were chilled for any reason, water vapor in the atmosphere would be lost and a chilling spiral would develop where the chill further reduced the greenhouse gas effect resulting in more chilling. The oceans would freeze right to the bottom. Life on Earth could not exist. It was absolutely necessary to have a warm ocean at the same time the land was chilled in order not to irreversibly lose the overall greenhouse gas effect. Otherwise the Earth would never again been suitable for animal life.

Both the heating factor and the chilling factor were necessary to have an ice age and together they were sufficient to have an ice age. Therefore an ice age happened. Further they were both necessary in order to retain the greenhouse gas effect or there would have been no recovery from the ice age.

7.3.2 Heating Factors

7.3.2.1 Heat Quantity

In order to build up the massive quantities of snow and ice of the Ice Age, a very large quantity of water had to evaporate. While much of the land had very recently been flooded (as one result of the impact of a shower of asteroids) with puddles and lakes being found everywhere, a much greater quantity of water than this was needed. Also there is always a certain amount of moisture in the air and recent volcanic activity would have added more, but such sources could only have supplied a very small portion of the total evaporant required. Only the ocean could have supplied the prodigious quantities of water that were needed. In fact, the amount of water which is still bound up in the glaciers of the world is always compared to a certain number of feet of ocean.

In both Greenland and Antarctica, the two major remaining ice-fields of the world, the ice is up to approximately two miles thick. If Antarctica alone melted, the level of the oceans of the world would increase by about 200 feet. (34) This means that in order to form only the Antarctica Icefield, a quantity of ocean equivalent to a layer of ocean water, 200 feet deep around the entire world, had to evaporate. When the water requirements of the Greenland glacier are added, another 28 feet of ocean are required. This great quantity of water evaporated, became moisture in the air and then fell as snow over Antarctica and Greenland until these great glaciers were formed.

During the Ice Age, other ice-fields besides Antarctica and Greenland were formed and they would have required an additional equivalent layer of ocean water of similar or possibly even greater magnitude. 'It is generally thought probable that the sea level has stood, during the glacial ages, perhaps 300 feet lower than now.' (35) Other commentators suggest that even more ocean water became tied up in the ice. 'An early and obvious theory was that … sea level was about 125 meters (412 feet) below present sea level.' (77) Based on these comments, 228 + 300 = 528 (or possibly 228 + 412 = 640) feet of ocean must have evaporated to form the glaciers. However, this would just be the ocean water that actually found its way directly into the massive ice sheets. Most of the water that evaporated would not have become glacial ice. Only a portion would have gone directly into the ice fields. Since the Earth is more than half covered by water, it would be reasonable to expect that more than half of the water that evaporated simply fell as rain or snow directly back on the ocean. It would also be reasonable to expect that only a portion of the precipitation over the land actually became part of the great glaciers. Some of it would have been rain and some of it would have been snow which did not become deep enough to press itself down into glacial ice. If only half of the precipitation over the land became glacial ice, this line of reasoning means that less than one-quarter of the water that actually evaporated became ice. (The surface of the Earth is about 70% water.) The total amount of ocean water that evaporated to form the great glaciers of the Ice Age would therefore have been at least four times the amount that became tied up in glacial ice. Therefore, the total amount of ocean water that evaporated was at least 528 x 4 = 2112 feet. Further, this estimate is clearly conservative. Any theory of an ice age must explain the source of the enormous amount of heat that was required to evaporate such an incredible amount of ocean water.

It is understood that a large amount of ocean evaporated but that all of this water was never missing from the ocean at any one time. Much of it evaporated and immediately returned to the ocean, either directly as rain and snow or indirectly from rivers (as it does at the present time).

7.3.2.2 Heat Source

If the Earth was universally warm prior to a major asteroid impact, as has been declared, (25) the oceans all over the world would have been warm. (26) Warm oceans mean that a great deal of heat has been stored and if the Earth chilled into the freezing zone, it would require considerable time for this heat to dissipate. The air temperature in the northern regions could fall below the freezing point and remain there for years before the temperature of a warm ocean would drop to freezing. Until the water reached the freezing point, a warm ocean would be an enormous reservoir of heat, which would be available for evaporation which would continually add moisture to the air to provide an ongoing source of rain and snow.

Just having been heated without having been initially warm, would probably not have been sufficient because any heating, which occurred, would not necessarily have included the high arctic where the islands, as well as northern Greenland, remained ice free during the ice age. (27) They remained ice free because the surrounding water was initially warm as well as deeper.

Some of the heat required for evaporation was available directly from the water because the water was warm – even in the high Arctic. More ocean heat was available from underwater sources of heat. The crust of the Earth had faulted or cracked by shock waves from the impact. These faults allowed volcanic activity to proceed in three different ways. First, molten material came directly up through the faults and formed ridges such as the mid-Atlantic Ridge. This activity would have been accompanied by much turbulence and the sea would have boiled on contact with the red-hot molten rock. Secondly we note that the major faults of the Earth are always accompanied by cone volcanoes. In fact the great majority of the volcanoes of the world occur within a few miles of a fault. It may be that the fault-causing shock and stress weakened the crust allowing hot material from within the Earth to ooze upwards bringing a lot of heat with it. Thirdly, the oceans are home to many Igneous Provinces. These are lava flows that have spread around on the ocean floor instead of piling up into mountains. (63) When all of this hot material became exposed to the ocean water, it too became hot.

7.3.2.3 Ongoing Heating

In fact, these volcanic sources of heat kept right on supplying heat even as the snow was falling and the great glaciers of the Ice Age were accumulating. In particular the ice cores recovered from both

275

Greenland and Antarctica show layers. (54) Since there are about 110,000 of these layers on Greenland and several times this number on Antarctica and they are quite thin (i.e. they only represent about ten inches of snow each), they could not possibly represent years as has been declared. They could however, indicate volcanic activity and that it was ongoing throughout most of the time that the ice was accumulating.

Other evidence supports this conclusion. 'The ice age volcanoes left huge deposits of ash. In the Western United States alone, more than 68 ash falls, coinciding with the Ice Age have been recognized. (Izett, 1981). The size of the ash beds indicates that some of these eruptions were gigantic. An exceptionally large ice age eruption was recently discovered in New Zealand (Froggatt et al., 1986). This eruption spread a distinct layer of ash over at least ten million square kilometers of the South Pacific Ocean. Based on correlations with modern volcanic eruptions, the dust and aerosol loading from the largest … was on the order of the worst nuclear winter scenarios. In these scenarios, almost all sunlight is blocked out for the entire world (Rampino et al., 1985; Froggatt et al., 1986, p 581).' (22) These reports appear to be concerning ash found on land. The source eruptions could have been on land or underwater. In any event, if there were numerous eruptions on land, one would expect that some would also have occurred underwater and been a source of heat for ongoing evaporation.

Initially warm water, (i.e. at 18 to 25 degrees Celsius) without additional heating, would not have been sufficient because of the incredible amount of heat required. It requires 540 times as much thermal energy to evaporate a quantity of water, as it does to change its temperature one degree. Therefore, cooling the entire ocean 25 degrees would not have been sufficient to evaporate the 2000 feet plus of ocean water required for the Ice Age. (Please refer to the Appendix for a detailed calculation.)

Further, the ocean had to evaporate at a certain critical minimum rate or the ice accumulation function would simply have terminated. This would have necessitated a certain associated minimum temperature which would altogether likely have been above room temperature. The maintenance of the required minimum cloud cover would have required a very high evaporation rate and room temperature water would not have been warm enough.

Not only was an enormous quantity of heat required but it was required at a rate which would actually cause the ice to build up. Things had to happen quickly. If the snow had not been produced at a prodigious rate, melting would have matched accretion, as it soon did anyway. During the time of the Ice Age, the Earth was cooling down toward a new climate equilibrium. The last of the dust clouds were dissipating and the Sun was shining through more and more. The snow clouds themselves kept the Earth cool enough a little longer but they could only shield the Earth from the Sun if they were thick enough and continuous enough. Reduced evaporation rates would have allowed the clouds to thin out and indeed that is how the Ice Age terminated. The cooling ocean simply could no longer supply cloud-making moisture at a fast enough rate. Ice accumulation therefore maxed out and the Ice Age went into its recession phase. It was the prodigious rate at which evaporation proceeded and continued to proceed due to ongoing volcanic activity coupled with the chill brought about by the sun-shielding clouds that enabled the Ice Age to happen in the first place. Without plenty of heat being added in a timely manner to an ocean that was already warm, significant quantities of glacial ice would not have formed before the ocean cooled down too much.

The Sun could never have been that source of heat because heat was required deep in the ocean. Further, solar heat would not have come quickly enough but when it did come, it was counterproductive and it melted the snow. On the other hand, underwater volcanic activity would certainly have followed a major asteroid impact but it had to continue long enough to provide the necessary heat in a timely manner. When the volcanic activity subsided, evaporation soon dropped to the critical level and ice accretion ceased.

7.3.3   Chilling Requirements for an Ice Age

The second group of conditions, which were necessary to bring about the Ice Age, caused and resulted in atmospheric cooling. It may initially seem contradictory to require both heating and cooling to effect an ice age but there is significant recognition that it has been far from simple and obvious how the great Ice Age occurred. Numerous theories have been advanced but none of them seem to have withstood serious consideration. In some cases, investigators have basically thrown up their hands in despair of finding an explanation. 'The cause of all these changes (i.e. the formation and disappearance of the ice fields), one of the greatest riddles in geological history, remains unsolved, despite the endeavors of generations of astronomers, biologists, geologists, meteorologists and physicists it still eludes us.' (28) Others are in agreement. 'Scores of methods of accounting for ice ages have been proposed … no theory is generally accepted … hopelessly in contradiction with one another and good authorities are arrayed on opposite sides.' (29) 'More than 60 different hypotheses of the origin of the Ice Age have been proposed and further suggested causes are published every year or two … Similar contradictions appear throughout the literature.' (30) The conditions for an ice age do not readily recur, and it has proven very difficult to postulate the actual environmental circumstances, which resulted in the Ice Age. There certainly is no ice age developing now, even though some ice fields are growing and some are diminishing. 'Actually the screening of the Earth by clouds of dust of volcanic origin was one of the theories concerning the origin of ice in the glacial epochs; however, like heat alone, cold alone would not have sufficed to produce the continental ice covers.' (70)

As mentioned a few paragraphs above, under Ongoing Heating, volcanic activity was still in progress, even as the snow was falling and the ice was accumulating. (22) While underwater volcanic activity continued to heat the ocean, the dust cloud from the volcanism continued to chill the land. As long as volcanic activity continued, both of these effects would be taking place. It was the combination of a warm ocean and cool land that enabled the great Ice Age to happen.

7.3.4   The Chill Factors

As the Ice Age progressed, there were several factors contributing to the chilling of the land.

7.3.4.1 Vapor Canopy Collapse

When the asteroid rushed through the atmosphere, it caused the destruction of the vapor canopy. The vapor canopy enclosed the oxygen-nitrogen layer of atmosphere completely. Locally, it was a layer of atmosphere floating on top of the oxygen-nitrogen layer. It had long-term stability because it was both lighter and warmer than the air underneath. However, when the asteroids came rushing through, stability was lost. The initial stability-destroying factor was the forced air movement directly caused by the asteroid. In front of the asteroid, air was forced aside. Behind the asteroid, air rushed back in. A short-term circulation pattern developed with air from the lower atmosphere moving vertically and then horizontally to fill the void left behind the speeding asteroid. This circulation pattern was localized and would have extended outward for several diameters of the asteroid itself. For example, if the asteroid diameter was 5 miles, the circular pattern of air movement may have extended outward for 20 or 30 miles. As this movement of air took place, the lower, oxygen-nitrogen layer was forced up into the H2O layer. Unfortunately, air cannot hold very much water vapor, which effect would now be referred to as relative humidity. When water is in vapor form, it can remain that way indefinitely as long as the temperature is high enough. However, when water vapor is mixed with air, there is a very definite limit to how much water can remain. The rest will precipitate out right away. The mass of water in the vapor canopy layer would have far exceeded the allowable upper limit for the mix. Therefore it would have rained. The rain would have been a downpour. Violent volcanic activity would have had a similar effect. Wherever a volcano blasted into the upper atmosphere, mixing of air and water vapor would have resulted and that portion of the atmosphere would then have been supersaturated with water. Nature would not allow this and rain would result. With the arrival of the asteroids the vapor canopy started to collapse. Collapse would have been augmented by the pressure waves propagating outward from both the asteroid's pathway as well as ground zero and whatever portion of canopy wasn't destroyed by one factor was probably destroyed by the other. As soon as any portion of the canopy disappeared, its insulating value was lost and local cooling resulted which would not have been reversible. Due to the world-wide development of volcanoes following the impacts of the asteroids, the entire vapor layer of the pre-existing atmosphere was destroyed resulting in the loss of its climate-moderating and heat-insulating characteristics. Extremes of climate for the whole world followed. Temperature reduction near the poles would have been most dramatic while the temperature near the equator would not have dropped at all and might, in fact, have risen.

7.3.4.2 The Dust Cloud

The impacts produced massive sunlight-blocking clouds, which were augmented by volcanic clouds. While volcanic clouds come with various characteristics, depending on the nature of the particular volcano, the impact activated numerous volcanoes so all types were involved. A major impact would stir up hundreds of volcanoes. The darkness produced by the impact cloud would consequently be reinforced. While it cannot get any darker than dark, the additional activity from the volcanoes would virtually guarantee that the whole world was included in the darkness. While total darkness would be most unpleasant, it could be survived until the interruption of photosynthesis killed the plants. However, darkness also means lack of sunlight which means cold and cold cannot be survived.

The aggregate cloud cover produced by the impacts and the resulting volcanoes would reflect some sunlight and block the rest. It is the constant inflow of sunlight toward the Earth that keeps the Earth warm. The drastic cooling would have soon caused all the water to freeze, cutting off drinking supplies Also, the plants would have frozen cutting off food supplies. All creatures need both food and water on a continual basis. Even short-term interruptions would jeopardize survival, but longer term interruptions would be fatal. How cold it would have gotten due to the great dust cloud is speculative but dropping below freezing would be expected to happen within hours. 'Without sunlight, the temperature of much of the Earth began to drop. On much of the land the temperature soon fell below freezing.' (11)

7.3.4.3 The Moisture Cloud

As the volcanic activity warmed the ocean, it commenced to evaporate. The evaporant formed great banks of moisture clouds or storm clouds which blocked the Sun as effectively as the dust cloud higher up. Any openings in one layer would have been covered by the other layer. The resultant darkness and cooling would have been world-wide.

7.3.4.4 CO2 Chilling

As discussed previously, the ancient Earth was universally warm. '... there is evidence that the Earth's polar regions were, at one time, a lot warmer than they are today. During his Antarctic expedition in 1907-09, Shackelton (1909, 2:314) found seven seams of coal, each between three and seven feet thick, while preserved remains of warm-water coral have been found at the watery northern pole within the Arctic Circle.' (36) There were even forests on Axel Heiberg Island, Spitzbergen Island and the New Siberian Islands. (12) If lush forest thrived in these places, one can only imagine what it was like elsewhere. The key to all of this plant growth was warmth. Since everything happens for a reason something caused it to be warm. The vapor canopy was a key ingredient in this arrangement but the extensive forest growth indicates that the level of CO_2 was probably much higher as well. (21) CO_2 also has a blanketing effect but would not have favored areas near the poles particularly. However, as the high level of CO_2 was reduced, its accompanying greenhouse effect would also have been lost. The CO_2 wasn't lost from the Earth but it was lost from the atmosphere by being absorbed into the ocean. At room temperature, CO_2 will dissolve in water at the rate of '0.9 volume of the gas dissolving in 1 volume of water at 20 C.' (38) However, as the temperature drops, water will dissolve more. 'At 0 C, 1 volume of water dissolves 1.79 volumes of the gas.' (i.e. CO_2) (39) In other words, water near the freezing point can absorb twice as much CO_2 as it can at room temperature.

After volcanic heating subsided, the ocean cooled due to its own evaporation. The earlier collapse of the vapor canopy caused the atmosphere to cool which in turn further helped to cool the ocean. The ocean started to take up CO_2. As the level of CO_2 was reduced in the atmosphere, its associated greenhouse effect was reduced. Thermally, this would have been similar to the loss of the vapor canopy but it would have taken longer. Consequently the atmosphere cooled a little more and the ocean cooled a little more, resulting in more CO_2 absorption into the ocean which resulted in further chilling of the atmosphere. This type of arrangement is called a positive feedback loop wherein the results of the activity make the activity happen even more. (Unfortunately the opposite is happening now where atmospheric heating is warming the ocean and forcing the CO_2 in the water to come out. This positive feedback will accelerate the warming of the Earth.) CO_2 was continually absorbed over the ensuing years as the Earth further settled into a lower temperature climatic condition.

7.3.4.5 Snow Cover

As discussed in a previous section, the post-impact winter which commenced within hours of impact would have quickly resulted in lowering temperatures over northern land areas to below freezing. Precipitation would soon have turned from rain to snow. The collapse of the vapor canopy would have resulted in a massive amount of rain. As the impact-produced cloud expanded and spread, larger and larger areas of the world chilled. The falling rain might have gone through a period of freezing rain before becoming entirely snow or it might have switched back and forth as the temperature drifted down past freezing. A post-impact winter developed and expanded as the cloud cover thickened and spread.

Even if it wasn't snowing over 100% of the land, wherever the post-impact winter developed, it would probably have turned the land white anyway due to hoar frost. When the temperature and humidity are just right, hoar frost will develop. Crystals of ice form right out of apparently clear air and a pattern of ice crystals could readily have formed on everything just as they do now during certain times in the winter. The result from a heat loss perspective would have been the same as if snow had fallen and covered everything. The ground would have been very reflective either way.

The thick, black, dust cloud was the initial cause of cooling because it prevented sunlight from getting through. Subsequently, the moisture-filled storm clouds had the same result. When the clouds eventually thinned out and some solar energy was able to penetrate, the snow cover would have reflected almost all of it away. As the snow and ice started to form, particles from the dust cloud would have been settling out, darkening the snow surface. (Please refer to the earlier references to ongoing volcanic activity.) The darker the surface was, the more heat it would have been able to absorb. However, as the dust and particulate matter in the cloud kept settling out, the snow cover kept building up. The Earth was continually being covered by bright fresh snow. As more and more of the dust settled out, the snow became less and less darkened by the dust. Eventually it would have been quite bright and would have assisted the cooling trend because it would have reflected most of any solar energy that penetrated the clouds. Reflection would have been most effective at the higher latitudes where the albedo may have resulted in three-quarters of incident solar energy being reflected. The albedo (i.e. the reflective factor) of fresh snow in an area free of trees and other vegetation, which can create an uneven surface, can exceed 0.7 which would leave less than 30% for any other purpose. (56) As the snow cover increased, vegetation would have been completely buried anyway and the air would have been increasingly free of dust so the reflectivity of the snow would have been near maximum. In this way snow tends to preserve itself because the air above an extensive snow cover is colder. Since atmospheric pressure decreases more with altitude in cold air an upper cold trough commonly develops above an extensive snow cover. (44)

7.3.4.6 Ongoing Chilling

These developments; vapor canopy collapse, CO2 absorption into the ocean and snow field development would ensure ongoing chilling after the initial chill producers, the clouds, had dissipated. The Earth would thereby have been prevented from returning to its pre-impact level of warmth.

The chill factors not only caused chilling, they also introduced different types of chilling. The dust cloud was a transient chill producer. The great dust cloud caused the Earth to chill for a few months. Then, as it dissipated, that particular chill factor was gone. The storm clouds were also a transient type of change but they might have lasted for several years instead of just a few months. The collapse of the vapor canopy brought a unilateral step-function change. This type of change happens once and remains in place. The circumstances producing the change do not go back and forth. Once this type of change is introduced, it remains that way for all future time. CO2 absorption, on the other hand, was an exponential type of change. This is actually the type of change that usually occurs in nature. In particular, when a change follows from an initiating event, the dependent change might follow an exponential relationship. This type of change is discussed in considerable detail in a subsequent chapter entitled; Evidence of Chaos and Catastrophe. With an exponential change, the same percentage of change takes place for every similar time period of interest. For example, if there was a 10% change the first month, there would also be a 10% change the second month. The difference is that at the beginning of the second month, there is less of the initial value to work with than there was at the very beginning. With such a constant percentage relationship, the parameter undergoing change, changes less and less all the time. For example, 10% of 10 is 1, which only leaves 9 to start the next period. Then 10% of 9 is 0.9, so the change the second month is less. Similarly, during the third month the change would be less again. CO2 absorption into the ocean would be expected to approximately follow an exponential type of change. Rather than being abrupt like the collapse of the vapor canopy, CO2 absorption settled down to its final state quite gradually many years later. It would remain that way until some distant future time when the oceans started to warm up again. The type of change brought by snow cover would be different again. In this case it would build up toward a maximum as the snow covered more ground and then it would taper off years later when the snow melted.

Both types of cloud caused cooling for as long as they existed but when they dissipated their cooling effect disappeared with them. The cooling effect caused by vapor canopy collapse would occur within weeks and be irreversible. Absorption of CO2 into the ocean would continue for many years and also be irreversible – at least in the short term. (i.e. hundreds to thousands of years) The chill factor introduced by snow cover would taper off as the snow melted.

7.3.5 Snow Field Development

7.3.5.1 Atmospheric Circulation

With the collapse of the vapor canopy, temperature discrepancies developed between the equator and the poles with the result that global air circulation patterns developed where previously there had not been any. Currently, global air circulation is understood as consisting of three major atmospheric circulation cells both above and below the equator. These are the Polar Cells, the Ferrel Cells and the Hadley Cells. (Please refer to the diagram entitled, 7.3 5.1 Atmospheric Circulation) The Hadley Cells operate near the equator. Starting at the equator, Hadley Cell circulation rises upwards until it approaches the tropopause which is usually about seven miles up. However, during the months immediately following impact, these air currents rose much higher because ozone had not yet formed in the upper atmosphere producing a tropopause. It had been prevented from forming because the water vapor in the canopy had prevented ultraviolet energy from reaching the oxygen and forming it. (60) (This is confirmed by the following. 'The amount of ultra-violet light that reaches the surface of the Earth depends on … water vapor. On heavily clouded days, very little ultra-violet light gets through to the ground.' (78)) The warm moist tropical air therefore rose quite high, released its load of moisture and drifted north. (south in the southern hemisphere.) The falling rain cleansed the atmosphere and kept right on cleansing it week after week. Cleansing from this factor was restricted to the extent of the falling rain which would have fallen out as soon as the air chilled enough with elevation. As the air moved north however, it descended and this brought the dust down to where it could be rinsed out of the atmosphere. Hadley cell circulation was most effective in removing dust from the atmosphere over the tropics, allowing the Sun to shine once again and heat the Earth. The Ice Age did not persist very long in the tropics.

While Hadley cell circulation is certainly still in existence, it cannot remove dust from the upper atmosphere anymore because its movement is restricted to the lower atmosphere below the tropopause. This restriction results from the presence of ozone. After the vapor canopy collapsed ozone started to accumulate and the atmosphere then became layered into the troposphere near the Earth and the stratosphere higher up with the demarcation being the tropopause. If this arrangement had been in place immediately following the asteroid impacts, vertical movement of air would not have been high enough to effect cleaning and the dust would have remained airborne much longer (as it does now following a major volcanic eruption).

Ozone development was delayed for many years. Ozone develops when UV radiation from the Sun strikes oxygen. However UV radiation is restricted by moisture. Prior to the first impact, the vapor canopy was in place and the oxygen in the atmosphere was effectively shielded from the UV radiation. Similarly when the massive moisture clouds were allowed to circulate to great heights, their moisture also shielded the oxygen and the development of ozone was delayed.

Instead of Hadley Cell circulation, the mid-latitudes have Ferrel Cell circulation, which proceeds in the opposite direction. In the Hadley Cells, before air moves away from the equator, it will have risen high above the Earth. The entrained moisture condenses and rain falls right away. With Ferrel Cell circulation, northward movement proceeds initially along the surface of the Earth. The moisture entrained in the moving air will therefore move along the surface and remain suspended in the clouds until an area cold enough to condense the moisture is reached. With this arrangement, moisture from warm ocean surfaces may be carried from lower latitudes to higher latitudes. This meant that during the Ice Age, moisture-laden air was continually carried northward where it cooled and dropped its moisture load. As soon as it reached areas below freezing, snow fell. The moisture-carrying clouds would also have helped keep the ground cold but some of the moisture would have fallen as rain before turning to snow farther north. It took years to cool the oceans and as long as they remained warm and the land remained cold most of the time, the snow would have kept falling. It would have been like a continuous conveyor belt of snow.

7.3.5.1 Atmospheric Circulation

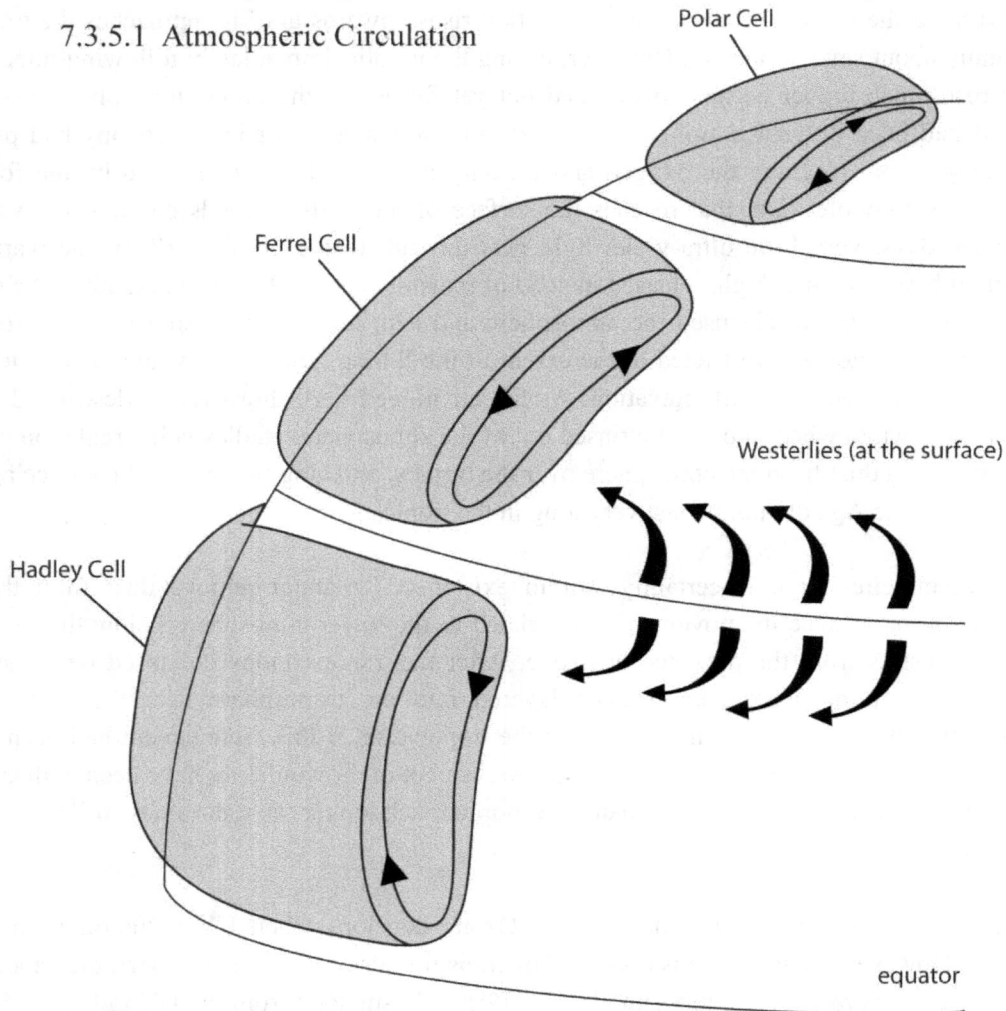

In particular, it is Ferrel Cell circulation that brings air from the Gulf of Mexico across the USA right into Canada. During the Ice Age, this pattern of air movement would have continually brought snow to the northern USA as well as the eastern half of Canada. The snow that dropped over the northern USA would have piled up and formed long tongues of ice, extending well down into the northern states. (This would be expected if the air currents repeatedly followed the same pathways.)

As storms from the Gulf of Mexico move north across the USA, they drift to the east. In this manner, moisture from the Gulf of Mexico would have provided the snow for the northeastern USA and eastern Canada. The great ice sheet covered more land than this however and extended further west than the moisture from the Gulf of Mexico could have supplied on its own. (62) However, it wasn't on its own. Sea level prior to the Ice Age was probably about 200 feet higher than it is at present. (48) The ocean therefore extended well up into central North America on either side of the present Mississippi River. Along with the Gulf of Mexico, this vast surface of warm water was available to feed the snow-producing engine and it remained this way until enough water became tied up in the snowfields to drop ocean level and thereby reduce the surface area available for evaporation. (In fact, from the Gulf of Mexico all the way to Hudson Bay, with minor exception, the elevation of the land is much less than 1000 feet above sea level at the present time. (64))

Moisture for glacier build-up also came from the endless puddles and lakes that were left over from the inundations resulting from the impact. One effect of a major impact is to produce massive waves which would have, in many cases, rolled right over the land. Most of this water would have returned to the ocean but any location which was the least bit depressed, would have been filled and stayed filled until the water either evaporated or left by some other means. At the beginning of the Ice Age the surface of the land included numerous lakes, ponds, and puddles and all of them were available to contribute to glacier accretion. Later when the glaciers were melting, some of these bodies of water reappeared.

One difference between present Ferrel Cell circulation and the circulation during the Ice Age relates to extent. At that time the Arctic Ocean was warm. Therefore there wasn't any Polar Cell in the northern hemisphere. Ferrel Cell circulation might have continued right out across the Arctic Ocean and the situation could have stayed that way for many years until the Arctic Ocean chilled and crusted over with ice. When this happened, the northern Polar Cell would have developed. At the South Pole, the Polar Cell developed much sooner because there is land at the South Pole which chilled as soon as the great dust cloud appeared overhead. This enabled descending air movement and the development of the South Polar Cell.

After cooling and dropping its moisture load, the northward-moving air continued out over the warm Arctic Ocean. There another load of moisture was picked up. Then, the flowing air mass would have risen and folded back over the land on its way to completing its Ferrel Cell loop. As on the way north, it cooled once again while passing back over the land on its way back south, so the second moisture load was dropped and snow fell across northern Canada and Greenland. In this manner, the Arctic Ocean contributed to ice-field accretion.

A similar return flow enabled the Arctic Ocean to contribute to the ice-fields of northern Europe and Asia. In the European case, the primary source of moisture was the North Atlantic, over which the air currents veer to the east the same as they do over North America. Eastern Siberia was spared from massive ice accumulation because it is too far from the north Atlantic and too far from the warm Indian Ocean to the south.

Above an ocean surface there is no restriction to air movement so the air is free to rise away from the surface and drift over the land. Any warm moisture-laden air, which passed over colder land, would chill. As soon as it reached its saturation temperature, precipitation would fall and would continue to fall for as far as the air from over the ocean was enabled to circulate over the land. The development of the Ferrel

Cell air circulation patterns was like having a great conveyor belt bring water, in the form of humidity, from the ocean and deposit it, in the form of snow, over the land.

7.3.5.2   Critical Evaporation Rate

The rate of water evaporation from a water surface is directly related to the temperature of the water. This phenomenon is always discussed with respect to vapor pressure. When water is at room temperature, its vapor pressure is four times as high as it is at freezing. (46) This means that warm water readily evaporates with water leaving the liquid surface and being forced into the air right above the surface. As water gets warmer, it evaporates faster. When the boiling temperature is reached, water will not get any warmer and will evaporate just as fast as its source of heat can supply the energy.

In order to form the vast ice-fields of the Ice Age, ocean water evaporated. However, for any of the evaporated moisture to contribute to glacial accretion, the ocean water had to evaporate at a rate that was within a certain critical range. There was both an upper limit and a lower limit to this critical evaporation rate. If evaporation was too fast, the air would have become super-saturated with water. Whenever the relative humidity goes above 100%, water is said to be supersaturated. The atmosphere never stays super-saturated. Whenever moisture is forced into the air above the saturation (or 100% relative humidity) level, rain will fall immediately and it will keep falling until the air is back below 100% relative humidity. This is just basic physics.

As earlier discussed, when the asteroids struck the Earth, numerous volcanoes erupted. Everything that came out of them was hot whether it was gas or ashes or molten magma. All of these hot materials caused the ocean to heat up and in many cases the ocean would have locally boiled. All of these activities would have sent tremendous amounts of moisture into the air but very little of it would have been useful for ice-field build-up. As soon as the hot evaporant rose above the ocean surface, it would have rained. There would have been a torrential downpour in the immediate vicinity of many eruptions. (This happens when the effluent from a volcano includes a lot of water.) None of the water involved in these downpours would have made it to the ice-fields and therefore none of it would have been useful for ice-field construction. Hence, there was an upper limit to the evaporation rate that would have been useful and that limit was just below the rate that would cause the air to become saturated. As long as the air did not become saturated and its relative humidity did not rise above 100%, the evaporated moisture might have contributed to glacial accumulation. Active under-water volcanoes would certainly have warmed the ocean. (In many cases, the ocean would have been above the boiling point, just as it is now at underwater vents where the temperature is commonly around 400F.) However the only portion of heat that might have contributed to forming snow was the heat that didn't force too much moisture into the air too fast.

There was also a lower critical limit to the required evaporation rate and it is even more important than the upper limit. While initial Earth cooling was caused by dust clouds, the snow-fields had to be kept cool long after the dust clouds were gone. Even at maximum evaporation rates, it would still have taken months to build up enough snow to cause glacial ice to form. Therefore the storm clouds had to take over the function of keeping the Sun from shining on the snow until enough had accumulated to form the glaciers (which were thought to be two miles thick, or whatever thickness actually resulted). Certain commentators declare that the ice became as thick as the present Greenland glacier. Others think it was actually much thinner. (80) The actual thickness does not concern us. The important factor is that the conditions had to remain near optimum for snow accumulation. As soon as the storm clouds thinned out, two circumstances counter to snow accumulation would have developed. Thinner clouds would have carried less snow and thinner clouds would have allowed the Sun to shine on the snow. As the clouds thinned out, snowfall dropped below the lower critical rate (which was simply the rate at which snow no longer accumulated). Thereafter, during every warm season the same amount of snow melted as had accumulated the previous winter. Just before this happened would have been the time of glacial maximum. Soon thereafter, more snow melted every year than fell every year.

In order to sustain adequate cloud cover, the ocean had to evaporate at or above the lower critical level. As soon as the evaporation rate fell lower than this the storm clouds would not have been adequate to sustain the snowfall fast enough and glacial maximum would have been reached. While the ocean started off warm and was warmed further by volcanic activity, it is doubtful that the critical minimum evaporation rate could have been maintained if the average ocean surface temperature dropped below room temperature although some commentators suggest a lower level. (68) While this temperature would be considered warm by today's standards it would probably have been barely warm enough for ongoing snow-field accumulation. (The reason for this suspicion is that evaporation is not directly related to temperature. As previously mentioned, vapor pressure for water is four times as high at warm-room temperature as it is at freezing. Half way between freezing and warm-room temperature, vapor pressure is less than one-half of what it is at room temperature. (46) In other words, water evaporates a great deal less below room temperature than it does above it.)

In general, the warmer the ocean was kept, the faster it would have evaporated. This means that the higher the volcanic activity in the ocean, the faster the glacial ice would have accumulated. The volcanic activity would have warmed the ocean causing greater evaporation. (87)

In order to form the great snow-fields of the Ice Age, the ocean had to maintain an evaporation rate that was above the critical rate (i.e. the rate at which ice would accumulate) and it had to maintain this rate until the glaciers had built up. Anything less and the snow would not have accumulated. Of course as water evaporates it cools itself and the massive evaporation that was happening would have been cooling the ocean at the same time. Once the source of heat was cut off, (i.e. because the volcanic activity died off) it was only a matter of time until the evaporation rate dropped below the critical level. After this, some snow would have continued to fall but even more melted and the great glaciers of the Ice Age thereafter continually reduced in size.

As a corollary, it is clear that the glaciers could not have been built up over a vastly extended period of time (e.g. 100,000 yrs.) because the critical evaporation rate could not have been sustained.

7.3.5.3 Modern Example

There is nothing surprising about the basic way that snow is produced. During winter, snowfalls are a common occurrence and they are always caused by the same circumstances. An open body of water evaporates moisture into the air above it. Air currents carry this moisture-loaded air to the land. If the temperature of the land is cold enough, snow will fall. The greatest snowfalls happen when the evaporating body of water is several degrees above freezing because the warmer it is, the higher its vapor pressure is with more water being forced into the air and being available to fall as snow. As a result of these factors, certain coastal areas commonly receive snowfalls of several feet from a single storm and these higher amounts occur during the fall before the water temperature approaches freezing. In particular, while not currently widespread, snowfalls of more than three feet can occur in one night. (45) 'The Buffalo area is periodically buried under what meteorologists call lake-effect snowstorms. These are intense local snowfalls to the lee of the Great Lakes that result when air picks up large amounts of moisture as it passes over the lakes. Single-storm snowfalls of three feet or more are common.' (55) While three feet of snow is a significant amount, it is equivalent to only about three inches of rain and it is not the least bit uncommon for a summer storm to bring three inches of rain in one day.

7.3.5.4 Volcanic Example

As with the very recent mini-nuclear-winter, which developed after the eruption of Tambora in 1816, the chilling factors did not distinguish between winter and summer. 'There was frost every month … A light snow in much of New England and Western New York State on June 6 was followed by moderate to heavy snow in New England on June 7 and 8. In Vermont there were drifts 18 to 20 inches deep.' (47) Similarly, as the Ice Age developed, it would have kept on snowing through June, July and August.

The deplorable situation in New England developed even though the ocean temperature was about the same as it is now. As the great Ice Age commenced however, the entire ocean was much warmer. The initial availability of moisture was therefore almost limitless and snowfall would not have tapered off until the temperature of the ocean had dropped significantly. In the meantime the blizzards raged on and the snow accumulated until it was thousands of feet deep.

7.3.5.5 The Lost Squadron

At the present time, even when the oceans of the world are cold, glacial ice can still accumulate locally quite quickly. This is at a time when the world is understood to be warming up. In fact, the warming trend will enable the ice in certain local areas to increase. Some of the glaciers of the world will get bigger. Any land area, which has an air temperature which is just below freezing most of the year and which has adjacent open water most of the year, will experience an increase in ice accumulation. As the world gets warmer the conditions in some local areas will more closely approach the ideal conditions for an ice age. However it is not going to happen on a large scale because these conditions will not be widespread. They will be local. The Sun will continue to shine in other areas and the temperature will be above freezing long enough to melt last winter's snow accumulation.

Current observations indicate that most glaciers are melting away but that at least one is increasing. In particular, the Greenland glacier is melting around the edges but getting thicker in certain inland regions. While there are few reliable observations to consider, the Greenland glacier is of particular interest because an event occurred in Greenland, which provides recent indisputable data on ice accumulation. With the availability of this data, the rapid buildup of ice, which occurred during the Ice Age, is more readily comprehended.

During World War II, a group of planes were being delivered from the USA to Europe by way of Greenland. At that time, fuel tanks were not large enough to allow direct flight, so the plan was to go from Greenland to Iceland and then to England. On the way something went wrong. These planes all came down on the Greenland glacier. Only one person was injured. After a hike to the coast, everybody was picked up and returned safely to home. These planes landed in 1942. (53) In 1992 one of the fighter planes was removed from the ice and brought back to the USA and restored. In the year 2000 this plane was flown. The ice and snow, which covered all of these planes, was more than 250 feet thick. Most of it was blue ice. Approximately one-third of it was packed snow. In order to remove the plane, which was later restored, the snow layer had to be stabilized so a shaft could be opened down through the snow and ice.

The time for the accumulation of the ice and snow covering the planes was fifty years. In this location, some snow melts during the summer but overall the glacier keeps getting deeper. It is impossible to determine how far the planes might have initially melted into the surface but it would be reasonable to assume that they did not sink the entire 250 feet. They probably only melted into the ice until snow cover shielded them from the Sun and prevented any further melting. Such melting into the ice would account for a certain amount of the total ice cover over the planes but not for all of it and probably not for more than a small percent of it.

The present day conditions in Greenland are certainly not ideal for the buildup of ice. The ocean around northern Greenland is certainly not warm and it isn't even open all year. Still, the ice, which covered these WWII planes, was 250 feet thick and it only took fifty years for it to accumulate. Under similar conditions the great glaciers that accumulated during the Ice Age, would only have required a thousand years to reach their maximum thickness. How long would it have taken if the conditions for ice accumulation had been ideal?

7.3.5.6   The Ice Cores

Certain ice cores recovered from Greenland have been named NorthGRIP, Grip2 and Grip. Also some have been recovered from Antarctica and have been named Dome Fuji, Vostok and DomeC. 110,000 layers have been counted in the Greenland Grip2 ice core down to the 2800 m level (9200 ft.). (54) Between the 2300 m level and the 2800 m level 25,000 layers were thought to be identified. Each layer is declared to represent one year. (54) The average ice thickness for each of the 25,000 layers would therefore only be 500/25,000 = 0.02 m or about two cm. (0.8 in.) Since it requires about ten inches of snow to make one inch of ice, the snowfall each year for 25,000 years would only have been eight inches. The upper 2300 meters included 85,000 layers which would represent 85,000 years. The average ice thickness for the upper region would therefore have been 2300/85,000 = 0.027 m or about 2.7 cm (a little more than one inch). Using similar reasoning the average snowfall every year for 85,000 years was only ten times one inch or about ten inches. The snowfall which covered the Lost Squadron, produced 250 feet of ice in just 50 years. Since five feet of blue ice will result from about fifty feet of snow, the average snowfall every year during those 50 years was about fifty feet. With this in mind, the conclusion that each of the 110,000 layers represents one year of snow and that snowfalls continued year after year for 110,000 years, producing eight or ten-inch snowfalls every year, does not hold any credibility whatsoever. Why would snowfall, during the present time, when circumstances are not even optimum for producing snow, result in fifty feet of snow in one year compared to ten inches? The only possible conclusion is that the layers do not represent time at all but only the conditions under which the snow was deposited.

As the land chilled, it started to rain. Very soon it was cold enough to snow. As with all ongoing blizzards, snowfall will not be constant day after day and week after week. There will be intense periods of snowfall interspersed with quieter periods. Neither is there reason to expect that volcanic activity had been constant or that the wind had always remained in one direction. The net result of all of these variations would be layers of snow. The layers would roughly correspond to the variations in activity. Also several periods of snowfall might not even be distinguishable if there wasn't some marker included. On the other hand, markers could show up any time – even several times in one month.

The evidence which is shown in the diagram, 7.3.5.6 'Greenland Grip2 Ice Core, Oxygen Isotope Ratio Variation' is more suggestive that the buildup of snow during the Ice Age, continued up to the 1500 m level. In would be reasonable to expect that volcanic activity would be generally quieting down and that soot and dust in the air would be disappearing by the time the ice had built up to this level. At this thickness of ice, (1500 meters (i.e. 5,000 feet)) 50,000 feet of snow would have fallen. This is a vast amount of snow under any set of circumstances, but under the favorable snow-producing conditions of the Ice Age, it was not too much to expect. Actually, a situation like that can be compared to the recent (fall 2008) snowfalls in south-western Ontario, where more than two feet of snow fell each day for two days in a row. If this type of snowfall continued and none of it melted, 50,000 feet of snow would fall in less than (50,000 ft./730 ft. per year) 69 years.

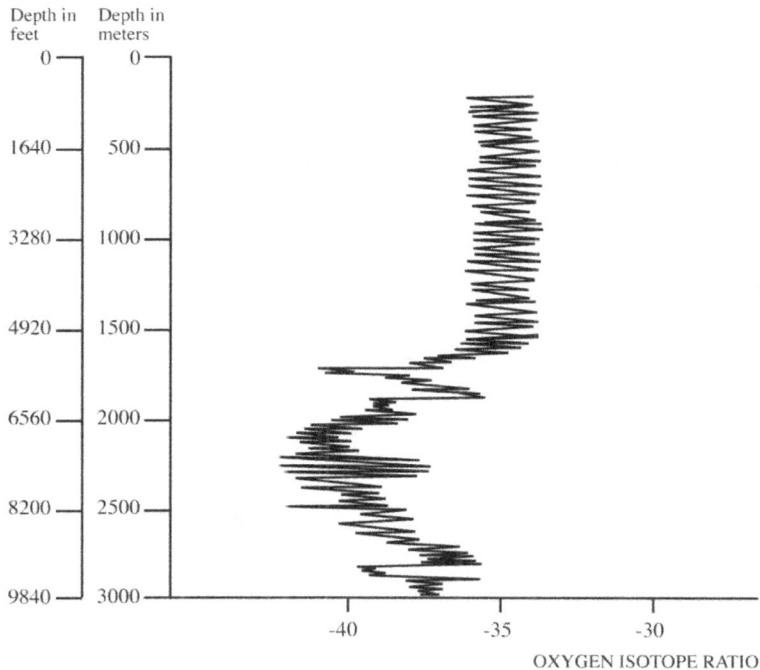

7.3.5.6 Greenland Grip2 Ice Core Oxygen Isotope Ratio
Variation With Depth

The variation of the oxygen isotope ratio with depth in the ice appears as
two distinct categories. The portion above 1500m appears to have been
deposited under one particular set of environmental circumstances, while
the portion deeper than 1500m appears to have been deposited under a
different set of environmental circumstances. This is suggestive that the
bottom portion was deposited during the Ice Age, and the top portion
since the Ice Age.

Oxygen isotope ratios were also measured throughout the depth of the entire ice core. As shown in the
diagram, these ratios have remained fairly constant from about the 1500 m level (5000 ft.) upward. Below
this level, the ratio was within a wider range of a slightly different constant. This is suggestive that
subsequent to the accumulation phase of the Ice Age, (i.e. the 1500 m level) circumstances changed.
Some circumstance introduced a step-function change so that the isotope readings from the 1500 m level
upward became much closer to the way they are at the present time.

7.3.5.7   Conclusion

More snow accumulated inland than near the ocean. Since the ocean water was still warm, many areas
near the ocean remained above freezing. The islands of the high Arctic, as well as parts of Greenland,
remained free of significant ice accumulation. (49) 'And what is no less surprising, the northern part of
Greenland, according to the concerted opinion of glaciologists, was never glaciated, "Probably, then as
now, an exception was the northernmost part of Greenland; for it seems a rule that the most northern lands
are not, and never were glaciated," writes the polar explorer Vilhjalmur Stefansson. "The islands of the

Arctic Archipelago," writes another scientist, "were never glaciated.' (83) In fact any area, which benefited enough from the moderating effect of the warm ocean did not experience the Ice Age. (If coastal areas were ice-free it may also have been partly due the fact that the ocean was initially deeper by both the amount that would soon become tied up in the developing glaciers as well as the shrinkage that would occur as the ocean cooled. These two factors would have caused sea level to be more than 200 feet higher than the present level.) This would have included certain coastal regions in the far north as well as mid-latitudes further south. The Ice Age was not necessarily a period of extreme cold. 'The warm ocean would have been a large heat source for the atmosphere. Winter temperatures over the ice sheets would not have been extremely cold, and areas south of the ice sheets would have been rather mild, mostly cloudy, and wet in winter.' (57) Everywhere that the ice fields formed had to be below freezing, but not necessarily extremely cold. Snow forms and falls more abundantly when the temperature is just below freezing than when it is much colder because very cold air cannot hold very much moisture or even pick it up in the first place.

While the extent of the Ice Age glaciers is commonly understood to have been restricted, excluding certain Arctic Islands for example, this does not mean that such areas were not snow covered. Glacial ice will not form until about 80 feet of snow has accumulated. Glacial ice may have been restricted to the areas commonly mentioned when these matters are discussed, but additional areas could also have been covered by up to 80 feet of snow or more. If this had been the case, there wouldn't have been any particular indicator left so the matter is inconclusive. However it would be surprising if the only snow that fell was deep enough to form glacial ice with nearby areas being left snow free. It is more likely that deep snow cover was very extensive and that it was simply deeper where glacial ice actually formed.

7.3.6 Climatic Transition Period

The ocean water was initially warm and readily evaporated. 'The temperatures of the oceans were warm … Warm ocean temperatures are often associated with increased rainfall over continental areas …' (50) Thus the great snow-producing engine was continually fed, enabling the blizzards to continue. However, as time passed, the temperature of the water in the oceans dropped and evaporation reduced. The colder water absorbed more CO_2. Any greenhouse effect of atmospheric CO_2 was thereby being reduced. All other effects of elevated atmospheric CO_2 were also gradually lost with the same time factor as the oceanic absorption of the CO_2.

It is suspected that as the ocean water cooled, atmospheric CO_2 was absorbed and the absorption generally followed an exponential relationship with time. Absorption would have been facilitated by movement in the cooling ocean. As the water cooled, it would start turning over. This would continually bring previously-unexposed water to the surface, exposing it directly to the atmospheric CO_2. The climate gradually became colder as this was happening until by the time most of the atmospheric CO_2 was absorbed, the temperature of the ocean had dropped to near its present level and the climate of the world settled out similar to the way it is now. CO_2 absorption tapered off as ocean temperature approached its present level (which is only a few degrees above freezing (52)) because the water had reached its CO_2 saturation level. It could not hold any more and the temperature could not drop any further (or the waterwould have frozen). The time period from impact until the pre-impact CO2 was absorbed down to a much lower level and ocean temperaturesettled out near freezing, was the climatic transition period interfacing the pre-impact climate with the present climate.

Fortunately, there is one set of ancient data available to us which appear to have this familiar exponential relationship and which therefore appear as a possible indicator that some significant factor of nature was being adjusted and that other factors of nature were tracking the adjustment – in this case that the climate was undergoing some dramatic transformation. These data have been included in the chapter entitled, 'Evidence of Chaos' and are shown on the diagram, 10.6.2 'Genesis Human Age Data'. These data do come close to an exponential relationship. It is certainly not likely that the recorders of this ancient data had any knowledge of exponentials. It seems that they were only keeping records. However, if these data were indirectly tracking the absorption of CO2 during and following the Ice Age, the implication would be clear that the Ice Age occurred during the time that humans were on the Earth and during a time when human life spans were dramatically shortening.

The understanding that the Earth cooled down significantly following a major impact is supported by evidence of cooling from plant remains and the oxygen-isotope method. Immediately prior to the Chicxulub impact, the Earth was warm. This was the Cretaceous Age of Chalk which is believed to have terminated when the Chicxulub impactor crashed into the Earth. This understanding is captured by the following comment which is quoted from Sir Arthur Holmes. 'The mean annual temperatures … were 25C…The general fall in temperature since the Chalk was deposited has been estimated from pollen and other plant remains as well as by the oxygen-isotope method … the cooling affected the bottom waters of the Pacific until they were reduced nearly to the freezing point … Today the oceans are cold … but at the onset of the Ice Age there was no melting ice to cool the oceans. Nevertheless cool they did.' (20) It is clear from these comments that the Earth underwent a major climatic shift from being universally warm to the climate that exists at present. To recognize that the entire world including the ocean to its very bottom, cooled right down to the freezing point is recognizing a major worldwide climate transition event.

7.3.7   Ice Age Duration

The Ice Age was of short duration. It is physically impossible for the great Ice Age to have lasted for more than a few hundred years because of the greenhouse gas effect.

As explained above in paragraph 3.15, Greenhouse Gases, it is the greenhouse gases that, along with the incoming energy from the Sun, keep the surface temperature around the world in the comfort zone. This cannot be violated. A worldwide reduction in the major greenhouse gas, water vapor, would result in the entire world going into the deep freeze and staying there indefinitely. In fact it would stay there until the Sun had increased its energy output enough to supply the warming effect that the water vapor had previously provided. According to nuclear physics theories this would be about another billion years. Who can wait that long? Having greenhouse gases in the atmosphere is not optional. It is absolutely necessary or the temperature on the surface of the Earth could not be within the range necessary for life to exist.

During the great Ice Age the temperature over the land was reduced which enabled snow to fall and ice to pile up. This reduction in temperature was caused by the two layers of cloud cover. One layer was a dust layer caused by the impacts and the volcanic activity. The second layer was moisture caused by the rapidly evaporating ocean. So the Earth chilled over the land but it did not chill over the ocean. Hence the average temperature around the world did not plummet. If it had plummeted all hope would have been lost and the water vapor levels in the atmosphere would have dropped and the associated greenhouse gas effect would have been lost with them. A deep freeze for the entire world would have resulted.

Instead the oceans remained warm and evaporated at prodigious rates providing snow for the ice fields. However within a relatively short period of time volcanic activity subsided. Thereafter the ocean cooled due to evaporation. As it cooled cloud cover became intermittent and the Sun shone through and melted the ice. The temperature over the land increased during this time. It was the combination of dropping ocean temperature and rising land temperature that kept the greenhouse effect on the average within the range which caused surface temperature to remain within the comfort zone. It could not have been any other way.

All of this means that the Earth has never suffered an ice age which lasted for thousands of years. If this had ever happened, the Earth would have locked up cold and would still be frozen solid. Without the greenhouse effect the energy from the Sun is simply not sufficient to keep it warm enough on its own.

The duration of the ice age was therefore directly tied to the temperature of the ocean. As the ocean temperature dropped the land warmed up and the greenhouse gas effect remained intact. How long did it take for the ocean to cool? As evidenced by the submarine canyons which are explained by a warm ocean, the ocean was still much warmer than glacial melt water at the time the glaciers were melting. However evaporation requires a great amount of heat and as it evaporated it chilled. It would have cooled substantially by the time the glaciers were melting but it was still warm enough to enable the melt-water to form the submarine canyons.

Volcanic activity would have subsided soon after the asteroids stopped falling and the great shockwaves stopped bouncing through the interior of the Earth and disrupting the crust. Thereafter the chilling effect of evaporating several thousand feet of water would have caused the temperature of the ocean to drop. The diagram shown in paragraph 10.6.2 is suggestive of a drop in CO_2 which would relate directly to a drop in ocean temperature. CO_2 is held in the ocean by temperature and a drop in temperature would enable the ocean to take up atmospheric CO_2 as it cooled. All through the entire event ocean temperature was balanced out by the rise in land temperature so greenhouse gas stability was retained and the average temperature of the Earth stayed within the habitable range.

To summarize, at the very most the Great Ice Age could not have lasted more than a few hundred years.

7.3.8 Summary of Main Events

The sequence of circumstances, which characterized the Ice Age, may therefore be summarized.

1. The Earth was universally warm, and the oceans were universally warm. Temperatures were between 12 and 25 degrees c. (54 and 77 degrees f) (51)
2. The first of a series of large asteroids crashed into the Earth.
3. A great deal of dust was projected into the atmosphere, which blocked solar energy from reaching the ground resulting in a serious chilling effect.
4. The protective atmospheric vapor canopy was destroyed with the result that heat was lost to space.
5. Hundreds of volcanoes erupted contributing to the dust cloud as well as heating the entire ocean until it was much warmer than room temperature.
6. The warm ocean evaporated water at a prodigious rate. Moisture clouds formed, helped to block out sunlight and carried thousands of tons of moisture landward.
7. Over the chilled land, snow formed in the moisture clouds and fell at very high rates.
8. As both the Hadley and Ferrel atmospheric circulation cells initially developed, they rose to very high levels in the atmosphere.
9. At these levels they entrained the dust, and brought it down to much lower levels where rain and snow removed it from the air.
10. The anticipated post-impact winter was thereby short-lived and ended much sooner than it might have otherwise.
11. The warm ocean surface continued to force prodigious amounts of moisture into the air.
12. Underwater volcanic activity, which was repeatedly reactivated for several months by ongoing asteroid impacts, kept adding heat to the water which caused the high evaporation rate to be maintained.
13. Snow continued to fall until huge mountains of it built up and under their own weight they pressed down and formed the great glaciers of the Ice Age
14. Widespread blizzards repeatedly developed and raged on week after week and month after month.
15. The snow continued to accumulate until the oceans cooled substantially and could no longer evaporate water at a fast enough rate to generate continuous cloud cover, thereby block the sunlight, keep the land chilly to enable ice-field accretion.
16. The loss of the vapor canopy, and the extensive snow cover caused the temperature of the polar regions to become and remain cold.
17. Ocean temperature dropped in response to a colder atmosphere, as well as the massive evaporation which had been going on for months.
18. As the ocean water became colder, it absorbed more CO_2 which further chilled the atmosphere, enhancing the further chilling of the ocean. (This could be called a negative feedback loop because the reducing result further diminishes the cause.)
19. Cooler oceans resulted in less evaporation, less cloud cover and less snow production.
20. Solar energy increasingly reached the surface of the Earth in the mid and upper latitudes.

21. However, since the moderating effect of the water vapor layer was lost, the greenhouse effect of CO2 was reduced and large areas were snow covered, atmospheric temperatures did not and could not return to their pre-impact levels.

22. Soon, the quantity of snow that melted during the summer matched and then exceeded the amount that had fallen the previous winter. Summers were quite short at first but quickly became longer causing the snow at the mid-latitudes to melt at high rates.

23. By this time, ice field accumulation had maximized and the great ice fields, which had formed in North America and Eurasia stopped accreting and started to diminish. While the northern continental Ice Age was over when the bulk of the continental glaciers melted, the Greenland and Antarctic Ice Ages are still not over and will not be over until these areas are also free of year-round ice.

24. Due to the large number of impacting asteroids, the Earth was nudged into a slightly higher orbit which changed the year from 360 days to 365 ¼ days. (see par. 13.20) This further reinforced the generally-cooler climate of the Earth.

7.4 Comparison: Post-Impact Winter and Ice Age

Characteristic	Post-Impact Winter	Ice Age
Climate	Cold	Mild, similar to presentwarmth at the equator, below freezing mid-latitudes, cold polar regions
	Dry	Wet, constantly raining orsnowing
Duration	Months, possibly years	Several years
Extent	The entire world	The entire world
Cloud Type	Dust	Moisture
Weather	Some snowfall	Prodigious snowfall
Ocean Evaporation	Negligible	More than 2000 feet
Survivable	No	Yes, in specific areas
Greenhouse effect	Lost	Retained
Heat-Cold balance	None	Retained

Even though there is some similarity between a post-impact winter and an ice age, it is clear, even from this brief comparison, that they are dramatically different phenomena.

7.5 Evidence

7.5.1 Post-Impact Winter

Evidence for any post-impact winters which occurred long ago will be difficult to find. Appeal could be made to the fossil record to support the conviction that many animals died. 'With photosynthesis at a standstill and the Earth plunged into deep cold, it is no wonder there was a great dying.' (11) However, the widespread and massively violent death witnessed by the fossils is equivocal because all of these unfortunate creatures could have died as a direct result of the impact itself. As we recall some of the results of the impact, it is clear that all aspects of nature suffered unimaginable trauma from the impact. 'Shock waves from the impact rattled the Earth … causing shockwaves that lasted for months. The giant wave circled the Earth many times. Some of the material ejected from the crater flew out of the atmosphere … rained havoc on distant continents.' (11) Clearly the entire world would have been involved in the aftermath of an impact and death would result for untold numbers of creatures. It would therefore be very difficult to identify any creatures that may have died due to the following cold.

7.5.2   Ice Age

An ice age, on the other hand, should leave evidence which is unequivocal and this is indeed the case. There are several lines of evidence supporting the existence of the Ice Age.

7.5.2.1   Present Glaciers

At the present time, two massive glaciers as well as numerous small ones, cover 10.4% of the Earth's land surface. At the maximum extent of glaciation during the Ice Age, it is thought that they covered 28% of the land. Thus they were almost three times as extensive as they are at the present time. (67) While this is a significantly greater area of land, such a conclusion gains some credibility from the very recent observations that even the present glaciers – especially in mountain areas, have been much more extensive than they are now. As an example, the melting of the Columbia Icefield in the Canadian Rocky Mountains has been measured for the last few decades and stakes have been placed to trace its retreat. Within the last one hundred years this glacier has melted back hundreds of feet. Also, within the same time period the massive ice shelves around Antarctica have been observed to break off in large chunks and drift away.

7.5.2.2   Isostasy

Isostasy refers to the general state of equilibrium of the Earth's crust. Both sinking and rising of the crust has been observed and directly related to changes in the amount of material sitting on the surface. For example, when Lake Mead in the USA was filled with water, the crust was measured to depress about 20 centimeters. (8 inches) (65) It seems reasonable to relate these types of observations to the changing land levels around both Hudson Bay in Canada and the Baltic Sea in Scandinavia. The shorelines of the Baltic Sea have been measured to be rising at about one centimeter every year. (66) Around Hudson Bay more shore area appears every year. Both Hudson Bay and the adjacent James Bay are very shallow so any change in elevation will be readily observed as increased shore area. The loss of water depth is also apparent miles from shore where sandbars almost break the surface making travel by canoe extremely hazardous. These bays are very large and storm waves are common. Recently, a heavily loaded canoe, being paddled by experts became lodged on one of the sandbars several miles from shore. The canoe was swamped by the waves and the entire party drowned.

The great glacier that existed in Canada would have involved the entire Hudson Bay area. Since this glacier has now gone, and the land is rising, it seems reasonable to claim this evidence as rebound due to the loss of the extra weight of the vanished glacier. Similarly, the Eurasian glacier would have covered the Baltic Sea area. Since it too is gone and the land is rising, it is reasonable to relate the two.

Land that is rising after apparently being depressed by extra loading is called rebounding land. While it is an assumption to claim that the land was originally higher and became depressed by the glacial load, the fact that the glacier is no longer there and the land is rising, is reasonable evidence that the glacier once existed.

7.5.2.3 Submarine Canyons

Submarine canyons are, with minor exception, extensions of large rivers. These canyons extend from the mouths of the rivers and carry on out across the continental shelves right to the abyss. Since, at glacial maximum, sea level was lower, it seems reasonable to conclude that water from each respective river simply flowed across exposed land as it made its way to the sea. This would readily explain submarine canyons in water which is shallower than the maximum depth that can be accounted for by water that was tied up in the glaciers. This depth has been assumed to have been about 300 feet or possibly 125 meters (412 feet). (24)

Other explanations have been offered to explain canyons which are much deeper than 400 feet (which all of them are). 'The major mechanism of canyon erosion is now thought to be turbidity currents and underwater landslides. Turbidity currents are dense, sediment-laden currents which flow down-slope when an unstable mass of sediment that has been rapidly deposited on the upper slope fails, perhaps triggered by earthquakes. Mass wasting, slumping, and submarine landslides are forms of slope failures (the effect of gravity on a hill-slope) observed in submarine canyons.' (24) In other words it is understood that there appear to be several contributing causes to submarine canyon expansion but none of these factors identify why canyons extend from existing rivers or why they were cut so deep in the first place. In recognition that these ideas are not really sufficient to explain the submarine canyons, 'Thornbury says after reviewing all the theories: The origin of submarine canyons remains a perplexing problem. The theory that they were cut by turbidity currents aided by submarine landslides, slumping, and creep holds a slightly favored position, not so much because it answers all the questions connected with them but because it encounters fewer difficulties than any other theory.' (74)

It has also been suggested that the canyons formed when great masses of water flowed off the continents and back into the ocean. (89) After the asteroid hit the Earth, water was in motion all over the world. This motion included massive waves and currents of every description. In some cases, water rolled right up onto continents and may have rolled right over them. The continental shelves consist of silt which appears to have been washed off of their respective adjacent continents by sheet water flow. Sheet flow however might not have always been maintained, particularly if some action of the flow or some other factor was able to channelize the flow. In such cases, the great masses of water leaving the continents might have become concentrated into channels which might have formed the rivers as well as extensions of the rivers well out into the sea. It must also be kept in mind that the unthinkable upheavals following impact would have temporarily emptied some ocean areas leaving the exposed bottoms vulnerable to channel cutting as the water returned. The major difficulty with this type of suggestion is the great depths to which some canyons have been formed. 'In a recent review of the problem, W. D. Thornbury, Professor of Geology at Indiana University, says: The difficulties encountered in explaining the lowering of sea level necessary for the canyons to have been cut by streams seem insurmountable ... If Tolstoy's conclusion that Hudson Canyon extends down to a depth of 15,000 feet is correct, the magnitude of lowering of sea level to permit subaerial (i.e. above water under the air) canyon cutting seems beyond any possibility of realization.' (75)

In spite of these difficulties a logical explanation for submarine canyons is possible as one consequence of the Ice Age. In order to have the Ice Age the ocean had to have been warm. In fact, the major contributing factor to glacial accretion was a warm ocean. As the ocean cooled, the evaporation rate dropped to the critical level where ice-field accretion was matched by ice-field melting. Soon thereafter the clouds thinned out, the Sun shone, and the ice-fields proceeded to melt.

However, the temperature of melt-water is always close to freezing and it would have been much colder and heavier than the ocean water, which was still cooling down and had not yet reached its present temperature. Therefore, the cold dense melt-water would have immediately sunk upon entering the ocean (even if the ocean had some salt in it) and it would have continued along the bottom until the deepest part of the ocean was reached. This type of flow is called 'density-induced underflows of denser water.' (90) The density difference can be caused by suspended debris, greater salt concentration or colder water. When the submarine canyons were formed, colder water was involved and it came into the ocean via the rivers direct from the melting glaciers.

An example of cold water underflow at the present time is in the North Atlantic at the terminus of the Gulf Stream. Here, between Greenland, Iceland and Europe, the surface water becomes chilled to the temperature where its density is the greatest, whereupon it sinks. Flow continues, but now it is deep in the ocean back toward the equator. (65)

The canyons are very large and therefore indicative of high flow rates, which in turn indicates, rapid melting. (e.g. the Astoria Canyon, the extension of the Columbia River in Oregon varies in width from 1.5 to 8.3 miles and has a maximum depth of 3000 feet (90))

Rapid melting and very high flow rates are also recognized from other evidence. 'J. D. Dana, studying the area of Lake Champlain and of the northeastern states in general, came to the conclusion that prodigious floods of almost unimaginable magnitude accompanied the melting of the ice ... floods rose two hundred feet above the present high-water mark.' (71) If flows from the melting snow and ice were many times higher than present river flows, this would explain the massive sizes of the submarine canyons. All of the rapidly-flowing, cold melt-water would have, as soon as it entered the ocean, headed straight for the bottom and would have left evidence of its passage on the way. The alluvial fans (i.e. a spread-out area of silt) at the terminus of the canyons (90) and the submarine canyons themselves are this evidence.

Evidence of rapid melting of the glacial ice is also available from 'Lake Agassiz, the largest glacial lake in North America. Study of its sediment, however, disclosed that its entire duration had been less than one thousand years, a measure of time unexpectedly short; this indicates also that the glacial cover melted under catastrophic conditions. Warren Upham, the American glaciologist, wrote: "The geologic suddenness of the final melting of the ice-sheet, proved by the brevity of existence of its attendant glacial lakes, presents scarcely less difficulty for explanation of its causes and climatic conditions than the earlier changes from mild and warm preglacial conditions to prolonged cold and ice accumulation".' (85) A warm ocean would have been part of the "warm preglacial conditions" and sudden melting of the ice explains the prodigious flow required to form the submarine canyons.

Even now undersea currents flow at various depths through the ocean. In particular in the Pacific Ocean there is a location where surface flow proceeds in one direction while lower down another current is flowing in exactly the opposite direction. Then just to confuse any theories that were being developed to explain such activity, a third current even further down was discovered flowing in the same direction as the surface current. One of the most exciting recent events in oceanography was the discovery of a powerful current running under the South Equatorial current but in the opposite direction. The core of the countercurrent lies about 300 feet below the surface, is about 250 miles wide and flows at least 3500 miles eastward along the equator at a speed of about 3 knots. (The speed of the surface current is only about one knot.) Beneath the swift-flowing eastward current was still another, flowing to the west. In only the uppermost half mile of Pacific equatorial waters, therefore, there are three great rivers of water, one above the other, each flowing on its own course independent of the other. (69)

Submarine canyons extending from major rivers occur all over the world and include; the Congo, the Amazon, the Hudson, the Indus, the Ganges. (24) This implies that all of these areas were also buried in snow which might not have been deep enough to form substantial glacial ice but plenty deep enough to produce massive amounts of melt-water when the clouds parted. Snow must increase in depth to about eighty feet before it becomes heavy enough to form ice. Much more than this could easily have accumulated without forming glacial ice thick enough to leave significant signs that it had existed. The glaciers of the great Ice Age were thought to have been more than one mile thick. (86) If a glacier had existed that was only one hundred feet thick, would it have left any evidence that it existed? If it is allowed that one foot of ice is formed by about ten feet of snow, it can be seen that significant amounts of snow could have fallen in places that are not usually considered as glacial regions. Snow still exists on some mountains near the equator but since submarine canyons exist in these areas, it is reasonable to expect that, during the Ice Age, it existed in the valleys as well.

7.5.2.4 Arctic Coastlines

Coastlines in the Arctic, both mainland and islands, are currently being washed away as permafrost melts. Both the melting and the disappearance are one-way, irreversible events. Such a conclusion has immediate ramifications. Firstly, the disappearing material must have been held in a continuously frozen state since it was either placed or exposed. Otherwise it would have disappeared long ago and would not be there to discuss at the present time. Secondly, this material must have become frozen immediately after it was either placed or exposed for the same reason. Disappearing coastlines are evidence that the great Ice Age developed quickly which, as discussed elsewhere herein, would have been one of the consequences of an asteroid shower.

7.5.2.5   Mammoth Ivory

The great Woolly Mammoth was similar in structure to present-day elephants which have large ivory tusks. The mammoth's tusks however, were even larger than an elephant's and could 'spiral to more than 13 feet', be as thick as a tree trunk and weigh 150 pounds. (91) Mammoth tusks are difficult to recover due to the hostility of extreme northern Siberia, but since it is becoming harder to get elephant ivory, (it is against the law), hunters venture into the Siberian wilds to recover mammoth tusks. Such ventures would be totally useless if the mammoth ivory wasn't fresh. This freshness is evidence of the great Ice Age.

As discussed above in paragraph 4.6, the mammoths lived in a warm climate and died very suddenly. The tusks which are currently being recovered provide evidence of this warmness or these great creatures would simply not be presently found in these regions at all. Clearly soon after they, or certain of their body parts, arrived where they are presently found, they were frozen in place and have remained frozen ever since. If this was not the case, the ivory would not be fresh but would, in fact, be useless. However, now that the permafrost is melting where the tusks are located, it is possible even for a single individual to dig them out of the ground.

The Earth was warm. Great Woolly mammoths roamed the land. An unimaginable catastrophe happened which killed and disarticulated them (along with numerous other species in the same location). Then the land became very cold and has remained cold right to the present time. The great Ice Age happened and because the land in the far north was so deeply frozen, it has taken longer to thaw out. But now that this is happening, it is clear that an ice age has occurred.

7.5.2.6   Lake Vostok

Lake Vostok is a large lake (about the size of Lake Michigan) about two miles beneath the surface of the ice on Antarctica. When a large borehole was opened in the ice to get to Lake Vostok, it was observed that the ice within about 600 feet of the water surface was muddy. (37) Mud was apparently distributed throughout the ice from that level all the way down to water level. This mud certainly did not work its way up through the ice. Its presence indicates that the water of the lake was disturbed (as it certainly would have been when tsunamis washed over Antarctica) and before it had a chance to settle down, it froze. It has remained frozen ever since. The events immediately preceding the Ice Age included not only great waves washing over continents but numerous violent earthquakes. Any water that pooled would have been agitated and muddy. Freezing quick enough to trap the mud indicates that the Ice Age proceeded immediately. The mud was trapped in the ice and remained there ever since. Since the ice is muddy from about 600 feet above the present lake surface, it can be concluded that the original lake was 600 feet deeper than the present lake. Otherwise if the water had been allowed to settle down first, the mud would not have become trapped in the ice as it formed and we would not know how deep the lake was originally. Only a violent catastrophe followed immediately by the Ice Age can explain why there is mud in the ice up to the 600 foot level above the present lake.

7.6 Repeatability

If a large asteroid crashed into the Earth, a massive dust cloud would certainly result. Numerous volcanoes would erupt and their production would augment this dust cloud. The overall result would be a world totally enveloped by a cloud. The immediate result would be lowered temperatures over the land. The volcanic activity would also heat the ocean and cause it to locally boil as hot lava entered the water. Wherever the water was heated significantly, it would evaporate. As this moisture drifted over the chilly land, moisture clouds would form. In summary, as discussed in detail above, the impact of an asteroid with the Earth would result in the formation of two types of cloud as well as a significantly warmed ocean.

7.6.1 Post-Impact Winter

In order to generate a post-impact winter, the chilling effects of an impact must be much more dominant than the heating effects. While the volcanic activity would put heat into the ocean, the ocean temperature would be starting off near freezing. (52) Secondly, a single impact would not activate as much volcanic activity as multiple impacts. As a result, the ocean would not likely become warm enough to evaporate significant volumes of water. Therefore the cloud-chill factor would dominate. Consequently if the Earth is hit in the future by either a large comet or an asteroid, it is quite likely that a worldwide post-impact winter would result. With a drop in temperature over the land that was not offset by an increase over the ocean there would be an overall reduction in the amount of water vapor in the air. Consequently there would be a reduction in the greenhouse gas heating effect of the water vapor and the Earth would spiral down into a deep freeze which would last indefinitely.

7.6.2 Ice Age

In order to recreate an ice age similar to the great Ice Age, a similar amount of ocean water must evaporate. More than two thousand feet of ocean had to evaporate during the Great Ice Age and it had to evaporate within certain critical rate limits. If the required evaporation slowed down a little too much, the ice would have stopped building up. Hence the evaporation rate must have been very high and it had to be maintained at a high rate in order for the great glaciers to accumulate. When the Great Ice Age started, the ocean was warm. Widespread volcanic activity warmed it even more until it was able to evaporate more than 2000 feet of itself and form the massive continental glaciers. Could this happen again?

If a large asteroid hit the Earth in the future a globe-encircling cloud would be produced and a worldwide chill would result. Would the ocean heat enough to evaporate the great quantities of water needed to build the ice fields? The worldwide ocean heating factor could only partially repeat because the ocean would be starting off cold and because a single impact event would not likely produce the amount of heating required. Therefore it would only be possible for a modest amount of snow-producing evaporation to happen before the ocean once again cooled down reducing evaporation to very low levels. A low evaporation rate would not enable the great snow accumulations of the Ice Age to redevelop so another Ice Age would not happen. From our understanding of greenhouse gases it is more likely that chilling the Earth with a world-wide cloud would produce a chilling spiral and it would simply freeze up.

8. Implications for the Windows of Life

8.1 Introduction

Stability is of primary importance for life to exist. Every particular entity of life simply must have environmental stability. Both its surroundings and life-support factors must operate reliably and predictably day after day for its entire lifetime. Even a rabbit cannot tolerate having its burrow collapse. The burrow must be predictable, reliable and structurally trustworthy.

Certain windows of life have a wide range. These windows will be more tolerant of disruption. Other windows are quite narrow and any disruption would mean disaster. The impact of even a relatively small asteroid of one or two miles in diameter would affect almost every window of life that one could imagine and the affect would be manifest over the entire Earth. However, if the impacting asteroid was large, or there was more than one of them, many windows of life would simply close.

8.2 Water

8.2.1 Sustenance Water

An asteroid impact, no matter where it might occur, is physically disrupting to the Earth in numerous ways. Therefore, there would be several ways in which drinking water or sustenance water would be adversely affected.

If the asteroid should impact on the ocean, the first disruption to the drinking water supply would be due to the water-splash. The ocean water would splash for a great distance. As the splash water descends and impacts the Earth, erosion would occur. Some of the splash water would fall as rain and wherever it was intensive enough, this rain would erode soil and thereby contaminate water. Some of the splash water would fall as globules, which would have effects similar to exploding bombs. This type of destruction would also contaminate previously safe sources of drinking water.

Also, if the asteroid impacted on the ocean, a giant wave would be generated. The wave would spread out and wash over the land wherever it comes ashore which would be disastrous for people in low-lying areas. As this type of wave comes ashore, it physically disrupts everything. Mud would be everywhere as well as anything else, which was dislodged by the wave. Sources of drinking water, which came into contact with the wave, could be polluted by several foreign substances. Salt and mud come to mind immediately. Human and animal waste would also be involved. Even though these substances may be unseen, it would only take a small quantity of any one of them to pollute a large volume of water. If the asteroid had been large, the wave would have been large. A large wave would travel inland for miles. In some cases, it would travel inland for hundreds of miles. All of the drinking water in these areas would then be unfit for drinking. Once it has become unfit, there is no way to tell how long it would take for it to become fit again.

8.2.2   Liquid Water

The impact of a large asteroid would result in major quantities of dust and ash being injected into the atmosphere. Sunlight would be blocked from reaching the Earth and chilly weather would follow. Open water would freeze. Depending on the extent of the snow and ice cover, liquid water might only be available in regions near the equator. Where it was not available, animal life would be dramatically restricted or eliminated altogether. In particular if the snow and ice of the subsequent ice age, covered most of the northern hemisphere, animal life would have terminated in these areas.

8.2.3   Ocean Level

The oceans of the world are a very important factor in the temperature stability of the world. The fact that there is more ocean surface than land surface, in the southern hemisphere, contributes to the fine tuning of temperature stability. Ocean water has a great ability to absorb heat without changing temperature very much. This helps to even out temperature variations on an annual basis. Without the extensive oceans, which we have, seasonal temperature variations would be much more exaggerated.

The predictability of the oceans is taken for granted most of the time and the location of the ocean surface is so anticipated that it is referred to as "sea level". The notion, which is inherent in this term, is that we really count on the oceans remaining in a predictable location. All other elevations around the world are referenced to "sea level". While the surface of the ocean continues to rise and fall with the tides, this too is very predictable and so is absolutely no cause for alarm. It is a characteristic of human nature that we expect things, which have always been there to always be there in the future.

If a large asteroid crashed into the Earth all of this predictability would be lost. Everything would change so quickly and so dramatically that it would be unbelievable. A serious upset in any one aspect of our surroundings is upsetting so it would not be surprising if attempts were made to ignore the whole thing. What response would be appropriate if the ocean suddenly surged up onto the land? Every reference point would vanish. Even swimming would not be possible because the moving water would undoubtedly be turbulent and it would also be moving at high speed.

If the oceans rose up and surged over the land, the primary disruption would be physical. Everything would be in devastating motion so all sense of orientation as well as reliable benchmarks would be lost. Until the oceans drained back and settled into a predictable location again, the situation would be untenable for life to continue. The "window of life" would therefore have been closed.

8.3   Rainfall

The rainfall resulting from an asteroid impact would be prodigious. Therefore, it would have a negative effect on the window of life. In fact, with the great amounts of rain, which would result from a large asteroid impact, life would simply not be possible in many areas and the window of life, in such areas, would therefore close.

There would be three main types of rain (or water dumping from the sky) resulting from an asteroid impact.

The first type of rain would be from the atmospheric shock-wave. As the shock-wave spreads outward from the path which the asteroid followed through the atmosphere, it would momentarily compress each particular point in the atmosphere as it passed. Compression would cause any water vapor molecules which were in the air, to coalesce or come together. Once water droplets are formed, they will not disperse immediately. Consequently, rain would fall. The amount of rain would be dependent on the amount of water, which had been in the air. Then, if the upper atmosphere had included a region, which was entirely water vapor, the atmospheric shock-wave would disturb it and it would be released from the sky and fall as rain. Actually it would fall more like a spreading wall of water. Such a layer of atmosphere might have contained up to the equivalent of an atmosphere (i.e. the weight of our present atmosphere) of water vapor. Since one atmosphere of water vapor is equivalent to about thirty feet of liquid water, this would have been an unacceptable amount of rain to deal with and neither trees nor houses could have withstood this much water plummeting from the sky. Everything would have been eroded. There would have been no place to stand. If one wasn't protected from this overpowering deluge, survival would not have been possible. However, if the impact site was far away, the shock-wave would be diminished and the associated rainfall would have been reduced. In these outer regions survival would have been possible but as long as the rain continued, shelter would have been mandatory.

The second type of rain would result from the water splash. This water would be falling from the sky and therefore might technically be called rain but there the similarity would end. It would be more like ponds or lakes falling. Some of the water splash would not have time to break up into droplets before falling back to Earth but instead would fall as large globules. The water aspect of these falling balls of water would be quite secondary to the mechanical crushing effect. Large globules of water, which could easily weigh several hundred tons, would crash back to Earth and crush everything on which they landed. Trees would be flattened. Dwellings would be crushed. Even structures, which were built of stone, would not be spared. Survival in these impact zones would not be possible. As the water crashed into the ground, the ground would respond exactly as if an object from outer space had landed. The difference would only be in the reduced speed of these water bombs but they would still create craters where apparently solid ground had just been. Then, the material from these secondary craters would splash and come back down as great mud-water deluges.

The third type of rain would have resulted from the chilling effect of volcanic dust. Volcanoes commonly produce a great amount of dust and as volcanic activity developed around the world following the asteroid impact, more and more dust would have filled the air. The first result of this dust cloud would have been to chill the Earth. Cooling would have caused condensation. Water, which had been in vapor form, would have turned into the liquid form. The liquid form could not have remained in the air so it would have come down. As this moisture condensed from the sky, it would have seemed like the atmosphere was collapsing. Unlike the atmospheric shock-wave which would cause the atmospheric water to fall out right away, the chilling effect of the dust would have caused atmospheric moisture to condense and fall over a longer period of time. If atmospheric moisture had been replenished during this time, rain (or snow) would have kept on falling and would not have stopped until either the chill factor was removed or the moisture-generating factor was removed In fact the combination of chilly land and a warm moisture-generating ocean would have been the appropriate conditions for an ice age in which case heavy precipitation would have developed and continued for months. Animal life could not have survived hundreds of feet of falling rain or snow, so the "window of life" would have closed under these circumstances.

8.4   Cloud Cover

A moderate amount of cloud cover provides welcome relief from the Sun and of course moisture clouds are necessary to bring the rain. However, the clouds resulting from an asteroid impact would be thick and extensive and envelope the entire world with a cover so thick that sunlight would not get through at all. The dust cloud produced directly by the impact would cover virtually the entire world but when the clouds resulting from the impact-initiated volcanic activity were also factored in, the world would not experience daylight at all for several months following an impact. The darkness produced by these clouds would be the darkness of a moonless and starless night and it would be persistent. Morning would simply not come.

The second type of cloud resulting from the volcanic activity in the ocean would have been moisture clouds. These would have generally spread out much lower down than the dust clouds and might even have extended right to the surface of the Earth. As soon as they drifted over cooler land it would have either rained or snowed virtually continuously for months and possibly for years - at least until the volcanic activity stopped warming the ocean water, causing it to evaporate.

Both of these types of cloud would be devastating to the "window of life". Since they would be continuous, thick and extensive, sunlight would not reach the Earth. Plants would die. Consequently plant-eating animals would also die. With sunlight blocked, the Earth would chill. Below-freezing days would extend right through summer so summer would not come. Crops could not be grown and the lakes would remain frozen. All of these effects would be detrimental to animal life and any animals that had survived the more immediate devastation of the impact would not survive the following months of below-freezing weather. For them, the "window of life" would have closed.

8.5   Snow Cover

The tranquil and beneficial environment of the pre-impact world was unilaterally disrupted by the impact of at least one asteroid and resulted in the great Ice Age. The first necessary condition for an ice age – chilly conditions over the land – was provided by the globe-enveloping sunlight-blocking dust cloud. The second condition – a world-wide ocean-heating mechanism – was provided by underwater volcanic activity.

As the Ice Age developed, snow accumulated. Accumulation was rapid resulting in several inches piling up every hour for weeks at a time. Occasionally it happens, even in this modern time that a snow-fall will be excessive for several hours. It has been reported from the Island of Newfoundland, that three feet of snow fell in a single night. The snow came down so fast that trees were constantly being broken and could easily have killed the witnesses to the event. The fire, which they had, did go out numerous times and it became harder and harder for it to be restarted. (1) Primitive shelters were no match for this type of storm. In this case, all hope would have been lost except that the snow stopped falling the next morning.

Even a snowfall of a few inches will be enough to upset the operation of a modern city. Work crews will spring into action as soon as the snow starts to fall and soon have the main roads open again. If several feet of snow should fall, it might take more than a week to restore a city back to full operation. But what would happen if several feet of snow fell every few days for the entire winter? Homes would soon be buried. Highways would be totally unusable and stores would run out of supplies. At first a few snowmobiles would be seen but if there is no replenishment of fuel supplies, they would soon run out of gas. People could still walk on top of the snow but where would they go? The stores would all be closed. Even if the stores could be tunneled into, the food supply would not last very long. As the snow continued to pile up it would soon reach the power lines. The power would go off. In order to keep warm, people could burn wood. However after both the wood supply and the chimneys are buried, wood burning would be no longer possible. Some of the houses, which were buried, would collapse. Some would not collapse but the occupants would soon run out of oxygen if they were buried under fifty feet of snow.

Certain evidence indicates that during the great Ice Age, ice covered much of North America to depths of hundreds of feet. Survival under these circumstances would have been impossible. Neither would travel to other areas have been possible because the falling, drifting snow would not have allowed this to happen. Even a snowfall of several feet would be detrimental to the "window of life" but the snow/ice cover of the great Ice Age would totally have closed the "window of life" and nothing could have survived.

8.6   The Crust of the Earth

The crust of the Earth is a thin layer of solid material, which is covering a great molten ball of liquid material. The crust is our home and without it life on the Earth would not be possible. However in order to be useful, it must remain solid and predictable. If it should start to move, pulsate, shake, upheave or crack, life will become much less tenable and may simply be terminated.

If a large asteroid should impact the Earth, the crust would move, shake, upheave and crack. Entire mountain ranges would form, escarpments would be thrust up and rift valleys would emerge as sections of crust sank. Wherever these things happened, the "window of life" would have closed.

8.7   Orbital Stability

The Earth enjoys a high level of orbital stability. In fact, the orbit of the Earth has been referred to as a "Goldilocks" orbit. It is just right. If the Earth was five percent farther from the Sun, it would be too cold and if it was five percent closer it would be too warm. (7) This is significant because there have been numerous asteroid impacts all around the world and several of them were upwards of 5 km. (3 miles) across. Apparently they were not large enough to significantly disturb the Earth's nearly-perfect orbit.

In the asteroid belt there are more than 200 objects which have diameters greater than 100 km (62.5 miles). Included among these monsters is Ceres, the largest, with a diameter of 933 km (583 miles). (8) Pallas, Juno and Vesta are also very large with Vesta being about 530 km (329 miles) in diameter. (20) (It must be noted that diameter is the right term because all of these objects are spherical in shape.) If one of these or a large comet should ever come to Earth, orbital stability would be a thing of the past and life on Earth would perish simply because the world would no longer enjoy an orbit which was appropriate for life to carry on.

It appears that Mars was not so fortunate. The three large impact sites on Mars indicate that three very large asteroids landed there. Since they all landed on one side, it would not be surprising if the orbit of Mars was affected. In fact, the orbit of Mars is quite elliptical and it wanders from about 206 million km to about 248 million km. (128 million to 154 million miles) from the Sun. (10) These distances translate to a deviation of plus or minus 9.2 % which is well outside of the five percent limit mentioned above. One can only speculate whether these large asteroids had anything to do with determining this non-circular orbit but since Mars only has about 11% of the mass of the Earth, (9) suggesting that they may have affected the orbit is quite realistic.

Mercury is also much smaller than Earth with a diameter of about 3000 miles but it has a very large impact site. The Caloris Basin is 800 miles across. (11) There is no doubt whatsoever that the impact of an object large enough to leave such a crater would have modified the orbit of Mercury. Possibly some portion of the non-circularity of Mercury's orbit could be attributed to the Caloris impactor. (The deviation of Mercury's orbit is approximately plus or minus 20 %) If a proportionately-large object hit the Earth, it would leave a mark about 2000 miles across. Suggesting that the orbit of the Earth would be modified by such an impact would be quite appropriate. While the objects which have hit the Earth have totally modified its surface features, at least the orbit of the Earth is still, fortunately, within an acceptable range. (However, the length of the year would have been changed.)

8.8   Earth Wobble

The impact of a large asteroid with the Earth would have an effect on rotational stability because the energy of such an object would be comparable to the rotational energy of the Earth. (2) While it would be smaller than the Earth's rotational energy, it would be well within the range where they can be compared. Therefore an impacting asteroid could put a wobble into the rotation of the Earth. The situation would be comparable to a BB hitting a large spinning top. The size of the BB may not seem comparable to the size of the top but its energy could be comparable to the rotational energy of the top.

An example of this type of upset is provided by the Chilean earthquake of February 2010. This earthquake was given a rating of 8.8 making it one of the most powerful earthquakes of all recorded time.

It is thought that it might have shortened an Earth day by 1.26 microseconds and moved the Earth's axis of rotation by 3 inches. Tsunamis were produced and recorded in far away New Zealand, with a surface waves reaching heights of between 3 and 7 feet. The water in Lake Pontchartrain north of New Orleans was caused to move. (19)

The Earth does have a wobble and the North Pole actually moves in a circle. For more than a hundred years scientists have been unable to explain the cause of the Chandler wobble, during which the North Pole describes a circle about 24 ft. dia. m and completes a single loop every 433 days. (4) In fact, there is more than one component to the wobble pattern with the others having much longer time periods. (4) The wobble of the Earth means that there isn't one position in the sky which is always above the North Pole. Even though the North Star or Pole Star is very close to such a position, it is always slightly offset and the extended axis of the Earth actually describes a small circle near it.

If a large asteroid hit the Earth, a wobble would be expected. It is fortunate for life on Earth that the wobble is small and doesn't have any noticeable effect on normal life activities. If it were noticeably larger, the seasonal pattern would be affected and larger areas of the Earth would experience total darkness every winter. If, for example, the axis became tilted more than the present 23 degrees and closer to ninety degrees, almost half of the Earth would experience winter which lasted for several months as well as summer which lasted for several months. Life under such circumstances would only be possible near the equator. It is fortunate that most of the asteroids which have hit the Earth (Sudbury excepted) came in almost straight down. Otherwise the wobble of the Earth would be much more severe and life on Earth would be much less viable.

8.9   Solid Fertile Ground

Solid fertile ground is an absolute necessity for the existence of life on the Earth. It might seem venturesome to get into a boat, eat raw fish and sail around the world but such activities do not have long-term viability. First, there must be a boat to get into and it must be equipped for an unending voyage. Both the boat and the equipment would require solid ground to prepare. Then, if the parents did survive a voyage of this nature for a few years, would the children be able to carry on?

Some people also live in swamps. However, even in these cases, there is still some solid ground or it would not be possible to remain.

If one or more large asteroids hit the Earth, the solid, fertile ground of the Earth would be inundated, over-run by water and washed away. While this was happening, life on Earth would be impossible and the "window of life" would close.

8.10   Carbon Dioxide – A Life Necessity

If a large asteroid should crash into the Earth, there would be several ways in which the carbon dioxide scenario would be upset.

The carbon cycle would be interrupted. The carbon cycle is a mental construct or framework, which is of assistance in understanding how carbon is utilized and transferred throughout nature. All plants and particularly trees, take in carbon dioxide, extract the carbon, and use it for construction purposes. The forests of the world are therefore of great consequence with respect to carbon dioxide because a very large amount of it is taken up and used by the forests. The loss of forest cover is therefore a serious concern. For example, the great rain forest of South America takes up a significant percentage of the CO_2 that is taken up by the entire world-wide biosphere so there would be world-wide implications for the CO_2 levels in the atmosphere if the rain forest was lost. The carbon dioxide, which is currently being produced all around the world, would not be taken up as quickly and more would be left to build up in the atmosphere. At first consideration, this might seem like a positive development because other plants would be able to grow faster and become larger. Unfortunately, this would not compensate for the loss of the rain forest and without it the atmospheric level of CO_2 would increase even more rapidly than it is at the present time.

It would be dramatically worse if all of the forests of the world were swept away completely. If the oceans were able to sweep over the land, either from a primary impact wave or from a remote tsunami, all forest cover would be swept away.

The problem with a major asteroid impact is that tsunamis would occur at many places around the world – even from a single impact. Both the major shock-wave, which would travel outward from the impact site, as well as the internal shock-wave, which would be travelling through and reflecting back and forth throughout the interior of the Earth, would cause tsunamis to occur at numerous places both near and far from the impact site. As these tsunamis propagated they might reinforce each other or succeed each other. With any of these scenarios, major water waves from the ocean would wash right over the land repeatedly. Forest cover would not survive this devastation because if any of the land could possibly erode, it would be gone. With an absence of forest cover, any CO_2 that was produced (i.e. due to rotting vegetation) would remain in, and keep building up in, the atmosphere.

While plant life might benefit from an increase in CO_2, animal life will not. As the concentration of CO_2 increases, more and more oxygen is displaced and long before the concentration reaches 10% of the atmosphere, breathing will be very difficult.

8.11   Carbon Dioxide – Temperature Regulation

Carbon dioxide is considered to be a major factor in both temperature level and temperature stability of the Earth. If the ocean were to undergo measurable temperature change, the carbon dioxide inventory would shift because the amount of carbon dioxide, which can be taken up by water, is temperature dependent. If the water cools, more carbon dioxide can be absorbed. If it should get warmer, the amount of carbon dioxide, which it already has, will be released. Fortunately water is like a great thermal flywheel and changing the temperature of the oceans can only happen very slowly.

An asteroid impact into the Earth, together with all of the physical actions and reactions, which accompany it, is a transient event. It is a singular happening. It happens and then it is over. The effects of the happening will certainly remain but the main event is not expected to recur. However because it would be such a major event, one might expect that some significant ongoing process would be modified. This could certainly be the case with the carbon dioxide inventory of the world.

It appears that, at one time, the Earth had a universally warm climate. Therefore the oceans would also have been universally warm. One effect of a major impact would be on the temperature of ocean water. If the internal shock-wave temporarily raised the pressure under the ocean crust until it broke, the molten magma from the interior would swell up through the crack. Locally, the ocean would boil. Water has an incredible capacity for absorbing heat. As the water boiled, most of the heat from the hot magma would be carried away as water vapor or steam. The oceans would therefore only boil locally so there would be some loss of carbon dioxide from this local part of the ocean. This heating effect would not have affected the entire ocean, partly because the rupturing and welling up of the ocean floor (caused by the upwelling shock wave) would have caused much of the ocean to race away towards the land. Even the evaporated water would cool down and precipitate as rain before very long so the heating caused by more carbon dioxide in the atmosphere would be partially offset. However an eruption of the ocean floor is a volcano. In this case it could be a long, narrow volcano running the length of the ocean. In addition to locally boiling the water and thrusting it into the atmosphere, the volcanic activity would also cause dust and ash to be thrust into the atmosphere. This dust and ash would not come back down very soon and would have the effect of blocking out incoming sunlight. This would cause the Earth, including the oceans, to cool. As cooling proceeded, the carbon dioxide, which was in the air, would become more and more dissolved in the ocean.

It will be recalled that carbon dioxide is a major factor in atmospheric heat retention. With more carbon dioxide in the air, there is a greater greenhouse effect. If the Earth should cool down, due to any reason whatsoever, this greenhouse effect would be reduced because some of the carbon dioxide would be absorbed into the ocean water. In this manner the transient cooling effect of the volcanic dust would initiate a general cooling cycle for the entire Earth. This cooling cycle would be self-propelled because as the ocean cooled, more of the carbon dioxide in the air would be removed from the air thereby further reducing its heat retention properties.

With such a large amount of volcanic dust being admitted to the atmosphere, it would be years before the associated cooling effect would diminish. By then, the oceans of the world would have cooled and much of the carbon dioxide, which was in the air before the impact, would have been absorbed into the ocean water. However, partway through this cooling cycle, another cooling factor would come into play.

Cooling would have been initiated because the heat from the Sun was blocked out. A great deal of the heat from the Sun would have been reflected from the cloud-tops but the clouds would have been so thick that that sunlight would not have reached the surface of the Earth anyway. Subsequently as soon as the warm Earth chilled to a temperature which allowed snow to form, any areas, which became covered with snow would also have reflected the heat from the Sun. The initial cooling effect of the cloud would thereby have been enhanced because the snow cover would have contributed to heat reflection.

In due course, the dust which initiated the whole cooling trend would have settled out of the atmosphere and the Sun would have shone again. However, by then there was less carbon dioxide in the air so the Earth stabilized at a lower temperature range. This was satisfactory because it was at a level, which was sufficient to support life over much of the surface of the Earth.

Therefore the overall effect of an asteroid impact on the temperature of the Earth was to reduce it. Through the agency of carbon dioxide, years after the impact, the temperature of the Earth settled out at a lower level but in the process, the window of life shifted towards closure, because the disparate temperature scenario of the present world results in a much narrower "window of life" than the pre-impact universally-warm scenario..

Fortunately, it did not close completely. Since the present trend for atmospheric carbon dioxide is towards increase, it would appear that the Earth is getting warmer again and that this aspect of the "window of life" is shifting further open. Hopefully it will not shift too far.

8.12   Ozone

When the ultraviolet light from the Sun acts upon atmospheric oxygen, ozone is produced. 'Atmospheric ozone is formed at altitudes above 20 km by the action on oxygen of short-wavelength ultra-violet light. This light is thus prevented from reaching the Earth's surface, where it would be harmful.' (12) 'The ozone absorbs still more ultra-violet light, of rather longer wave-length. The energy of this absorbed light is transformed into heat in the air...' (14) In this manner, ozone provides a shield from the damaging ultraviolet rays of the Sun, which can cause skin cancer. 'One of the most interesting and most publicized harmful effects of ultra-violet light is its role in the cause of cancer of the skin. Fair-skinned people and those who sunburn easily and do not tan well are especially susceptible to ultra-violet light induced cancer of the skin.' (13) From a "window of life" viewpoint, it is therefore important that a layer of ozone be present in the upper atmosphere at all times.

Ozone is easily destroyed by chlorine. The freons – a class of popular refrigerants for many years - include chlorine in their long and complex atomic structure. Since some types of freon break down in the upper atmosphere and release chlorine atoms, they have been identified as hazardous material and are being phased out of production and use.

Chlorine is also injected into the atmosphere by volcanic activity which may propel it high into the upper atmosphere where the ozone is being produced. It only requires a very small amount of chlorine to destroy a lot of ozone. Therefore widespread volcanic activity would result in a significant reduction of ozone which would result in an increase in the amount of ultraviolet light which reached the surface of the Earth

A great deal of volcanic activity resulted from the asteroid impact. Fortunately, from a "window of life" viewpoint, most of the volcanic material would have settled out of the atmosphere within a few years, and the negative effect which it would have had on the "window of life" would have been correspondingly short-lived.

8.13   Wind

Light winds are rather pleasant but heavy winds are hazardous and make life more difficult. As wind speed increases, trees will not grow and soon nothing will grow because the soil will all be gone. However, up to a certain speed, wind is not a disadvantage to life.

If a large asteroid hit the Earth, there would be local serious disruptions to the wind but of the many life-threatening hazards, which would develop, wind would be very low on the list. However there would be a long-term effect. The volcanic activity initiated by the asteroid, would have been effective in precipitating the protective vapor canopy from the upper atmosphere. Therefore, the universally warm climate of the Earth would have been terminated. Consequently, the northern regions became colder and a significant difference in temperature developed between the equator and the poles. This resulted in the development of world-wide wind patterns where there had not been any before. These winds move heat from the tropics to the middle and high latitudes whereas prior to the impact of the asteroid, this would not have been the case. Occasionally these new winds – in particular tornadoes and hurricanes - are hazardous to life but as long as they remain tolerable, there is little effect on the "window of life". (This situation appears to be changing as the Earth warms up producing more powerful thunderstorms, more energetic tornadoes and more massive hurricanes – all of which are more hazardous to life and thereby narrow the "window of life".)

8.14   Vapor Canopy

The vapor canopy was viable as long as the atmosphere was stable. Its heat distribution function would have operated continuously and its light propagating function would have illuminated areas of the Earth which were beyond the horizon and which would otherwise have not received enough light to enable plant growth (e.g. Axel Heiberg Island in the high Canadian Arctic). The arrival of an asteroid would have upset the stability of the atmosphere and it would have destroyed the vapor canopy. As it collapsed, the associated rainfall would have been intensive. While this would have been locally and immediately detrimental, its effects would have been short-lived.

The longer-range implications of canopy loss are more important. With its loss, most of the solar heat received near the equator will remain there. At the same time, polar areas will have lost their insulating covering. As a result of these two changes, temperature discrepancies have developed between the equator and the poles. This has caused global air circulation patterns to develop in the atmosphere and long-range currents to develop in the ocean but neither of these energy transfer mechanisms are able to efficiently redistribute the abundant solar energy which continuously pours down on the equator. Mid-latitude regions have become chillier and polar regions have gone into the deep freeze every winter and

313

barely come out of it during the summer. Both plant growth and animal life in the high Arctic and Antarctica has been dramatically reduced. While the "window of life" did not totally close in these areas, as a result of the asteroid impact, it became seriously curtailed and virtually closes every winter.

Another long-range implication of the vapor canopy loss involves the exposure of the Earth to ultra-violet light. The moisture of the canopy would have blocked ultra-violet light completely thereby preventing its harmful effects from reaching the ground. When it was removed, ultra-violet light could reach the ground bringing its detrimental effects with it, including skin cancer. The loss of the vapor canopy would therefore have been very similar to the loss of the ozone layer as discussed above where it was noted that 'One of the most harmful effects of ultra-violet light is its role in the cause of cancer of the skin.' (13)

8.15   Magnetic Field

As discussed earlier in the chapter entitled, Some Windows of Life, the magnetic field of the Earth is very necessary for our protection. However, it is weakening and at the present rate of decay it will be gone within a few hundred years. The Earth's main magnetic field, its dipole field, is decreasing at a rapid rate which is "unparalleled" by any other geophysical phenomenon. The rate of decay is several few percent per century. (5) This rate of decay was confirmed by the Magsat project as a 'fraction of 1 per cent per decade. (5) Other reports agree with these conclusions. 'Scientists after Gauss continued to make global measurements of the field. ... covering the whole period from 1835 to 1965. They drew a startling conclusion: during those 130 years, the Earth's magnetic dipole moment (i.e. the main magnetic field) had steadily decreased by over eight percent!' (16) Depending on how such data are extrapolated into the future, the magnetic field will approach zero in between 1200 years and 2000 years in the future. When that happens, the Earth will be totally exposed to incoming cosmic radiation with its attendant hazards. In fact, the consequences of the decaying magnetic field are astonishing. It is heading quickly toward a lethal environment of cosmic ray bombardment. Cosmic rays are very high speed charged nuclei coming in from outer space with high enough energies to change elements. (5) 'Many types of radiation can cause cancer.' (6) Cellular function may be temporarily or permanently impaired... or the cell may be destroyed. Exposure to doses too low to destroy cells can induce cellular changes...' (15)

The data not only indicates that the strength of the main magnetic field is decreasing, it also indicates that secondary fields are increasing in strength. While there continues to be a net loss overall, the non-dipole components have increased in strength quite significantly during the period from 1970 to 2000. (17) It appears that the magnetic field is not only becoming weaker but that it seems to be breaking up into higher-order fields like eddies in a stream. Whether this development is directly linked to the arrival of the asteroids, or not, is speculative. However, their arrival completely disrupted the tranquility of the Earth including any current that might have been circulating in the interior. If such a current was caused to break up into secondary currents, the main current might die off. There was so much upset accompanying the asteroids' arrival, that the turbulent activities within the Earth could also have included perturbation of the currents which cause the magnetic field. (18)

8.16   Summary and Conclusion

All of the "windows of life" must simultaneously and continuously remain open within a fairly narrow range, in order for life on Earth to survive.

When a large asteroid - one that was several miles in diameter and big enough to punch right through the crust of the Earth - impacted the Earth, a devastating chain of events followed. Many "windows of life" closed immediately and some of them remained closed for several years.

As the asteroid passed through the atmosphere, an enormous shock-wave formed which flattened everything for many miles around the impact site. This spreading shock-wave squeezed the moisture right out of the atmosphere producing an expanding curtain of rain. The vapor layer in the upper atmosphere collapsed and the resulting rainfall amounted to several feet of water. The Woolly Mammoths were killed by this shock wave as well as numerous other animals.

The impact splash also produced rain but more importantly, it launched enormous globules of water into the air which crashed back down and flattened everything, which was not already flattened by the atmospheric shock-wave. Both mud and rocks were included in the splash material just for good measure.

The ground shook. The initial horizontal displacement near the impact site was thousands of feet before the first bounce-back occurred. The back and forth shaking lasted for weeks before dying out. This earthquake also included an up and down component. Sometimes a ring of mountains formed around the impact site while regions outside of the crater area experienced both up-rifting and down-rifting. Valleys formed. Escarpments rose. Volcanoes erupted.

The horizontal displacement of ground adjacent to ocean caused world-encircling tsunamis to be generated. In the case of the Vredefort impactor in South Africa every coast of Africa shifted and generated tsunamis. Similarly when Chicxulub in Mexico was formed a great series of tsunamis left the west coast of Central America and travelled across the Pacific Ocean. Even the Sudbury and Manicouagan impactors in North America would have caused tsunamis to leave every shore of North America. There are many other large impact sites around the world and every one of them would have generated numerous world-encircling tsunamis. There was no refuge whatsoever from any of these waves which kept coming repeatedly for weeks.

Great shock-waves spread throughout the interior of the Earth and as they came up under the crust on the far side of the Earth (i.e. the antipode), there were more earthquakes volcanoes and tsunamis. One of these shock-waves stressed the Atlantic Ocean floor into failure mode and it split right down the middle. An enormous, upwelling, gushing wall of water, steam, dust and ash surged up into the atmosphere and formed the longest volcano the Earth has ever seen. The water in the ocean rose up on the upwelling ocean floor and rushed towards the land. This wave was initially very high – in fact, thousands of feet high. When it reached the land it washed right over it causing widespread sheet erosion and displaced everything right down to the bare rock. Great floating masses of vegetation were swept along with this wave and the eventual pileups were, in some cases, thousands of feet deep.

The interior shock-wave wasn't finished after one transition through the interior of the Earth but traveled back and forth several times. More tidal waves and tsunamis resulted from each pass and all patterns of movement in the interior of the Earth were disrupted. This may have initiated the breakup of the main magnetic field of the Earth into secondary and tertiary fields with a resultant overall reduction in main field strength and the accompanying cosmic-ray shielding effectiveness that the main field provided.

Included in the effects of the asteroid impact was the production of an enormous amount of dust —both directly from the impact as well as from the resultant volcanic activity. These developments initiated a chilling cycle over the land. As soon as the temperature dropped below freezing the great Ice Age was underway. Snow production was fueled by the heated ocean water, which supplied the moisture for making the snow. It snowed all summer. The snow piled up until it was several thousand feet deep.

All of this happened as the consequence of a single asteroid impact. However there was more than one impact. In fact, it was an asteroid shower and all of the above-noted effects – in particular the prodigious snow-making activity were augmented and caused to continue for several years. The asteroid shower devastated the surface of the Earth and totally rearranged all of its surface features. (Similar devastation visited both the Moon and Mars as the result of asteroid showers.) Things did not settle down again for months. The interior shock-waves bounced back and forth through the interior of the Earth for weeks and the giant ocean waves went right around the world several times before being interrupted by the newly-formed mountain ranges.

Would any form of life have been able to survive such devastation? Could any form of life have survived the impact of even a single large asteroid? What if there were several large impacts, several medium impacts and numerous small impacts. Neither animals nor people could have survived. Birds could not have survived. Possibly a few swamp creatures could ride it out and float up onto the masses of vegetation. Many fish would have died as the ocean boiled. Larger animals had less chance of survival. Insects might have attached to some floating vegetation and make it through. However, for most of the life-forms on the Earth, many of the windows of life closed. With several windows shut at the same time, there was no hope at all that any dependent life-forms survived.

9.   Evidence of Asteroid Impact

9.1   Introduction

Asteroids are also called minor planets and like the major planets, they orbit the Sun. Most asteroids have orbits in the region of the solar system, which is between Mars and Jupiter. The orbits of many asteroids are reasonably circular but there are also many which have elliptical orbits, which may take them out past the orbit of Jupiter or in closer to the Sun than Mars. There is also a group of asteroids, which have orbits, which are well within the orbits of the four inner planets including Mercury, Venus, Earth and Mars. These objects are of particular interest and any, which cross the orbit of the Earth, are potential threats to the Earth.

When they are compared to the major planets, asteroids are not very large. However, as possible objects of destruction, their relatively small size becomes of much greater interest. The largest asteroid is called Ceres and it was discovered by Giuseppe Piazzi in 1801. It is approximately 680 miles in diameter. (1) Asteroids range in size from Ceres all the way down and there really isn't a lower limit to the size of an asteroid.

Asteroids continue to be identified, numbered and tracked. Included in the new discoveries are those with orbits which are within the orbits of the four inner planets. Of particular interest, of course, are those that cross Earth's orbit. Only a few years ago this number was set at forty. As instruments improved and more objects were identified, the number grew. It continues to grow and it is now in the hundreds. Improved instrumentation and observation techniques enable more and smaller objects to be seen. The beneficial aspect of this is that the newly identified objects are smaller. The discouraging aspect is that there are so many of them.

Since asteroids are not very large when compared to other celestial objects, it is easier for their orbits to be modified. Asteroids, which presently orbit beyond Mars, could be drawn in closer by the gravitational effects of other objects in their region of space. Orbital changes could also be brought about by any large object, which might pass through the inner solar system, come near an asteroid and thereby attract it out of its orbit into a more dangerous orbit from the perspective of the Earth. (2) Occasionally large comets pass through the asteroid belt on their way in from the far reaches of the solar system. An asteroid could be dislodged from its orbit by a passing comet and its orbit could be further affected by the gravity of both Mars and the Earth. This means that even though it may have a stable orbit between Mars and Jupiter for many years, this does not mean that its orbit will always be stable and not a concern to the Earth.

As discussed above, observations confirm that many asteroids are known to be potential threats to the Earth. However, in addition, some of those, which are currently far away, could be drawn from their distant orbits and brought within range of the Earth. Since asteroids are not very large, they can even affect each other's orbits. (3) The summation of all of this is that it is totally impossible to really determine how many asteroids might be a threat to the Earth. In confirmation that this is the case, asteroids have recently passed close to the Earth and were not detected until they had actually passed.

317

Currently, it is thought that there may be a near flyby every week. (4) The volume of space is just too great for anyone to ever be able to completely identify how many objects may actually become a threat.

The total number of asteroids might never be known either. New ones are identified and cataloged at the rate of many thousand every year. There is no end in sight. To suggest that there might actually be several million asteroids would not be an exaggeration in the least. (5)

9.2   Impact Site Expectations

The expectation that asteroid more impact sites will be found on the Earth is quite realistic in part because there appear to be numerous impact sites on the Moon, Mercury and Mars. If these objects of the solar system have received this type of attention, then why wouldn't it have happened on the Earth as well? The search continues in an ongoing effort to identify any places where an asteroid might have come down but because the Earth is generally covered with vegetation it usually requires considerable effort to locate an impact site. Also, the Earth has an active environment. The effects of wind and rain could modify evidence of impact and erosion could even wash some evidence away. The Earth also has a great deal of water which can simply hide evidence. Anytime the water was in motion, other materials could have been relocated into a crater thereby obscuring it and making identification more difficult. The Earth is very large when compared to either Mars or Mercury or the Moon, which raises the expectation that the Earth should show at least as many impact sites as either of these other heavenly bodies and there might even be more. In at least partial satisfaction of these expectations, many impact sites have been found and new ones continue to be identified. There is no doubt that as time goes by, even more sites will be found.

There is a second major category of reasons why impact sites are difficult to identify, which relates to the structural nature of the Earth. The Earth is liquid. When a rock is thrown into a pond, within a few minutes it is not possible to identify any trace that it happened. The pond is liquid and simply fills in again and pretty soon it appears as if nothing happened. The situation is improved if the pond has a thin crust of ice. The rock would fracture the ice and leave a hole. The evidence is now a hole in the ice as well as possibly fragments of ice on the surface. However, the water in the opening will soon freeze over again and the most obvious evidence is lost. The freshly frozen part might not appear quite like the rest and so suspicions may be aroused. Such an anomaly could justify more detailed and careful study.

An ice-covered pond has similarities to the crust-covered Earth. If a large object strikes the surface of the Earth, it might go right through the crust into the interior. The major opening will disappear but there should be some indication that an opening had been made and more detailed study of such a site would be justified.

In recognition of all of the above, there is another overwhelming reason why more asteroid impact sites have not been found. Since the Earth is mostly covered by water, the impact of a large asteroid would be a catastrophic event of global proportions. Included in the effects of a major impact would be giant tsunamis and worldwide earthquakes. The geology of the entire Earth would be modified by a single major impact. In this manner an asteroid might be able to obliterate any obvious evidence that it had arrived. Included in the destruction could be destruction of the impact site itself. This reality is well demonstrated by the Chesapeake Bay example as this site is totally buried and has only recently been identified (following the study of suspicious drill cores).

9.3   Asteroid Size

Whenever a crater is discovered on Earth, there is always an attendant effort to determine how massive the incoming object would have been. While this may be an interesting exercise, it remains that it is not possible to make such determinations. There are several reasons why this is the case. First, the damage, which is done by the object, is not only a function of its mass but is a function of its speed as well. Therefore if the damage has been estimated, the mass can only be estimated by making an assumption concerning the speed. The energy of the object is the product of the mass and the speed but the speed factor is multiplied twice. This is called a square law. It means that the speed is much more influential on the resultant energy than the mass and a relatively small difference in speed will have a disproportionate influence on the energy. For example, if the mass were doubled the energy would be doubled. However if the speed were doubled, the energy would be quadrupled. It has been repeatedly stated in numerous articles that the speed of an asteroid can be in the 10,000 to 100,000 miles per hour range. This is a ten-fold range of speed and translates into a 100-fold range of energy.

Secondly, asteroids land in various types of terrain. If an asteroid lands where there is plenty of overburden, the relatively loose unconsolidated material, which is resting on bedrock will be displaced and form the classic bowl-shaped crater. The amount of material, which is displaced, could therefore give an indication of the asteroid's energy. Then, by arbitrarily setting the speed, the size can be determined. One good estimate deserves another.

If an asteroid lands on exposed rock, the classic type of crater will not be formed. There might be a rim around the crater but the center will not be proportionately depressed. Since the impact site will not be crater shaped, using the missing material to determine the incoming energy will not be possible.

The fourth factor, which bears directly on any determination of size, is whether or not the object was an appreciable fraction of the thickness of the crust of the Earth. If for example, an incoming object was ten miles in diameter, it's diameter would be 50% of the thickness of the Earth's crust. Why then would not an object this large simply smash right through the crust and dive into the interior? If this occurred, the impact site would be totally different than if all of the incoming energy were dissipated during removal of material from the site. As the size of the incoming object gets larger and surpasses the thickness of the crust, a smaller percentage of its energy will be dissipated on the surface and more and more of its energy will be directed into the interior. Since the energy of a single large asteroid is more than enough to build a mountain range, having some of the energy dissipated away from the site seems like a good idea. This does mean, however, that the size of the crater will become proportionately smaller as asteroid size increases. Therefore, while an object, which is only one hundred feet in diameter, might make a crater, which is more than one mile in diameter, an object which is fifty miles in diameter could make a crater which was only a few times that large.

9.4   Evidence from the Moon

The Moon has 200,000 impact sites, which are greater than one km. in diameter. (25) Some of these are craters, which are many miles across. They are so large that if a person stood in the middle of one of them, he would not be able to see the rim. There are also large areas on the Moon, which are called seas. These appear as areas with very few crater marks and are suspected to be lava, which poured out onto the surface as a result of a major impact. One of these, Mare Orientale is 560 miles in diameter and ringed by mountains. Another one is Clavius, which is 211 miles across. These features of the Moon indicate that it has been a place, which has experienced numerous impacts by very large objects. As it would on Earth, each major impact would have generated several life-endangering events. All major impacts would cause the ground to shake. Earthquakes or in this case, moonquakes would have been just as devastating on the Moon as similar events would have been on Earth. There would have been a great splash of material. Rocks and boulders would have showered down over a wide area. Of course there wasn't anybody there to experience the danger but from what can be observed from Earth, there would have been a great deal of danger to anybody who had been there.

A further factor of interest in the case of the Moon is the observation that while the near side has evidence of major impacts, the far side has a bulge (called highlands) where the terrain is uneven with jagged ridges and the general appearance of having been disturbed – possibly from within. If large objects have impacted the near side of the Moon, it would not be surprising if the far side showed evidence of upheaval. The shockwaves from the impacts would have gone right through and when they came up, the surface would have been thrust upwards and broken and highlands would have been formed.

9.5   Evidence from Mars

Mars has many features, which are evidence of asteroid impact and the three largest can be seen from one side. These impact sites are named Isidis, Hellas and Argyre. The largest of these, Hellas, is variously given as 1430, 1304 and one thousand miles across. The other two are smaller. Argyre is 1120 miles in diameter and Isidas is 680 miles in diameter. None of these impact sites are very deep in comparison to their respective diameters. This suggests that the impacting objects were swallowed by the planet and then the molten interior enabled the crater floor to reestablish near the original level under the influence of gravity. The impactor could have been very large but since there isn't a classical bowl-shaped crater but only a relatively shallow basin, using the displaced material as an indicator of its size is not appropriate. Recourse must be made to speculation. Whether it was a comet or an asteroid, a major shock wave would have been propelled through the interior and would have come up under the far side in just a few minutes. Chaos resulted. Massive volcanoes erupted and the surface structurally failed and collapsed. (Further comment is included in the section 'Evidence of Chaos).

Shock-waves from these impacts would explain the elevated area on the other side as well as the adjacent Valles Marineris, the immense canyon system. With any one of these impacts, the surface of the planet would have been violently shaken. It would have been impossible to stand and if people had been there, they would have been tossed around so much by the Marsquake, that they would not have survived. There would also have been a shower of rocks and boulders, which would have further endangered any life-forms had any been present.

9.6   Evidence from Mercury

Mercury is the smallest planet in the solar system. However it has a feature which indicates that it has taken a major hit from an asteroid. The Caloris Basin of Mercury is greater than 800 miles in diameter and it is ringed by mountains. (27) This feature is believed to be the result of an impact. The diameter of the planet is just over 3000 miles so the impact of an object large enough to form a crater 800 miles in diameter, would have been most catastrophic.

9.7   Evidence from Tethys

Tethys is a moon of Saturn and has an impact feature approximately 250 miles in diameter. The diameter of this moon is only 660 miles. (26)

9.8   United States of America

9.8.1   Chesapeake Bay

One of the most recent discoveries of a major impact site has been at Chesapeake Bay in the eastern United States. The Chesapeake Bay area has been well populated for more than two hundred years. However it is only recently that a major crater has been identified. The difficulty in making the identification relates to the presence of both water and the soil, which has partially refilled the crater. Apart from the fact that the crater is completely buried, it is very large and includes part of the bay as well as land areas on either side. It was therefore not at all obvious that an impact site was located in this area. Once the discovery was made, a great deal of effort has been expended to identify the location of the crater rim as well as material, which would have been ejected from the ground when the impact occurred. The impact site involves a lot of unconsolidated material, which is relatively easily moved and some of this material appears to have slumped back in. (6) The diameter for this crater is approximately fifty-three miles. The object, which made it would, of course, have been much smaller and probably less than five miles in diameter but as discussed above, it is not really possible to identify how large it was.

The Chesapeake Bay crater continues to be studied. These studies include the drilling of a deep bore-hole in an effort to actually find the asteroid. This type of activity is usually not successful and it probably will not be in this case either. The incoming object would have had a diameter, which was a significant fraction of the thickness of the crust of the Earth. It is therefore very likely that it punched right through the crust and came to rest someplace deep in the molten interior of the Earth. Therefore, the asteroid will not be found because it would have been assimilated into the molten interior and become totally unidentifiable, even if its actual location was found.

9.8.2 Manson, Iowa

Where the town of Manson, Iowa now stands is an impact site approximately twenty miles across. An object possibly 1.5 miles in diameter (ten billion tons) created a crater 3 miles deep, which was subsequently filled with gravel leaving no visible evidence on the surface at all. (7)

9.8.3 Yellowstone National Park

It has recently been identified that the entire area of Yellowstone National Park is a crater. It has been suggested that it is the crater of a very large volcano, which has only recently crusted over. (26) The entire park is geologically active. The ground shifts and water shoots up. A small lake in the park recently tipped up and poured its water out the low end. While popular thinking suggests that it is a volcanic crater approximately forty miles across, there really is no way to distinguish a large volcanic crater from an impact site. Yellowstone National Park might very well be the site of an asteroid impact, which would also result in volcanic activity making definitive identification more difficult.

9.9 Canada

Numerous impact sites have been identified in Canada and several additional sites are suspected of having craters buried just out of site. There are more than a dozen, which are classified in the 6-60 mile diameter range. (8) There are two which are in the over 60 mile diameter range. In the identification of craters, size is not an advantage. If a crater is relatively small - less than three or four miles in diameter it might be reasonably obvious. If however it is more than fifty miles in diameter, it could be too large to be seen from any particular position. Even if a person were to stand on the rim on one side and look across, the far rim would be over the horizon and not at all obvious. Without the help of satellites, topographical maps and bore-hole data, large impact sites are almost impossible to identify.

9.9.1 The Sudbury Crater

In Northern Ontario there is a very large crater. The Northern Ontario crater is oblong in shape and lies in a north-west to south-east direction. The shape and direction are suggestive of an object coming in at an angle from one of these directions. It is also quite curious that many of the mines of Northern Ontario are located along the rim of this crater. It is probable that the incoming object disturbed the crust of the Earth so seriously that the minerals melted and pooled and have now become more available for excavation. One provincial park, Killarney, is located on the rim of the crater. Within this park there are numerous trails, which are appropriate for hiking and one of them goes right up to the top of the rim. Unfortunately, the crater is so large that even when a person is standing on the highest part of the rim, there is really no way to identify that it is a rim at all. Even ignoring the forest cover and local ups and downs of the terrain, the far side of the rim is so far away that the curve of the Earth makes it impossible to see. Satellite photographs, topographical maps and detailed studies of the terrain are required to actually locate the entire rim.

The Sudbury crater is approximately 150 miles across the long axis and about 100 miles across the short axis. It is the largest crater in Canada and to date the largest one to be identified in the western hemisphere.

9.9.2   The Manicouagan Crater

In the Province of Quebec, there is a large crater, called the Manicouagan crater and it is located approximately 110 miles NW of Sept Iles, which is on the Saint Lawrence River. This crater is about sixty miles in diameter, which means that it was also made by a large object. (9) The interior of the crater includes a peak more than 1600 feet high indicating that the impacting object punched right through the crust, molten material refilled the opening and because of the major disturbance to the underlying magma, the surface actually bulged up a little. Another curiosity in this case is that the entire surrounding region appears to have been elevated. It has the characteristic of tundra, which is usually located further north. The area supports a small herd of caribou. According to reports from airline pilots, the crater outline – because it is a circular lake - can be clearly seen from the air.

9.9.3   Others

While relatively small craters like the Manson, Iowa crater, are not of particular interest for this discussion, it is worth noting that Canada has approximately twelve impact sites, which are between six and sixty miles in diameter. The objects, which made these craters, would certainly have caused trouble in Canada, but all of the other impact sites listed herein would have caused worldwide trouble.

9.10   Mexico

The Chicxulub crater is located partially on the northern coast of the Yucatan Peninsula of eastern Mexico. The other portion of this crater is located under the waters of the Gulf of Mexico. As with the Chesapeake Bay crater, the Chicxulub crater was not at all obvious. It was only identified during a study of the geology of the region using information from well-drilling records.

This crater is approximately 112 miles in diameter and is currently being given credit for the extinction of the dinosaurs a long time ago. (10)

9.11   South Africa

The Reitz (or Vredefort) multi-ringed crater of South Africa is between 186 miles and 312 miles in diameter, depending on which geological feature is recognized as the rim. (11) The asteroid, which would have made this crater, would have been unimaginably large and would exceed the usual suggestion that an asteroid is probably less than 5% of the size of the crater, which it produces. The reason for this is that large asteroids like this would certainly have enough energy to punch right through the crust of the Earth and transfer most of their energy to the molten interior. A classic bowl shape would consequently not be available to enable energy (and hence size) estimates to be made. While it will never really be possible to determine just how large the asteroid was, it could easily have been greater than ten miles in diameter.

9.12   Kazakhstan

The Ishim impact crater in the Taniz Basin of Kazakhstan has a diameter, which is estimated to be from 220 miles to 450 miles. (12) From satellite imaging, this crater not only has broken through the crust of the Earth but has fractured the crust in the entire surrounding area. As asteroids increase in size, the craters, which they produce, become proportionately smaller. Most of the energy of these large asteroids would still be with them as they punch right through the crust into the interior. The crust of the Earth would have little effect in slowing them down. As their speed is retained, so their energy is retained and instead of the energy being dissipated by breaking the crust, the great energy of such asteroids would be transferred to the molten interior. There, they would generate shock-waves, which would distribute their energy to the far side of the world. For this reason the very large asteroids will have been disproportionately larger than their craters would indicate. If this reasoning is correct regarding the Ishim situation, the actual asteroid could have been several tens of miles in diameter. (This is probably too large to be acceptable for popular thinking but never-the-less this is what the evidence indicates.)

9.13   Australia

9.13.1   Acraman Crater

The Acraman crater of south central Australia is similar in size (i.e about 55 miles in dia.) to the Chesapeake Bay crater of the United States. One edge of it is about one hundred miles from the ocean and the crater can be identified by the lakes - including both Acraman Lake and Lake Gairdner - which lie within its boundaries. (13)

9.13.2   Woodleigh Crater

The Woodleigh crater of Western Australia is about 25 miles in diameter. It is very close to the west coast of Australia and also very close to the town of Wooramel. (14)

9.13.3   Yarrabubba Crater

The Yarrabubba crater is also in Western Australia about 300 miles further inland and a little south of the Woodleigh crater. Its diameter is given as 19 miles. (15)

9.13.4   Tookoonooka Crater

The Tookoonooka crater is located further east of the above two craters, is about 35 miles in diameter and lies just west of the Grey Range. (16)

9.14   Russia

9.14.1   Popigai Crater

The Popegai crater of Russia is about the same size as the Manitouigan crater of Quebec, Canada (i.e. about 62 miles in diameter). It is located about 200 miles from the Arctic Ocean in Eastern Siberia and about 300 miles west of the Lena River. (17)

9.14.2   Puchezh-Katunki Crater

This crater is about 50 miles in diameter and is located about 200 miles due east of Moscow. (18)

9.14.3   Kara Crater

The Kara crater is located within 100 miles of the Kara Sea and a similar distance south-west of the town of Kara on the northern Russia coast. Its size is given as 40 miles in diameter. (19)

9.15   Europe

The border of the Czech Republic is a circular arrangement of mountains making that country a possible asteroid impact site. (20)

9.16   Possibilities

9.16.1   China

In the north-western part of China there is a geological feature which includes an almost perfect circle of mountains surrounding the Takla Makan Desert. This feature appears very similar to features which can be seen on the Moon and which in those situations are referred to as impact sites. The circle of mountains is approximately 800 miles across from north to south and approximately 1600 miles across from east to west. If this feature was formed by an asteroid, it would have been several tens of miles in diameter. It is more than a bit curious that the elevated Plateau of Tibet is immediately south of this feature and that several mountain ranges including the Himalayan Mountains are immediately south of the elevated plateau and concentric about it.

The massive Deccan Plateau of India was formed by an outpouring of lava about two miles thick covering 200,000 square miles. (27) Also in the Indian Ocean south of Ceylon, the ocean floor is formed of lava where attempts to take bottom samples from this plain had little success, for the corers were broken repeatedly, suggesting that the bottom was hardened lava and that the whole vast plateau might have been formed by the outpourings of submarine volcanoes on a stupendous scale. (31)

This type of outpouring is called a Large Igneous Province. These types of formations include continental basalts, associated intrusive rocks, volcanic passive margins, oceanic plateaus, submarine ridges, ocean basin flood basalts, and seamount groups. They represent major global events. Preceding major volcanic activity, the uplift above the area would have been considerable. Maximum uplift can reach thousands of feet and uplift would have been rapid. (31) This is exactly what would be expected if there been an increase in pressure deep in the Earth and what would be expected from an asteroid impact at Takli Makan. Such an event would have caused Asia to bump into India rather than the other way around.

9.16.2   Africa

In south central Africa there is a feature called the Congo Basin. This formation is approximately 1000 miles in diameter and almost completely surrounded by mountains. If this feature is due to the impact of a large asteroid, it would have been even larger than the one which made the Kazakhstan crater.

9.16.3   Madagascar

On the sea floor north of Madagascar a feature has been identified which could be an impact site. This feature is about 200 miles in diameter and will no doubt be the subject of further investigation.

9.16.4   USA Eastern Seaboard

Off the eastern coast of the USA a feature has been identified which could be a large impact site. Certain nearly circular patterns in the seafloor indicate that a crater might be involved. This feature is distinct from the Chesapeake crater, which has been well investigated and established as an impact site.

9.16.5   Canadian Shield (Laurentian Plateau)

The Canadian Shield is approximately centered on a depression (i.e. Hudson Bay), is reasonably circular, includes "hundreds of extinct volcanoes" and the Mackenzie dike swarm "which is the largest dike swarm known on Earth". Many mines, including diamond mines centered on Kimberly pipes, are spread throughout the entire region. It is also very large (3,000,000 sq. mi.) extending from Southern Ontario to the high Arctic Islands and Greenland and is the scene of unimaginable chaos. It is tempting to think that this feature was formed by a very large asteroid.

9.17   Speculation

None of these possibilities have been generally accepted identified as impact sites by the geological community. Whether they will be or not will only be confirmed or denied by careful examination of these areas by competent people. Since very large asteroids would punch right through the crust of the Earth and only leave a perimeter mark or hardly any mark at all (as in the Sudbury, Chiczulub and Chesapeake Bay cases) easily identifying an impact site is not possible. Further in the Quebec case, the Manicouagan crater is only evidenced by a relatively shallow circular depression. Both the Chinese feature and the

African feature meet the expectation of at least a partial perimeter mark because they both have a nearly circular surrounding formation of mountains. The Canadian feature is nearly circular and has clearly been the scene of violent disturbance. Prior to confirmation, through detailed investigation, all of these possibilities remain speculative. The fact remains however, that in the future, more large impact sites will be identified. (In keeping with this expectation, an impact site has been very recently discovered on the floor of the Indian Ocean. It has been named the Burkle Crater and is located 900 miles south-east of Madagascar at about 33 degrees south latitude and 58 degrees east longitude. While it is only 18 miles in diameter, it would have been made by an asteroid in the serious-trouble range.) Deciding where to look will probably be initiated by factors, which are initially remote from the subject but somehow indicate that an impact site investigation is warranted.

9.18 Terrestrial Maria

While all of the above discussion relates to direct evidence of asteroid impact, there are two other features of the Earth, which are indirect evidence of asteroid impact.

9.18.1 Overthrusts

On the Moon there are large surface features called maria. Maria is the Latin word for seas. Apparently at the time this designation was first used, it was thought that the dark regions on the surface of the Moon were seas. There is some justification for such thinking as the areas being referred to are darker, flatter and more devoid of impact sites than the rest of the lunar surface. The contrast is actually quite striking as the Moon, generally speaking, is riddled with craters while the maria are almost void of craters. This leads to the reasonable conclusion that these features were formed quite recently. They have not had time to accumulate very many impact sites. Since the oceans of the Earth would appear dark from space, it would have been reasonable to conclude that the dark areas of the moon were also seas. If this were the case, an impact site would not show even if it had happened yesterday. While it is now perfectly clear that the lunar maria are not water seas, they might have very recently been lava seas. One could even wonder if they were lava seas at the time that they were first called seas. This, of course, is speculation but the fact remains that the areas in question are evidence of catastrophe because something very significant caused the lava to flow and apparently caused it to flow recently. In further support of the speculation that the maria are indicative of impact sites, is the fact that located beneath most of these features are mass concentrations. These mass concentrations, or mascons, are responsible for keeping one side of the moon constantly facing the Earth. (21)

The Earth has features, which are geologically very similar to the maria of the Moon. While there is plenty of water surface on the Earth, there are also large areas indicative of lava flow over the surface. The usual practice is to refer to such areas as thrusts. Unfortunately, many of the terms used in modern geology are loaded and the term thrust is such a term. The implication and intention of the designation is that the areas in question have been forced up on top of the original (if there is such a thing) surface and have slid along until they finally came to rest. The problem is that there is no evidence of sliding. Thrust formations are often very thick and very heavy. If there had been sliding action, whether it was slow or fast, there would be some indication of such action at the interface. Unfortunately there is no such evidence and the interface sites are sometimes not even smooth. If these areas did not over-slide, then it is reasonable to conclude that the material on top flowed into position and cooled. The fact that gravel, as well as artifacts, are found underneath some thrusts, (23) is conclusive that these surfaces were once exposed ground and that they weren't seriously disturbed or totally destroyed by the arrival of the new material. Gravel certainly cannot be squeezed in between huge slabs, which extend for hundreds of miles or even for one mile. People do not drop their tools underneath a surface. They drop them on the surface. Of course, tools or other artifacts can intentionally be buried but the lack of any evidence of burial together with the great depths involved rule out that option. Thrusts were not forced over in a solid state but simply poured over in a liquid state. In this case, it is reasonable to conclude that something caused the material to flow and because it is so extensive in volume and area, whatever it was that caused such flow was overwhelmingly catastrophic. If an asteroid should impact the Earth, the accompanying catastrophic events would include molten material from the interior of the Earth being forced up and caused to flow across the surface. Neither plant nor animal life could ever withstand a catastrophe of this nature.

The Lewis Thrust of North America extends over an area of approximately 30,000 sq. km. from Alberta down through Montana and is bounded by the Rocky Mountains on the west. The upper layers have been dated at a billion years and are resting on material dated at a million years. (23) Conventional understanding suggests that when the mountains were formed, old material was forced up on top of young material. The thickness of the 'older' upper portion exceeds a kilometer. What incredible and mysterious force could move this extensive volume of material and place it on top of much younger material without leaving overwhelming evidence of movement? Why wasn't the gravel on top of the 'younger' layer underneath, scraped away and ground into powder along with the mortar and pestle found in the gravel? (24) The horizontal travel required is several hundred kilometers. How could the upper slab have been placed without suffering buckling. The force, which would be required to move a great slab of material such as this, would have exceeded the ability of the rock to remain flat, horizontal and intact.

If, however, an asteroid impacted the Earth, material from the molten interior would be expected to come to the surface and flow out over it. In such a case, since sliding would not be involved, evidence of sliding would not be observed and indeed it is not observed. Such an area should more properly be referred to as terrestrial maria, since, when it happened, the area would have appeared as a great molten sea of lava and when it cooled, wherever it was still exposed, it would appear like the great maria on the Moon.

9.18.2 Igneous Provinces

An igneous rock is a type of rock which forms when magma cools. If it is found spread over an area of the Earth, that area will be referred to as an igneous province. In this respect an igneous province is very similar to an "overthrust". While the popular notion of an overthrust is of a large area of solid rock being forced to slide on top of another large area of rock, the upper layer was actually placed as a liquid and would therefore have flowed into place. Igneous provinces would have formed by lava flowing into place as well.

There are numerous igneous provinces on the Earth and some of them are quite large. Their formation indicates that molten material oozed out of the Earth, spread around and cooled into solid rock. Many igneous provinces are found beneath the sea. Many are also found on land. The Deccan traps of India are recognized as an igneous province and they cover an area of at least 200,000 square miles. (28) All of these formations are indicative of the mammoth pressure waves which would have propagated throughout the interior of the Earth following an asteroid impact and they therefore provide evidence of such impact.

10.   Evidence of Chaos and Catastrophe

10.1   Carbon14 Evidence

10.1.1   Introduction

Carbon14 came to the attention of the public during the mid-nineteen hundreds through the work of a man named Willard F. Libby. He identified a procedure, which was supported by solid theoretical work, to determine how long ago an object had been made. This procedure was primarily applicable to objects, which were made of wood and assumed that the manufacture of the item had taken place soon after the wood had been harvested. Measurements and calculations would determine the time which had expired since the wood stopped living. Other previously-living plants could also be used but since wood was very common, wood was most often used.

As long as a plant is living, it will interact with the atmosphere. The interaction of interest in this case concerned the respiration of the plant - in particular the take-up of carbon dioxide from the air. All plants take up carbon dioxide and use the carbon as a structural component. All plants, including trees, are built of carbon. The carbon is taken in by the plant and used for construction purposes. Therefore, when the plant stops living and is harvested, no more carbon will be taken in. The carbon, which is already part of the plant, is expected to simply remain in place indefinitely. This expectation, on its own, was not of much use in determining how long it had been since the plant died until it was realized that there are three different types of carbon and that one of them could be a time indicator.

Most of the carbon, which is found in nature, is carbon12. There is some carbon13 and a very small amount of carbon14. Carbon13 is a little heavier than carbon12 and carbon14 is a little heavier than carbon13. However, carbon is still carbon and any interactions, which carbon has with the rest of nature, will be similar, no matter which type of carbon is involved. Therefore, when carbon reacts with oxygen and forms carbon dioxide (CO_2), all three types of carbon will be involved expectedly in the same ratios in which they would normally occur in nature. While this doesn't always happen, the expectation that it will happen is the basis of the idea that carbon14 should be useful as an indicator of time.

Neither carbon12 nor carbon13 change with time. They always stay the same no matter how much time is involved. However, carbon14 does not stay the same but gradually changes into nitrogen. Therefore there will be less and less carbon14 as time goes by and the diminishing amount of carbon14 will provide us with an indicator of how long it has been since the plant (i.e. usually wood) was alive.

Carbon14 is said to be radioactive. As it changes from carbon back to nitrogen, a small part of the carbon atom is lost. Radioactive changes are predictable and fortunately, all radioactive changes follow a similar pattern of change. This pattern of change is called the half-life law. No matter how much material there was at the beginning of the period of interest, there will only be one-half of it left after one half-life has passed. Of course, different radioactive materials have different half-lives. In the carbon case, the half-life of carbon14 is of particular interest and it has been determined to be approximately 5730 years. (28) Since 95% of a sample will be gone after about five half-lives, a half-life of 5730 years is of very great interest as a dating technique for artifacts, which might be up to a few thousand years old. If the half-life were only 100 years, the active ingredient would be virtually all gone after 1000 years and so would not hold the potential to be useful for dating something that was several thousand years old. Since the history of humanity has occurred in the last few thousand years, the development of the carbon14 dating technique aroused a great deal of interest in the scientific community. Since the middle of the twentieth century, this technique has been applied to many samples to determine how old they were. However the use of carbon14 as a time indicator has a serious limitation.

10.1.2 (1) Time Constant Curve and Half-Life Curve

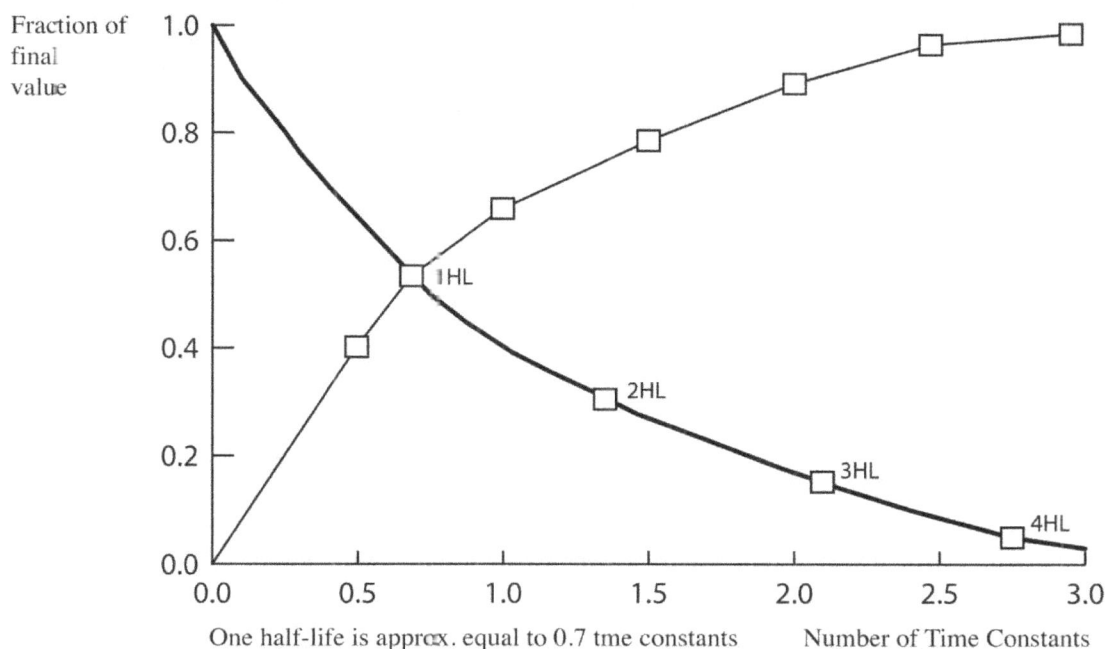

One half-life is approx. equal to 0.7 time constants   Number of Time Constants

A half-life curve is usually shown dropping off because the material remaining is in reduction. A time constant curve may be shown increasing (as above) or decreasing. The shapes of these curves are exactly the same and they follow an exponential relationship.

10.1.2   Buildup and Decay

In nature, responses to change often follow the half-life relationship. For example, with processes such as heating a kettle, filling a tank or charging a capacitor, the parameter of interest will very likely follow the half-life response curve. In most of these cases, a half-life parameter will not be measured but another idea will be introduced, which is called a time constant.

The time constant relationship is shown in the diagram, 10.1.2 (1) Time Constant Curve, which in this case is showing how the voltage across a capacitor varies as the capacitor is charged. After the amount of time, which is called the time constant, has passed, the capacitor will be approximately 63% charged. When two time constants have passed, the capacitor will be approximately 86% charged. After three time constants have passed, the capacitor will be 95% charged.

A very similar situation develops when a leaky bucket is being filled with water. As shown in the diagram, 10.1.2 (2) Leaky Bucket, water is being poured into the top of the bucket. In this example, the bucket has a lot of small holes in it and so as the water is being poured in the top, some of the water will leak out through the holes. After a while when the water level has risen, there will be so many holes which are leaking that the leak rate will equal the fill rate. When this occurs, the water level in the bucket will not rise any higher. The parameter, which is of particular interest in this example, is the water level change with time. As time passes, the water level gets higher and higher but as it gets higher, the leak rate increases and the rate of water level increase slows down. The increase in the water level will keep getting slower until it actually stops. This occurs when the leak rate is the same as the fill rate. The change in water level will follow the same pattern of change as the voltage on the charging capacitor. The curve, which describes the charging capacitor, could be placed right on top of the curve for the filling bucket and it would be found that the two curves had exactly the same shape. It is very curious that this would be the case but nature has many other examples of changes, which follow this very same pattern.

10.1.2 (2) Leaky Bucket

The holes in the bucket cause the water to leak out. The more the water level rises, the more quickly the water will leak out of the sides. The increase in water level will follow a time-constant or exponential curve.

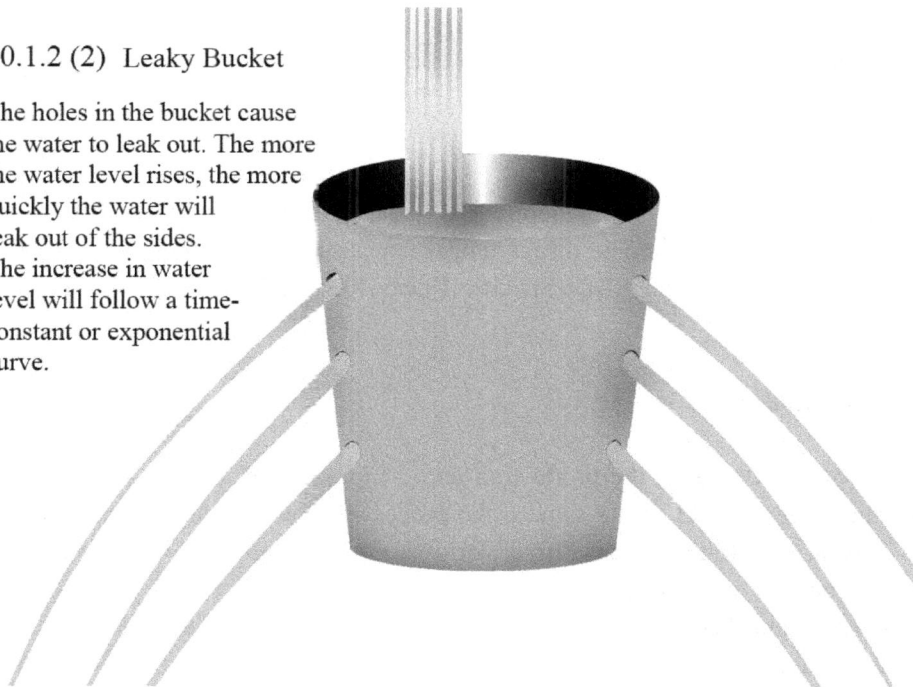

Another example of this pattern of change occurs when a kettle is being heated on a stove. As it is heated, the temperature of the water will increase. The temperature will not jump up suddenly. As the water gets hotter, further increases in temperature will take longer because as the kettle gets hotter, it loses heat at a faster rate. A time will come when the kettle will not get any hotter. At that time, the heat, which is being put into the kettle, will be exactly the same as the heat, which is leaving the kettle. When the heat, which is going in, is equal to the heat, which is leaving, the temperature will just stay the same. If the change in temperature were measured up to the time that the kettle did not get any hotter, a pattern of change would be found, which would be similar to both the capacitor and the bucket.

This type of change has a very particular mathematical description which is called an exponential. This type of pattern of change is found a great many times in nature. When either the half-life of a radioactive material is known or the time constant of a capacitor circuit is known, as well as either the starting point or the finishing point, both the future and the past of the particular parameter of interest can be determined. These well-known physical relationships will be particularly helpful in determining the validity of the carbon14 dating procedure.

10.1.3 C14 Production

The atmosphere of the Earth is approximately 80% nitrogen and 20% oxygen. In the upper atmosphere, a very small amount of the nitrogen is continually being changed into carbon14. Energy is required to make this change. The energy is supplied by incoming cosmic radiation, which liberates neutrons, which become captured by nitrogen making the nitrogen heavier than usual. Soon, this heavy nitrogen loses a proton and carbon14 results. The production rate of carbon14 was determined by Libby to be 18.8

SPR(Specific Production Rate). Others disagreed with this value and placed it at 27. (2) Unfortunately, there is no way to determine if the rate has been constant during the period of interest.

10.1.4 C14 Decay

Carbon14 is radioactive. Therefore it is continually disappearing. Actually it does not disappear but simply changes back into nitrogen. The disagreement concerning production rate means that the decay rate, which is usually given as 16.1 SDR (Specific Decay Rate) is between 20% and 40% less than the production rate. (2, 34)

10.1.5 Carbon14 Buildup

Therefore, on a world-wide basis, the production and decay rates of carbon14 are not the same. They should be the same. They should be the same if the process has been in operation for more than eight or ten half-lives. This puzzled Libby because ifthe world was actually millions of years old, the SDR should have long since come into equilibrium with the SPR. (2) This situation can be understood from the chart entitled Carbon14 Buildup. The decay rate should gradually build up until it is the same as the production rate. The increase in decay would be the same as the increase in voltage of a capacitor, which was being charged. As shown above, when a capacitor is being charged, the voltage gradually increases. It will follow a pattern of increase, which is called an exponential. The shape of the exponential curve is the same as the shape of the half-life curve, which indicates that the two different processes will have similar patterns of change. As discussed above, this pattern of change will also be similar to the change in water level, which occurs with the leaky-bucket example. When a kettle is heated on a stove, the temperature of the water will change in a similar way. This type of change is very common in nature and is in fact the way nature usually responds to a change in circumstances.

With the leaky-bucket example, it is clear that as long as the quantity of water which is leaking out of the bucket is not as much as the quantity which is being put in, the water level in the bucket will keep on increasing. The example of the charging capacitor is very similar. As long as the voltage across the capacitor is less than the source voltage, it will keep right on increasing until it matches the source voltage. Voltage measurements would determine if the capacitor was fully charged. It would be expected to keep on charging until the source voltage had been reached and it could not be charged any further.

The situation with radioactive material is very similar. Once the process has begun, the decay rate should build up and keep right on building up until it matches the production rate. Therefore if a situation is found where the decay rate is less than the production rate, it is clear that the decay rate is still building up. This means that the process has only recently begun. (i.e. It only began within a few half-lives ago.)

10.1.6 The Present Time

The chart entitled, 10.1.6 Carbon14 Buildup, shows how the decay rate of carbon will relate to the production rate. This chart represents a considerable amount of time involving several thousand years. In the situation, which is shown by the chart, the decay rate starts at zero and then gradually builds up. At the right side of the chart the decay rate almost matches the production rate. This chart would be very useful to us if we knew where on the chart, the present time was located. What location on the chart represents the carbon14 situation in nature at the present time? What location on the curve represents the present time? From the 'present time' position, the part of the chart to the right would represent the future and the part to the left would represent the past.

The 'present time' location can be determined by knowing what the current level of decay actually is at the present time when compared to the production levels at the present time. Fortunately, for carbon14, (as discussed above) both production and decay have been measured. It has been determined that at the present time, the decay rate of carbon14 is about 20% less than the production rate. When this position on the curve is found, it will represent the present time. Unfortunately all measurements in the world of science do not agree and some estimates of the decay rate place it 40% less than the production rate. The present time position for the 40% case, on the chart is therefore a little further to the left of the 20% case.

10.1.7   Carbon14 Startup

The time since the process began, can be determined from two factors. The half-life of the material must be known and the difference in production and decay rates must be known. When these two factors are known, the time since the process began, can be calculated. On the Carbon14 Buildup diagram, we would then simply follow back down the buildup curve until the zero line is reached. Since there is scientific disagreement on the level of production, both possibilities will be examined. First we locate the 20% difference location on the chart. (See location A) Then by following the radioactive half-life curve back down to zero, the time since start-up can be determined. On the diagram this is approximately 2+1/4 half-lives or about 2+1/4 x 5730 years (from 10.1.1 above) = 12,000 years. (see location A) Secondly, the 40% case will be located and once again the curve will be followed back to zero. In this case, the time since startup is approximately 1+1/3 half-lives or about 1+1/3 x 5730 years = 7,000 years. (see location B) From these determinations, the time since carbon14 started to be produced in the atmosphere of the Earth, is between 7,000 and 12,000 years ago.

10.1.6 Carbon 14 Buildup

How long ago did the Carbon 14 production process begin?

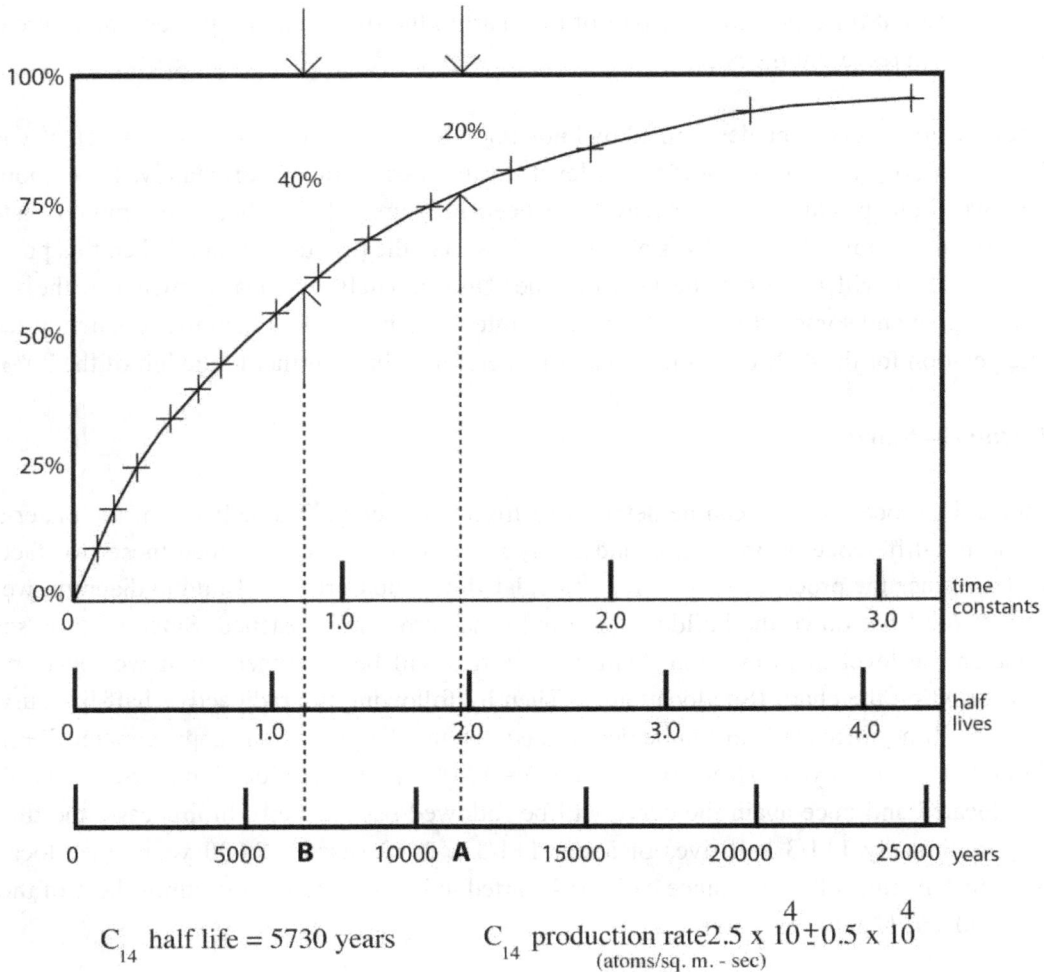

C_{14} half life = 5730 years C_{14} production rate $2.5 \times 10^4 \pm 0.5 \times 10^4$
(atoms/sq. m. - sec)

10.1.8 Partial Startup

The situation is modified if it is assumed that the process was always in operation at a low level and then suddenly increased to a higher level. If, for example the process was formerly operating at about one-half of the present level, the time since the increase, would be much shorter. Suppose that the process was formerly operating at the 50% level and then suddenly increased to the present level. In this case we only follow the curve back down to 50% of the present level. When this line of reasoning is followed, the time since startup is reduced to 6,000 years for the 20% case and to about 1,000 years for the 40% case. Similarly, if the process were operating at 25% prior to a sudden increase, the time since that increase would be approximately 9,000 years for the 20% case and 5,000 years for the 40% case. All of these determinations can be made directly from the Carbon14 Buildup diagram.

10.1.9   The Half-Life Curve

It is a curious reality of the half-life curve that no matter where we are on the curve, the shape of the curve is exactly the same. This feature is what makes the above discussion valid. If we are in the middle of the curve and only chose to use 10% of it, the shape of that 10% would be the same as the whole curve. Even if we are at one end and extended the curve further, the shape would be the same as the original curve. It might be that the beginning, (of whatever portion it is that we want to study), is part way up from the beginnings, which are shown in the above diagrams. The discussion can still proceed because the shape of part of the curve is the same as the whole curve. In fact the half-life curve can be extended both ways indefinitely and this whole new curve would still have the same shape as the original section.

10.1.10   Catastrophic Implications

The time since the process either began or suddenly increased is not nearly as important as the fact that the process did begin and the fact that it had a definite beginning. If it began, something caused it to begin.

The production of carbon14 is understood to occur in the upper atmosphere when incoming cosmic radiation collides with nitrogen atoms and causes them to change into carbon14. Therefore, since the process had a beginning, either the cosmic radiation began or the nitrogen was suddenly exposed to the cosmic radiation. Prior to such an event it must have been shielded from the incoming cosmic radiation. If this were the case, the catastrophic event was the removal of the protective shield.

A layer of atmosphere, which consisted of water vapor, could have provided a partial shield to incoming cosmic radiation. Therefore, the catastrophic event could have been the removal of this water vapor layer. As soon as that happened, carbon14 production would have increased to the present level and the decay rate would then have proceeded to build up towards the present level.

10.1.11   Theory Correlation

The earlier discussion (in the chapter entitled The Vapor Canopy Hypothesis) identified several factors, which indicated that there might have been a water vapor layer in the upper atmosphere at one time. This is supported by the above deduction that the carbon14 process had a definite and not-too-distant-in-the-past beginning. If there was a protective water vapor layer and it was suddenly lost, the carbon14 production process would have started. Alternatively, if the carbon14 production process was already underway, it would have increased to a higher level. With either of these possibilities, the idea that a pre-existing water vapor layer was suddenly destroyed and the conclusion that there was a definite beginning to the carbon14 process, correlate each other.

The conclusion that the carbon14 process had a beginning or at least a significant increase in activity quite recently also agrees with the observation that coal has a low level of carbon14 activity. (i.e. 50,000 yrs. (1) Whereas if coal is as old as claimed for the carboniferous period – 100 million years – it should not have any carbon14 count at all!) If coal consists of the plants, which existed prior to the recent impact of a large asteroid, these plants would have been growing in an atmosphere, which was partially protected from incoming cosmic radiation and one which had very little carbon14 production in process. Hence the carbon14 count would indicate that coal is older than it is in reality.

10.1.12   Carbon14 Dating Process Upset

Unfortunately, if the carbon14 process had a beginning or even a significant increase in activity within the last few half-lives of carbon14, the use of carbon14 for dating ancient artifacts is compromised. Any dates, which are determined will be in error and will indicate more time than is actually involved. A radioactive process such as this could only be expected to give valid results if the process was in equilibrium and the production and decay rates were the same. When they are not the same, it means that the process is still ramping up and is not yet well established throughout nature. Even though it appears that several thousand years are involved, it is clear that steady state has not yet been realized and the process will not really be very useful until this happens.

An upgrade to the use of carbon14 as a dating procedure could be realized if a time dependent correlation or adjustment factor were included. Even though the originator of the carbon14 theory recognized that production and decay were not in balance, (35) no adjustment factor was ever introduced.

10.1.13   Catastrophic Atmospheric Change

In order for carbon14 production to either have been initiated or to have undergone a major increase in activity, there must have been a catastrophic irreversible change in the atmosphere of the Earth. If the previous atmosphere included a water vapor layer, this is exactly what would be expected if a large asteroid crashed into the Earth.

10.2   Evidence from the Rivers

10.2.1   The Hydrologic Cycle

Water constitutes a major portion of the material of the Earth. There is water in the air in vapor form. There is water in the soil. There is water in underground streams and reservoirs and there is a great amount of water in rivers, ponds, lakes and oceans. Water in its solid form, ice, covers most of Antarctica as well as Greenland. With the exception of some of the ice, all of this water is in motion. The source of energy to move the water is the Sun. As the Sun heats the Earth, water evaporates from everything that the Sun shines on including trees, grass and open ground. Lakes, rivers and oceans evaporate as the Sun heats the surface layers of water. It requires a lot of energy to evaporate water but as the Sun shines on it, the temperature increases and evaporation results. In particular, as the Sun shines on the oceans in the tropical regions, the ocean water evaporates into the air leaving behind any salts or minerals, which were in the water.

This evaporated water then becomes part of the atmosphere, mixes with the other components of the atmosphere and is carried along by the moving air. As the weather patterns of the Earth shift and change, the temperature of the atmosphere will occasionally drop. Rain will result. Some rain falls right back on the ocean but a lot of rain will fall on the land. Some of this rain will supply the trees and crops with needed moisture and some will either fall directly on the lakes and rivers or percolate through the soil before either evaporating again or trickling into streams and lakes.

The water of the Earth is always on the move. This great pattern of movement is referred to as the hydrologic cycle. The hydrologic cycle involves a pattern of water transfer from the oceans to the land and then back to the oceans again. The heat from the Sun evaporates the water in the ocean. Clouds are formed. The water vapor in the clouds is transported over the land by the great wind patterns of the world. When the clouds pass over the land, rain falls on the land. Some of this water soaks into the ground. Some of it runs off right away into streams and ditches. Lakes and rivers become filled with water, which runs off the land. Most rivers and lakes empty into the ocean. A few do not reach the ocean but depend directly on evaporation for the release of water. All of the oceans evaporate and a few bodies of water, like the Dead Sea, which do not empty into the ocean, also evaporate. Either way, the water, which previously fell as rain, is recycled into the air before travelling back over the land to fall again as rain.

This cycle of recirculation is quite predictable. Most rivers do not undergo a significant change in level year after year. In fact, the water levels are so predictable that bridges are built and boats operate on regular schedules. It is also true that some rivers go into excessive flood mode quite often. The Amazon River is a good example of this characteristic. Water levels in the Amazon River can change dramatically overnight. This happens when there is a heavy rainfall in the mountainous regions where the river originates. This unpredictability means that bridges cannot be built and any activity along the shores must recognize that serious flooding can occur at any time. The general cycle of activity, however, is the same as with the more predictable rivers. Rainfall is the source of water for the river, which returns the water to the ocean and the cycle just keeps on going. The evidence shows, however, that this was not always the case and that the entire process had a recent beginning.

10.2.2 River Deltas

Many of the major rivers of the world, which flow either into the ocean or one of its adjacent bays or estuaries, have a buildup of material at their mouths. These buildups are called deltas. Deltas are formed from material, which the river carried from locations upstream. When the mouth of the river is reached, the silt drops out of the water because the water slows down and cannot carry the silt any further. As time goes by, the silt accumulates and forms an extension of the land. Fortunately, the rate of buildup of all of the major deltas of the world, have been calculated. For example, The Mississippi is pushing its delta into the Caribean Sea at the rate of a mile every 16 years. The conclusion is obvious: at this rate it represents an advance of 250 miles in only a few thousand years so the river cannot be older than that. (6) This determination assumes that the flow rate has always been the same (as it was up until quite recently when some of the flow was diverted to the west). Such a short time frame is quite alarming but even more alarming is the notion that the entire process had a beginning, which beginning was when the silt of the delta started to be deposited.

The time since the rivers started to flow seems to be quite short. From another viewpoint, it can be deduced that if these rivers had been flowing for several hundred thousand years, their deltas would be much more extensive by now. In the case of the Mississippi River delta, at the current rate of 2 million tons per day, it would only require a few million years to fill the entire Gulf of Mexico. (6) Similarly, if the Amazon had been flowing for a long time, the Amazon River delta would be much further out into the Atlantic Ocean than it is now.

10.2.3   River Startup

The river deltas of the world have a very definite and measurable size. It must therefore be concluded that they had a definite and relatively recent beginning. If the deltas had a beginning, the rivers must have had a beginning.

10.2.4   Niagara Falls

The Niagara River flows between Lake Erie and Lake Ontario. The elevation of Lake Erie is much higher than Lake Ontario and most of the elevation change occurs at the Niagara escarpment. At this location there is a sudden drop from the higher elevation region which includes Lake Erie to the lower level which includes Lake Ontario. Both Niagara Falls and the American Falls occur where the Niagara River plunges over the escarpment.

Every year the Niagara River works its way back further and further into the rock and the falls move a little further upstream continually. Observations have been made concerning the rate at which the river erodes the rock where it plunges over the edge. Reports are also available from the early observers to the area. These data along with the variations in the thickness of the various rocks along the Niagara gorge enable estimates to be made concerning how long the river has been flowing over the edge of the escarpment. The town of Queenston is located right at the escarpment and is actually partly on top and partly at the bottom. Important historical battles were fought in this area and this is the place where the Niagara River started to work its way back into the rock. Queenston is now several miles downstream from the falls. This distance is therefore a time indicator. When this distance, the observed and reported erosion rates as well as the variations in the type of rock along the way are considered, it can be determined that the water started to flow over the edge of the escarpment approximately 7000 years ago. (7) Further, since flow was greater at the end of the Ice Age, the time might have been even less. '… it was concluded, seven thousand years may constitute "the maximum length of time since the birth of the falls." In the beginning when immense masses of water were released by the retreat of the continental glacier, the rate of movement of Niagara Falls must have been much more rapid; the time estimate "may need significant reduction," and is sometimes lowered to five thousand years.' (26) Whether the time was five thousand or seven thousand is secondary to the notion that the water started to flow. It had a definite beginning.

10.2.5   Evidence Correlation

The notion that major rivers, in particular the Niagara, had a definite beginning correlates with the notion that the deltas of the great rivers of the world also had a definite beginning and started to form at some particular time in the past.

10.2.6   Weather Pattern Startup

Rivers are fed from rain. If there is too much rain, rivers will flood. If there is not enough rain, rivers diminish and dry up. The further conclusion is therefore that the rainfall patterns, of the world, were also started at the time that the rivers started to flow. Rain is produced by the great weather patterns around the world. It must therefore be concluded that these weather patterns came into existence at the same time as the rivers started to flow and the deltas started to accumulate.

10.2.7   Catastrophic Geological Change

If there were no flowing rivers prior to the beginning of delta formation or the beginning of the migration of Niagara Falls, the geology of the Earth must have been dramatically different from the way it is at the present time. Therefore it must be concluded that there was a world-wide catastrophic change in the geology of the Earth. This is what would be expected if a large asteroid crashed into the Earth. There would be major changes in the geology of the Earth, in the weather patterns of the Earth and in the location and indeed the actual existence of the great rivers systems of the Earth.

10.3   Evidence from the Glaciers

10.3.1   Glacial Retreat

There are two great glaciers in the world and numerous smaller ones. The largest one is on Antarctica, where the ice is more than two miles thick in places. The great glacier, which covers most of Greenland, is also approximately two miles thick. Other glaciers or significant accumulations of ice occur in various mountain areas around the world.

These glaciers are the remnants of much larger accumulations of ice, which are thought to have covered much of both North America and Eurasia at one time. It isn't really possible to determine how extensive the ice was but it is readily possible to conclude that it was much more extensive in the past than it is today. The reason for this is because there have been numerous eyewitnesses to the fact that even within the past two hundred years, the ice accumulations which still lingered in the mountains were much more extensive. For example, photographs have been taken of the Columbia Icefield (which is located in the Rocky Mountains) when it extended all the way across the lowest region of the valley and part of the way up the other side.

The great ice shelf, which extends around Antarctica, is part of the great buildup of ice which occurred during the Ice Age. In the very recent past, this ice shelf extended much further out into the ocean and even within the last few years, very large sections of the ice have broken off and drifted out to sea.

10.3.2   Transient Weather Change

The great ice fields of the Ice Age formed and then, with the exception of Greenland and Antarctica, they melted away. In order for this to have happened, there must have been a major shift in the weather patterns of the whole world. It was a temporary or transient shift. Prior to the Ice Age, the weather did not produce ice. Then, for a short period of time the weather produced ice in vast quantities which was followed by the present weather patterns which do not produce significant ice either.

This evidence from the glaciers correlates very well with the expectation that an ice age would occur if several large asteroids impacted the Earth. Then, also as expected, the weather patterns would slowly shift closer to the way they are at the present time (i.e. warm summers, cold winters) and the accumulated ice would melt.

10.3.3   Lake Vostok

Lake Vostok provides evidence that an ice age happened and happened quickly. If the Ice Age had developed slowly over a period of years, the mud in the ice above Lake Vostok (see section 7.5.2.6) would have had a chance to settle and this ice would be mud-free. Since it isn't, the logical conclusion is that an ice age indeed happened and that it developed quickly. It is both logical and consistent to think that the Ice Age once it commenced, would have proceeded quickly on Antarctica because Antarctica is a land mass that receives very little solar heat at the best of times. When the Earth became cloud covered due to the impacting asteroids and the volcanoes, Antarctica did not receive any heat from the Sun at all. So the temperature dropped quickly and the entire continent went into the deep freeze. Lake Vostok froze down to the level where heat loss through the ice to the atmosphere matched the heat gain from the interior of the Earth. At this level the water did not freeze any deeper and remains in thermal balance to the present time.

10.4   Geological Evidence

10.4.1   Adirondack Rock

The rock, which is shown in the photograph, 10.4.1 'Adirondack Rock' was found in the Adirondack mountain range in upper New York State. While there really are a lot of rocks in the Adirondack mountain range, this one is unusual because it is resting very close to the very peak of a mountain. Its actual location is approximately one hundred feet horizontally and five feet vertically from the highest point of Mount Colden, which is located ten miles southeast of Lake Placid.

While resting on a slope of approximately twenty degrees, the rock is overhanging a greater slope of about forty-five degrees. If Mount Colden should shake too much during an earthquake, the rock could be dislodged and plummet into Lake Colden below.

There are very few ways to explain how it arrived at its present position. It does not appear to be the result of the erosion of surrounding material because it is well differentiated from the mountain on which it is resting. Neither does it appear to have come from the mountain. There isn't any place higher up where it could have come from. All of the sides of Mount Colden are steep. It is therefore difficult to imagine how it might have been pushed up the side of the mountain almost to the very peak. Further, it is certain that it wasn't carried up by any one person or even a large group of people. Never-the-less there it sits, but quite possibly, it will not sit there very much longer. Its precarious position means that there is no long-term stability for it and since these mountains do shake every few years, the next shake could be its last.

This rock has company on Mount Colden. A little further from the peak on the other side there are several other rocks, the positioning of which is similarly difficult to explain. One of them is possibly ten times as large as the rock, which is shown and it is resting in a manner which is suggestive of someone having thrown it up from the valley below.

10.4.1 Adirondack Erratic Boulder

The situation speaks to both catastrophe and recency. It would have required a catastrophe to place it in its present position. Also, the placement event must have been recent. Since it is so close to the actual peak, and since the mountain shakes so frequently, (which shaking would cause it to migrate closer to the edge), it can be concluded that it has not been up there very long. Neither will it be there much longer.

Therefore, whatever catastrophic event caused it to rest on a mountain-top, took place in the not-too-distant past.

If one or more large asteroids should crash into the Earth, rocks will be thrown high into the air. Some of them would land great distances from their starting point and some would be thrown so high that they would almost go into orbit. The impact of an asteroid would also cause mountain ranges to form. The rock in the picture might have been one of the many, which were thrown up into the air or it might have ridden the mountain up as it was being formed. In either case it is evidence of a recent great catastrophe, which was accompanied by a lot of chaos.

Geologically the Adirondacks are included as the southern extremity of the Laurentian Plateau (or Canadian Shield). This vast 3,000,000 square mile formation includes most of Canada, all of Greenland as well as the north-eastern United States. Does the youthfulness of this Adirondack Rock suggest youthfulness for the Laurentian Plateau?

10.4.2   Faults, Rift Valleys and Escarpments

Faults are cracks in the crust of the Earth. These cracks are found all over the world. In some cases they are geological curiosities but in other cases they are of great concern for human safety.

Within the last few years, a tunnel has been excavated under the English Channel. While the total length of the tunnel is less than thirty miles, the rock through which it passes has been cracked in several places. These cracks are nearly vertical and separate one section of rock from the next section. The tunnel walls are reinforced through these weak places in the rock but the concern remains that if any of these rocks should shift, the tunnel would not be usable. Flooding could result from even a minor shift in any of the rock sections but modest flooding can be handled by pumps. However there isn't any way to handle a major shift. Even if only one section of rock shifted up or down by a couple of feet, the utility of the tunnel would be lost. Utility might be regained by doing some modifications but there would be no guarantee that further shifting wouldn't happen.

Faults are never welcome. A fault indicates a place where, any movement, which is going to happen, will happen. Solid rock, which does not have any cracks in it, is much more welcome. Whenever a large building is to be built, it will never be constructed across a fault line. It would be considered terribly foolish to ever build a building across a fault line.

Similar reasoning would be used for any major public work. If, for example, one end of a bridge was placed on one side of a fault, and the other end was placed on the other side of the fault, this would be considered an open invitation for trouble.

Faults are very common and have been found all over the world. Faults indicate that the crust of the Earth has been stressed into failure mode by forces, which were so great, that they cracked the very bedrock of the Earth.

If an asteroid should impact the Earth, the crust of the Earth would crack. It could crack hundreds or thousands of times and due to the great extent of the displacement and shaking from the impact, cracks in the crust of the Earth could be formed all over the world - even from a single impact.

Faults or cracks in the crust of the Earth are evidence that the world has experienced a period of extreme stress. The stress was so great that it cracked the crust of the Earth for miles and miles both vertically and horizontally. Thousands of cracks have been formed. Accompanying many of these cracks there is displacement of the rock on one side or the other and this displacement also indicates that some extreme forces have been involved. In many locations, two or more cracks are separated by a section of rock, which has either risen or fallen compared to the surrounding rock. Both escarpments and rift valleys have thereby been formed. How much force would it take to raise a section of the Earth's crust, which has become an escarpment? How much force would it take to depress the crust of the Earth to form a rift valley?

If a large asteroid should crash into the Earth, the crust of the Earth would crack. Rift valleys would form and escarpments would form. The faults of the Earth provide evidence that there has been a period of extreme chaos and catastrophe on the Earth and thereby provide evidence that one or more large asteroids have impacted the Earth.

10.4.3   Mountains

Mountains are obvious evidence of catastrophe. There are sharp edges, cracks and sheer vertical surfaces everywhere. Mountains are often arranged in long groups, which run parallel to each other as well as to an ocean coastline. It appears as if energy was somehow transferred from the ocean floor into the land and the land folded and crumpled under the stress.

Most mountains are formed from sedimentary rock. The material for this type of rock has been placed by moving water. However, in mountainous areas, this rock is found thrust up into the sky. If everything had been peaceful on the Earth it would have remained flat, undisturbed and under water. As it is found thrust up in the air instead, it is obvious that some extraordinary force became involved and jammed the sedimentary layers up out of their flat position into jumbled regions of chaos.

In some areas of the world the mountains are in the form of a circle. If such circles were found on the Moon, they would be identified as impact sites. If they are found on the Earth, they should also be considered as impact sites. (e.g. the border of the Czech Republic)

In some areas of the country, elevated areas are referred to as mountains, which might be more properly called escarpments. There is usually a relatively flat, elevated area with one or more sharp sides. It appears as though these elevated areas have been thrust up by some vertical force. If the Earth had always been tranquil, these elevated areas should be level with the lower surrounding areas.

Most mountains are formed from sedimentary rock. The curved arrangement of the layers of rock into anticlines and synclines is suggestive that they were formed between the time that the sediments were deposited and the time that the sediment hardened into rock. This in turn leads to the conclusion that the mountain-building process (for the whole world) occurred over a very short period of time.

Further, since sharp edges and craggy peaks are common, the soft sedimentary material must have been held in position immediately or it would have simply crumbled away. The agent for its retention was the post-impact winter and the following ice age. The crumbly material at the peaks cooled quickly to below freezing. This development enabled the sharp stark features of mountains to be retained until the material hardened. They all remained cold until the waning days of the Ice Age and by that time the sedimentary layers had hardened into rock. From this requirement for immediate stability it is clear that the Ice Age followed quickly after mountain formation.

Mountains and mountainous-looking escarpments are evidence of unthinkable catastrophe. If an asteroid should smash into the Earth, mountains would form and the crust of the Earth would be compressed, displaced and greatly distorted from tranquil, level, peaceful-looking plains into mountains, escarpments and rifts.

10.4.4 Submarine Trenches

At various places in the ocean there are deep trenches running parallel to continental coastlines. These trenches appear to be regions of the ocean floor, which sank down from some original level location. If an asteroid should smash into the Earth, the ocean floor would be displaced horizontally towards the continents. Something would have to give. Either the surface layers of the continent would buckle and compress into mountains or the ocean floor would ride up onto the continent, crushing and folding as it went or the ocean floor would fold downward and form a trench. Then, when the force which caused the displacement was reduced, there could have been a partial rebound effect, which would enable some displaced material to slump partway back. For all of these events, the results are catastrophic and when they occurred, the whole region would have been in utter chaos.

10.4.5 Gravel Deposits

In many areas of the world gravel is found. Gravel is so important to any modern society that the modern industrialized world cannot thrive without it. Gravel is commonly found right on the surface of the ground. Mining is carried out by simply scooping the gravel up. In some areas the gravel is several hundred feet deep.

All gravel is composed of rounded stones. Rounded stones are expectedly formed by water, which has been moved horizontally. For example, exposed gravel is commonly found near rivers and always on the inside banks where the river turns. The lateral secondary currents are understood to have placed the gravel, which is found in these locations and the stones, which compose the gravel deposits are invariably rounded. In these situations, it is understood that the gravel, (even though it may weight thousands of pounds), has been placed by the force of the moving water.

Terminal moraines represent another type of gravel. Terminal moraines are the piles of material, which are found at the terminuses of mountain glaciers. The material in a terminal moraine however, is not rounded. A great number of sharp-edged stones are included. Apparently the stones which are pushed along by a sliding glacier do not roll very much.

On the Canadian prairies, straddling the border between Alberta and Saskatchewan, there is an interesting formation, called the Cypress Hills, which is an area that is elevated above the surrounding prairie by several hundred feet. Trees grow there as well as grass and the entire area appears quite different from all of the lower, treeless, surrounding landscape. The Cypress Hills are capped by an 82 foot-thick layer of rounded stones. The stones appear mixed with other material and there is enough topsoil to enable the grass and trees to flourish. The Cypress Hills are not a terminal moraine. It is a rounded formation, which appears right out of place on the prairie. The roundness of the stones indicates that a great current of water was involved in their placement as well as in the removal of similar surrounding material to leave only the Cypress Hills. (29)

If an asteroid should impact the Earth, there would certainly be vast and incredibly violent currents of water flowing right over the continents. The Cypress Hills are evidence that such great chaos and catastrophe once visited the area.

In many places where gravel deposits are found, the surface of the land is rolling mounded hills. These mounds may have local elevation variations of fifty or one hundred feet or even more but the depth of these deposits often extends for several hundred feet. Water made these placements. The water was on the move as witnessed by the roundness of the stones. If the blade of a giant bulldozer had made these placements, or the front of an advancing ice sheet had made these placements the stones would not be rounded. In either of these cases the stones might not be rounded at all. They would be basically the same as they were before the movement occurred.

Gravel deposits are evidence that moving water was involved. If a large asteroid should impact the Earth, the water from the oceans would pour right over the continents and all of the loose material, which was on the continents, would be displaced. The speed of the water would determine whether the material would keep moving or if it would drop out of the current and settle in one place. As the great volumes of water from the ocean, from the water splash and from the rain, swept back and forth over the land the velocity of the currents would vary with both time and location but they would eventually diminish. It would require an unbelievable amount of power and energy to place a major gravel bed. If an asteroid should impact the Earth and cause the ocean water to pour over the land, the energy, which was transferred to the moving water, would enable the gravel deposits to accumulate. The gravel deposits of the world are therefore evidence of a time of unthinkable chaos and catastrophe.

10.4.6   Rim Gravel

There is one type of gravel, which deserves particular attention. It is referred to as rim gravel because it is found at elevations of 6900 to 7900 feet above sea level on the Mogollon Rim in northern and central Arizona. (9) Much of this gravel lies on top of an erosion surface. It therefore appears that the sequence of events included: A. Surface is eroded. B. Gravel is placed. C. Gravel is partly eroded. The gravel is an assembly of rounded rocks indicating that it was formed and placed by water. The underlying surface appears to have been overrun by water. These deposits are at a considerable elevation. The reasonable conclusion is therefore that the area was swept by a great flow of water, which was interrupted and then another flow brought the gravel. Finally a third event of flowing water swept away much of the gravel deposits leaving only what remains today along the top of the rim. It is evident that a great catastrophe occurred involving a massive flow of water because this type of stone cannot be moved at all unless the water is flowing at high speed.

10.4.7   Folded Rocks

In many locations around the world, rock formations are found, which consist of wave-like patterns of material. There are usually several layers and the layers appear like some great wave became frozen as it moved along. It also appears that the layers were originally placed in a horizontal position. Then, after numerous layers had been laid down, the whole area was subject to some compressive force, which reduced the horizontal extent of the structure. However simply squeezing the formation laterally would not achieve the smooth folded effect. It appears that a combination of forces was involved one of which produced a sinewave effect in the material and the other force achieved some horizontal compression. The displacements became permanent so the rolling wave remained in the rock as it hardened. (30)

The rolling wave-like shape of some of these folded rocks appears very similar to the wave-like effect, which is produced in pliable ground during an earthquake. During an earthquake, waves sometimes form in the surface of the ground and move horizontally away from the epicenter. Unconsolidated material is ideal for this type of energy transmission. If the regions where these waves were occurring was also being compressed, the land would not be able to settle back to its original position. Folded rocks would result. Whether this was the mechanism, which caused these rocks to appear this way or not is secondary to the fact that they consist of wavelike formations which have been formed from a pre-existing flat layered arrangement. This would have required very great forces and a lot of energy. These rocks provide evidence that chaos was involved in their formation - at least chaos with the magnitude of an earthquake and possibly even more. If an asteroid should impact the Earth, there would be adequate energy to cause these folded rock formations to come into existence.

As with mountains, which have been formed from sedimentary rock, folded rock formations must have occurred between the time their sediments were deposited and the time that they hardened into solid material.

10.4.8   Erratic Boulders

Erratic boulders are boulders which are found in locations where there is no similar parent material. Erratic boulders appear to be out of place. They are far from home. Sometimes the nearest similar material may be more than one hundred miles away. Erratic boulders are often very large and one of the most spectacular ones on Earth is 'a mass of chalk stone near Malmo in southern Sweden, which is "three miles long, one thousand feet wide and from one hundred to two hundred feet in thickness"..' (10)

If a large asteroid should impact the Earth, erratic boulders would be one of the results. A major impact would throw rocks so high into the air that some of them would be temporarily in sub-orbit. They could easily come back down many miles from where they started. Erratic boulders are direct evidence of chaos. When they were coming down or being rolled along, the "window of life" would have been narrowed and there are no forms of animal life, which could have survived their decent.

10.4.9   The Grand Canyon

The Grand Canyon is a well-known and well-photographed feature of the Arizona landscape. Due to its great size and colorful features, many people travel for considerable distances just to spend a few hours enjoying the view.

There is a river at the bottom of the Grand Canyon, which also provides a means of exploration. The river, however, is too small to explain the size of the canyon. While the river occasionally approaches the steep banks of the canyon it only does this in a small percentage of its total distance through the canyon. There are large flat areas on the floor of the canyon where there is enough room to graze animals and carry on with a farming operation. If the present river did not form this canyon, what did form it? Actually the river may have formed this canyon but not in its present form. The flow would have had to have been greater by several magnitudes, which would make comparisons with the present river unrealistic.

The canyon is evidence of a period of massive high velocity erosion because the material, which is obviously missing, is nowhere to be found. The material, which once filled this canyon, would require a very large volume to accommodate. However, the missing material is not found near the canyon or downstream of the canyon. A catastrophe of unthinkable magnitude carried away the material from the Grand Canyon with such velocity that it cannot be found.

10.4.10   The Badlands

The Grand Canyon has similarities to the Red Deer River valley of southern Alberta, which is the primary location of the badlands of Alberta. In this case as well, it is obvious that the missing material is nowhere to be found and has gone completely away from this location as well as any locations downstream. The present river therefore did not form the Red Deer River valley. It must have been formed by a flow of incredibly larger magnitude.

10.4.11 Underfit Rivers

Rivers, which are much too small to explain the canyons or valleys they flow through, are called underfit and most of the rivers of the world fall into this category. This means that they are much too small to account for their respective valleys and that some other explanation is required. Underfit rivers are therefore evidence of catastrophe because a much greater water flow was required to form the valleys, which they occupy.

10.4.12 Drumlins and Eskers

Drumlins are mounded features on the surface of the Earth with the shape of an inverted bowl of a spoon. Eskers are ridges of gravel, which wind across the countryside going up and down hills. Both of these features are commonly offered as evidence of glacial movement but are not of much support if the glaciers did not actually move. However both of these features could be the result of water movement. Eskers have been explained as having been formed by water, which flowed through melting glaciers but since eskers go uphill, this explanation is not compelling. Drumlins are said to have been formed as the great glacier advanced. However, drumlins consist of material which would be relatively easy to move and some of them have fairly sharp tops. Since they are sitting on overburden, one wonders why this overburden wasn't moved by the glacier. Drumlins could have been formed by water, which moved over an obstacle creating a downstream eddy into which silt and gravel were deposited. Eskers could be markers of where water flowed through a glacier or where other massive flow patterns were caused to deviate and stall, dumping their loads.

Both drumlins and eskers could be the result of flowing water. Flow of the required magnitude would have been catastrophic.

10.4.13 Volcanoes

Volcanoes are evidence of chaos. In 1943, at Paricutin, Mexico, a small volcano developed on a property, which was being farmed. Initially, the farmer noticed some smoke coming from the ground. Within a week there was a mound 500 feet high and within two years there was a mountain 1400 feet high and half a mile across. This mountain of material from deep in the Earth made the land useless for farming. Fortunately, no one was injured. (4)

The inhabitants of Pompeii were not nearly so fortunate. When nearby Vesuvius erupted in AD79, the towns of Herculanium, Pompeii and Stabiae were destroyed. So much material was discharged that Pompeii was completely covered. Many people died and were quickly buried - hopefully in that order. 18,000 more died in 1631 as the result of another eruption. These events were chaotic disasters. (11)

Some volcanoes, like those on Iceland, spew out volumes of molten rock, which flow down the side and harden within a few days. The red-hot flowing rock will completely destroy any biological material in its pathway.

From a study of Yellowstone Park, using high altitude photos, it has been concluded that the entire park was once a crater more than 40 miles across and is thought to be the cauldron of an extinct volcano. If indeed it was a volcano, the chaos which would have followed its eruption would have involved the entire Earth.

In geology very few things are known for certain and this is also true for Yellowstone. It might not be a volcano. It might be an impact site. While the evidence indicates that the entire park is actually the floor of an old volcano, this would be exactly the description for an impact site if an impacting asteroid was large enough to punch right through the crust of the Earth. After such an object had gone into the interior, there would be a large area of the Earth exposed and connected to the molten interior of the Earth. Consequently, molten material would well up and fill the opening and volcanic activity would be expected. Then, due to gravity, the area would solidify with the general shape of the spherical surface of the Earth. Years later it would not be possible to determine what had actually happened? Even with evidence of a rim, it would be impossible to determine whether the feature was primarily a volcano or an impact site, which included volcanic activity after the asteroid had gone far into the interior. It will be recalled that the largest impact site on Mars is only recognized because there is a circular formation of relatively low mountains. If evidence such as this was destroyed at the time of the disaster, it would be impossible to determine later if a volcano had exploded or an asteroid had smashed through into the interior. Either way it would have been an event of unimaginable chaos.

The third possibility is that the feature resulted from an upwelling pressure wave from the interior. If an asteroid smashed into the Earth, on the opposite side, (i.e. the antipode) a pressure wave would surge up from the interior and lift entire continental regions up with it. If the crust locally was not able to carry such a stress load, it would fracture or blow out, resulting in volcanic activity. Since this appears to be what has happened on Mars, it would not be surprising if something similar happened on Earth.

If a large volcano was active and then settled down, it would cool and crust over. The crust would increase in thickness as time went by but the region could remain active for hundreds of years. The crater at Yellowstone Park is certainly active. New hot water springs show up regularly and geysers are always blowing off. Areas of the surface occasionally blow right out and earthquakes are so frequent that it is almost a continuous activity. Recently a portion of the park swelled up and dumped water out of the lake. Then it dropped back a few inches but reports indicate that it may be rising up again. All of these activities indicate recency but just how recent no one can say. It is certain, however, that the area is active and that molten material is not very far below the surface. (5)

When the massive asteroids smashed into the Earth long ago, many hundreds of volcanoes erupted all over the world. The local chaos and catastrophe at the location of each eruption would have been overwhelming enough but the collective effect would have made life on the Earth almost impossible.

10.4.14   Crater Rims

Impact crater rims have elevated areas - otherwise they would not usually be recognized as craters - as well as depressed areas. These rims include evidence of horizontal as well as vertical disruption. Perfectly continuous circular arrangements are not found. Instead, there are elevated ridges interlaced with lower mounds as well as trenches and any combination of trenches, hills and valleys, which can be imagined. Such formations are evidence of unimaginable chaos. Just being in such an area at the time of impact, is unthinkable. Even being well outside of the rim would not have been much better as evidence of disruption can be found in these areas as well. In the Chesapeake Bay crater example, disruption outside the rim includes material, which appears to have been ejected from the crater area. The rim itself seems to have been raised up prior to slumping back down again. All is chaos.

10.4.15   Arctic Islands Formation

As evidence that the Earth is warming, it has been repeatedly pointed out that the permafrost in the northern regions is melting. Not only is it melting but the upper regions of the still-frozen areas are warming up. Where this is occurring on islands, which depend on the frozen soil to resist erosion, the ocean is now eroding the land away. This implies that these islands would have been eroded and would have disappeared long ago if they had not been continuously frozen since they were formed. It may therefore be properly asked; what was the unimaginably chaotic event, which formed these islands and then caused them to become frozen and remain frozen, right up, in fact, to the present time?

10.4.16   The Canadian Shield

'The Canadian Shield …is the Earth's greatest area of exposed Archaean rock. (i.e. old rock, in this case Precambrian rock, the oldest rock of the Earth) It … has been repeatedly uplifted and eroded… with much volcanic activity … some of the oldest (extinct) volcanoes on the planet … over 150 volcanic belts …making the tally of volcanoes reach the hundreds. …The Canadian Shield also contains the Mackenzie dike swarm, which is the largest dike swarm on Earth.' (37) A dike is an 'intrusion of igneous rock across strata. In other words, the material of the Earth cracked and molten material from further down came right up to the surface a great many times. The Canadian Shield also includes '… the Sudbury basin an … impact crater (as well as) … the nearby … Temagami Magnetic Anomaly … striking similarities to the Sudbury basin. This suggests it could be a second metal-rich impact crater … (as well as) the giant Manicouagan … one of the largest known meteorite impact craters on Earth.' (37) The Canadian Shield is a story of chaos beyond anyone's imagination.

10.4.17   Antarctica

The geology of Antarctica provides direct evidence of chaos and catastrophe. In particular in the area referred to as the dry valleys, there is evidence of unimaginable water flow. Deep canals cut 'several hundred meters deep' have been cut into the surface. Very large boulders have apparently been tossed around like matchsticks. All of this chaos is blamed on an explosive surge of water which gushed into the bedrock tearing away chunks of rock the size of refrigerators and tossing them down Antarctica's Wright Valley.(39) Other areas of Antarctica have suffered similar chaos suggesting that catastrophic floods were not rare in the continents geologic history (39).

10.5   Biological Evidence

10.5.1   Fossils in Rock

Fossils are formations in rock, which have been caused by some type of life-form. Occasionally, there are impressions in rock, which look like part of a life-form, but very often the impression will be of a complete creature. Eyes, legs, antennae, as well as numerous other parts of a body may be identified. Details are often so evident that not only species but subgroups within a species as well as intricate eye details may be identified. (38)

While occasionally single individuals are found, most often numerous fossils are found closely packed together in the rock. Hundreds or even thousands of fossils are commonly found all thrown together with no particular regularity at all. Large numbers of many different types of creatures are packed and mixed in together in the same volume of material with no preference shown for any particular type to be in any particular position or orientation. Some of these fossil beds are hundreds of feet deep. (12)

Most fossils represent creatures which are extinct, but it is not uncommon to find fossils of extant creatures as well.

It has been declared that a fossil would form if a hapless creature, either simply died or became caught somehow and could not escape and so perished and became a fossil. There is a recent example of this type of declaration where a rhinoceros became stuck in the mud near the shore of a small lake in Africa. Several people who were in the area to make a movie decided that this creature should be rescued before it died and became a 'fossil'. Indeed it might have died but it would never have become a fossil. In fact it would not have become a fossil any more than any one of the forty million buffalo, which roamed the North American prairies (until the white man came with his firearms and exterminated them) did.

Fossils are not being formed at the present time. Whenever a creature dies, nature always recycles it before it can possibly be buried by any natural means. Usually, anything which dies will simply be eaten by other creatures. Any number of recycling agents will appear without warning or invitation. When they set to work, it might only require a few hours until there is very little left to identify. Then, any remains will just be absorbed into the local environment.

The fact that fossils are found at all is direct evidence of catastrophe. Further, the way that fossils are arranged is evidence of chaos. Fossils are never found neatly arranged. They are always packed together in a disorderly manner. (13) Occasionally, fossils are found in a layered arrangement. Declarations are then made that each layer represents a period of time in the history of the Earth when only creatures of the type in that particular layer existed. Within these layers however, disorder reigns. Then to make matters worse, creatures from other layers are so often found mixed in that any claim that a layer represents a particular period of time becomes completely meaningless. (14) Disorderly layers, layers in the "wrong" order or the inclusion of fossils from either an "earlier" time or a "later" time are evidence of chaos.

In order to produce a fossil, the object, which is to be fossilized, must be isolated from the atmosphere. It must be quickly and totally buried. Otherwise the oxygen, which is part of the atmosphere, will react with the creature causing it to degenerate, rot and disappear. Since fossils are found in groups, where they often number into the millions, there must have been a terrible and extensive catastrophic event, which not only terminated these creatures but broke them into numerous pieces and then buried them as well. Since there are a lot of fossils, there must have been a lot of chaos.

When fossils are found, they are commonly found as part of an extensive formation, in which they are all broken, mixed and jumbled up together. It always appears as though the whole formation had been tumbled and rolled along for some distance before coming to a stop. Parts of creatures are found in every conceivable position and pressed in so close to other creatures that there isn't even room for any fill-in material. (15) These arrangements are direct evidence of chaos.

If a large asteroid should crash into the Earth, energy would be available to break, gather, roll, and tumble large numbers of creatures together into totally disarrayed formations. Water could have been the agent for this. If a large asteroid should impact the Earth, water at high speed would have washed over the continents. In so doing it would not only have uprooted the trees and totally rearranged the geology of the Earth but at the same time, it would have destroyed, broken, gathered and deposited large numbers of creatures as well.

10.5.2 Frozen and Compressed

Biology provides another type of evidence of chaos, which has similarities to the more classic fossil rock formations. In this category there are two different examples. The first example comes from Russia and in particular the islands to the north of Russia in the Arctic Ocean. Liakhov, Stolovoi, Belkov and the New Siberian Islands consist mainly of two things - bones and sand. (16) Many of the bones are the bones of the great wooly mammoth. Apparently there are so many bones that the islands appear to be structurally composed of them. It would not be possible to estimate the number of creatures, which contributed to these assemblies, but to suggest that it would be in the millions is realistic. Something caused these creatures to be in this state and the state, which they are in, is a state of total and utter chaos. Only a great catastrophe could have caused this to happen. As with so many other fossils, since there are so many creatures represented, there must have been a lot of chaos.

The second example comes from Alaska. Alaska is a region, which includes permafrost. Permafrost is a type of soil, which remains frozen all the time. It never melts. It seems to have been frozen for a long time already and the reason it never melts is because the winters are too cold and the summers are too short and not warm enough. Permafrost can be excavated. One way to do this is by using a high-pressure jet of water. The water will not only melt the frozen soil, but it will wash it away at the same time. In Alaska while excavating permafrost, the remains of animals have been found. So-called ancient creatures as well as modern creatures have been found and commonly found in the same formations. This type of find is, of course, of interest but the way in which these creatures are arranged in the soil is also of interest. They are always mixed up in a totally chaotic manner with not only other different kinds of creatures but with rocks and parts of trees and mud. All of this is called muck. (17) It is a jumble of everything, which nature found handy. It is chaotic to say the least. A great catastrophe occurred, the ground thereafter froze and the evidence of the catastrophe is now available for us to consider. Something caused a great jumble of trees and rocks and soil and creatures to be rolled and tumbled and compressed together in great piles. A great catastrophe caused this to happen.

In the Yukon Territory of northwest Canada, large numbers of bones have been found. Along the Yukon River and its tributary the Porcupine River as well as the Whitestone River, Old Crow River and Timber Creek, bones from ancient creatures have been found in abundance and continue to be found right to the present time. Many of these bones have been identified and include; Giant Beaver (similar in size to a black bear), American Lion (25% larger than present day African lion), Jefferson's Ground Sloth (size of an ox), Steppe Bison (slightly larger than modern bison), Scimitar Cat (about twice as big as a modern mountain lion), Woolly Mammoth (ten feet tall), Short-Faced Bears (tall as moose), Camels (one-fifth larger than modern breeds), Hyenas, Hunting Dogs, Voles, Shrews, Lemmings, Mastodons, Horses (a small breed), Caribou and Giant Pika. Approximately 60 mammal species, seven fish and 33 bird species have been identified. Only a small portion of the discoveries were fully articulated skeletons. In certain locations along Timber Creek, the ground seemed paved with bones. (18)

Occasionally the bones of an animal will be found in the wild but usually when a creature dies, the bone material is recycled through other creatures or the environment and the material in its original form will not be found. Included in the Yukon finds were complete animals. One of the horses had a long blond mane and whitish body hair and appeared to be unmodified from the way the animal would have appeared when it was alive. Some of these finds were not actually fossils but bones from which DNA could be extracted. Many bones were found on or right at the surface and some were found through mining. One can only wonder how many might actually be still in the deposits available for discovery. (A reasonably similar group of fossilized animals existed in the upper Mississippi River valley. See Appendix)

If a large asteroid should impact the Earth, there would be plenty of chaos. Energy would be available to create the muck of Alaska and The Yukon. Untold numbers of animals would perish and be swept away.

10.5.3   Northern Coal

There is an island in the far north called Spitsbergen Island. This island is located directly north of Norway and is almost as far north as the northern coast of Greenland. It does warm up in the summer but as with all northern regions, summers are short. Winters are always long. Coal (as well as fossil pines, firs, spruces, cypresses, elms, hazels, and water lilies) has been found at Spitsbergen. This material either grew at this location sometime in the past when the Earth was warm or it arrived there some time in the past when the Earth was warm. (6)

Such material from the far and frozen north is evidence that there has been a major shift in the climate of the Earth. At some time in the past, the climate of the whole Earth dramatically changed. Prior to this change the northern regions were much warmer. In order for this to have been the case, the atmosphere must have been constructed differently. The present atmosphere certainly cannot sustain or support a warm Earth. Even with the present warming trend, it is not very likely that the regions of the far north will ever be warm enough to enable coal producing plants to flourish. Some dramatic event caused a major and sudden (otherwise the plants would have rotted and disappeared) shift in the climate of the entire region.

10.5.4 Polystrate Fossils

In the world of biology, there are many more plants than there are animals. It would therefore be expected that the fossil record would include plants as well as animals and this is certainly the case. In particular, fossils of trees have been found, which are called polystrate because they extend through several layers (or strata) of rock. These fossil trees are sometimes two or more feet in diameter and extend vertically for twenty or thirty or even fifty feet. Within this vertical distance, there may be upwards of a dozen different layers of material. (19)

Fossils of this nature are difficult to explain using conventional arguments because it is clear from our present knowledge that trees are made of wood and dead wood cannot be exposed to the atmosphere for very long before it will weather and rot. The layers through which these fossils pass are usually declared to have taken a great many years to develop. Explanations such as this are therefore not valid. Either environment-proof trees were involved or the layers were deposited within a fraction of the time it would have taken for the trees to rot in which case the placement of material around these polystrate trees was a catastrophic process.

Fossil trees are also found extending for several feet through several layers of coal. For example, polystrate trees are found in the coal mines in Alaska. (19) In the case where coal is involved, extension of a fossil tree through any more than a few inches of coal is sufficient as evidence of catastrophe because the coal-forming material must have been placed within the "rot time" of the tree.

Some of these polystrate fossils pass right through several layers of coal as well as several layers of rock. If the long-time-frame scenario is accepted, a relatively tranquil scene might be envisaged. However if these layers were deposited within the degeneration time for an exposed dead tree, a much more dynamic situation was involved. It appears as though these fossil trees were standing upright, anchored in a layer of material when another layer of material was deposited right on top of the base layer. Then another layer of material was deposited on top of that one. The trees would be very well secured in their places by the three layers and so they continued to stand straight up. Soon more layers were deposited until they were covered and buried forever. This second scenario speaks of catastrophe. A layer of coal testifies to a mat of plants being swept into place. Then, possibly, a layer of inert material was washed over the plants. The next wave brought a second batch of plants and so it continued until numerous layers were in place. A great catastrophe was in progress. These polystrate fossils are direct evidence that a catastrophe was taking place, which was great enough to uproot trees as well as wash away great amounts of rock-forming material. Polystrate fossils are direct evidence of unthinkable catastrophe.

If an asteroid should impact the Earth, there would be energy available to uproot trees, deposit them elsewhere as well as wash away the overburden and redeposit it elsewhere.

10.5.5   Dinosaur Eggs

In the foothills and mountains of Montana, a great many dinosaur eggs and dinosaur footprints have been found. Both of these are evidence of chaos. The footprints are always found in a straight line, which is not the way we would expect to find them. Creatures like this would be expected to meander. There are such a large number of footprints that it appears as though a herd of creatures passed through the area in a great hurry. Something was causing them to hurry. Similarly the eggs, which are strewn haphazardly about, have not been placed in the expected regularity of nests. In fact, there aren't any nests in the area but occasionally there is an assembly of eggs, which on casual inspection appears like a nest. However, on closer inspection it is clear that the eggs in these assemblies, were not even deposited on the same level. One such assembly appears to have been deposited on three levels spaced apart vertically by a significant fraction of the size of each egg. (20) It appears that the mother paused to lay her eggs and that even while she was laying them, the area was being inundated by silt-loaded water. Most of the eggs do not occur in any type of group but are just scattered about randomly. Occasionally a juvenile is found in the area suggesting that the inundation was too much for the smaller ones. A great catastrophe was in progress.

10.5.6   Arctic Islands Forests

The islands of the Canadian Arctic have a great amount of mummified, sawable, burnable, apparently ancient wood on them. Ellef Ringnes, Amund Ringnes, Axel Heiberg and Ellesmere Islands have accumulations of fossil wood, some of which appears as the remains of in-situ trees. What incredible catastrophe caused this vast forest to be leveled, and turned these islands into barren wasteland? It also caused all of this wood to be preserved without rotting, right to the present time. Alternatively, it is conceivable, that these great masses of wood were floated into place. A great floating log boom on the ocean would have been immediate evidence of unthinkable catastrophe. The flotation means would then have had to recede, immediately prior to the setting in of the deep freeze preservation phase. It is clearly more probable that all of these trees originally grew in the vicinity where they are presently found. Further, if a glacier had moved across these islands and flattened the trees, it would also have scraped them off! It is therefore clear, that the demise of these island forests was not the result of a moving glacier.

10.5.7   Woolly Mammoth

As well as countless Woolly Mammoth bones there have been discoveries of virtually complete animals. They have been found entombed in ice as well as in the permanently frozen soil of Siberia. One animal in particular – named the Beresovka Mammoth – was complete with stomach contents and eyeballs. Its meat was edible without causing harm to the recipients. (31)

It is completely and utterly impossible for such a state of preservation to be realized unless the animal had been quick frozen immediately after dying and then kept frozen ever since. Even one brief warm period would have been too many. It is therefore instructive to ask what catastrophe caused the death of these animals and then sent the area into the deep freeze for such an extended period. Clearly a complete climatic change occurred to the extent that even the food that these creatures ate cannot now be found for hundreds of miles. The evidence, including their very sudden death (food was still in their mouths) and the fact that they suffered major internal hemorrhaging, (32) could be explained if they were killed by a shockwave. An incoming asteroid would have had an accompanying shockwave which would have killed everything for a thousand miles. Russia has numerous major impact sites which were made by objects which would have produced shockwaves of the type that could have instantly killed the entire mammoth herd as well as sweep most of them away (their tusks were broken off) and bury them safe from the atmosphere so they would not decompose before being frozen. The immediately-following post-impact winter and subsequent Ice Age produced the deep freeze which has persisted right to the present time.

If a shower of asteroids hit the Earth, the accompanying shock-waves would have killed all of the Woolly Mammoths of Siberia and elsewhere. (as well as the elephants, hippopotami rhinoceroses, horses and mastodons (33)). Other effects of such a catastrophe would have buried most of them.

10.6   Human Records

10.6.1   Genesis Flood Report

The Book of Genesis is understood to have been written by Moses more than three thousand years ago. It is thought that it was written with recognition of verbal accounts, which had already been preserved for many years by the people of that time. The origin of the record is therefore impossible to date but it was a long time ago and may even have been among the very earliest records, which we have of ancient times. The Book of Genesis is recognized as a religious book by three mainline world religions. The reason for this is because of the frequent references to the Supernatural throughout the book. There are therefore two ways to read a book of this nature. One way is with acknowledgement of the Supernatural and the other way is without acknowledgement of the Supernatural. In this second mode of reading, the text may simply be considered as a record of events. Records of this nature are of particular interest because there are so few documents, which are available from such an ancient time. They are also of interest because of the possibility that they were written much closer to when the recorded events took place and when they were still fresh in the collective memory of the people. There are four entries in this ancient record, which are of particular interest with respect to evidence of catastrophe.

The first of these entries is from Gen. 8: 3. 'And the waters returned from off the Earth continually'. A paraphrased translation would be 'and the waters repeatedly swept back and forth across the Earth'. There was a going followed by a coming. Water was coming and going repeatedly. This reference is speaking of water waves, which were coming and going. The Earth was being swept by great waves which kept coming and going repeatedly for 150 days. For five months, great waves swept back and forth across the Earth. This report is therefore of a catastrophe of unspeakable proportions and one, which would have made survival impossible.

The second entry which is also of interest is from Gen. 7: 11. 'All the fountains of the great deep burst forth'. The deep refers to the oceans. The word fountain has a dual meaning and both of them apply in many cases. The first meaning is reservoir, as in fountain pen. A fountain pen is a pen, which has a reservoir of ink inside of it. This innovation was a major step forward when it was introduced but now the ballpoint pen has completely taken over. (And it also has a reservoir of ink inside of it.) Fountain also means welling up or surging up. Many fountains push water up into the air from a reservoir. The water then falls back into the reservoir and is continually pushed back up into the air again. The text is therefore explaining that the water in the oceans was thrust up into the air and possibly out onto the land as well. The further implication of the text is that the water was in great commotion and it was making a lot of noise. This ancient text is obviously describing a great catastrophe.

The text in Gen. 7: 12 refers to rain. However in ancient Hebrew, there are two words for rain. The rain in this case is a violent heavy rain. This is the type of rain, which would be expected during a hurricane. It has been reported that when a certain hurricane came ashore in Georgia several years ago, it rained at the rate of six inches every hour for six hours. This would make a total of three feet of rain. This is the type of rain, which is being discussed in the text with the difference being that the report refers to the rain lasting for more than a month. At the present time there are no known ways that a rain this violent can continue for that long. The text is clearly referring to an unusual catastrophic event.

The comments in the Genesis text refer to the great flood event as destroying everything and everybody in the whole world. The text uses the word 'all' repeatedly. In addition 'every' and 'everything' are also used numerous times. This report is clearly referring to an event of catastrophic proportions - one, which was affecting the entire world.

10.6.2   Genesis Human Age Data

The book of Genesis also includes a lot of numbers, which include the ages of the people who were in the direct line of descent to Abraham. These were the important people of the time and their lifespans were noted as well as their ages at the time that their oldest sons were born. From this information, a graph can be developed. As the diagram, 10.6.3 Genesis Human Age Data shows, there are two periods of particular interest. For the first period of time, an average age can be determined for all of the people who are listed. With the exception of the one named Enoch, this average is 912 years. While these are a very long life-spans when they are compared to life-spans at the present time, rather than dismiss this type of comment outright, it would be instructive to try to understand how such long ages could actually happen.

The second set of data, which is included on the diagram, shows that the ages started to change at the time of the reported flood. From that time onward, the ages were reduced. However the reduction pattern is of particular interest. The ages did not just drop off suddenly. They dropped off in a manner, which appears to be following a particular pattern. In fact the drop-off relationship is very close to a half-life curve or an exponential type of curve. This is the way that nature usually responds to a change in some circumstance. In this case, age is changing but because it is changing in a very particular manner, the situation deserves more consideration. While the data do not follow a time-constant or half-life curve exactly, a mathematical relationship can be offered which would follow the actual data very closely. For this reason it is strongly suspected that some aspect of nature is changing - in the way that nature usually changes - and that human lifespan is tracking the change.

10.6.2 Genesis Human Age Data

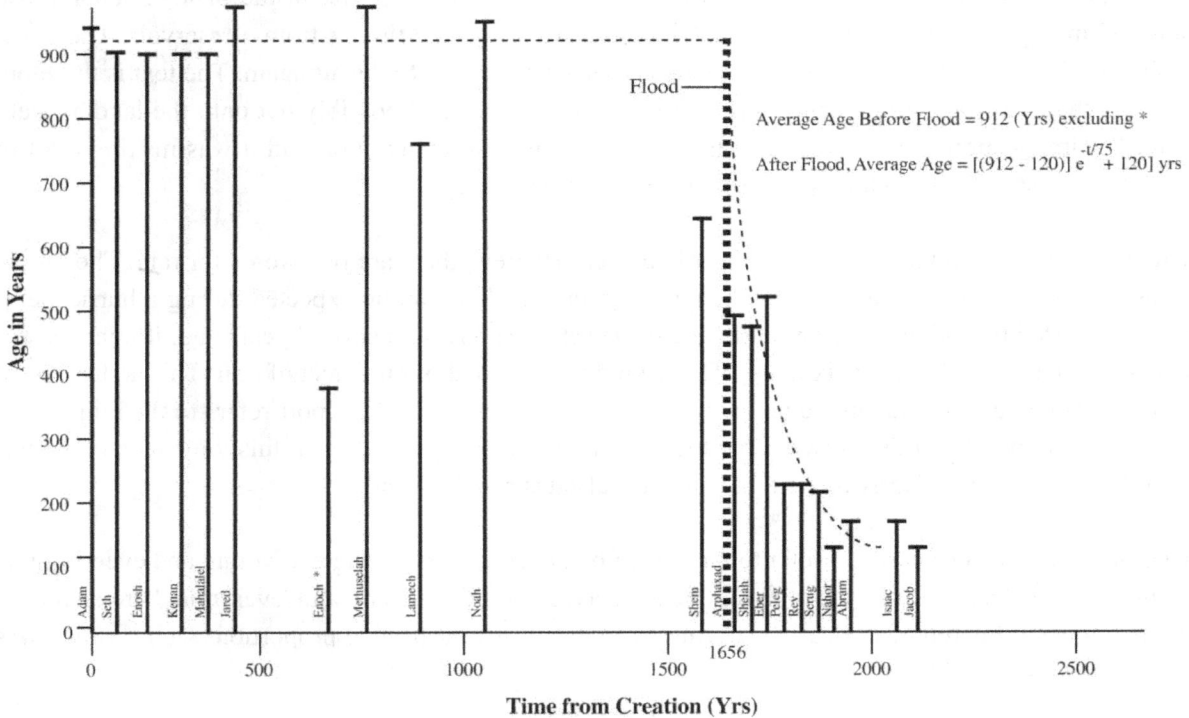

Flood

Average Age Before Flood = 912 (Yrs) excluding *

After Flood, Average Age = $[(912 - 120)] e^{-t/75} + 120]$ yrs

Age in Years

Time from Creation (Yrs)

10.6.3   Theory Correlation

A theory has been offered in the literature, wherein the lifespan for humans would be considerably increased if the atmospheric concentration of carbon dioxide was increased in the presence of an increase in the partial pressure of oxygen. This theory has been called The Hypothalamus Disregulation Theory. (33) Higher concentrations of carbon dioxide result in vasodilation of the blood vessels in both the skin and the brain. Blood flow is greater and together with a greater partial pressure of oxygen, cell electrosensitivity would be improved. Therefore the hypothalamus, which is part of the brain, would be less active and aging would be delayed.

If, prior to the impact of a large asteroid, there was a water vapor layer floating on the present atmosphere, the partial pressure of oxygen would have been greater. If the world was universally warm at that time, the water in the oceans would also have been warm. With these relationships in place, the

359

exponential drop in lifespan following the flood report in Genesis, might actually be tracking the decrease in the temperature of the oceans of the world. This was the time of the Ice Age. By the time the temperature of the oceans had dropped, there was a significant reduction in human lifespan. Also, as the temperature dropped, there wasn't enough heat to sustain evaporation and prolong the Ice Age. The cloud cover from the original catastrophe had cleared away by now so the ice melted off again but the temperature profile of the Earth settled out much colder and much closer to the way it is at the present time. The ice and snow did not melt in the arctic regions because they were now much colder than before. In fact, because the heat from the Arctic Ocean had now been spent, evaporating water for the Ice Age, the arctic became much colder than it was before. The temperate latitudes warmed up again during the summer and thereby shed most of their ice and snow. This would have caused more flooding but at least the great ice fields were gone. Atmospheric carbon dioxide would have been taken up by the cooling ocean water while the Ice Age was in progress. At the present time, the average temperature of ocean water around the world is close to freezing (21) and consequently the oceans have absorbed and are holding a great amount of carbon dioxide. This is presently a cause for concern because if the world is warming up, this carbon dioxide will come out of the oceans and further enhance the warming trend. This would become a positive-feedback situation for which the terminal condition is unknown.

At the very least, the particular way that the human age data were changing as reported by the referenced text is symptomatic of a major change, which was taking place in the natural world. Since these data correlate well with the conditions, which are required for an ice age, the suspicion is that the age data are tracking some basic factor, which was changing while the ice age progressed. With the removal of the pre-impact water layer, the Earth would not have been able to warm up again to the levels it had prior to the catastrophic event and the climate of the Earth would therefore settle out with a lower set of temperature conditions.

10.6.4 Primitive Tribes

Primitive tribes are increasingly difficult to find. With the desire to find treasure in far off places and the presence of communication satellites, there are not very many places left where the people have not been exposed and over-exposed to either European or North American culture. However until quite recently, it was possible to speak of groups of people who were primitive and who had not had very much, or possibly, any contact with the so-called civilized world. It was from these groups that curious and very similar reports were recovered concerning the occurrence, in some far off time, of a great flood, which came upon the Earth and destroyed almost everybody. This type of report would have been passed along from one generation to the next - probably as these groups gathered at their central meeting places. While modern civilization does bring a lot of advantages, sometimes things of value are lost and may never be recovered. The retention and conveyance of a report of this nature is curious on two counts. The report apparently was very similar from one tribe to the next even though these tribes had not been in contact with each other within memory. Secondly, it is of interest because it testifies directly to a catastrophic event of world-wide proportions and of devastating effect.

10.6.5 Epic of Gilgamesh

The Epic of Gilgamesh is an ancient Babylonian document, which is named after the hero of the narrative. Evidently Gilgamesh was a king who lived at Erech many years prior to the time of the Babylonian king Assur-bani-pal who had included the Epic in his royal library during his reign from 668 - 626 B.C. This period of time predates most of the history of people who lived around the Mediterranean Sea but it ends at the beginning of the time of the ancient Hebrew prophet Jeremiah. The Epic includes several stories of the adventures of Gilgamesh as well as other narratives and appears to be partly factual and partly fanciful. In the course of the entire collection, reports of two floods are included. One of these reports appears to be similar to the flood, which is reported in Genesis and the other might have been of a flood, which was confined to Mesopotamia. Both of these reports are of interest to us.

The Mesopotamian flood appears to have affected most of the valley surrounding the Tigris and Euphrates rivers as is witnessed by a thick layer of silt which has been found throughout this region. In fact, the entire area is underlain by a thick layer of silt. In addition, the basement rock includes two dominant faults, which run for several hundred miles up the valley. If these cracks resulted in the vertical downward displacement of the region between them, a suitable receptacle would have been formed to receive the great silt deposit, which is currently resting there. The depression might have enabled the sea to inundate the valley bringing another load of material, which formed the current flood-indicating layer. With any of these scenarios, it is clear that the region has undergone at least one great catastrophe.

The other flood report is also of interest because it mentions a great vessel, which provided protection for the only survivors of the flood. This vessel, unlike the one, which is reported in Genesis as being oblong, was cubicle in shape but of similar dimensions to the one in Genesis. The implication is clear that the flood was so serious that the only possibility of survival was by way of this vessel.

There is a possibility that these two reports are referring to the same event. From our observation point several thousand years later, it is completely impossible to draw any reliable conclusion. However it is clear that both reports are referring to some catastrophe which was of overwhelming importance to the people who are making the report as well as to the entire surrounding area.

10.6.6 The Great Dilemma

A great dilemma accompanies reports of overwhelming catastrophe such as the Genesis report. In particular the Genesis report includes the following points:

a. Hurricane-like rain for more than one month.
b. Mountain-overwhelming waves coming and going repeatedly for five months.
c. All animal life died.
d. Earth was uninhabitable for more than one year.
e. Dramatic 13:1 reduction in human lifespan.

The narrative is clearly reporting the results of an asteroid shower. If a person simply intended to write a report of an asteroid shower all of these points would be included. The dilemma arises because the report has been made by intelligent human beings. How long have intelligent report-making human beings been on the Earth? The corollary follows immediately; how could the present diverse cohort of animal life have evolved within the time span that intelligent report-writing human beings have been on the Earth?

10.7   Evidence from the Solar System

10.7.1   The Moon

There are three features of the Moon, which are indicative of catastrophe. These include the craters, the mountains and the maria.

There are estimated to be 200,000 craters on the Moon, which have diameters greater than one kilometer. A few craters are very large. Clavius is 211 miles across. Other large craters include Copernicus and Tycho. Oceanus Procellarum may be hiding a crater and it has an area greater than the Mediterranean Sea. (22)

The maria (Latin for seas) cover one-half of the visible side. The maria do not have any crater sites of significance. This probably indicates that these areas are comparatively recent features, which have not had time to accumulate very many impact sites. The maria are thought to be the result of molten material from the interior rising to the surface and spreading around before hardening. Whatever enabled molten material to reach the surface must not only have broken through the crust but must have disturbed the interior causing it to well up and pour out across the lunar surface. A large object like an asteroid crashing into the surface and punching through to the interior would have enabled this to happen. The second way these features could have formed would happen if a large object crashed through the crust and generated a major shock wave, which propagated through the interior to the far side. When this shock wave came up under the crust, the crust would have been stressed into failure mode, broken and hurled upward from the surface. Molten material would fountain up and spread around. Even if such a shock impulse did not hurl crustal material high into the air, the rising pressure wave could have caused the crust to fail and material to surge up through the failure region. In either case the original causal surface feature would probably have been obliterated by the molten material.

The Moon has many mountains. The 'mountains of eternal light' near the North Pole only become darkened when there is a lunar eclipse. Some of these mountains are higher than the highest peaks of the Himalayas.

The Moon always has one side facing the Earth. This indicates that the material of the Moon is not evenly distributed and that the Moon is 'unbalanced' with a greater concentration of mass on the side facing the Earth. The asymmetrical distribution of mass was confirmed when satellites orbited the Moon and deviated slightly from their expected orbits. (24) These mass concentrations are located beneath the maria.

The far side of the Moon has a bulge. 'Since the Apollo 15 laser altimeter experiment, scientists have known that a region of the lunar far side highlands is the highest place on the Moon. Additionally, the far side has only highlands and no maria. (40)

It could be that the mass concentrations, the maria, the highlands and the bulge are causally connected. The maria could have been caused by large impacting objects, which fractured the solid upper part of the Moon enough to allow molten material from the interior to migrate to the surface and spread around. This would not necessitate that the entire interior of the Moon be molten but would require that some portion of it was and that the impacting objects allowed pourable material from the molten region to rise up and fill the depressions caused by the impacts. The shockwaves produced by the impacting objects would have propagated through to the other side creating the 'chaotic' highlands and the bulge. Subsequently the tidal effect of the Earth's pull would have stabilized the bulge in place where it remains to this day.

All of these lunar features are indicative of catastrophe. Craters always form suddenly and violently. Lava flowing for miles across the surface is certainly not an appropriate scenario for a quiet Sunday afternoon, particularly if it was enabled to flow by some large impacting object such as an asteroid or comet. When the mountains were thrust many miles up from the surface, it would have been a good time to have been elsewhere.

As mentioned earlier, the Moon has more than 200,000 impact sites, which are greater than one km. in diameter. Half of the visible side of the Moon is apparently the result of massive lava flow. Utter chaos has taken place on the Moon. The entire surface is covered by features, which can only be explained by some major chaos-causing event. If asteroids hit the Moon, utter chaos would result and extensive evidence of chaos is exactly what would be expected.

10.7.2 Mars

Mars has three major impact sites. Hellas, the largest, is 1240 miles across. The other two, Isidis (680 miles across) and Argyre (1120 miles across) are also very large by any comparison. When any one of the objects, which formed these craters, hit the surface, it would have been most catastrophic. Even though the atmosphere of Mars is very thin (about 1% of Earth) and any atmospheric effects would not have been nearly as pronounced as they might have been on Earth if similarly-sized objects arrived, all of the other results of a major continental impact would have occurred. The surface would have been displaced and shaken most violently. There would have been a major shower of debris and a great amount of dust would have been stirred up. As with a major impact on Earth however, the direct impact effects would have been just the beginning of trouble. A shock wave would have been initiated in the interior and it would have traveled right through to the other side within a few minutes. When it came up on the far side, the second round of chaos would have resulted. This, in fact, appears to have been what happened on Mars.

The Tharsis Bulge is a huge swelling in the Martian surface, almost directly opposite Hellas, the largest impact site. This elevated area is approximately 6 1/2 miles higher in the center than the average Martian

363

surface and it extends over an area of 3000 square miles. (23) Included in the elevated region are the four largest volcanoes in the solar system. Immediately adjacent is Valles Marineris, the immense canyon system, which varies between 50 and 75 miles across, between 3 and 4 miles deep and which runs for several hundred miles. Radiating outward from the main trench are large arrays of tensional faults (graben). (i.e. the ground was stretched)

The highest of the volcanoes is Olympus Mons. Its elevation is variously given as 75,000 feet, 88,600 feet, 27 kilometers and 37 kilometers. The base is 335 miles across. The opening at the top is 53 miles by 37 miles. This volcano is so large that there is no particular place on it where a person could stand and see all of it because of the general curvature of the planet's surface. If a person stood on the highest point, the slope would extend all the way to the horizon and the lower part of the slope would be beyond the Martian horizon. It is about the same size as the state of Missouri where it is, of course, not possible to see the far side of the state from any particular position along the border, or even from the top of a tall building. Where the base meets the surface plain, the edge of the volcano goes straight up for about four miles. Confronting any would-be adventurers crossing the Martian surface would be an escarpment four miles high. The slope of the volcano is quite gradual, however, so if a person could get up the escarpment, hiking up the rest of the way would be relatively easy. A mountain such as this on Earth would allow hikers to actually hike right into space.

Three other volcanos; Arsia Mons, Pavonis Mons and Ascraeus Mons are nearby and within the Tharsus Bulge region. While not quite as large as Olympus Mons, they are also much larger than anything on Earth. (8)

The Hellas side of Mars shows almost no other evidence of chaos but the volcano side shows total chaos. The surface has boiled up into four giant volcanoes and the surface has been split wide open. It appears that the Hellas impactor generated a major shock wave, which traveled through the interior and erupted out the other side. As this shock-wave welled up under the surface opposite Hellas, the surface of Mars was forced upward. However then it would actually have been the surface of a larger sphere. Therefore the surface crust had to split open because there wouldn't be enough surface material to enable this larger surface to remain continuous.

The escarpment surrounding Olympus Mons indicates that the entire volcano has been thrust upward by an increase in internal pressure. As the pressure wave forged upwards from the interior, the area was lifted and a few regions which could not carry the shear stress were lifted even higher (i.e. about four miles in the case of Olympus Mons). In several locations the crust gave way completely and material from the interior poured out onto the surface forming classical-looking cone volcanoes on top of an area, which was already elevated.

While the overall area is elevated as the Tharsis Bulge, the region immediately surrounding Olympus Mons is depressed. The volcano rises from within this localized depression. Since it appears that a great amount of material has been ejected from the interior onto the surface to form the volcano, it might be that this left a void, which caused this area to slump back down after the upward force subsided.

When the shock wave energy had dissipated and the surface settled down again, there were four huge volcanoes belching out material, an extended fracture in the surface as well as gigantic ridges and dislocated blocks. The entire side of the planet shows evidence of utter chaos. However utter chaos is exactly what would be expected from a major asteroid impact.

10.8   Summation

The evidence of chaos around the world is well summarized by Velikovsky in his book entitled Earth in Upheaval which is itself an excellent description of what would happen on Earth if it was struck by one or more large asteroids. 'Whenever we investigate the geological and paleontological records of this Earth we find signs of catastrophes and upheavals, old and recent. Mountains sprang from plains, and other mountains were leveled; strata of the terrestrial crust were folded and pressed together and overturned and moved and put on top of other formations; igneous rock melted and flooded enormous areas of land with miles-thick sheets; the ocean bed flowed with molten rock; ashes showered down and built layers many yards thick on the ground and on the bottom of the oceans in their vast expanse; shores of ancient lakes were tilted and are no longer horizontal; seacoasts show subsidence or emergence, in some places over one thousand feet; rocks of the earth are filled with remains of life extinguished in a state of agony; sedimentary rocks are one vast graveyard, and the granite and basalt, too, have embedded in them numberless living organisms; and shells have closed valves as they do in a living state, so unexpectedly came the entombment; and vast forests were burned and washed away and covered with the waters of the seas and with sand and turned to coal; and animals were swept to the far north and thrown into heaps and were soaked by bituminous outpourings; and broken bones and torn ligaments and the skins of animals of living species and of extinct were smashed together with splintered forests into huge piles; and whales were cast out of the oceans onto mountains; and rocks from disintegrating mountain ridges were carried over vast stretches of land, from Norway to the Carpathians, and into the Harz Mountains, and into Scotland, and from Mount Blanc to the Juras, and from Labrador to the Poconos; and the Rocky Mountains moved many leagues from their place, and the Alps traveled a hundred miles northward, and the Himalayas and the Andes climbed ever higher; and the mountain lakes emptied themselves over barriers, and continents were torn by rifts, and the sea bottom by canyons; and land disappeared under the sea, and the sea pushed new islands from its bottom, and sea beds were turned into high mountains bearing sea shells, and shoals of fish were poisoned and boiled in the seas, and numberless rivers lost their channels, were dammed by lava and turned upstream, and the climate suddenly changed; tillable land and meadows turned into vast deserts. Reindeer from Lapland and polar fox and arctic bears from the snowy tundras and rhinoceroses and hippopotami from the African jungles, and lions from the desert and ostriches, and seals, were thrown into piles and covered with gravel, clay, and tuff, and the fissures of multitudes of rocks are filled with broken bones; regions where the palm grew were moved into the Arctic, and oceans steamed, and the evaporated seas condensed under clouds of dust and built mountainous covers of ice over great stretches of continents, and the ice melted on heated ground and cast icebergs into the oceans in enormous fleets; and all volcanoes erupted, and all human dwellings were shattered and burned, and animals tame and fierce and human beings with them ran for refuge to mountain caves, and mountains swallowed and entombed those that reached the refuge, and many species and genera and families of the animal kingdom were annihilated down to the very last one; and the earth and the sea and the sky again and again united their elements in one great work of destruction.' (copied from Earth in Upheaval by Immanuel Velikovsky as published by Random House Publishers in 1955 and used here with the kind permission of Rafael Sharon, his grandson) (25)

11. Predictions for the Future

11.1 Life in the Universe

Based on the 'Window of Life" concept, it will become increasingly frustrating to realize that there is no life elsewhere in the universe. This prediction is based on the following three factors.

i) The discovery of an actively temperature-stabilized planet (such as the Earth) in the thermally-habitable zone of a 'hot' sun such as the Sun, accompanied by an appropriately-sized and distanced moon and appropriately-distanced orbital stabilizers like Jupiter and Saturn, will not happen.

ii) The simultaneous existence of a complete set of necessary life-enabling factors will not be found.

iii) The increasingly-recognized complexity of life, will reduce any hypothetical discovery, from some finite chance, to virtual zero.

11.1.1 Temperature Stability Criteria - Part 1

The Earth enjoys a very high level of temperature stability. This stability is due to a multitude of factors, which are part of the structure, environment and processes of the Earth. It is also due to the fact that the Earth has both a stable, nearly-circular orbit around a sufficiently-quiet singular Sun. From Earth, it is not possible to study environmental processes on distant (i.e. beyond the solar system) planets but it is possible to predict whether or not a distant planet could possibly have a stable temperature. For example, a planet might have a stable temperature if it had a steady and predictable circular orbit. It would not necessarily have a stable temperature because of a circular orbit but a circular orbit is a minimum, necessary condition for temperature stability. Conversely, if a planet had an orbit, which was not circular, without knowing anything else about that planet, it could be concluded that it would not have temperature stability and consequently could not be a place where life could thrive.

In order to have a circular orbit, a planet must be involved with a singular star. Unfortunately, observation indicates that singular stars are in the minority. 'A few years ago Levy and I decided to determine more carefully the frequency of doubles and other multiple stars among sun-like stars. Levy and I observed all 123 sun-like stars visible to the unaided eye in the Northern hemisphere. Some of the systems observed were triple or even quadruple. Among the 123 primary stars 57 were found to have one companion, 11 to have two companions and three to have three companions. Thus more than half (57%) of the stars have at least one companion. The preceding inventory is only part of the picture, because we are unable to detect all the companions that are likely to be present.' (1) These comments were made more than thirty years ago and since then the percentage of multiple stars observed has increased. 'In fact, … single stars are rare; in spite of the appearance of the night sky to the unaided eye, most stars are actually binary….Our Sun, as a single star, is part of a minority population.' (2)

Other commentators offer similar or higher estimates. 'About 75 percent of all stars are members of a binary system.' (19) In addition to the binaries there are triples and quadruples. When these are also included, the proportion of multiple stars is upwards of 80 percent.

It is an absolute necessity that a sun be singular before there is any possibility of an associated Earth-like planet having a circular orbit. (Massive spheres of gas such as Jupiter do not qualify in this case as Earth-like). Most stars are binaries. Instead of just one glowing mass of hot gas, there are two. The two stars revolve around each other and from a great distance it appears that there is only one object. It requires the optical advantage of a telescope with its associated instrumentation to determine that a star, which appears like a single object, is really two objects circling around each other. In some cases, there are three objects circling around each other. It is reasonably understandable that two objects could circle each other for a period of time, but it is not really understandable how three objects could do this for any period of time without their orbits becoming unstable. However, whether this happens or not is quite secondary to the fact that two or more massive objects, (like our Sun) which circle each other, have, associated with them, a gravity field, which includes massive fluctuations. These fluctuations guarantee that no nearby object, like a life-enabling planet, could ever have a circular orbit. In fact it could not have any orbit at all unless it was close enough to one of the stars so that star's gravity clearly dominated the planet's orbit. It is more likely that ' … untold numbers of worlds have … fallen into their suns or been flung out of their systems to become "floaters" that wander in eternal darkness.' (32)

Unfortunately, for those who prefer to imagine a universe, which is teeming with life, most stars are either binary or triple. Some estimates have placed the portion at ninety percent. (3) Examples of binary stars within the near portion of the Milky Way Galaxy are; Eta Cassiopedia, Procyon, Sirius, (brightest star in the sky) Ross 614, G1 166, G1 702, G1 783 and Groomsbridge 34. Examples of triples are; Alpha Centura A and B along with Proxima Centura, (the nearest star to the solar system) G1 166, G1 1245, G1 570, G1 663 A and B along with G1 664 and EZ Aquarii. (4)

It has been speculated that a binary star system would have a habitable range wherein life could exist. The nearest star system to the solar system happens to be a triple but it is still offered as a possible energy source for a life-inhabited planet. 'The good news for believers in alien life forms is that a binary system could harbor life but the stars would have to be a fair distance from each other, in a stable orbit…..As it happens, there is a binary system that fills the bill….Alpha Centauri A and B….move around in an orbit that takes them as close as 11 AU from each other (roughly the distance from the Sun to Saturn) and as distant as 36 AU (as far as Neptune). Each star could have its own system of planets, but they'd need to be in relatively close (i.e. to their host star) so as not to be disturbed by the gravitational pull of the other star – roughly 3 AU. In our solar system four planets reside within 3 AU of the Sun – Mercury, Venus, Earth and Mars.' (20) Actually, this is not good news at all. Having a second star in the solar system wondering between Saturn and Neptune would cause any planet positioned between Mercury and Mars to have an elliptical orbit. The energy received from the host might be suggestive of enabling a habitable zone but the energy level would not be constant. Only a constant energy supply is permissible. Even the Earth, with all of its temperature regulating mechanisms cannot tolerate a 5% orbital deviation. (21) Further it is now known that Alpha Centura includes a Red Dwarf star which actually does have a planet in orbit around it. However any planet that orbits a Red Dwarf has no hope of being able to support life because the planet would have to be very close to get enough energy. Being close however is intolerable because the planet would not be able to rotate and properly distribute the incoming energy properly across its surface. For these reasons Red Dwarfs are not considered possible candidates for a life-enabling planet.

It may subsequently be argued that ten percent, or more, remains, which is still a very large number of stars. However, most of the stars, which do not qualify as binaries, have one or more massive dark objects nearby. In fact, some observations, together with the associated calculations, have determined that there must be three massive dark objects orbiting the Sun in question. Upsilon Andromedae is a star in the constellation Andromeda and it has been determined that there are three massive objects orbiting around it. One planet is about three-quarters the size of Jupiter and orbits every 4.6 days. The next one is twice the size of Jupiter and about as far from the host star as Venus is from our Sun. The third object is even larger at four times the mass of Jupiter and completes an orbit in 3.5 years. (5) In many cases, these dark objects are much larger than Jupiter and orbit their host sun much closer than Jupiter orbits our Sun. 'Instead of finding the Jupiter-sized planets in orbits like those of Jupiter and Saturn, the two largest planets in our solar system, the planet hunters were stunned to find that all of the new planets orbit closer to their parent stars – and most of them are much closer. Eight of the 17 planets circle their stars at less than the distance of Mercury, the closest planet to our Sun. The other nine range from Mercury's distance to roughly three times Earth's distance from the Sun.' (22)

While this report was discouraging enough when it was given, more recent reports (i.e. late 2009) have not really offered a better scenario. 'To date, astronomers have identified more than 370 "exoplanets", ... Many are so strange as to confirm the biologist J.B. S. Haldane's famous remark that "the universe is not only queerer than we suppose, but queerer than we can suppose". Included among the findings is a planet with more than 5000 times the mass of the Earth (CT CHA B) and another one with almost 8,000 times the mass of the Earth (HD 43848 B). In fact, of the first 370 planets to be discovered only 13 have been identified with less than 10 Earth masses. (As the gravity of a planet increases it becomes more difficult for water to evaporate.)

At one time, objects in the universe were very clearly placed into one of two categories. (neglecting comets) Self-illuminating objects were stars and the rest were planets, either large or small. This view has changed and there is now no clear demarcation between the two.This view results in a continuum from high-mass stars, to low-mass stars, to high-mass brown dwarfs, to low-mass brown dwarfs, to high-mass planets to low-mass planets. As long as the definition of brown dwarfs, stars and extrasolar planets remained theoretical constructs, stars and planets remained distinct objects. However, with the recent discovery of brown dwarfs and extrasolar planets within the gap, the distinction between stars and planets is now murky. (6)

The gravity perturbations of all multiple-mass stars leave no doubt that any planet such as Earth would not be able to remain in the vicinity. It would be accelerated into a trajectory, which could easily include having it directed into the central mass or flung completely away from the entire system. 'None of Upsilon Andromedae's planetary companions are like Earth, and would not be hospitable to earthly life. And even if an Earth-type planet existed in this system, it might risk being obliterated by its huge sister planets....which have elliptical orbits that could cross the circular orbit of an Earth-like planet, if one existed, which could fling this new world into space.' (5) Another commentator offers a similar conclusion. 'Earth could not exist around any of the stars that harbor the newfound planets. The gravitational influence of the big planets would not permit a stable orbit for an Earthlike planet. Smaller planets would be tossed completely out of their orbits and off into deep space.' (22) Any Earth-like planet that managed to exist in a system which included monsters such as those discovered so far, would certainly not have a circular orbit. Therefore there would be no possibility of life being sustained because the temperature would always be deviating too much.

While the great majority of stars are either known or expected to be multiple-mass systems, the primary driving motivation of the planet-search activity is to find a star with an Earth-size planet. One of the main detection procedures involves monitoring the pathway of the host star. If the pathway deviates from a straight line in a rhythmic or cyclic manner, there is probably a planet nearby, which is causing this deviation. Observation and calculation can then enable how many planets are in orbit. The faster the deviation, the closer or larger the planet which means that detecting small ones (i.e. like Earth) will be very difficult anyway. The second, apparently most promising detection procedure, involves scanning stars for minute changes in light output. If the change is periodic, expectedly a planet is causing it.

It is highly probable however, since both instrumentation and technique are constantly improving, that Earth-size planets – possibly many - will be discovered. In order to qualify as a base for life, such a planet must have a constant temperature, which is in the appropriate range. Therefore it must not be too close to its sun. Herein lays the first problem. In order to be detectable, the planet must be close enough to its sun to cause it to deviate either fast enough or often enough to be detectable by telescopes controlled from Earth. Our Earth orbits our Sun once a year. It therefore causes our Sun to deviate very slightly from a perfectly straight path with a one-year cycle. While this makes detection virtually impossible, an Earth-size planet is too small to be detected anyway. 'A planet the size of Earth would not exert enough gravitational force to cause a discernable wobble, the astronomers said.' (5)   More probably, any deviation pattern of a distant star will be detectable because it is much less than one year, indicating that the orbital path of the associated planet is quite close to its host sun. If, for example, it was as close as Mercury is to the Sun, (which is 36,000,000 miles), the orbital period would be similar to Mercury's (i.e. 88 Earth days). This would be detectable, as would longer periods, if the associated mass was large enough to cause a detectable wobble. While it is unlikely that an Earth-size planet will be detected this way, the chances would be improved if the orbital period was much less than one year. However, Mercury is much too hot to enable life to exist. Such would be the case for any remote planet orbiting close to a star, which was similar in size to the Sun.

It is understood that if the host star were larger than the Sun, the preferred distance for a planet would be greater than the Sun-Earth distance. Conversely, if it were smaller, the preferred distance would be less. It remains however that the orbital period would be similar to an Earth-year. Detecting a one-year period wobble in the pathway of a remote star will not be easy, because such a wobble would be much less than the diameter of the host sun.

The next part of the problem is that the wobble must be symmetrical indicating that the orbit of the planet is circular. If the orbit is not circular, temperature deviation will be too great, causing the possibility of life to be nonexistent.

Another aspect of temperature stability relates to size. A planet larger than Earth would have stronger gravity. Water would therefore not evaporate as easily and not only would plants (which required water) be parched but the energy transfer and distribution function provided by evaporated water would be reduced. For example, a planet has recently been discovered in the Gliese system and it is 'just twice Earth's mass' (31) Rivers and lakes would not easily form with a planet this large (particularly since its host star is 'dimmer than our Sun' (31) like 95% of all stars) and any hydrologic cycle would be less active than Earth's so the temperature stabilizing function these factors would provide would be reduced. The situation would be even worse with the "most promising" find. 'Indeed, the most promising planet yet found – the "super Earth" Gliese 581 d, has seven times Earth's mass. (33)

On the other hand a planet smaller than Earth would have difficulty holding an atmosphere. Even if the orbit of an Earth-like planet was "just right", in order to have temperature stability, the size must also be "just right". Will a rocky planet with a circular orbit at just the right distance ever be found? If and when it is found, the host of life-enabling necessary factors discussed herein would also have to be present to enable a Window of Life to exist.

11.1.2   Temperature Stability Criteria - Part 2

While the possibility of finding a singular star appears minimal, the possibility of finding a singular star with long-term stability is even more remote. Unfortunately, this includes our own Sun, which may not have long-term stability either.

Nova events are very common throughout our Milky Way Galaxy. While nova does mean new - ancient observers evidently thought that a new star was being formed - the nova event is actually a brightening event. It was, until fairly recently, understood to be the termination of the star. It has now been observed that an existing star may suddenly brighten before slowly fading back to its original brightness. Within the Milky Way Galaxy approximately thirty such events are observed every year. Unfortunately, the degree of brightening is not exactly the same for each star. The flare-up for some stars is much less than the flare-up for others, which leads to the possibility that numerous flare-ups occur which are neither observed nor reported.

When events such as these are observed, a theory is sure to follow. In this case it is theorized that the flare-ups are the result of accumulated hydrogen fusing into helium. We understand this to be similar to the reaction of a hydrogen bomb. When the flare-up event is over, the star continues to shine as before. However the accumulation of material is expected to continue. This means that periodically, the star will flare up again. Long term temperature stability for any nearby planet would therefore not be possible.

11.1.3   Tidal Problem

The Sun is included in a very small group of stars (approx. 5%) which are bright. This means that the heat produced enables an Earth-like planet to be a considerable distance away and still receive enough energy to enable life to exist. At these greater distances, the tidal effect of the Sun is reduced and, in fact, barely noticeable. On Earth for example, the tidal effect of the Sun on the ocean can only be measured in inches.

Unfortunately, most stars are low-mass stars and are dimmer than our Sun. This means that a life-enabling planet must be much closer in order to receive sufficient heat. However closer means that tides will be higher. Science Daily reported that "Tides can render the so-called habitable zone around low-mass stars uninhabitable." Astronomers at the Astrophysical Institute Potsdam studied the effects of tides on planets around low-mass stars (the most numerous stars in the galaxy) and found that the ... increased volcanism... make them...uninhabitable. "I think that the chances for life existing on exoplanets in the traditional habitable zone around low-mass stars are pretty bleak, when considering tidal effects" lead researcher Rene Heller remarked. "If you want to find a second Earth, it seems that you need to look for a second Sun." (34) In other words, in order to avoid the tidal problem of having a planet near a dim sun, an Earth-like planet must orbit a bright sun. This reduces the number of stars as possible energy suppliers for Earth-like planets to less than 5% of all existing stars.

11.1.4 Life-Enabling Factors

A great many factors must simultaneously be present for life to be possible. These include all of the 'Windows of Life" discussed earlier, four of which are mentioned here.

A planet with a stable nearly-circular orbit is the minimum necessity for temperature stability but it is only one of the numerous factors, which are required.

The structure of such a planet must include a relatively thin crust and a molten interior. If the crust is too thin, it will not be structurally sound enough and the surface will include too many hot spots. If the crust is too thick, surface temperature will deviate into the too-hot and too-cold categories, too much of the time.

Next, because the Sun only shines directly on one area at a time, the surface of the planet must include large bodies of water. It takes more heat to change the temperature of water than it does to change the temperature of rock. Large bodies of water therefore act as thermal flywheels and help to regulate the temperature, not only from day to night but also from summer to winter. In this regard it is fortuitous for the Earth that its largest surface areas of water are facing the Sun when the Earth is closer to it. They soak up the extra heat and release it later thereby helping to stabilize the temperature. If it were the other way around, the surface of the Earth would experience greater extremes of temperature.

A magnetic field is important because it intercepts damaging incoming radiation. If the Earth were to lose its magnetic field, the incidence of cancer would be higher and many more people would die from it.
An atmosphere is beneficial. In addition to being an aid to breathing, it helps to transfer moisture to the land and heat from the tropics to the poles. It is correct that Venus has an atmosphere. Unfortunately it consists of carbon dioxide under very high pressure. (28) This would be a death trap for virtually all forms of life. In order to have an appropriate atmosphere, a planet must be within a fairly narrow size range. If it

is too small, its gravity will not be able to hold an atmosphere. If it is too big, the temperature of the vapor point (or boiling point) of water will be too high and life will not be possible (i.e. water will not vaporize and thereby transfer moisture to the land). The atmospheric mix is of great consequence. Oxygen is absolutely necessary but too much would result in uncontrollable forest fires. The oxygen must be diluted with an inert gas. Nitrogen is good choice because it does not readily react with anything. There must also be a small amount of carbon dioxide. If there isn't enough, plant life will not thrive. If there is too much, it becomes difficult for people to breathe.

While the probability of finding an appropriately-sized planet is vanishingly small, the probability of finding one with even one or two of the above four factors in their respective life-enabling ranges, is just too small to be realistically possible. If a probability were computed for this situation, it would be found to be one in a number which is greater than the estimated number of atoms in the visible universe. For those of us who are more mathematically challenged, 'it didn't happen'.

11.1.5   Life

A continually increasing amount of information is available concerning the great complexity of life. The human body or any animal body includes numerous complex molecules, transfer systems, communication systems and numerous different kinds of fluids. So much information is available on the complexity of an animal body that no one person or professional specialty is able to comprehend it all. The question must be properly asked; how did things become so complex? Could it ever have happened by chance? The short answer is obviously 'no'. The long answer must include a probability discussion.

Since it will not be possible to deal with the entire animal body at this time, a discussion of the improbability of one aspect will have to suffice. Proteins are found in all animal bodies. Proteins are molecules, which are necessary for all of life's processes. In any animal body there are numerous different types of proteins. (approximately 100,000 in a human body)

Proteins are formed from amino acids of which there are twenty different types. Amino acids are the basic building blocks of proteins and are themselves small molecules consisting of several atoms connected together in a particular sequence. Proteins are formed by connecting the amino acids together end-to-end in very particular sequences. If the right acid is in the wrong place even once, the protein is useless or damaging to its host. Any particular protein will include several different amino acids and it is the type (of acid) and the particular sequence of assembly that makes each protein unique.

A thought experiment can be formulated in which all twenty amino acids are available, in unlimited quantity, to form a protein. Then one could ask; what is the probability that any particular protein will be formed? (We assume for a minute that the amino acid to amino acid bonding process is able to proceed somehow.) Each location in the chain (i.e. the protein is like a chain of amino acids connected end to end) could possibly be filled with any one of the twenty possibilities. Therefore, the probability of forming a protein - any protein - is one divided by 20 x 20 x 20 x … for the number of positions in the protein which is usually about 300 but some proteins have 3000 acid positions. Therefore, the probability is one divided by 20 x 20 x 20 x 20 x .... 300 times. This is one divided by an extremely large number. In plain English, the resulting probability is zero. The possibility or probability of 300 connections happening in the proper sequence is so vanishingly small that it is outside the range of reality. It is so small that there is only about one chance of a useful protein forming in a mass of useless proteins as large as the Earth. 'The number of possible proteins of a given length of 'n' amino acids is 20(n), since there are 20 different amino acids possible for each link in the protein chain. The number of variants for a protein containing 100 amino acids would be 20(100). This means that there are more theoretically-possible proteins that can be formed from 100 amino acids connected together than there are particles in the universe. (i.e. the visible universe) (7) This very improbable result is obtained even when the soup has been carefully prepared but in reality this would not be the case. The number of molecules in a mass as big as the Earth is incredibly large. The number is so large that it does not have any practical everyday meaning and neither does the number of particles in the universe. Therefore, in plain language, there is no possibility that proteins were formed by chance or by themselves, but if proteins had never been formed, life would never have happened.

If the discussion were to continue from this point, it must include the (virtually) innumerable other molecules as well as the numerous interactive systems and procedures, which are going on within the cell all the time. These activities number in the billions and they continue hour after hour, day after day. Molecules are moved to their required positions by little built-in outboard motors. Gates open to allow certain molecules through. Other molecules switch back and forth from a double state to a triple state as energy is stored, transferred and released. The probability of all of this happening (and if it didn't, the viability of the host organism would be either compromised or terminated) is zero point zero to millions and millions and billions of decimal places.

11.1.6   The Drake Equation

A discussion of this nature would be incomplete without mentioning the Drake Equation which was originally formulated by a person with that name. There are usually several factors included with the first one being the estimated number of stars in a galaxy, which number is commonly given as 200 billion. The actual number is not known but 200 billion represents a reasonable starting point for this discussion. The second factor is the fraction of stars with habitable planets. This is a much more difficult factor to determine because of the numerous factors involved. For example, some regions of a galaxy are recognized as unsuitable. Far out in the galactic disk, the rate of star formation is too low to make many heavy elements, so stars in the distant suburbs are probably planet free. Close to the galactic center, there are lots of stars forming to make planet-building material, but the supernova rate is so high that it may be hazardous to evolution. (30) In other words, a galaxy might have many stars but most of them, due to their location, are not suitable. Even in the acceptable zone, many more – in fact most of the rest – must also be eliminated. We must ignore stars so massive that they live fewer than four billion years as well as stars so much less massive than the Sun that their habitable zones are very narrow. (30)   Further, as mentioned earlier in this chapter, in order for a planet to have a constant temperature, its orbit must be circular which necessitates that its sun be singular and have a constant heat output. Also it must not have any large companions, which would prevent an Earth-sized planet from having a stable circular orbit.

With these factors alone in mind, the number of possibilities has already been reduced to very few but several additional factors must also be recognized.

As mentioned above in 11.1.3, the candidate star must not be a low-mass star and must have its appropriately-sized planet (In 11.1.1 above it was identified that a planet which is too large will not allow its water to evaporate properly and one that is too small will not restrict evaporation properly. Therefore, the planet of interest must not be much larger than the Earth or much smaller.) at exactly the right distance. (Recall that the Earth has a "goldilocks" orbit which would result in disaster if it varied by 5%.) The planet must have a Moon-sized moon at just the right distance for axial stability as well as modest tidal production. Further it must have an appropriate atmosphere as well as a large ocean surface. Both a strong magnetic field and an ultraviolent shield are absolutely necessary. As this list continues, the conclusion that must be reached is that the Drake Equation, as given in most of the usual forms (29) is naïve, because it does not recognize any of these factors.

11.1.7 Life in the Universe Conclusion

From all observations to date, it appears that virtually all of the stars in the Milky Way Galaxy, are either double, triple, or are closely associated with one or more massive objects, most of which are orbiting at a fraction of the distance that the Earth is from the Sun. This makes virtually all of the stars in the Milky Way galaxy unfit as sources of energy for a life-sustaining planet.

In order to avoid tidal problems, an Earth-like planet must be a considerable distance from its host sun. This eliminates dim stars which includes 95% of the stars of the galaxy.

As discussed in the Windows of Life chapter, the Moon has a significant influence on the stability of the Earth's orbit as well as a stabilizing influence on the tilt of the Earth's axis. Further, it appears that the orbital stability of the Earth is enhanced by the position and mass of Jupiter. The possibility of finding a planet with these additional stability factors is very very small.

Each of the categories discussed above have associated with them a probability of occurring which is vanishingly small. Unfortunately, probabilities must be multiplied. It is correct that we are in the realm of the unreal and as long as that is the case, one more surreal item will not be out of order. When several vanishingly small probabilities are multiplied together the result is even smaller. It is in the unthinkably-small category. In everyday terminology it must be concluded that it certainly did not happen. As time goes by and more and more knowledge in each of the above categories is gained, it will increasingly be concluded that life is no accident, did not spontaneously happen and does not occur throughout the universe.

Stars have been likened to snowflakes from the perspective of variety. It has been suggested that there is no such thing as two identical snowflakes because after observing a very large number of them no two were found alike. A similar situation appears to be the case with star systems because the structural variety appears endless. Therefore, it is predicted that it will be concluded that the probability of finding a life-enabling Earth-like planet is less than one, in a number, which will exceed by orders of magnitude, the number of stars in a typical galaxy (i.e. 300 billion).

11.2   Asteroid Impact

Is it possible that another asteroid will strike the Earth? It has been claimed elsewhere in this document, that a shower of asteroids hit the Earth at a previous time and in so doing completely reshaped its surface. Could it happen again?

Based on current evidence, it is certainly possible that the Earth could be struck by another asteroid. The present capability of technology has enabled thousands of asteroids to be cataloged and tracked. Based on this information it is known for certain that numerous asteroids regularly cross Earth's orbit. Just a few years ago, only a few were known to cross. However, every year a great many more asteroids are identified and added to the already long list and there is no end in sight. The only positive aspect is that improved technology is able to identify smaller and smaller objects so the ones that are being added to the list, are usually not very large. Even as this is happening, however, speeding asteroids regularly appear near the Earth on their way by and sometimes they are not observed until they are already past. It is common for an asteroid to pass by the Earth closer than the Moon. This is close. Could one of these asteroids hit the earth? Yes it could. Will one of them hit the Earth? It is quite realistic to expect that one will. It is a matter of time but not the times usually offered. If one passes by about once a week, in one thousand years, fifty thousand asteroids would have come very close to hitting the Earth. After all, a large asteroid was observed to hit the Moon in 1178 AD. (23) 'Statistical analysis indicates there are some 1,500 dangerous rocks – half a kilometer wide or larger – not yet located and another 100,000 or so big, but less threatening NEOs (near Earth objects) undiscovered. Any one of these would produce a cataclysm.' (24) Authorities are concerned. 'All this has prompted NASA, the US space agency, to establish a committee to seek expert advice on how to use existing telescopes to map the orbits of the estimated 2,000 asteroids and comets larger than one kilometer in diameter that could potentially threaten Earth with widespread destruction.' (27)

Smaller objects have not been completely listed and probably never will be listed. The problem relates to the difficulty in actually recognizing smaller objects as well as the fact that smaller objects may move in less predictable orbits because their orbits are more easily modified by other objects in the solar system. 'Over time, the gravitational influence of Jupiter gradually modifies their orbits, occasionally shifting them into trajectories that carry them inward to where the Earth circles the Sun.' (26) The expectation from all of the evidence is that it is highly probable that an asteroid in the house-size category will hit the Earth within another few hundred years.

The greater question is; what damage would be done? The answer is that a lot of damage would be done and the amount of damage would depend of where the object landed. A house-size asteroid could devastate an entire state or province. However, a more subtle complication relates to the fact that the Earth has already been impacted and the results of those impacts are still here. They include a cracked or faulted crust. This means that the next impact does not have to crack the crust but only disturb the present cracks. Most volcanoes occur near one of these faults. (25) It is therefore predicted that the next asteroid to strike the Earth will disturb and reactivate a lot of volcanoes. The results of such activity are well understood and include colder than usual summers with accompanying crop failures. Near the impact site, many people would die. However, far from the impact site, a lot more would die of starvation because of the crop failures caused by the chill brought by both the volcanic and impact clouds.

11.3 Carbon Dioxide

The amount of carbon dioxide (CO_2) in the atmosphere will continue to increase. In fact it will continue to increase even if no more carbon fuel is burned. Even if the entire world stopped burning gasoline, diesel fuel, fuel oil, natural gas, propane and all the other types of fuel, which contain carbon, the amount of carbon dioxide in the atmosphere would continue to increase and there are several reasons why.

Carbon dioxide is credited with being a greenhouse gas. The presence of carbon dioxide in the atmosphere helps to retain the heat from the Sun and thereby increase the temperature of the Earth. If the Earth gets warmer due to CO_2 increase, not only the land areas become warmer but the oceans will also become warmer. However, the oceans contain a great amount of CO_2, which is held in the ocean water because it is cold. If the ocean water was warmer, it could not retain this CO_2 and some of it would come out. (8) As it comes out, it will add to the greenhouse effect and cause the Earth to become warmer still. Then the oceans would become warmer and release more CO_2. Then the atmosphere would become warmer and further warm the oceans. This type of activity is called positive feedback. This is a type of arrangement where the results enhance the cause, which further enhances the results. Such a situation can only be stopped if some part of the feedback loop is interrupted. If, for example, the increase in temperature resulted in an increase in cloud cover, less sunshine would reach the Earth and the temperature would not continue to increase but would level off instead. In the case of the Earth, no one knows for sure if this will happen and if it did happen, what temperature would be reached before it leveled off? Before leveling off however, it could easily exceed the range, which is acceptable for human life.

The combustion of fossil fuel certainly aggravates the problem but only represents a portion of the total amount of CO_2 that enters the atmosphere every year. Much CO_2 comes from natural sources including the breath, which all animals and people exhale, as well as rotting and burning vegetation. When these natural sources of CO_2 are recognized together with the various positive feedback loops already operating, the conclusion is reached that even if no more carbon fuel was burned, the level of atmospheric CO_2 will continue to increase.

11.4 The Temperature of the Earth

The Earth will continue to get warmer for the next few decades and probably for much longer. There will be numerous effects of a warming Earth and most of them will be negative.

11.4.1 Ocean Level Increase

The level of the water in the ocean will rise. Sea level will change. The change of such a basic reference is psychologically upsetting enough but the physical effects will be quite measurable. The change of sea level will be due to two main factors: a) expansion due to temperature increase (i.e. warm water needs more space) and b) melt water from the melting ice-fields.

As water gets warmer, it expands. This effect holds for water, which is warmer than four degrees Celsius. From this temperature upwards, water expands. Therefore as the Earth warms and in particular as the oceans warm, the water in the oceans will expand. The resulting change in sea level will be the same as if more water had been added to the ocean.

Warming the Earth will result in melting the two great glaciers of the Earth. If both the Greenland glacier and the Antarctica glacier melted completely, ocean level would be higher by more than two hundred feet. (9) It is obvious that inland continental glaciers and parts of the two great glaciers are melting. In fact, some inland glaciers such as the Columbia Ice-field in the Canadian Rocky Mountains are melting very quickly and will be completely gone within another century. (This conclusion is reached by observing the stakes that have been placed in the ground to follow the retreat of this glacier.) The Greenland glacier is reducing overall, but building up in some inland areas. Warm water adjacent to a chilly inland area is the perfect recipe for ice buildup. The great continental ice shelf around Antarctica is melting and breaking away in large amounts.

As the great glaciers melt away, ocean level will rise and by and by the great glaciers will no longer exist.

11.4.2 Permafrost Destruction

The destruction of permafrost is a one-way street. Once it melts it will not reform. In order to form permafrost in the first place, an extended period of time was required when the weather was below freezing much more of the time than it was above freezing. This drove the frost line very deep and a layer of frozen soil was formed. When summer came, only a few feet of soil near the surface had time to melt. During winter the cold frozen soil continued to conduct heat to the cold atmosphere and the depth of the frost line just kept getting deeper.

Something similar can and does happen at the present time if ice arenas or frozen food storage facilities are not designed properly. Heat will conduct from the unfrozen ground under the cold area, into the cold area. From the cold ice slab of the arena or the frozen room of the food storage facility, heat is removed mechanically to the outdoors. The ground underneath will freeze and the region of frozen ground will continue to deepen as long as the thermal conductivity of the frozen region - be it the ice of an arena or the low temperature of a frozen locker - exceeds the thermal conductivity of the unfrozen ground. On the other hand, if the facility is designed properly, a layer of insulation will be included beneath the area to be cooled. With this arrangement, the ground will not freeze because the insulation reduces the rate at which heat can transfer from the ground through to the chilled area. The more remote ground can then supply heat fast enough to keep the ground, adjacent to the insulation, above freezing.

As the Earth warms up, the permafrost areas will melt to a greater depth every summer. The following winter the ground will refreeze, but not as far down as the permafrost region, which then becomes isolated from the annual freeze-thaw cycle. The permafrost region will subsequently slowly melt because heat from deep down in the ground will flow up into it. Once it becomes melted, and since it is isolated from the freeze-thaw cycle, it will not refreeze. Milder winters prevent the frost line from going deeper than the summer warming depth, so any newly-frozen ground will completely thaw prior to the next cold period.

Permafrost is definitely melting. There was a time in the recent past when permafrost was recognized as a structural support for buildings, roads and pipelines. This situation is changing and structures, which were once quite solidly placed, have shifted badly. Permanently frozen soil also prevented erosion. When the permafrost melts, the structural value of the soil is lost. It is anticipated that permafrost will continue to be lost and its structural benefits will continue to be lost with it.

Along with the loss of certain structural benefits, melting permafrost will also release its methane load. Permafrost in some areas of the world, particularly Russia, retains a huge amount of methane. There are an estimated seventy billion tons of frozen methane gas in the west Siberian bog alone – and it began to melt in 2005. (10) Methane, as a greenhouse gas, is about twenty times as potent as carbon dioxide. (The permafrost also includes CO_2 which will also be released as it melts. (14))

11.4.3   Desert Formation

As the Earth warms, some areas will become too warm for either continual human occupation or for plant growth. As the temperature goes up, moisture is driven away. If it is not replenished, the ground dries out and plants will not grow. When plants are lost, the water table sinks. When the water table sinks, the ground dries out. This is a type of positive feedback cycle and is very difficult to reverse. It is therefore anticipated that as the Earth warms up, present desert areas will get larger and new deserts will form. (11) In particular, the state of California is becoming desert and will not be able to supply the vast quantities of food that it does now.

The loss of plant cover is also a factor augmenting the atmospheric buildup of CO_2. Plants, in order to build their structures, extract CO_2 from the atmosphere. As plant cover is lost, less CO_2 is removed from the air which reinforces the warming trend. If the warming trend results in plant cover loss, another type of positive feedback cycle comes into operation.

11.4.4   Arctic Seaway

As the Earth warms up, the amount of ice in the Arctic Ocean will diminish. In fact the thickness of the pack ice in the Arctic has reduced considerably in the last one hundred years. Also, the southward-drifting pack ice also melts much sooner leaving seals with no place to rest and polar bears with no seals to eat.

(12) The bears therefore go to shore sooner and do not become as heavy as they did a few decades ago. (17) More importantly, as the ice melts, ships will be able to navigate the Arctic waters longer every summer season. Trial runs of large ocean ships are already planned for Hudson Bay and this type of activity will only increase in the coming years. Soon, it will become commonplace for large ships to go through the Arctic Ocean to the Pacific Ocean instead of through the Panama Canal.

11.4.5 Temperate Zone Shift

For climatic discussion purposes, the Earth is commonly identified as having various zones. The one with generally the most pleasant climate is called the Temperate Zone. In this zone the temperature during the winter may dip below freezing but it will not remain there for very long. Summers are expected to be warm and may even be hot but not as extreme as tropical areas, which are closer to the equator. As the Earth becomes warmer, the temperate zone will shift northward. Spring will come sooner and fall will not be as cold. The moderate climate of this zone will then be enjoyed by people living further north. Crops will be harvested from the warmer northern regions and second crops will increasingly be viable from the present temperate zones. (15) All of this seems like a positive development except that the regions, which were once temperate, will gradually become sub-tropical and that may not be quite so desirable. If the formerly sub-tropical areas become tropical and the formerly tropical areas overheat, either populations must make major shifts northward or they will simply not survive. It is anticipated however, that major population shifts will not occur simply because masses of people cannot easily move. Therefore, since the formerly tropical areas are shifting toward desert, large populations of people will expire. (13) The people of Africa cannot move en masse into Europe. The people of India cannot easily cross the mountains and displace the people of Central Asia. As the Chinese wheat crops increasingly fail, great hordes of Chinese will not be able to occupy Siberia as easily as they might wish. It is therefore anticipated that some populations of the world will be decimated by a continually-warming Earth.

11.4.6 Flora and Fauna Shift

As the basic climatic zones of the Earth shift, so too will the types of plants and animals, which live in any particular area. For example, as winters become less severe, animals, which once lived further south, will be found in areas, which were previously too cold for them. An example is the opossum, which is now found across southern Ontario. When winters were more severe, the opossum could not survive in this area because its furless ears and tail would freeze and cause its death. A few had always survived in barns where they could benefit from the protection, which the barn provided. However, as the weather warms, they can survive with less protection and consequently will expand their range northward. Also there are numerous other examples of insects, birds and animals being spotted further north than they would have been expected, only a few short years ago.

11.4.7 Snowfall Changes

Snowfall is certainly a sign of cold weather. However it is not a sign of extremely cold weather. There are two conditions, leading to maximum snowfall which are; temperatures near freezing, and open water. In fact maximum snowfall will occur when the temperature is just below freezing. The reason is simply that warmer air can hold more moisture. As air becomes colder, it holds less moisture. This is, in fact, why it rains or snows in the first place. Above freezing, it will rain. Below freezing it will snow. Just below freezing, maximum snowfall occurs. It follows that as the weather warms, regions where the temperature is just below freezing will experience more snowfall than regions, which are further north and consequently colder. The second condition for maximum snowfall is open water. When a lake or river forms a crust of ice, it can no longer evaporate moisture into the air. Subsequently, there will be a reduction in snowfall over adjacent land areas.

An example, where both of these factors are relevant, is the city of Buffalo. Buffalo is situated at the eastern end of Lake Erie. Before Lake Erie freezes in the fall, Buffalo experiences large snowfalls. Open water is also indicative of modest temperatures. When the air temperature is just below freezing, more moisture is retained and more snowfall results than when it is colder. The great snowfalls, which regularly occur in Buffalo, are the result of both of these factors at work.

It is therefore anticipated that as the climate warms, certain lakes will remain open all year and adjacent land areas will experience greater snowfall than they did when these lakes were frozen for much of the winter. The greater snowfall will be augmented by the milder temperature, which will enable the air to hold and carry more moisture from the open water to the land. Milder winters, which result in more snowfall is, in some ways, the opposite of what one might expect. However, this is what will happen, and such regions will develop farther northward with a warming climate.

11.4.8 Hurricane Increase

As the Earth warms, there will be an increase in the intensity and frequency of hurricanes. (16) This will simply be the result of the heat from the tropics being transferred in greater quantities towards cooler regions. Whenever an area of the world becomes warmer than another area, heat will flow from the warm area to the cooler area. As the Earth warms, the great oceans in the tropics will get warmer. Since polar areas will still be colder, heat will be transferred in greater amounts to these colder areas. Hurricanes are very similar to thunderstorms and essentially form in the same way. As warm moist air rises, it rotates. Over large open areas, the rotational action from separate rising air columns becomes coupled and soon a large portion of atmosphere is rotating as one system. The power to drive this rotation is provided by the heat in the warm water. A warmer Earth will result in warmer water and warmer water will provide more energy to power more and stronger hurricanes. Hurricanes will therefore become more frequent, larger and more destructive.

11.4.9 Water Vapor Increase

As the air gets warmer, it can hold more moisture and moisture is also a greenhouse gas. Just like CO_2, moisture traps heat in the atmosphere and prevents it from escaping, further enhancing the warming trend.

11.4.10   Albedo Effect

A fourth type of positive feedback loop involves the albedo effect. The albedo is the fraction of sunlight which reflects from the Earth. Sea ice, which is covered by snow, has a very high albedo of 0.8 or 0.9. This means that almost all sunlight will be reflected. However, as the sea ice melts, more open water appears, which has a much lower albedo of perhaps 0.1 or 0.2, indicating that most of the Sun's energy is being taken up by the water. The water will therefore warm up. As the water warms, more sea ice will melt producing more open water. Then, the open water will warm up and melt more ice. The open water will release more moisture into the atmosphere and cause the local climate to be more moderate. A warmer Arctic Ocean is a warmer world.

11.4.11   Antarctic Ice Shelf

The Antarctic ice shelf almost completely surrounds Antarctica. Just a few years ago it not only surrounded Antarctica but in many places extended seaward for many miles. The ice shelf consists of freshwater ice and its relatively level upper surface is several meters above sea level. A further result of a warming ocean will be the loss of the ice shelf and it is currently disintegrating very rapidly. Even a few years ago it was much more extensive. During recent years very large sections have broken off and drifted away. As an example, in February, 1997 a vast crack opened in the Larson B ice shelf. 'A 27-mile by 3-mile chunk of the shelf later broke away ' (18) The disintegration of the entire ice shelf will continue as the ocean warms.

11.5   Polar Cell Demise

The atmosphere of the Earth includes three atmospheric circulation patterns called Hadley Cells, Ferrel Cells and Polar cells. Included in the North Polar Cell is a column of frigid air from up near the tropopause which descends and flows southward across the frozen Arctic Ocean and all surrounding lands as far south as the Canada-USA border. However, the Earth is warming up. The Arctic Ocean becomes more ice-free every summer. As it warms, ocean currents from farther south will increasingly find their way into the Arctic Ocean further augmenting the warming trend. When the warming process has proceeded so far that the Arctic Ocean is ice-free for most of the year, the Polar Cell will start to peter out. When the Arctic Ocean remains open for even longer, the flow of very cold air from high in the atmosphere over the North Pole, will be opposed by the much warmer air near the surface. One result of the partial stalling of the Polar Cell is that another positive feedback loop will have developed, (i.e. the more it stalls, the more it will be caused to stall.) further enhancing the warming trend with the result that the stall period will become progressively longer every summer season. The final result will be that the Polar Cell will not exist at all and that the Ferrel Cell, which currently ascends further south near the Canada-USA border (along with the Polar Cell) will continue much farther north before ascending.

There will be several implications to this development. First, the flow of warm air from the southern USA will continue farther and farther north until it includes most of Canada. This will cause Canada to experience more and longer heat waves and periods of drought every summer.

Secondly, the Northern Polar Jet Stream, which occurs where the upward atmospheric flow of the Polar Cell and the Ferrel Cell diverge, (35) will shift further north. All airline traffic which currently takes advantage of the jet stream's powerful flow will no longer benefit from the free ride. It is predicted that this situation will continue until the Northern Polar Jet Stream no longer exists.

When this happens, the Polar Cell atmospheric circulation pattern will have met its demise and the Ferrel Cell will simply continue further and further north until it is well out over the Arctic Ocean before rising.

The Polar Cell and the asteroid shower/ice age are directly connected. Prior to this unthinkably disasterous event, the Arctic Ocean was ice free all the time. Air that drifts over a warm surface tends to rise and this would have stalled any tendency for the Polar Cell to develop. Polar Cell circulation involves the air in the upper troposphere sinking from up near the tropopause down to the surface prior to drifting south again. Since that probably wasn't allowed to happen, any atmospheric circulation at that time would have been dramatically different and there would have been neither Polar Cell circulation nor Northern Polar Jet stream movement. The asteroid shower/ice age changed all this. First the land and sea became dark. While the darkness caused a chill over the land the suddenly over-heated ocean proceeded to evaporate prodigious amounts of water. At first it simply rained right back down on the Arctic Ocean. The warmth of the water ensured that the nearby Arctic Islands would receive rain instead of snow so most of them remained ice free throughout the ice age. However the dark persisted. This together with the high evaporation rate caused the ocean to cool down. As volcanic activity abated the water was allowed to cool down even more. Soon a crust of ice formed. When this happened, atmospheric temperature plummeted. The snow had already piled up to tremendous thicknesses on the land which caused the situation to worsen. As the darkness persisted for several years the Arctic Ocean became ice-covered all year. This allowed and caused the atmosphere over the massive ice surface to drop and thereby established the Polar cell. It has remained in place ever since but now that the Earth is warming up it will not persist. It will peter out and Canada will suffer more prolonged heat waves every summer.

11.6 Pacific Impact Sites

There are two reasons that major asteroid impact sites are expected to be found on the Pacific Ocean floor. The first is simply that the Pacific Ocean is very large and would reasonably be expected to be the recipient of a reasonable portion of all of the asteroids which have struck the Earth. The second is that both North and South America have mountains which run parallel to their Pacific Ocean coastlines from the far north to the extreme south. These mountains were probably formed by shockwaves which traveled partway across the Pacific Ocean floor before slamming into continental material. While the Pacific Ocean is home to several antipodes including the Tanitz Basin in Kazakhstan, Vredefort Crater in South Africa, the 200 mile diameter impact site surrounding The Czech Republic, and the Kara impact site in Russia, the fact remains that the shockwaves which would have emanated from a ground zero would have always been much greater in magnitude than the shockwaves from an antipode. It is therefore anticipated that major asteroid impact sites will be found on the Pacific Ocean floor.

11.7 Positive Feedback Loops

There are several positive feedback loops caused by and accelerating the warming trend.

i.   Warmer air will result in warmer oceans, which will release more CO2 causing more warming.

ii. Warmer air holds more water vapor, which traps more heat causing more evaporation.

iii. Open sea water absorbs more of the Sun's energy than sea ice. The warmer water melts more ice resulting in more open water which heats up even more.

iv. The warming Earth is melting the permafrost, releasing methane, a greenhouse gas twenty times as effective as CO2 in warming the atmosphere. More warming = more melting = more methane release = more warming.

v.   The warming Arctic Ocean will stall the Polar Cell. As it starts to stall, the Arctic Ocean will be allowed to warm even more. As it warms the polar cell will stall more and more. In this manner the warming contributes to the stalling and the stalling contributes to the warming.

vi. All of these sequences are called positive feedback loops because the result of each activity causes that activity to increase. They are also called viscous cycles because the worse they get, the worse they get

11.8 Axial Tilt Increase

Current wisdom advises that the Earth's axial tilt will decrease in the future before increasing again with a 41,000 year cycle. This is expected to happen because of minor changes in the Earth's orbit and the Moon's orbit. Unfortunately measurements of the Earth's axial tilt have only been made for a small fraction of that period of time so that firm conclusions based on measurement are not available. However whether or not the predictions are valid, the expected variation will be over-powered by the decelerating Earth. The Earth is slowing down. (36) Due to this slowdown, it is predicted that the Earth's axial stability will be completely lost within a time-frame of much less than 1,000,000 years. (This matter is discussed in detail in 'Fairytales for Adults' and 'The non-Myths of the Bible'. When that happens the Earth will not be habitable at all!

12. Conclusions

12.1 Introduction

While the life processes of each and every type of life on Earth are extremely complex, the complete set of circumstances, which enable life to continue and thrive, are even more complex. Each of these circumstances can be compared to a window, which lets light shine through. On either side of the window the light does not get through. As long as the complete set of life-enabling circumstances is present and each one is within its appropriate range, the conditions will be satisfactory for life to continue. Collectively, these conditions are like a set of windows, which are lined up behind each other in such a way that light can shine right through all of them. However if any one of them were to shift one way or the other, the light would not get through and the necessary set of life-enabling circumstances would be incomplete. Life would therefore terminate.

Several of the windows of life are one-sided. In this case if there is a shift one way it will not really matter but if there is a shift the other way, it will matter and the requirement for completeness will be broken. Whether the windows are one sided or two sided, is of little consequence to the main requirement that all of them must simultaneously be within their respective life-enabling ranges or life will not carry on but, will instead, be extinguished.

12.2 Asteroid Energy Distribution

A large incoming asteroid has associated with it, multiple life-destroying effects. Since it would be approaching at very high speed, (popular knowledgeable opinion places incoming asteroid speeds in the 25,000 to 100,000 miles per hour range), as it passed through the atmosphere, a very high-pressure atmospheric shock-wave would be generated. This wave would be powerful enough to totally flatten entire forests for many hundreds of miles from the impact location and exterminate vast herds of animals instantaneously.

If the asteroid crashed into an ocean, a water splash would be generated. The high speed means that, as above, water would also leave the impact site at a high speed. Some of it would be going fast enough to leave the Earth completely. Most of it would come back down however, and this could occur anyplace on the surface of the Earth. Some of it would come down as rain and some of it would come down in great life-destroying globules as devastating as exploding bombs.

The rapid passage of the asteroid through the ocean would result in a very large temporary opening in the ocean, all the way to the bottom – a blast effect. The ocean floor would be laid bare for several asteroid diameters in every direction.

Penetration into the ocean floor would result in a mud-rock splash following the water splash. As the asteroid crashed deeper into the ocean floor, the material, which was directly in front of it would be forced sideways. Some of it would turn upwards and form the mud-rock splash. Further out, material would move laterally, elevating the surrounding ocean floor.

Asteroid contact with the solid crust of the Earth would result in the production of innumerable flying boulders. Some would be material from the fracturing asteroid. Some would be broken pieces of Earth-rock. The scenario would be similar to driving a giant hammer, as fast as possible, into a pile of gravel. The gravel would splash outwards in every direction for distances which were much greater than the diameter of the hammer. Similarly, rock which was forced into the air by a crashing asteroid would fly outwards for many miles. One of the largest craters in the world is near Sudbury, Canada. This crater is 250 km. across which means that rocks flew into the air from the entire area defined by this very large crater rim. How far did they go? Some would fall within another 100 km. but some would have gone much farther and appear completely out of place where they landed. Erratic boulders are found all over the world and asteroid impacts could account for all of them.

Two types of water wave would result from ocean impact. First, there is the primary wave, which is the result of the asteroid pushing water out of its way. All of the water, which would be pushed aside and which does not contribute to the water splash, would pile up in a great circular mountain of outward-moving water. This mountain of water would be augmented by the secondary effect of the ocean floor being elevated under it and the elevation factor would contribute to the overall energy of the wave thereby increasing its destructive force. The size of this wave would be commensurate with the size of the asteroid. Therefore it could be several miles high. Most continental land masses would not be able to stop it and it would wash right over some continents and almost totally inundate others. Unconsolidated material such as topsoil, subsoil, sand and gravel would be stirred up and relocated. No vegetation would be able to withstand the force of this flowing water.

When the asteroid hit the ocean surface and then the ocean floor, sound waves would be generated which would move outward through the air, the water and the ocean floor. These sound waves would all be travelling at different speeds and would therefore arrive at remote locations at different times.

The lateral displacement of rock from the impact site would be the beginning of an earthquake because the crust of the Earth would be forced sideways. Then after the asteroid had punched right through into the interior, the crust would rebound. Thereafter, for an indefinite period of time, which could be weeks, it would continue to vibrate back and forth. The shaking would be so violent that trees would be shaken right out of the ground. Boulders would fly into the air and the crust of the Earth would be cracked into numerous pieces. These stress-relieving cracks would permit some regions to dislocate vertically upwards and some areas would dislocate vertically downwards. Rift valleys would thereby be formed as well as escarpments, plateaus and mountains. An entire continent would shake. Any continental dislocation, which was adjacent to water, would produce tsunamis. These water waves would then propagate outwards and would travel across entire oceans and continue right over low-lying land.

The energy of a large high-speed asteroid is extremely great. (30) If the diameter of the asteroid was comparable to the thickness of the crust of the Earth, it would punch right through the crust and penetrate deep into the interior. It would eventually come to a stop, but its effect on the molten interior of the Earth would continue. A high-pressure wave in the magma would be generated and it would propagate outward from the asteroid pathway in all directions at high speed. In this manner the asteroid's energy would be absorbed by the Earth and spread throughout the entire sphere of the Earth which would therefore be less likely to displace from its orbit.

The interior pressure wave would spread throughout the interior of the Earth in every direction. It would propagate outward from ground zero and travel under the crust. The high pressure of this wave, would stress the crust at the same time that the earthquake was shaking the crust. This combination would result in three-dimensional displacement of the crust and include, as before, mountain building, escarpment building and rift valley formation. Within minutes, the pressure wave would come up under remote continents on the far side of the Earth. These too would be uplifted and mountains, rifts and escarpments would form.

The interior pressure wave would also come up under remote oceans. Since ocean crust is not as strong as continental crust, it would be uplifted even more. It would also be easier to crack and if the greatest concentration of stress was at the middle, half way between two continents, it would crack and displace upwards at the middle. The resulting uplift would generate another water wave, which would travel towards the continents in both directions. Magma would be forced up through the opening. In the Atlantic Ocean case, a great ridge would be formed right down the middle.

The upward displacement and sudden fracturing of the ocean floor would also result in a horizontal displacement. The ocean floor would be smashed against the continents, with a high-energy impulse similar to a giant sledgehammer hitting a steel bar. If the surface layers of the continent were recently deposited sedimentary materials, they would be quite weak. Such continental material would therefore compress, fold and buckle and mountains would be formed. If the ocean crust dislocated downward, a trench would be formed parallel to the continental shorelines. In the very minimum, the basalt layer forming the ocean floor would buckle, possibly far from shore, but also near the shore wherever the compressive stress was too much for the ocean floor to withstand.

The upwelling pressure wave would reflect from the underside of the crust and then travel back through the interior. It might travel back and forth several times. Wherever this pressure wave contacted the underside of the crust it would stress the crust and additional faults could form. Every time the interior pressure wave approached one of these faults, material from the interior would explode or ooze up onto the crust and some would escape into the atmosphere. This volcanic activity along the edges of the massive plates formed by these faults would be reactivated every time an interior pressure wave surged up under a plate interface. The plates of the crust of the Earth would be jostling on a pulsating molten interior much the same as pack ice is jostled by a passing water wave. This pulsating and jostling could continue for months and possibly not settle down for years.

In these ways the great energy of a large asteroid or comet striking the Earth would be distributed throughout the entire Earth.

12.3   Major Impact – No Survival

The evidence shows that the conditions for life to exist on Earth at one time were ideal. The Earth was universally warm. Atmospheric pressure was greater. Plant growth was abundant. Swamps, forests and oceans were teaming with life. The surface of the Earth was protected from incoming cosmic radiation as well as ultraviolet light. The evidence also indicates that the life-spans of both animals and people were greater than they are at the present time and that many types of life were very robust. The "windows of life" were all lined up well within their respective life-enabling ranges.

The evidence also shows that the Earth has been through unbelievable chaos. This evidence is so very widespread that there is no location on the Earth, which is more than a short distance away from some significant evidence of chaos. The total number of creatures, which have been killed and immediately buried, as evidenced by the fossil record, is impossible to number. Something unthinkable happened to the Earth.

It has been estimated that the total devastation from a single asteroid, approximately ten miles in diameter, would be sufficient to decimate life in an entire hemisphere. (7) This is suggestive that if an even bigger one struck the Earth, the devastation would be sufficient to decimate life on the entire Earth. Instant local death would result from the immediate effects of the impact. Soon-to-follow death would result from the spreading blinding flash, the high-pressure freight-train effect of the atmospheric pressure wave, the earthquake, the water wave, the shower of rocks, and the mountain-forming crustal deflections. A massive cloud of dust and debris from the impact would rise high up into the atmosphere and start to spread out over the Earth. The impact cloud would be augmented by volcanic clouds. Soon the entire Earth would be enveloped in thick choking cloud. Death during the next few days would result from choking on an atmosphere which was full of dust and particles of rock. There would not be any clean air to breath over the whole Earth. A post-impact winter would set in right away. The cloud would bring darkness and because the energy from the Sun would be blocked from reaching the ground, it would become cold. The cold would kill plants and freeze drinking water as well as any unprotected animal life. Starvation would follow. Knowledgeable commentators state that between 75% and 96% of all plant and animal life on the Earth would die following the impact of a single large asteroid. (1) (2) For example, the Chicxulub asteroid impact event is claimed to have resulted in 'mass extinction, which included the dinosaurs.' (9) Or 'This is exactly what seems to have happenedwhen the dinosaurs were extinguished along with 70 percent of all other animal and plant species.' (18) On the other hand, from the expected results of the Chicxulub impact, it is a wonder that any life is left on the Earth at all. Between the direct devastation of the impact and the following post-impact winter, how did anything survive? A survival scenario following a major impact has never been offered. Is therefore concluded that, as a result of one major impact with the earth, animal life would be annihilated.

12.4   Multiple Impacts - Single Event

12.4.1   One Mass Extinction

Even small asteroids are feared much more than their size would suggest. The impacts of objects larger than 1 kilometer would have global consequences including climate change. Most inhabitants of the planet would be killed. (19) Asteroids which were over one kilometer in size, and therefore potential causes of mass extinctions, are expected to strike at least every few hundreds of thousands of years. (20) If the generalization that there is a ten to one ratio between asteroid diameter and crater size, (which is probably reasonable for small asteroids at least) the evidence included in the Earth Impact Database (EID) indicates that the Earth has experienced at least 70 impacts involving objects greater than one km. diameter. (29)

With respect to the above declarations, an estimate can be carried out to determine how many animals would survive the repeated impacts that the Earth has endured. For the sake of discussion, if it could be allowed that there would be, more conservatively, only an 80% extinction accompanying all asteroid impacts which produced craters over 250 km. in diameter, there would have been four (reading from the EID (19,20)) 80% extinctions. Further, if a 50% extinction accompanied all asteroid impacts which produced craters between 150 and 250 km. in diameter there would be another mass extinction. (This is conservative in consideration of the 70% quoted above.) If all eleven of the craters between 50 and 150 km. diameter indicate 30% extinctions and the 17, between 25 and 50 km. diameter indicate 15% extinctions, the estimate can be carried out. The Manson, Iowa, crater is in this category at 35 km. diameter, and is considered to have been formed by a stony meteorite about two kilometers in diameter. (10) 'An electromagnetic pulse moved away from the point of impact at nearly the speed of light, and instantly ignited anything that would burn within approximately 130 miles of the impact (most of Iowa). The shock wave toppled trees up to 300 miles away...and probably killed animals within about 650 miles.' (11) This description, while consistent with others, does not mention any fatalities resulting from ongoing effects such as the dust cloud. Also, it is highly probable that there are crater sites which have not yet been discovered. Further, the Earth Impact database indicates that there are at least 40 more with diameters between 5 and 10 km. The following calculation will not include any entry to recognize these last two factors, which would only make things worse.

The oceans of the world account for approximately 70% of its surface area. Therefore it would be reasonable to expect that impacts would also have occurred in the oceans. In particular a crater was recently discovered under the Indian Ocean. The Burkle Crater is only 18 miles in diameter and is well buried below the bottom of the sea but it makes one wonder how many more there are on the ocean floor. (12) Further, the ocean floor includes numerous areas called igneous provinces, many of which are much larger than the largest craters so far discovered. These are places where molten lava from the interior has come up and spread around. (13) Do any of these provinces cover an impact site?

While any of these additional impacts would also have resulted in much destruction, the following calculation will ignore them. The portion of surviving species from an 80% extinction is 20%. Similarly

389

from a 50% extinction the surviving species are also 50%. From 30%, 70% remains and from 15%, 85% remains. Therefore the calculation appears as follows.

> 20% (Tanitz Basin, Kazakhstan, 350 km.)
> 20% (Sudbury, Canada, 250 km.)
> 20% (Vredefort, South Africa, 300 km.)
> 20% (Czechoslovakia, 320 km.)

These four events would therefore leave 0.2 x 0.2 x 0.2 x 0.2 = 0.0016 or 0.16%. This means that only a fraction of 1% of all animal life on the Earth would be left as a result of the four largest impacts. Between 50 and 150 km. diameter there is only one event. Therefore from the above assumption of 50% survival for this category, only 50% of all animal life would be left after the asteroid in this category hit the Earth.

> 50% (Chicxulub, Mexico, 170 km.)

It is noted that this particular event is credited with completely wiping out all dinosaur life. '…a city-size meteorite slamming into the region … generating a global cataclysm. Giant waves inundated shorelines, and fine dust blotted out the Sun and cast the world into darkness. Most scientists now accept … mass extinction, which included the dinosaurs.' (9) The 50% being used for the calculation is therefore very conservative, as mentioned.

There are twelve craters between 50 and 150 km. diameter, to which a 30% extinction was assigned. Therefore, for this part of the calculation we have the following.

> 70% (Acraman, Australia, 90 km.)
> 70% (Beaverhead, USA, 60 km.)
> 70% (Charlevoix, Quebec, 54 km.)
> 70% (Chesapeake Bay, 90 km.)
> 70% (Kara, Russia, 65 km.)
> 70% (Kara-Kul, Tajikstan, 52 km.)
> 70% (Morokweng, South Africa, 70 km.)
> 70% (Puchezh-Kalunki, Russia, 80 km.)
> 70% (Tookoonooka, Australia, 55 km.)
> 70% (Manitouigan, Quebec, 100 km.)
> 70% (Popigai, Russia, 100 km.)
> 70% (Ries, S. Bavaria, 80 km.)

These twelve events only leave 0.7 x 0.7 x 0.7 x 0.7 x 0.7 x 0.7 x 0.7 x 0.7 x 0.7 x 0.7 x 0.7 x 0.7 = 0.0138 which is only 1.38% of all animal life remaining. In a similar manner, the 18 impact craters between 25 and 50 km. would only leave a small portion surviving.

There are also eighteen craters between 25 and 50 km in diameter.

85% (Araguainha, Brazil, 40 km.)
85% (Carswell, Canada, 39 km.)
85% (Clearwater East, Canada, 26 km.)
85% (Clearwater West, Canada, 36 km.)
85% (Kamensk, Russia, 25 km.)
85% (Keurusselka, Finland, 30 km.)
85% (Manson, USA, 35 km.)
85% (Mistastin, Canada, 28 km.)
85% (Mjoinir, Norway, 40 km.)
85% (Montagnais, Canada, 45 km.)
85% (Saint Martin, Canada, 40 km.)
85% (Shoemaker, Australia, 30 km.)
85% (Slate Islands, Canada, 30 km.)
85% (Steen River, Canada, 25 km.)
85% (Strangways, Australia, 25 km.)
85% (Woodleigh, Australia, 40 km.)
85% (Yarrabubba, Australia, 30 km.)
85% (Sahara, Eygpt, 31 km.)

Therefore, these 18 events would only leave 5.28% of all animal life remaining.

When all of the above events are considered together, only 5.28% of 1.38% of 50% of 0.16% would remain. This is approximately 0.00005%.

In everyday language this is less than one ten-thousandth of the original number of creatures. If this calculation is one thousand times too pessimistic for whatever reason, it would still leave only 0.05%. Since there were at least four impacts even greater than Chicxulub, the above calculation does seem to line up with the following comment. 'Every 300,000 years, a one-to-two-kilometer-wide asteroid crashes, initiating a short-duration global winter. Every 100 million years, a bigger asteroid hits, producing the world-wide calamities that have punctuated evolutionary history. There are 14 known mass extinctions in Earth's geological record.' (2) One is inclined to ask how many times this sort of thing can happen and still have anything left. Of course, a different set of assumptions could be made and a different result would be obtained. It could be argued that some of these impacts occurred prior to any animals being on the Earth. One commentator offsets this idea by declaring that 'evidence gathered from around the world related to earlier (i.e. earlier than Chicxulub) mass-extinction episodes, 230, 365, and 445 million years ago.' (1) This clearly implies that there were enough creatures here 445 million years ago to realize a mass extinction if they were all or mostly killed. On the other hand, even more impacts than those listed above were probably involved because it is presumptuous to think that all impact sites are currently known. In particular, the Earth is mostly covered by water. While a few craters have been found underwater, why wouldn't there be even more underwater than there is on land? In addition, more craters will certainly be discovered on land. In particular, the Chiczulub Crater and the Chesapeake Bay Craterare very recent discoveries. Unknown sites, like these two, may be so completely buried that they will not be found until some totally unrelated activity leads to their discovery, even though they may be as large as Chicxulub. No matter how many craters were actually involved in mass extinctions, the fact will still remain that following every major impact there would be very few, if any, survivors. Then, when the next impact happened, only a small percentage of that number would be left and so on. The so-far-unmentioned problem is that the surviving cohort would not include any large animals but would be dominated by vermin and other small creatures which, simply because of their size, were better able to avoid destruction.

In the subsequent paragraph 12.5.1, it is concluded that the world-enveloping dust cloud resulting from every major asteroid impact must have been brought down in a timely manner or life would not have survived at all. It is difficult to hypothesize how this type of dust cloud could have been brought down, let alone in a timely manner. If even one dust cloud was terminated quickly, it would have been unusual and unexpected. How can the unexpected termination of dozens of dust clouds be explained?

Since the Earth has experienced several impacts, which are in the very large category, life must have been extinguished or virtually extinguished, several times. This is suggestive that life did not really result from evolution because there would not have been enough time for a diverse range of species to have evolved prior to the first major impact. Then, starting with the surviving cohort from the first impact, another diversity of species had to be in place prior to the second impact and so on. Mass extinctions every 50 or 100 million years would have made it very difficult for evolution to proceed properly, partly because, after every wipeout, a different mixture of survivors would have been the starting point for the next period. On the other hand, if life on Earth was originally created, the setback of numerous wipeouts would have required virtually total creation all over again after every wipeout. Whatever explanation is accepted must account for the present diversity of life on the Earth. In fact it must account for the known diversity which was even greater in the immediate past because of the numerous species that are known to have recently become extinct.

There are several very large impact sites on the Earth. With each of these impacts, the "Window of Life" would have closed, if not from the impact events directly, at least from the subsequent post-impact dust cloud. Could it have closed repeatedly? The probability that a diversity of life could have spontaneously developed prior to the first impact, is mathematically virtually zero, so the possibility that it could redevelop after subsequent impacts is just another zero. Any declaration that life repeatedly redeveloped into a multitude of different species from a few vermin-like and other stragglers, is preposterous. Life is just too complex. Could a great diversity of life forms have been repeatedly created? While this certainly is not very likely, neither is there any evidence to support such a hypothesis. **Therefore, it is concluded that; there has only been one mass extinction of life on Earth.**

12.4.2   The Great Ice Age

The Earth has experienced an ice age. In order for this to have happened, possibly 500 feet of ocean water evaporated and became tied up in the massive glaciers of that time. As discussed in the chapter entitled,

Post-Impact Developments, in order for 500 feet of ocean to become ice, much more than this actually evaporated. Obviously not all of the evaporant and its subsequent precipitant would have become ice. It would be generous to suggest that 25% of it did. This conservatively means that more than 2000 feet of ocean evaporated in order for 500 feet of it to become ice. This is a very large amount of water. In order to evaporate such a large amount of water, a correspondingly large amount of heat would be necessary. The required heat came from volcanic activity which appeared in several forms including; igneous provinces, underwater rifts and underwater cone volcanoes. Impacting asteroids caused all of this to happen. Would enough volcanoes be activated from the impact of one asteroid, to heat the ocean enough to evaporate more than 2000 feet of it? No they would not. It would require a multiple-impact event to produce enough volcanic activity to evaporate the 2000+ feet of ocean required for the Great Ice Age. Conversely, in order to have a world-wide ice age it would be necessary to cause the near-simultaneous eruption of hundreds of volcanoes which would require a multiple-impact asteroid shower.

12.4.3   The Greenhouse Gas Effect

A single major asteroid impact would have produced a globe-encircling cloud. It is entirely likely that it would also have stirred up several volcanoes the clouds from which would have augmented the impact cloud causing further cooling. This would have been a recipe for disaster because the cooling effect of the clouds would have reduced the overall amount of water vapor in the air. With the loss of this important greenhouse gas the Earth would have chilled to well below freezing within a short time. (Please refer to Climate and the Oceans by Vallis p. 165) While any underwater volcanic activity would have warmed the ocean partially offsetting this loss, it is not likely that a single impact would have been enough. Therefore, the Earth would have spiraled into a deep cold and simply frozen solid. While the cloud cover would have blocked the Sun, more cloud cover would not block the Sun more because once it is blocked, more cloud will not make much difference. Therefore multiple impacts would not have had much more of a cooling affect than a single impact. However multiple impacts would have had a significant effect on ocean heating. In fact an asteroid shower would have caused the ocean to boil in places and this great heating effect would have not only provided moisture for the great ice fields but it would have provided an abundance of water vapor over the ocean to offset the chilling effect of the loss of this important greenhouse gas over the land. Therefore, on a global basis a balance would have been retained and the Earth would not have gone into the deep freeze as is obviously the case. A multiple-impact event would have provided enough heat to offset the chilling effect (on a world-wide basis) so the Greenhouse Effect could be retained to prevent the Earth from going into a one-way deep-freeze.

12.4.4   An Asteroid Shower on Earth

From the above lines of reasoning, the conclusion is reached that; The totality of major asteroid impacts on the Earth was a singular event. It was an asteroid shower, which might have occurred over a period of several months, but never-the-less it happened as a singular all-inclusive event and then it was over. This asteroid shower is referred to in the literature as the Period of Heavy Bombardment. (31) The Earth and its climate were dramatically changed but when everything settled down, life on Earth was again possible. The "Window of Life" reopened, albeit much narrower than before. Since even a single impact would have been sufficient to change the topography and environment of the entire Earth, we are fortunate that the changes which occurred ultimately resulted in a set of circumstances, which re-enabled life to exist.

The above singular-event conclusion is paralleled by other singular-event conclusions, concerning the solar system.

12.4.5   Mars

On one side of Mars there are three very large impact craters. The Hellas Impact Basin is the largest with a rim diameter of about 1300 miles. The Argyre Impact Basin is smaller at 1120 miles across, but still very large compared to anything on Earth. The third largest is the Isidis crater with a diameter of 680 miles. While these three large craters are significant because of their size, the remaining craters in the same hemisphere are significant because of their numbers. The "Hemisphere of Craters" includes 91% of all of the craters on Mars which are 20 miles in diameter or larger. There are 3068 craters on the "Hemisphere of Craters" side. Directly opposite these significant features, at the antipode, there is a group of four volcanoes on an elevated region called the Tharsis Bulge, a huge swelling in the Martian surface. 'It is approximately 6 ½ miles high in the center relative to the general crust of Mars and it is about 3000 miles in diameter.' (14) The four volcanoes located on the bulge are equally impressive in magnitude. Olympus Mons is the largest and rises 85,600 feet above its base. The escarpment defining its outer edge is 20,000 feet high. (15) The other three; Arsia Mons, Pavonis Mons and Ascraeus Mons are less than half as high but still very high compared to anything on Earth. Extending away from the volcano area is a trench 4000 km. long flanked by side trenches and features that appear similar to dry river beds. 'Extending eastward from Tharsis is an immense canyon system, Valles Marineris, while radiating outward from it are large arrays of tensional faults (graben); both are presumably related to the formation of Tharsis itself. North of the canyons are numerous fascinating outflow channels.' (16) The impact basins on the other side also have river-like features including Surius Valles, Dzigai Valles, Palacopas Valles and Nia valles. All of these features are declared to have been formed 'during the Heavy Bombardment period.' (15) Incidentally, the crater count for the volcano side is only 292, which is about 9% of the number on the "hemisphere of Craters" side. (28)

Three massive craters as well as more than 3000 others are on one side of Mars. On the other side there are less than 300 but there are four huge volcanoes and miles of trenches large and small, none of which have any significant evidence of being impacted at all! None of this evidence is coincidental.

The conclusion is therefore drawn that: **The significant surface features of Mars were all formed during a singular all-inclusive event.**

12.4.6   The Moon

Similar evidence is available from the Moon. All of the Maria on the Moon are on one side. The Maria are large "seas" of lava, most of which are accompanied by mass concentrations (mascons) beneath the surface. (17) This evidence leads to the conclusion that the Maria are impact locations and that the impacting objects are still in the vicinity. Each asteroid was large enough to punch through the solid surface and right into an interior molten region.

These lava flows are virtually void of craters, which is the reason that the Maria were noticeable in the first place. If the impacting objects had been randomly spaced out over time, the Maria would show various degrees of cratering. They do not. They appear fresh in the geological sense without craters of any significance. If the arrival of the Maria-causing objects had been spread out over eons of time, a much more random Maria pattern on the surface of the Moon would be expected. Instead, the massive Maria structures are all located on one side.

Further it must be noted that the Maria half of the Moon faces the Earth. The reason given for this phenomenon is that the geographical center of the Moon is offset from its gravitational center. The Maria face, (i.e. the near face) of the Moon has therefore become locked toward the Earth. The cause of the imbalance is the lunar mascons which are located beneath the Maria (6) as well as the bulge on the other side This evidence further supports the notion that the Maria were formed by the impact of very large objects.

It may therefore be concluded that; **The Moon was pummeled by an asteroid shower, which caused both the maria and the bulge to form and which was over before the Moon could rotate and allow the impacts to be spread out over the entire lunar surface.**

(It may very well be that the multiple-impact event on the Earth, the Moon and Mars was actually three aspects of a singular event in the Solar System. If there was an asteroid shower on either the Earth, the Moon or Mars, neither of the other two would likely have escaped a similar fate.)

12.5   Post-Impact Dust Cloud

It is reasonable to expect that the last major asteroid impact would have terminated life on Earth. Further, since there have been several very large asteroid impacts, life should have been terminated every time. Plants must have light. If they do not have light they will shrivel and die. If a cold period accompanies the darkness produced by the impact cloud, they will shrivel and freeze and die. If, after several months or years, the cold is relinquished and the light returns, some plants might rejuvenate but the animals, that needed these plants for food during the cold period, will not rejuvenate.

12.5.1   Its Timely Demise

In order for life on Earth to have any possibility of rejuvenation after the development of a massive dust cloud is for the dust cloud to be brought down in a timely manner. If it is not brought down, light starvation will terminate plant life and dependent animal life will not survive. It is absolutely essential that the dust cloud be brought down without delay. Therefore, it may be concluded that; **following every major impact in the past, the resulting dust cloud was brought down in a timely manner.**

12.5.2 Termination Agency

A dust cloud will not come down quickly on its own. In fact the dust cloud from even a single volcano can stay aloft for years. With a little reflection, this is what one would expect. Since the atmosphere is layered and there isn't any significant interaction between the layers, dust which is hoisted high into the stratosphere or beyond will stay airborne indefinitely. As an example, the dust produced from the volcano Tambora, which erupted in 1816, created a thick layer of dust 'that girdled the globe and lasted for years.' (Please refer to the discussion in the chapter entitled Post-Impact Developments) Therefore it can be concluded that; **in order to bring down a post-impact dust cloud within the time required for plants and animals to have any hope of survival, an appropriately-timed cloud-termination agency is absolutely necessary.**

12.5.3 Vapor Canopy Existence

A Carolinian Forest once existed on Axel Heiberg Island in the high Canadian Arctic. Since this island is within eight degrees of the North Pole, it does not receive light from the Sun all year. In fact there are always several months during the winter when it receives no significant light at all. This makes it impossible for trees to grow there – even if other circumstances such as heat and nutrient supply were adequate.

If light from the Sun is to reach this area it must travel past the shadow line formed by sunlight as it comes straight from the Sun. In fact to reach Axel Heiberg Island at noon at winter solstice (i.e. the shortest day of the year when the Sun is lowest in the sky) sunlight must be bent or reflected around the Earth for about 15 degrees. (recall the definition of twilight in par. 5.14) Current phenomena such as noctilucent clouds only reflect sunlight past the shadow line for about 9 degrees. In this case, however, only simple reflection is involved. If a combination of reflection and refraction were involved, sunlight would travel further past the shadow horizon and the Earth would be illuminated by the Sun over much more than 180 degrees of its surface.

A vapor layer above the oxygen-nitrogen atmosphere would have provided the required combination of reflection and refraction to enable sunlight to reach Axel Heiberg – even at winter solstice. Since the trees of Axel Heiberg (and Antarctica) certainly did exist and since they certainly did require adequate light it can be concluded that; **during the time that trees existed on Axel Heiberg Island, a vapor canopy existed above the present atmosphere.**

It was warm on both the islands of the high arctic as well as on Antarctica. This means that the overall water vapor content of the atmosphere was greater with its associated ability to cause warming on the surface of the Earth. Since the average temperature on the surface is currently close to ideal (as evidenced by the concern over global warming by even a few degrees) it would seem that the increased warming effect would have meant disaster. However, with a vapor canopy in place this would not have been the case because it would have not only distributed energy better it would have intercepted more incoming solar energy before it reached the surface of the Earth thereby preventing overheating at the surface. (see 5.8 (2)) In fact, it was necessary for this to happen. It may therefore be concluded that; **during the time that trees existed on Axel Heiberg Island, a vapor canopy existed above the present atmosphere.**

12.5.4   Post-Impact Dust Cloud Termination

Water is the best agency to remove dust from the air. If the dust-carrying air can be submitted to precipitating water, the dust will be removed from the air and taken to the ground. The vapor canopy provided the required dust-cleaning function in two ways. Firstly, since the canopy was in fact water, when its stability was disturbed it precipitated out of the atmosphere and took a certain amount of dust to the ground with it. Secondly, when it vacated its position above the oxygen-nitrogen layer, this lower layer was allowed to expand but because ozone had not yet been produced, there wasn't any temperature inversion. Therefore as atmospheric circulation patterns first developed, they included the upper atmosphere thus including the dust-entrained regions. Therefore, it can be concluded that; **collapse of the Vapor Canopy enabled the atmospheric circulation to include very high elevations which resulted in the Post-Impact Dust Cloud being terminated in a timely manner.**

12.6   One Ice Age

From the above discussion, there are two basic conclusions which indicate that there has only been one Ice Age. From the certainty of the existence of the trees of Axel Heiberg Island and from the evidences given in the chapter entitled; The Vapor Canopy Hypothesis, it is clear that a vapor canopy once existed. Unfortunately it was destroyed by the disturbances associated with the multiple-impact asteroid event. However, its destruction allowed atmospheric circulation to extend high above the Earth, entrain the dust from the light-interrupting dust cloud and bring it down to where it could be rinsed from the air. This singular development truncated the period of darkness and saved the Earth from death due to light starvation. While the Earth can do without sunlight for a brief period, such a circumstance cannot be allowed to continue indefinitely. Unfortunately, the vapor canopy could only be brought down once and its associated dust-removal function could therefore only happen once.

Secondly, it was necessary that a major multiple-impact event to have occurred or the ocean could never have been warmed enough to evaporate the required 2000+ feet of ocean water at or above the critical minimum rate that was absolutely necessary to maintain the moisture cloud cover and thereby protect the developing ice-fields from melting.

A multiple-impact asteroid event in association with the collapse of a vapor canopy will never again occur. Therefore it can be concluded that;there has only been one Ice Age and there will never be another Ice Age.

12.7   A Terminal Event

The trauma caused by a major asteroid strike would destroy life on the Earth. Whenever a major-impact event, such as Chicxulub or Chesapeake Bay is discussed, mass extinction is always included. 'When Canadian geologist Alan Hildebrand identified the Chicxulub Crater off Mexico's Yucatan coast as the precise impact point of a 15-kilometer-wide object, scientists began re-evaluating the role NEO's (Near Earth Objects) play in planetary evolution. Evidence suggests that 65 percent of the Earth's animal species – including dinosaurs – were wiped out in that cataclysm 65 million years ago.' (22) Although this is a very sobering evaluation, it is not singular. 'Hurtling at velocities up to twice those of asteroids, a comet the size of Haley's – about the width of a small city – could hammer the Earth with such force that most of the life forms on the planet would be annihilated. This is exactly what seems to have happened years ago when the dinosaurs were extinguished along with 70 percent of all other animal and plant species.' (18) Some commentators are even more pessimistic, if that is possible and have declared that the collision of the earth with an object 15 km. in diameter would have made the Earth's crust move up and down for hundreds of feet 1000 miles from ground zero. When the greatest of the mass extinctions occurred between 75 per cent and 96 per cent of all species were exterminated with only one in fifty types of species surviving. (1) Small size would have been an advantage. Therefore mice and rats would have stood a better chance than cattle, dogs and sheep.

As discussed earlier, volcanism would certainly accompany a major impact. Asteroid-impact and volcanic-eruption scenarios are closely related. Large impacts would cause volcanic eruptions. Volcanic ash would produce darkness and cooling. Also the Deccan Volcanoes were active at the time of the Great Extinction. The volume of their lava flows exceeded two million cubic kilometers, and billions of tons of sulfuric acid were injected into the atmosphere. The consequences would have included acid fog and acid rain. (1) While discussing the mass extinction which is thought to have occurred 250 million years ago, volcanic activity was also recognized. The volcanoes of the Siberian traps erupted massively at about this time, spreading lava over several million square miles. (25)

The diagram entitled, 6.8.5 (2) 'Fault Diagram – Tectonics' in chapter 6, illustrates the pattern of faults that exist around the world. These faults are cracks in the crust of the Earth. Something caused these cracks. In a previous chapter it was suggested that the pressure wave (or shock wave) generated in the interior of the Earth by a major impact would cause the crust to fail. It certainly has failed. Volcanoes are commonly associated with these faults. While not literally right on a fault, most volcanoes are located within a few miles of a fault. (There are also volcanoes many miles from any fault.)

If another asteroid should strike the Earth another massive pressure wave would propagate back and forth through the interior of the Earth. Whenever it surged up under the crust, the crust would be agitated and displaced. Volcanic activity would be sure to follow. Even as volcanoes were associated with past events, they will also be associated with future events. Volcanic clouds of dust and ash would augment the destructiveness of any impact. Death during the subsequent post-impact winter would surely follow.

While there seems to be agreement that destruction of all forms of life would occur and extinction is repeatedly acknowledged, it is a wonder that anything survived at all. Most of the effects of an impact would involve the entire world. The blinding flash would travel outward in a straight line and hence not shine on the far side, but all of the other aspects of disaster would reach the whole world. If the shock wave from Krakatoa could encircle the Earth numerous times (23), the shock wave from an asteroid impact might go around a dozen times. How far would a wave ten miles high travel? How many times would the interior shock-wave reactivate volcanic activity? Most forms of life – whether plant or animal – would not survive. Among those that did survive, smaller sizes would dominate. 'It was a miracle that any of the higher life forms made it through. Indeed, almost every land animal that weighed more than about 50 pounds had been extinguished, probably because they were most vulnerable in their high position in the food chain.' (26) If smaller was better, rats and other vermin could have become numerically dominant, while horses, cattle, sheep and goats would have been reduced to few, if any, survivors.

Under normal circumstances, animals are commonly found in particular ecological niches or locations rather than spread evenly over the whole Earth. If any such areas were spared from the worst of the results of a major impact, the animals normally found in these areas, would have had a better chance of being spared. This factor together with the advantage of small size would guarantee that the surviving cohort would not include a wide diversity of species.

However, why is any survival expected? The cloud cover would naturally have persisted longer than the time required for most types of plants to die from light starvation or cold. Even now, plants must grow during the growing season. If even one growing season is missed, many plants will not start up again the following year. Most plants, that were not killed when the post-impact winter arrived, would not have survived if the next growing season was delayed. Also, how long would it normally take for "billions of tons" of contaminant to leave the atmosphere? Unless something equally far reaching and spectacular cleaned the atmosphere and washed out the bulk of the contamination, the Sun would not have shone through soon enough to have mattered. If the atmosphere had not been cleaned in a timely fashion after past impacts, no life of any consequence would have survived, let alone the 4% to 25% as suggested. The basic marvel is not that 96% were destroyed but that 100% were not destroyed. One commentator has declared that bacteria would have survived a mass extinction as witnessed by the recent discovery that living bacteria were found entombed in the abdomens of bees thought to be 30 million years old. By using techniques that precluded contamination colonies of bacillus sphaericus (removed from the fossil bees preserved in amber derived from Dominican Republic strata thought to be between 25 and 40 million years old.) was cultivated. (24) However, it is hereby submitted that bacteria, cockroaches, and vermin do not constitute survival!

As discussed in the chapter entitled; Post-Impact Developments, it would require a very large amount of ongoing volcanic activity to cause the ocean to warm enough for it to evaporate 2000+ feet of water to form the glaciers of the great Ice Age. Further, it was concluded above that a multiple-impact asteroid shower would have been necessary to enable this to happen.

Moreover, a single impact would produce a world-encircling dust cloud which would be augmented by all of the volcanic activity that the impact stirred up. Such a dust cloud would cause the Earth to chill but the volcanic activity from a single impact would not produce sufficient warmth to offset it. Further the oceans are presently cold. Volcanic activity would warm them slightly but not into the temperature range required for the massive evaporation required by the Great Ice Age. The cold which would be produced by the dust cloud from the next major impact would dominate and the world would be subjected to a drawn-out world-wide post-impact winter.

Therefore, it is concluded that; **the next major Earth impact involving either an asteroid or comet greater than five miles in diameter will terminate life on the Earth.**

12.8 Mountain Formation

Most of the continental mountains of the world are formed from sedimentary rock. In many places this rock has been deformed into mounds and trenches called anticlines and synclines. This could only have happened if the material had been pliable enough to bend a great deal without breaking. Sedimentary rock certainly cannot be bent these days or it would simply fracture. It is concluded that; **all of the mountains of the world formed in between the time that their sedimentary layers were deposited and the time that they hardened.**

Could there have been a case where the sediments of Asia were in a pliable, malleable state during which time the mountains of Asia formed and then many years later the sediments of Europe were in a pliable, malleable state during which time the mountains of Europe were formed? Then at some other completely different time, the sediments of South America were pliable and malleable and the mountains of South America formed. At still some other time, the sediments of North America were in a soft, pliable, malleable state and the mountains of North America formed. Could the sediments on the various continents have been placed at wildly different times?

It would have required a great deal of energy to deposit the sediments. Their existence in a soft "wax-like" state would have been a short-term situation. This means thatduring a very short period of time, a massive energy source appeared and placed the sediments. Then before they hardened, another similarly-powerful energy source appeared and distorted them into mountains. Could this type of sequence have occurred on various continents during widely-separated periods of time until all of the mountains of the world were formed? Could a similar sequence happen again at some time in the future?

It is completely unthinkable that thousands of feet of soft sediments could have appeared in one area of the world to be followed eons of time later by a similar development in another part of the world. This leads to the conclusion that; The sediments on every continent of the world were placed nearly simultaneously as a singular event.

The reality that the sediments could only have remained in a soft wax-like state for a short period of time leads to the conclusion that; all of the mountains of the world were formed within a short period of time in a near-simultaneous event.

Mountains consist of alternating anticlines and synclines. Some have remained in a curved shape and with some the domes of the anticlines have broken open. On a much smaller scale, earthquake waves present very similarly with domes and troughs with the domes often broken open. It is therefore concluded that; mountains are giant earthquake waves which formed as a shockwave passed and the material has remained in a chaotic state ever since.

Mountains are found around the entire world and they were all formed within a short time period. It would have been impossible for a single asteroid to have provided the energy to form all of the mountains of the world.

It may therefore be concluded that; an asteroid shower occurred and it provided the energy to form all of the mountains of the world within a short period of time.

Mountains are formed from sedimentary rock which must have been in a soft, pliable state or else the folding and bending in the mountain layers would not have occurred. Neither would the mountains have retained their sharp-edged features unless the soft crumbly material had been stabilized until it could harden. The post-impact chill and the succeeding ice age provided the necessary stabilizing factor. By the time the Ice Age was over years later, the sedimentary material had hardened into rock.

It is therefore concluded that immediately after the mountains were formed, they froze and it was the reduction in temperature that enabled them to retain their sharp features until they hardened.

12.9   Vapor Canopy Eliminated

While the high arctic islands have been described as once being 'hot and steamy' it is also quite common for this phrase to be used whenever the prehistoric world is described. However a 'hot steamy world' would have been impossible without a vapor canopy interrupting a portion of the incoming solar energy. Otherwise the greenhouse effect of such an abundance of water vapor in the atmosphere would have caused over-heating. The current situation is alarming enough and it only involves a very small change in the world-wide average surface temperature. If the entire world or even a significant fraction of it were 'hot and steamy' the average surface temperature would have been too hot for animal life over most of the world. The only way that this over-heating could have been avoided is if there had been some means to interrupt a large portion of the incoming energy from the Sun.

It is therefore concluded that prior to the asteroid impact event, the world was enveloped in a blanket of water vapor which the incoming asteroids totally destroyed.

12.10   Window of Life Closure

The Moon is receding from the Earth. As it approaches its maximum distance from the Earth, one night on Earth will be three weeks long and one day will be three weeks long. The temperature extremes resulting from this arrangement will prevent life on the Earth from being viable. Fortunately, there will be a few hundred million years left before this situation develops.

The magnetic field of the Earth is becoming weaker. As it weakens, increasing amounts of cosmic radiation will reach the surface of the Earth and an increasing number of cancers will develop. Within about one thousand years, the strength of the field will have been reduced significantly and all associated negative developments will be manifest.

The mutation load among humans is increasing. Due to advances in modern medicine, more and more mutants survive to maturity and pass their mutation to the next generation. Currently there are more than 5000 diseases associated with mutations. What will life be like when there are 50,000?

The ozone layer is seriously threatened and appears to be thinning out. The stratosphere will chill as thinning proceeds with the result that storm clouds will be able to rise higher and higher. Thunder storms will consequently become increasingly ferocious and hurricanes more powerful.

Salt continues to enter the oceans from all major rivers of the world. While some components of the salt can be useful to certain small creatures of the ocean as construction material, if the saltiness exceeds the level at which these creatures can survive and therefore use some of the salt, the buildup of salt will proceed unrestricted. The oceans would then eventually become as salty as the Dead Sea and just as void of life. Fortunately, it will be several million years before this happens.

The Earth is losing more heat than it gains. This is known simply because interior temperatures are higher than surface temperatures causing heat to flow from the interior outward. Consequently, the solid crust will get thicker and thicker. (This assumes that all original internal heat generators have petered out.) The interior of the Earth is currently molten, which is due to original heat, nuclear heat generation or some other heat production activity. All of these heat-generating activities will die out eventually resulting in a colder planet with a thicker crust. As a result, the long-term climate of the Earth will gradually become cooler. The only way that this procedure can be offset is if the current atmospheric warming trend heats the atmosphere until it is as hot as the interior of the Earth. Otherwise the Earth will eventually cool off. Fortunately, this is a long-term effect.

Atmospheric CO_2 levels are currently increasing at a rate exceeding 2 ppm per year. (3) This is a short-term phenomenon with several results which are not beneficial. More solar energy will be retained in the atmosphere with the result that both the surface and atmospheric temperature will rise. As it rises, more

CO2 will be released from the ocean compounding the rate of increase. As the ocean gets warmer, its water will expand inundating some coastal areas. As the weather gets warmer, all of the glaciers around the world will melt. Ocean level will rise another two hundred feet from this effect alone. (4) There will be widespread crop failures, starvation, and greater attempts among starving people to migrate. As the atmosphere gets warmer, it will hold more moisture. This too is a means of trapping solar energy and will accelerate the warming trend. Where the temperature increases will level off is not known, but anything approaching either body temperature or the temperature at which seeds cannot germinate, will make the Earth uninhabitable.

'What we are seeing in the climate domain is not just increases in the average surface temperature of the Earth. We are at risk from the climate from heat waves, from floods, from droughts, from wild fires, from heat waves, from sea-level rise. …. We are seeing changes in circulation patterns, changes in storm tracks, increases in flood intensity and frequency, increases in drought intensity and frequency, more and stronger heat waves, more powerful tropical storms – right across the board, everything that is expected to result from global climate change driven by greenhouse gases is not only happening, but it's happening faster than anyone expected.' (This comment was apparently made by one John Holden and is most appropriate to the world's current situation.)

Terrible things are starting to happen but the worst one may be the effect on food production. Higher temperatures will have a disastrous impact on food crops in parts of the world where the average temperature during the growing season is already close to the critical maximum. This is the uppermost temperature where plants can germinate. Any higher than this and the plant cannot reproduce and will effectively be sterile. Many tropical countries already experience extended periods during the growing season when the temperature is almost at that level. In particular rice yields fall by 15% for each degree in temperature rise during the growing season. (27)

The more complex something is, the easier it is for it to be compromised. The human body - or indeed the body of any creature of Earth - is extremely complicated. Thousands and thousands of components and circumstances must be in place and operating properly or it will not work at all. If, even one of these many thousands of factors were to be seriously compromised, the particular body affected could not continue living.

Similarly, water is easy to contaminate but hard to purify. Even a very small amount of contaminate will compromise a large amount of water and do it irreversibly unless some extraordinary measure is taken. As an example, once salt entered the oceans of the world, they would forever after be salty. There isn't any process, which can make them fresh again.

The final conclusion is therefore reached. THE "WINDOW OF LIFE" WILL CLOSE. A great flurry of activity will precede the closing. Even now a significant amount of activity is directed toward keeping the window open. IT WILL NOT STAY OPEN. Even as water is easily polluted, so the "Window of Life" is easily closed and in fact it will close. WHEN IT CLOSES, INTELLIGENT LIFE IN THE UNIVERSE WILL BE GONE AND WILL NEVER AGAIN EXIST.

13. A Theory of the Earth Based on Asteroid Impact

Explaining the Trees of Axel Heiberg

13.1 Tranquility

The Earth was universally warm as evidenced by; the forest on Axel Heiberg Island in the High Arctic of Canada, the trees piled up on the New Siberian Islands north of the Siberian mainland, the coal of Spitzbergen Island north of Norway, the tropical trees of Antarctica and the coal of Antarctica. The oceans were also warm. The temperature of both water and land at the higher latitudes was in the range of 15 to 25 degrees Celsius. The temperature near the equator was similar. The temperature of the entire Earth was within the comfort zone for humans and winter never came. (i.e. Similar to the Faroe Islands in the North Atlantic at the present time – only warmer.)

13.2 The Vapor Canopy

Solar energy was distributed around the entire Earth with a very efficient energy transport mechanism which, because it had a high thermal capacity and was in continuous motion, kept the Earth wrapped in a blanket which worked so well that the entire Earth enjoyed temperatures within a very narrow band. Not only did the Earth travel around the Sun on a "Goldilocks" orbit, it also had a "Goldilocks" energy transport mechanism. A layer of water vapor existed above the oxygen-nitrogen atmosphere, its stability being dependent on the lighter weight of H_2O compared to O_2 or N_2. Its preferential heat-absorbing characteristic enabled it to absorb and distribute solar energy equitably around the Earth. The light reflecting and refracting characteristic of its water droplets and tiny ice crystals higher up, caused the light of the Sun to propagate far beyond the horizon and illuminate the high Arctic similar to the way that atmospheric ice crystals currently produce sundogs and to the way that the noctilucent clouds currently reflect solar light. The vapor canopy distributed both heat and light in a way that resulted in a world that was universally warm and illuminated over much more than 180 degrees of its surface. With this arrangement the polar regions were never really dark.

13.3 Atmospheric Circulation

The present atmospheric circulation patterns - in particular, the Hadley Cells which distribute solar energy from the equator to regions both north and south – did not exist. Without the dry descending air of the Hadley Cell loop, the band of deserts which currently encircles the world at 20 to 30 degrees of latitude would not exist. In particular, the Sahara Desert wasn't there and in its place there was a land of rivers, lakes and forest.

13.4 The First Impact

A large asteroid hit the Earth. It was so large that the atmosphere did not slow it down at all and it hit the Earth so hard that it completely disappeared into the interior of the Earth. World-wide chaos followed immediately. The blinding flash ignited fires within the line-of-site of the impact. The atmosphere contorted as the expanding pressure wave flattened everything with the efficiency of an expanding steel plate. The ground convulsed in an unusual earthquake which resulted in the ground moving back and forth for hundreds of meters as well as up and down for hundreds of meters. Nothing within at least one thousand miles had any chance of survival. Tsunamis, initially several miles high, propagated outward from the impact site as well as from suddenly-displaced land along coastlines and they travelled completely around the world repeatedly. The interior pressure wave which was generated by the asteroid as it dove into the depths of the Earth, travelled right through the Earth and when it came up on the other side of the world, the crust of the Earth cracked open. Lava poured out and spread around creating Igneous Provinces covering thousands of square miles. The Mid-Atlantic Rift was formed at this time. Other mountain ranges were formed as the crust of the Earth was shocked sideways. Volcanic activity developed around the entire world.

13.5   Woolly Mammoths

The high-pressure atmospheric shockwave produced by the incoming asteroid(s) (possibly Popigai, Tenitz, Kara or Puchezh-Katunki in Russia) as it plummeted through the atmosphere, killed the entire Woolly Mammoth herd instantly.

13.6   Erratic Boulders

Erratic boulders are found all over the world and were produced from fragments of the crust of the Earth as the asteroid smashed into the crust and scattered debris for hundreds of miles in every direction. More debris was generated at the antipodes when the upwelling pressure wave surged up from the impact site on the opposite side of the world.

13.7   Mountain Formation

Mountains consist largely of wave patterns in the rock consisting of both domes and troughs (anticlines and synclines). Earthquakes produce waves which have been observed to be several feet high. Mountains are giant waves – the giant waves of a giant earthquake – which were produced when asteroids smashed into the Earth.

13.8   The Dust Cloud

Effluent from both the impact site and the volcanoes produced a massive dust cloud which spread out and covered the entire world with a layer of material so thick that the Sun could not reach the surface at all. The entire world became as dark as night. Solar energy was completely cut off and as soon as the tsunamis stopped rolling over the land, the air chilled and the temperature dropped below freezing, killing any remaining plants and animals as well as freezing small bodies of water.

13.9 The Vapor Cloud

The ocean was warm prior to the impact and close to room temperature all over the Earth. As a result of the impact, underwater volcanic activity including igneous provinces, rifts and cone volcanoes warmed the ocean even further until, on the average, it was much warmer than room temperature. Warm water evaporates and the unusually warm ocean evaporated at a prodigious rate. A massive, extensive vapor cloud was formed. Wherever this vapor cloud chilled, precipitation resulted. Over warm ocean and warm land areas it rained in torrents. Over the land that had been chilled by the dust cloud, it snowed so heavily that several inches often fell every hour for days at a time. This abundant precipitation continued as long as the ocean evaporation rate was high enough to maintain continuous cloud cover and prevent the Sun from melting the accumulating snow.

13.10 Vapor Canopy Collapse

The violent inrush of the asteroid and the elevating power of the mushroom cloud which developed from the impact as well as the violent up-thrust from numerous volcanoes disturbed the stability of the vapor canopy, causing it to collapse. Removal of the vapor canopy caused the associated temperature inversion to disappear. Secondly, the ozone layer had not yet formed (the water in the vapor canopy had blocked ultraviolet radiation from reaching the oxygen layer and forming ozone) creating the current temperature inversion which restricts upward movement. Initially, part of the dust cloud was removed directly by the collapsing vapor canopy. Subsequently, dust was removed on an ongoing basis by the developing Hadley and Ferrel atmospheric circulation cells which were initially allowed to rise to very high elevations (and entrain the dust) because there was no temperature inversion to restrict their upward movement. Removal of the dust cloud truncated the post-impact winter which would otherwise have continued for years. However the chill which the dust cloud initially created continued as long as the vapor cloud cover remained continuous enough.

13.11 Species Elimination

Numerous additional impacts occurred over a period of several months (This was the actual Period of Heavy Bombardment referred to in the previous chapter but not the one that is the result of a computer program.) all of which produced more cloud and furthered the collapse of the vapor canopy. The vapor canopy could only collapse once. Hence, its collapse could only have allowed atmospheric circulation to extend into the stratosphere once. Therefore it could only contribute to the truncation of one post-impact winter. If the asteroid impacts had been separated by any number of years, repeated post-impact winters of unlimited duration would have resulted because there would have been neither a collapsing vapor canopy nor unusually-high atmospheric circulation to clean dust from the air before the Earth spiraled into a deep cold and stayed there. Life on Earth would have been terminated every time. Even if this had not happened, multiple time-separated impacts would have reduced the numbers and types of animal life to less than 1/1,000,000 of 1% of entire original cohort. Even if repeated impacts had been separated by millions of years and the Earth had not gone into the deep freeze, they would have precluded any redevelopment or evolution of any significant number of new replacement species because there would

407

just not have been enough time. From the current great diversity of species and from the expectation that a single major impact would eliminate virtually all of them, it is clear that there has only been one mass-extinction of life on the Earth. (The question concerning how any life was able to survive at all is beyond the scope of Part 1 but will be discussed in Part 2, Surviving the Ice Age.)

13.12   Coal Bed Formation

The coal beds formed from material which included numerous kinds of non-swamp trees as well as trees which might have been growing in water. (1) The continent-crossing tsunamis which were generated by the asteroid impacts removed and relocated all of this material and formed massive mounds which rolled and tumbled before coming to rest. Successive water movement brought other types of material which buried all of this vegetation and packed it down. The type of material that was relocated determined the type of coal that resulted.

13.13   Cloud Transition

It is widely and justifiably expected that any of the major asteroid strikes which the Earth has experienced would have resulted in a post-impact winter which would have been caused by the massive dust cloud resulting from the impact. Worldwide volcanic activity would have added to the dust cloud ensuring that all land areas would be cold as well as dark. It would have stayed this way for years killing virtually all forms of life that survived the immediate devastation of the impact. However, within weeks after the asteroid struck, the post-impact winter was truncated and the great Ice Age developed instead. This transition happened because of a change in the nature of the cloud cover. The dust cloud was brought down. As the vapor canopy collapsed, it brought down dust directly. Then, without the temperature inversion caused by the vapor canopy, the developing Hadley and Ferrel atmospheric circulation cells were allowed to rise to unusual heights, entrain more dust and bring it down to lower levels where it was washed out of the air by precipitation. The unusually warm ocean evaporated at prodigious rates, moisture clouds formed and these same atmospheric circulation cells brought the moisture (storm) clouds from the ocean to the land.

13.14   The Ice Age

Prior to the first impact the oceans were warm and then a great deal more heat was added by volcanic activity raising the average temperature of the ocean to a much higher level. Warm water evaporates and the unusually warm ocean evaporated water in prodigious amounts which caused the resulting moisture clouds to also form a thick sunlight-blocking blanket. These massive moisture clouds from the oceans were carried over the chilly land, (chilled by the sunlight-blocking action of both the dust cloud and the moisture cloud) so it started to snow. It kept on snowing for months and months until the oceans cooled down and evaporation dropped below the critically high rate which was required to sustain moisture cloud continuity and hence enable glacier accretion.

Cloud continuity and the continually cooling ocean (which absorbed increasing amounts of CO2) contributed to maintaining the chilling effect. If a little sunshine did get through the clouds, it was reflected by the snow. The temperature at the North Pole dropped slowly until the Arctic Ocean crusted over with ice -then it plummeted.

When the ocean cooled to a certain critical average temperature, its reduced evaporation could not sustain continuous storm cloud cover. Subsequently, the storm clouds began to break up. Snowfall became intermittent and then it stopped altogether during the summer months and summers became warm enough to melt the snow from the previous winter. With this development, accumulation phase of the Ice Age was over.

13.15   Glacier Volume

Glacier volume is always referred to by the amount of ocean water it would take to make the amount of ice of interest. For example, the great glacier on Antarctica is referred to as being equivalent to 200 feet of ocean. The volume of the Greenland glacier is only about 10% of the Antarctica glacier. During the Ice Age it is believed that an additional 300 feet of ocean was transferred from the ocean and became tied up in glacial ice. With respect to these estimates and opinions, it is obvious that a great amount of ocean water was involved in forming the great glaciers of the Ice Age. As discussed in the chapter entitled, Post-Impact Developments, only a portion of the actual water that evaporated actually became tied up in glacial ice. Therein it was suggested that at least "four times as much water evaporated as became ice" which means that at least 2000 feet of ocean had to evaporate for the Great Ice Age to have happened. This is a very large amount of water and there are very few possibilities available to explain how such a vast quantity of water could evaporate at all. It follows immediately that this process was a very rare and unrepeatable event.

13.16   Singular Ice Age

The absolutely incredible amount of ocean water that had to evaporate to form the great glaciers of the Ice Age was caused to evaporate by world-wide volcanic activity which included hundreds of underwater cone volcanoes as well as igneous provinces and rifts (underwater mountains split down the middle). In turn, this high level of volcanic activity was caused by multiple asteroid impacts.

In addition to heating the ocean and producing a great dust cloud, the atmospheric upheaval brought down the vapor canopy removing its insulating and temperature-regulating characteristic. Its demise fortunately removed much of the dust from the air thereby truncating the developing post-impact winter. The heat released from the interior of the Earth provided one of the **necessary** conditions for an ice age. The chill caused by the world-enveloping cloud and the loss of the vapor canopy provided the second **necessary** condition for an ice age. These two **necessary** conditions provided **sufficiency** for the Ice Age to happen. **Therefore, the Great Ice Age happened.** The uniqueness and rarity of these two conditions appearing simultaneously meant that only one Ice Age could ever have happened. Another reason involves the stability of soft mountain material.

13.17   Mountain Stability

Since mountains have been formed from soft sedimentary material, this material must have been stabilized immediately or it would have crumbled. The chill that followed their formation stabilized the material until it hardened. Therefore, in order to have mountain ranges there must be: a, the deposition of soft layers of material; b, the deformation of these layers into mountain waves and c, the stabilization of the deformed layers until they hardened.

13.18   Submarine Canyons

Once the snow started to melt, it melted at alarming rates and the major rivers became swollen with melt-water. Since the melt-water was close to freezing, (all melt-water is always close to freezing) it was colder than the water which was still in the ocean and which had not yet completely cooled down. (In order to have the Ice Age, the ocean had to have been warm – actually very warm.) Therefore, as the colder, denser melt-water entered the ocean, it headed straight for the bottom forming submarine canyons on its way.

13.19   Climate Transition

The Ice Age was the interface transition period during which time the pre-impact universally warm climate was changed to the warm-at-the-equator and cold-at-the-poles climate that the world has now. A post-impact winter would not have produced such a transition because it would have been reversible. While a post-impact winter would be the kiss of death for most life surviving an impact, it would not cause a permanent climate change. When the dust settled out, whether it was years or decades later, temperatures would have returned to their previous levels. In order to transform from one type of climate to another, something irreversible must happen. The irreversible factor was the collapse of the vapor canopy. With its collapse, everything changed. The Earth still received the same amount of heat from the Sun but now it mostly impinged on the Earth near the equator and there wasn't any mechanism to efficiently transport it to higher latitudes. Atmospheric storms and ocean currents soon developed due to the temperature gradients between pole and equator, but their energy transfer capability was no match for the high thermal capacity and transfer efficiency of the vapor canopy. Hence the new climate would be characterized by disparity of temperature between the equator and the poles.

Our current atmospheric circulation patterns developed during this transition time. Hot air near the equator rose, traveled northward and descended to feed the westerlies. The great band of deserts that currently encircle the world about 25 degrees above and below the equator, were the result of this new circulation system.

13.20   The 360 Day Year

As mentioned in Par. 6.6.5 and 6.6.7 above, the expected results of a large asteroid impacting the Earth include giving the Earth an impulse or a push. This impulse would affect the speed of the Earth in its orbit or possibly alter the direction of the orbit. It follows that numerous impacts as part of a singular event (i.e. an asteroid Shower) would have an integrated result with respect to the orbit of the Earth.

When the asteroid shower occurred on the Earth it happened within a reasonably short time frame of probably not more than a few months and possibly much less. The reasoning which supports this conclusion includes the effect which the swarm of asteroids had on both the Moon and Mars. In both of these cases the great majority of impacts happened on one side. Neither the Moon nor Mars had time to rotate and spread the impacts more evenly over their respective entire surfaces.

Of course the rotational speed of neither the Moon nor Mars would have been known prior to impact but since the vast majority of impacts have occurred on one side only it appears that the event was quite short in duration. On the other hand the rotation of the Earth is reasonably rapid and only requires 24 hours. It is therefore reasonably easy to imagine the impacts being spread around the entire Earth. We understand that the rotation of Mars is also quite rapid currently being very similar to the Earth's rotation. However it is also true that Mars is much closer to the orbit from which the asteroids originated. This orbit is even further out than Mars where it has variously been postulated that either a collision occurred between a small planet and something else (i.e. a comet) or the small planet simply exploded for some other reason. In either case, the products of the mishap are the asteroids, many of which are still in orbit around the Sun. If a swarm of these newly-formed objects drifted closer to the Sun (just as some of them have drifted further out to the orbit of Jupiter) they would have encountered Mars, possibly years before encountering the Earth-Moon system. During the intervening time they would have had time to spread out much more.

In any event, a swarm of asteroids impacting the Earth would have affected its orbit. A swarm of asteroids coming in from one direction, (as they would have because they orbit the Sun in the same direction as the Earth) would have pushed the Earth a little faster, hence into a higher orbit - possibly a more elliptical orbit but any higher orbit would have resulted in a longer year.

In order to change from a 360 day year to a 365 ¼ day year only requires about a 1.4% change. This appears to be a reasonably achievable result considering the number of large asteroids involved, each one of which would have knocked the Earth forward slightly in its orbit. Further, more than the known number actually impacted because it is a virtual certainty that more impact sites will be identified in the future. These could possibly include the Congo Basin in Africa and the Takla Makan Basin in China.

It must also be noted that the planet Mercury has suffered a similar disruption. The monstrous impact site on Mercury, Caloris, is given as 800 miles across. A very large object caused this feature as well as the chaotic 'weird terrain' on the opposite side. We also notice that Mercury has a very elliptical orbit. It is suspected that these two features of Mercury are related.

It is therefore offered as one of the results of the great asteroid impact event that that the orbit of the Earth was changed and that the length of the year was increased. This conclusion agrees with Velikovsky (2) and certain others (3) who have commented on this matter.

13.21 The Final Result

When the Arctic Ocean cooled enough it became ice-covered. Prior to cooling substantially, it would have kept some land areas in the high Arctic ice-free. When the Arctic Ocean froze and both land and sea became covered with snow, the reflective characteristic of these areas enhanced chilling. This in turn, caused the development of the Polar atmospheric circulation cell which includes a great downwash of frigid air from up near the tropopause (Air at this elevation is commonly minus 100 degrees F.) to flow across all Arctic regions. The loss of the vapor canopy's ability to carry sunlight to these high latitudes resulted in several months of darkness every winter. The loss of warmth and the loss of light removed any possibility that the high Canadian Arctic would ever again be suitable for the growth of Carolinian forests. CONSEQUENTLY, THE TREES OF AXEL HEIBERG ISLAND WERE FROZEN IN THE DIRT AND WOULD NEVER GROW AGAIN.

Part 2 - Surviving the Ice Age

Foreword

Any theory of the Earth, no matter what the particular subject is, must recognize that numerous large asteroids have impacted the Earth. There are theories explaining how mountains were formed, how submarine canyons were formed and how volcanoes erupt. There are theories to explain how fossils occurred, how gravel deposits and drumlins were formed and how submarine trenches occurred (not the same as submarine canyons) and how underwater rifts developed. Very very seldom do any of these commentators mention asteroids. The subject just doesn't come up. This is most unfortunate because every few years another major asteroid impact site is found. One of the most recent is Chesapeake Bay on the eastern seaboard of the USA. Also, erratic boulders are found all over the world and their presence is usually attributed to ice movement but ideas such as this do not stand up to very much scrutiny especially since many of these boulders weigh several tons and are found at elevations higher than their origins.

All theories of the Earth involving either geology or geography simply must recognize the effects of asteroid impact because it happened. When most of the popular science versions of theories in these categories were first brought forward, asteroid impact sites were unknown. That is not the case now however and therefore many existing theories of the Earth must be set aside or reworked in recognition of the effects of these asteroids.

We do recall that Aristotle who lived more than two thousand years ago held many ideas that were far from valid. Unfortunately, people clung to his ideas up until just a few hundred years ago. This was done in spite of the fact that Aristarchus, who lived at the same time as Aristotle, held valid ideas but was simply ignored by the mainstream of popular thought. Then we had Galen (a medical practitioner of the 2nd century) who had a host of valid ideas but also many that were not the least bit valid. Unfortunately, some of the invalid ones held sway for more than one thousand years in spite of correct knowledge being available. We recognize that misconceptions such as these were not very productive but should be hesitant to judge because the present generation is doing no better. Contrary evidence to numerous popular theories continues to pile up as the result of ongoing research but many old invalid ideas are still being held as absolute truth when they should be discarded. The discard pile should include the Laplace Theory of Solar System Formation, (which died when extra-solar planets were discovered) the Theory of Evolution (which should have died when the DNA molecule was discovered) and String theory (which has no evidence whatsoever to support it). Most ice age theories should also be set aside because both heat as well as cold are required to have an ice age and they have not recognized this basic criteria nor how such an apparent contradiction occurred. The ideas in this part, as well as in Part 1, might not become popular very soon. However, the contradictions that continue to develop with the ongoing accumulation of knowledge will not be ignorable forever and sooner or later, will have to be faced. The present discussion will contribute to this reality.

In summary it must be recognized that the multiplicity of theories of the Earth are simply not cohesive. One of the most glaring examples is the idea that the dinosaurs lived one hundred million years ago. This idea remains entrenched in spite of the fact that a mummified fully-articulated carcass of a Hadrosaur (Dakota) was discovered just a few years ago. Another example relates to the problem explaining how the great diversity of animal life on the Earth made its way through the numerous global-generating disasters which immediately followed every major asteroid impact. Even the arrival of the Burkle impactor in the south Indian Ocean would have been followed by world-wide flooding and volcanism and another mass extinction. Life in any form is fragile and the Window of Life is easily closed. How did anything survive the Ice Age? The idea, by the way, that the Ice Age mammals survived it, is not valid. Surprisingly on this point no survival explanations have been offered. There isn't any animal that survive in ten feet of snow let alone in a thousand feet of it - especially if it should remain in place for several years! Suggesting that many animals survived this event is not the least bit credible and only provides another diversion from coming to terms with the survival issue. In fact, in order to provide the credibility that is so desperately needed all theories that are offered must contribute to an overall picture that is completely cohesive. Herein a coherent argument is offered while drawing attention to the unexplained survival problem.

1. Introduction

The Window of Life concept recognizes that life is very complicated and that the circumstances required to enable and support any life-form are equally complicated. Even a caterpillar requires food, protection, air to breath and an environment which has a particular not-very-wide temperature range. If any one of these necessary life-enabling factors is removed, the caterpillar's life will terminate. Each of these factors may be compared to a window which allows light to shine through. When all of the windows are lined up the light will get through but if even one of them moves to far to the side the light will not get through at all. Neither is it optional that all of the factors be lined up, rather, it is absolutely necessary. All forms of life are like this. Even a dandelion requires water, nutrient-loaded soil and just the right amount of sunshine in the proper temperature range. In fact in biology, the water, oxygen, nitrogen, carbon, phosphorus, and sulfur cycles are of particular interest because they are all integral to the cycle of life. It is therefore recognized that all forms of life require a very large list of factors to be present in their respectively appropriate ranges. Collectively, such a list is herein referred to as the Window of Life. Unfortunately, any set of circumstances with such a complex multiplicity of factors could be easily upset. This truth underlies the current concern about the environment of the Earth. In this case the extensive amount of attention that is being paid to the apparent change in the average temperature of the Earth underlies a concern that the world is becoming less hospitable for life in general. Such concern is well founded. What if the temperature in the tropics and the sub-tropics increased until it was above the germination temperature for our food supply? If such an unthinkable thing happened life on the Earth would simply dwindle away. Alternatively, what if a large asteroid crashed into the Earth?

The destructive power of a large incoming asteroid has been recognized by numerous commentators and comments are usually offered regarding the effects it would have on all forms of life on the Earth. Estimates are even assigned with respect to the percentage of the total living creatures that would be annihilated if a large asteroid should strike. One commentator stated that if an asteroid in the neighborhood of ten miles in diameter were to hit the Earth, all forms of life in an entire hemisphere (i.e. half of the surface of the Earth) would be decimated (which means that 90% would be killed). (1) Another suggests that 'there would be a great dying.' (2) Others suggest that between 75% and 90% of all life – including plants and animals – would be killed as the result of a single large impact. (3) The Chicxulub impact on the Yucatan Peninsula in Mexico would have resulted in a great mass extinction which included the dinosaurs. (4) In support of this assertion we have 'This is exactly what seems to have happened … when the dinosaurs were extinguished along with 70% of all other animal and plant species.' (5) In fact there do not seem to be any dissenters who disagree with these seemingly extraordinary claims. After all to kill most of the life on the Earth is – to say the least – a major undertaking.

While there is a reasonable amount of discussion regarding mass extinction there is none regarding survival. How did anything survive? Suggestions are simply not offered. Perhaps it is incomprehensible that everything could be killed. Since we are still here it may therefore be unthinkable or unallowable that everything could be killed. Why is this? If 90% could be killed from a single impact what if there were several nearly coincident impacts? How would anything survive? No theory of survival has ever been offered and it is not likely that it ever will be. The task is simply too great and credibility would be strained. It is much easier to suggest that 10% remained so the question can be avoided altogether. However it cannot be avoided because the evidence shows that there have been several very large impacts and there is no doubt that more impact sites will be confirmed in the coming years. Neither can the question be avoided because if it is allowed that 10% survived, how could all of the present diversity of creatures on the Earth have evolved from that surviving 10%?

The subtitle, "Surviving the Ice Age", is related to the impact of the asteroids because the two phenomenon are directly connected. In fact it was the impact of the asteroids that caused the Great Ice Age. Many, possibly more than sixty, explanations have been offered to explain the Ice Ages but none of them stand up to very much scrutiny. The basic reason for this is that both a major world-wide chill factor as well as a major world-wide heat factor would have been required to produce an ice age. Cold alone would not be sufficient and of course heat alone wouldn't do much either. You obviously cannot have an ice age if you cause to Earth to be heated! But heat was certainly necessary to cause ocean water to evaporate. In fact it had to evaporate at or above a necessary critical rate to enable ice accumulation to exceed reduction. Further, it is obvious that you cannot form vast quantities of ice if the water does not leave the ocean. This is clear by recognizing the equivalent amount of ocean that is still tied up in the glaciers of the world. It has been estimated that the Antarctic glacier has enough water in it to raise the level of the entire ocean about 200 feet. (7) The other glaciers including Greenland, would add about another 20 feet. During the Ice Age it has been postulated that there may have been 120 m (400 ft) (8) more of ocean water involved. Also, since all of the water that evaporated would not have actually formed ice, much more than the suggested total of 500 or 600 feet of water must have left the ocean in order for the glacial ice to have built up to these vast thicknesses. In fact if we conservatively suggest that four times these amounts evaporated, (i.e. making a total of 2000 to 2400 feet) it is clear that a great amount of heat was required. In review, Part 1 above includes an extensive discussion explaining how the Great Ice Age and the impact of the asteroids are connected as well as how the Ice Age developed from the results of the impacts.

How would all of the heat and all of the evaporation as well as the unthinkable blizzards, that must have raged on week after week, have affected survival? It is hereby offered that even a rat would have had trouble coping with such a scenario. The Great Ice Age would therefore have narrowed and probably have closed the Window of Life on its own. If it had been immediately preceded by the hazards which would certainly have accompanied any major asteroid impact, the problem of survival becomes much clearer. Then, if it had been an asteroid shower instead of a single hit, the survival problem becomes so unthinkable that it doesn't even seem sensible to contemplate it.

In fact an asteroid shower is what actually happened. The reasoning that led to this conclusion involves the unsolved problem of survival and the regeneration of life across the whole world. No commentator has ever offered an explanation of how any remnant of life was spared from the destruction resulting from an impact and indeed it would be very hard to explain. Instead of explanations declarations are made. For example; '… an asteroid … about 5 miles in diameter … energy released was greater than 100 million megatons of TNT … it is no wonder that there was a great dying.' (2) The commentator in this case stops short of saying that everything died but at the same time doesn't offer any explanation regarding survival.

This appears to be the general opinion among those who comment on the situation at all. Another example is a little more explicit and declares that 96% of all species died. (10) How do we know that it wasn't 98% or 99%? What miraculous coincidences occurred to preserve a sufficient diversity of creatures to enable the Earth to be repopulated with the present diverse cohort? Evolution would have to be working overtime to achieve such a feat and some explanation of how this occurred should be forthcoming. However this has never happened. It isn't as if the impact of an asteroid was a local event. It was not. Every major impact was a global event. In fact a global event would have been the norm right down to incoming objects which were even less than one mile in diameter. If one should review the Earth Impact Database (which is available on line) he/she will note that there are numerous impacts where the impacting object would have been larger than this. In support of the opening statement above, a review of the hazards to life developing from an impact should therefore be carried out.

It is unfathomable how any form of life, whether plant or animal, could survive these hazards. In fact a survival theory has never been offered. With all of this in mind, it seemed appropriate to recognize and discuss the problem in more detail. That is what the following discussion is all about. How did life survive the great asteroid-shower/Ice Age event?

2.   The Hazards of Asteroid Impact

2.1   The Extent of the Hazards

The hazards of impact are numerous and many of them would involve the entire world. In fact there are even some that would develop as the asteroid plunges through the atmosphere. As it nears the surface of the Earth more life-threatening situations would develop. Actual contact with the Earth, whether it was land or sea, would involve several more and penetration into the Earth would extend the list. As the asteroid disappeared into the Earth and was swallowed out of site it may be out of site but the devastation list is still near its beginning. The interior of the Earth is understood to be molten and would absorb the asteroid or its remains but the energy associated with such a heavy fast-moving object would continue to spread far into the interior and the material in the interior would be the medium for this energy to spread throughout the entire sphere of the Earth. The remains of the body of the asteroid would be expected to come to a halt deep in the Earth but the energy would not come to a halt. Similarly, a storm on the ocean may peter out but the waves which it generated would continue on and may completely cross the ocean and cause damage on a beach a thousand miles from the original storm. In this case the ocean water is simply a means of transmitting energy. The molten material in the Earth's interior would also be a means of transmitting energy and would transmit the asteroid's energy throughout the entire world. When the energy-carrying shockwaves reach the underside of the Earth's crust on the far side of the world, more disruption and life-threatening events will develop. When a large asteroid hits the Earth there really is no place to hide either on the impact continent, the ocean or the far side of the world.

An asteroid would enter the atmosphere at very high speed. Using passing asteroids as a reference, the speed of one that was approaching the Earth could exceed 50 miles/sec. (180,000 miles per hour) (12) If we allow that the atmosphere is 20 miles deep, an asteroid moving at such a speed would only require about 0.4 seconds to reach the surface. The air directly in its pathway would not have time to completely move aside. The air that did move aside would be compacted into a thin walled cylinder of very high pressure. This shockwave would be thrust outwards at a speed commensurate with the asteroid's speed and would have an effect similar to an expanding steel plate. Everything within its danger zone would be blown away. The remaining air which did not have time to move aside would be increasingly compressed in front of the asteroid and its temperature would rise. It would soon become hot enough to glow. This is called the blinding flash and would be similar to looking at an exploding atomic bomb. These two atmospheric effects would be devastating to any form of life within their respective danger zones.

Contact with either land or water would generate another series of events. First of all a splash would form. Some of the material at ground level, whether it was water or ground, would be splashed away. Water impact would result in great globules of water arcing outwards in all directions. They would land with the effect of exploding bombs. Land contact would not be any better. A mud-rock splash would develop and be projected outwards similar to the water globules. Asteroid crater rims vary with the size of the asteroid, of course, but if the crater rim was 200 miles across we could expect broken rocks to be projected outwards for several times this distance. In this manner erratic boulders would be spread over the countryside for hundreds of miles.

As an asteroid presses on into the Earth more material would be forced aside. This would be the beginning of a massive earthquake. The displacement of the ground would not be the few inches to the few feet that we experience with most earthquakes but would be several hundred to several thousand feet. Such displacements would diminish the farther the distance was from ground zero but some residual effect could easily be felt all the way around the world. When this ground-travelling shockwave reached an ocean a great tsunami would develop and race away at several hundred miles per hour.

Deep in the interior the asteroid or its remains would come to a halt. The pressure wave which formed ahead of it would not come to a halt. It would travel right through the Earth and come up under the far side. This location is called the antipode. At this location the crust of the Earth would be thrust upwards. The crust would crack and indeed it is cracked. The upwards displacement could be thousands of feet and the cracks or faults could extend for thousands of miles. Several more hazards would accompany these antipode developments including the projection of hot material from the interior into the atmosphere or onto the ocean floor. As the ground swelled up under the ocean the ocean water would be caused to run down off of the bulge that was formed. Sheet flow of ocean water would result and it would race across the top of undisturbed ocean water and would not be easily stopped by any shoreline. Therefore being on the opposite side of the world would not avoid the hazards of survival. In fact it might even be worse.

Many other woes would accompany an asteroid impact event and all of them would be multiplied if the event was an asteroid shower instead of a single impact.

2.2 Primary Atmospheric Activity

2.2.1 The Blinding Flash

2.2.1.1 The Cause

Any asteroid, meteor, meteorite, comet or even a piece of so-called space junk would enter the atmosphere of the Earth at high speed. Commentators seem to agree that comets would be moving the fastest. An extreme example of comet speed has been provided by the one named Howard-Koomen-Michels 1979X1 which closed on the Sun at 150 miles per second. (540,000 miles per hour, (1)) Part of the reason for such an extreme speed was the Sun. The gravity of the Sun is very high and would have accelerated that particular comet continually as it approached its fate. The gravity of the Earth is, of course, much lower but never-the-less a comet could plunge into the Earth's atmosphere at a speed which was in the range of 10 to 50 miles per second. (36,000 to 180,000 miles per hour, (1)) Asteroids also move at very high speed by any comparison and their speed would be in this same range. (2) The atmosphere is not very thick when such speeds are being considered. If it were allowed that the atmosphere was 50 miles deep, an incoming threat which wasn't slowed down would reach the surface of the Earth in a time period of just a few seconds. Of course there wouldn't be any expectation that such an object would slow down. Neither would there be any expectation that it would reveal itself with a fiery streak. A large object, because of its mass, would not heat up very much. However, the air in front of it would be compressed and simply because of this compression, would heat up. In fact it would become red hot or even white hot. A large incoming object could be 5 miles in diameter. In order to get out of the way of an asteroid this large, the air that was in the center of its pathway would have to move 2 ½ miles to the side. Some air away from the center of the trajectory would slip off to the side. The air near the center is expected to be trapped – partly because the leaving air would have trapped it. As a high-pressure cone of air builds up in front of an asteroid, it would actually provide a better means for air deeper in the atmosphere to be deflected out of the way. However, in spite of any such development, a significant quantity of air would remain trapped, heat up until it was white hot and thereby produce a blinding flash. One commentator has offered that the first suggestion of catastrophe would have been a flash of the brightest and most blinding light ever witnessed by a human. (48)

2.2.1.2 Comparisons

This flash can be compared to an explosion. In 1914, near the beginning of World War I, a munitions ship exploded in the Halifax, Canada harbor. (3) A blinding flash was produced and contributed to the blindness inflicted on numerous spectators to the event. This type of calamity happens very quickly. If one senses that something is wrong and simply looks up, there would not be time to change your mind. The damage would have been inflicted instantly.

An even better example is the explosion of an atomic bomb. The bomb could explode many miles away but if any living creature was exposed, blindness could result. Further the flash would be just the visible portion of the damage. A flash of light is an electromagnetic phenomenon. Included with the visible energy there would also be infra-red energy (i.e. heat energy). This blast of heat would cause the instant combustion of everything in its pathway that was flammable. A house with wood siding would be instantly set ablaze. A forest would ignite. Anything damageable by such a blast of heat and light would be damaged. One report from Japan, immediately after one of the atomic bombs was dropped right near the end of WWII, spoke of a horse walking by in front of some observers. All of the skin on one side of this animal had been burned off. The entire muscle arrangement on the whole side of the animal was visible. There would be no saving an animal in such a case. In fact the most merciful thing that could happen at that point would be for it to succumb quickly to its injuries.

A burst of electromagnetic energy of this nature would also include energy with lower frequencies than either light or heat. A pulse of energy in the lower frequency range could damage both electrical and electronic equipment. This has always been a concern in the event of a nuclear war. The explosion of either an atomic or hydrogen bomb could disable the control circuitry in an intercontinental bomb-carrying missile even if there wasn't a direct hit. In order to offset this expectation the storage silos have all been "hardened". In this case hardening is, in part, referring to protecting everything in the silo from the high energy electromagnetic pulse which would be produced when a nuclear bomb explodes. After all if there was a nuclear attack, retaliation would seem to be in order. We can all be thankful that there seems to have been a shift in thinking in recent years after it was realized that there was so much killing power available that there might not be any survivors anyway. But prior to this, much effort was expended in trying to protect these weapons from being disabled even before they left their storage locations.

In the event of a wide-spread nuclear war, electromagnetic blinding flashes of energy would have been all over the world with the result that there would be blindness, fires and disabled electronic equipment all over the world. The only favorable comparison to the blinding flash from an incoming asteroid at this point is that in the asteroid case the damaging effects would have been restricted to a roughly circular area around ground zero with a diameter of possibly 1000 miles (Please refer to the Appendix for this calculation.) that is until the atmospheric shockwave spread the fires out even further.

2.2.1.3   The Hazard

There isn't any form of life that could have survived the blinding flash. The temperature … would rise to 60,000 Kelvin … everything including people, houses, factories and cars – would vanish like cellophane in a flame. … every standing thing … would be on fire or would be dead. (48)

2.2.2   The Atmospheric Shockwave

While the blinding flash would be a serious threat to the survival of animals upon which it was able to shine, the atmospheric shockwave would be a threat to survival for all animals directly exposed or not. While the blinding flash would be an electromagnetic (i.e. light, heat and electronic) hazard for hundreds of miles and cause fires for hundreds of miles, the atmospheric shockwave would be hazardous for many miles further - even for thousands of miles.

2.2.2.1 The Cause

As an asteroid plunges through the atmosphere the atmosphere is modified in three basic ways. The air which is directly in its pathway is of the most interest because it is the effect of the asteroid on this air that generates two of the atmospheric hazards that threaten all forms of life in the area. The first of these, the blinding flash, develops because some of the air becomes so highly compressed that it heats up until it is visible. Fires result. Air which is directly in front of the asteroid will compress and ride with the asteroid. It is unable to get out of the way of the asteroid simply because there isn't time for it to move aside. The total time available to do anything at all is dictated by the tremendous speed at which all asteroids would have as they approached the Earth. 'Most of these objects are hurtling at such high velocities – 11 to 100 kilometers per second … ' (5) In miles per hour this speed range translates to approximately 20,000 miles per hour to 200,000 miles per hour. (Please refer to the appendix for this calculation.) Neither would they slow down as they approached the Earth with its tremendous gravity helping them along. While the atmosphere of the Earth really doesn't have an outer limit, for practical purposes we can recognize that there is hardly any air more than twenty miles from the Earth. Therefore at fifty kilometers per second (100,000 miles per hour) there will be less than one-half of a second for an object to reach the surface. In other words there would be less than one-half of a second for any air in front of the asteroid to get out of the way. The net result is that some of it will not get out of the way but will simply be compressed and ride in front of the asteroid all the way to the surface creating the Blinding Flash mentioned above. The second hazard is also the result of air being compressed. Some of the air will not ride with the asteroid but will move aside. The air that moves aside will also be compressed but rather than stay trapped it will slide off to the side and develop into a cylinder of highly compressed air with an initial diameter the same as the asteroid. This cylinder will not be very thick and neither will it be standing still. It will be moving outward from the asteroid's pathway at high speed. A spreading shockwave is thereby formed and while it consists only of air, it presents a major hazard!

2.2.2.2 Comparisons

A shockwave of this nature might be compared to other shockwaves which have been experienced in recent history. Krakatoa, the reputedly greatest explosion of the eighteen hundreds, included many hazards one of which was an atmospheric pressure wave which was heard many miles from the site of the explosion. It was heard in both Japan and The Philippines, both of which are thousands of miles from the Sundi Straight (located between Java and Sumatra) where the explosion occurred. It was also heard on Rodriguez, an island east of Madagascar and 3,000 miles from Krakatoa. In all of these cases, a distant rumble was audible to the residents who obviously would have wondered what was happening. (6) Atmospheric pressure instruments were in use at that time and when the records were assembled much later it was clear that the atmospheric pressure wave had actually encircled the Earth three times before dying out. (4) Krakatoa was a very large explosion. While the hazards to life from the shockwave were, in this case, restricted to a short distance from the actual explosion, the disturbance to the atmosphere was detectable all the way around the world.

A further comparison could be made to the Halifax explosion which produced a pressure wave as well as a blinding flash as discussed above. While the blinding flash was a threat to unprotected eyes, the shockwave which the explosion produced was a threat not only to animal life in the vicinity but also to all of the structures in the neighborhood. Buildings were blasted away from the exploding ship and everybody in the vicinity was either killed outright or seriously wounded. The high pressure shockwave expanded outwards pushing everything that was movable ahead of itself. This included the ocean water around the ship. Witnesses reported that a temporary opening appeared in the ocean water revealing the very bottom of the sea. (7) Atmospheric pressure did this. It is understood that the pressure wave was mostly air and other gaseous products of the explosion but it could readily be compared to an expanding steel plate. There isn't any form of life that can withstand a force of this nature. As with the Krakatoa explosion, the structural devastation was restricted to the area immediately around the blast site but the expanding pressure wave continued for many miles further.

A third and even more appropriate comparison to an asteroid shockwave can be made to the Tunguska event. Tunguska is a river in a remote part of Siberia. In 1908 an explosive force of 50 megatons destroyed a large area of Siberian forest about 100 km in diameter. (9) No one knows exactly what happened but reasonable explanations have been offered based on an atmospheric explosion of an incoming object such as a comet or asteroid. 'The explosion knocked over an estimated 80 million trees over 2150 square kilometers (830 square miles). It is estimated that the earthquake from the blast would have measured 5.0 on the Richter scale, which was not yet developed at the time. An explosion of this magnitude is capable of destroying a large metropolitan area.' (8) Since an impact site does not exist, it means that the damage was effected by an atmospheric shockwave. If an asteroid had been involved, expectedly pieces of it would have been found. On the other hand since comets are understood to be mostly frozen water anything that remained would have melted and evaporated or simply left the area as a trickle of water leaving no evidence of its former presence.

While the exact cause of the damage does have academic interest, the effect of the blast is of more practical interest. With both the Krakatoa explosion and the Halifax explosion, material was dislodged and blown away from the area for either hundreds or thousands of feet. With the Tunguska event, material, in this case mostly trees and other movable surface material, was blown away from ground zero out to a radius of about 30 miles. (9) Out to this distance the trees have been completely removed and swept away. Beyond the 30 mile region, the trees are still there but have been flattened away from ground zero as if a giant roller rolled them all down. How much force is required to break off a tree and throw it for 30 miles? What would have happened to a deer or a bear caught in the area? What would have happened to a groundhog down in its burrow? And what would have happened to a human being? 'The sounds were accompanied by a shock wave that knocked people off their feet and broke windows hundreds of miles away.' (10) All of this devastation was caused by an expanding curtain of compressed air produced by the explosion of an object possibly 50 feet in diameter!

2.2.2.3   Effects of the Asteroid Shockwaves

2.2.2.3.1   Trees Piled Up

As mentioned above, trees were broken off and piled up as a result of the Sandusky event. Therefore if piles of broken trees were found elsewhere it could be suggestive of something similar having happened. Great mounds of broken trees have been found in northern Russia. In this particular example some of the piles are reportedly 300 feet high. (14) Trees do not just pile themselves up! In this example the trees are all broken and piled up and mixed with the remains of animals. Some enormous force would have been required to form an assembly of this nature and the expanding high-pressure shockwave from an incoming asteroid would be a serious candidate.

Another example of broken trees and broken animals is found in the muck of Alaska. Muck is simply frozen ground which includes mud, trees, (i.e. parts thereof) animal remains (commonly broken) stones and anything else that was in the vicinity when this event occurred. 'This muck is a frozen mass of animals and trees. … torn trees are piled in splintered masses.' (25) It is also curious that in this case as well as the previous one, both the trees and the animal remains have been preserved right to the present time. Animals buried in the ground are expected to decay before very much time elapses. Actually decay would only occur near the surface of the ground whereas with animal remains that are buried deeper, other underground processes would be involved and before very long it would be difficult to identify any remains at all. The fact that all of the constituents of the muck are identifiable implies that the ground froze soon after the event and has remained in a frozen state ever since. The enormous force that placed the muck was therefore followed by continuous cold weather. The impact of an asteroid would not only be upsetting for the geology of the Earth but would be expected to have climate altering effects as well – particularly if the atmosphere of the Earth was different prior to the event. While further comment on this topic is well warranted and will be included in a later chapter, the present dilemma remains as to how all of these great masses of broken trees occurred. The expanding high pressure shockwave from an incoming asteroid would be a serious possibility just as it was with the Sandusky event. One can only imagine how hazardous it would have been for a person to have been in any area when huge masses of broken trees were being swept along to their final resting place. Survival simply would not have been possible.

2.2.2.3.2   Disarticulated Animals

A disarticulated animal is an animal which has been broken apart and its pieces might not even be in the same vicinity. In many of these cases not only are separate pieces found but the pieces themselves are often broken. 'Mammal remains are for the most part dismembered and disarticulated, even though some fragments yet remain, in their frozen state, portions of ligaments, skin, hair and flesh.' (21) As evidenced by the unsolvable problems (i.e. even the bones they found were broken) encountered when they tried to reassemble "Lucy", (20) identifying the overall shapes and sizes of animals found in this way simply isn't possible. It must be even more frustrating to find an entire animal minus the head because now we would almost have a good report but the missing head prevents the possibility of satisfactory conclusions. Such is certainly the case with disarticulated animals of any kind and this problem becomes of much greater interest when it is realized that this state of affairs has been found all over the world. If animal bones were separated from each other, it would be reasonable to expect that the same thing would have happened to a human being caught in the area. Something very hazardous to life caused this to happen.

2.2.2.3.3 Woolly Mammoth Extinction

The great woolly mammoths have all been killed! Every last one of them has died. They certainly did not die because they were not fit. They were very fit. They were large and powerful animals and there would have been very few predators with the courage or confidence to attack one of them. To make it even worse, they were grazing animals and appeared to have lived in herds which would have made an attack even more hazardous. However there doesn't seem to be any suggestion that they died from attack but there is very explicit comment that their death was so mysterious that it is unexplainable. " ... one of the most perplexing mysteries of the northern tundras: the existence of thousands of frozen animal remains." (11) One thing that is clear is that they died very suddenly. "Well-preserved food fragments were found in the mouth and between the teeth of the mammal. This could only mean that the animal (referring to the Beresovka Mammoth) could only have met with a sudden death and did not have time to swallow its last meal! ... Further indication of sudden death was found in blood which was collected in great masses due to hemorrhage. ... Bits of frozen blood were found which, when heated, turned into dirty red spots indicating that the oxygen content of the blood had not been fully extracted. This condition frequently indicates a sudden death." (11) Not only did they die suddenly but they were frozen solid very soon thereafter. It is well known that ivory, to be useful for carving, must be fresh. The mammoth ivory which was found was fresh as new, indicating that it froze quickly. (12)

As mentioned above, there were "thousands of frozen animal remains". Other commentators confirm this fact. "... Such was the enormous quantity of mammoth remains that it seemed ... that the island was actually composed of the bones ... full of mammoth bones." (13) Therefore the great herd of woolly mammoths can be readily compared to the great herd of buffalo (which is thought to have numbered 40 million) that once roamed the North American central plains. It is therefore apparent that the termination event must have been of great magnitude.

The evidence can therefore be summarized as follows; Millions of healthy, giant, fit, woolly mammoths died so suddenly that they could not even swallow their last bite nor could their metabolisms utilize the oxygen that was already in their blood. Their blood vessels had burst open and they had hemorrhaged internally. As the high-pressure shockwave hit these animals their bodies were compressed. This would be like laying them down on the ground and running over them with a heavy contour-following roller. Every part of their bodies was squeezed. As the bones were squeezed some of them would have broken. As their muscles were squeezed the blood would have been forced out. As their blood vessels were squeezed they ruptured and the blood collected into various pools. The skin would have broken open but it was the skin that was doing the squeezing. So the blood just remained inside.

An atmospheric shockwave did this. The atmospheric shockwave produced by an incoming asteroid would have had the power to kill every animal for a distance of several thousand miles from ground zero. Neither were the woolly mammoths favored by these shockwaves. Many other animals (so-called modern) died and were buried with the woolly mammoths. (19) To make matters worse, Russia, where many of these animal remains have been found, has been hit by several large asteroids. Included are; (at least) Kara, (65 km.crater dia.) Kamensk (25 km. dia.), Popigai (100 km. dia.), Puchezh-Katunki (80 km.dia.), and Ishim (very large, in the 300 to 700 km. dia. range also called Teniz Basin, actually in Kazakhstan) as well as many smaller ones. (15) ("At least" is included because very large impact sites continue to be discovered around the world every few years and it would not be surprising if new ones were found in Russia as well.) So the mystery of why these vast numbers of animals died so suddenly is not really a mystery at all. They died because they were struck by the expanding high-pressure atmospheric shockwaves produced by several incoming asteroids. If one of these shockwaves couldn't quite do the job the one coming from the other direction could. There was no escape and there was no time to escape. Further, the fact that the entire herd numbering into the millions died suddenly the same way is evidence supporting the idea that an asteroid shower was involved instead of a single hit. (Further comment on this factor will be included in a later chapter.) Supposedly a woolly mammoth could have withstood much more trauma than a person could have but the Woolly Mammoths all died. Therefore it would be safe to assume that wherever a woolly mammoth was when it died would have been an unsafe place for any people to have been because they also would have died.

2.2.2.3.4 Dinosaur Extinction

Many other animals have also died in vast numbers in the past and there is an air of mystery attached their deaths as well. This becomes most apparent when the death of the dinosaurs is considered. At least in the dinosaur case, popular science has tied their deaths directly to a particular asteroid. (16) Chiczulub is an impact site located partially on the Yucatan Peninsula of Mexico. The site is about one-half on the mainland and one-half in the water. As with many discoveries of asteroid impact sites, Chiczulub was found when exploratory drill cores were examined. The crater outline is included in the "very large" category at 170 km. (EID, 106 Miles) Various commentators give it credit for the extinction of the dinosaurs as well as nearly everything else. "… such an impact killed the dinosaurs … virtually every plant and animal had died." (19) One is inclined to notice immediately that dinosaurs once existed all around the world and they existed in both the very large category as well as quite small (i.e. chicken size) Never-the-less they have all perished. Dinosaurs are not found at the present time and while many other species have been lost in recent history, it does raise a question as to why every last one of them would have been killed by a single event. The first thing that is made clear by this declaration is that Chiczulub is understood to have been a world-wide disaster and it certainly would have been. Then we must ask if it resulted in the death of very fit animals both large and small on a world-wide basis, how could anything else have survived? How could any creature have found shelter from a disaster that was killing every last dinosaur over the entire world? The first part of the answer is simply that untold millions of other creatures must have died at the same time from the same cause. Whether one is accepting of the asteroid/dinosaur-death hypothesis or not it does remain as a reasonable assumption that such an event would have killed nearly everything whether one likes such a conclusion or not. Some commentators have admitted to this reality. "About one-half of the genera living at that time perished, including marine and flying reptiles, microscopic floating animals and plants…" (18) Again it must be asked, how could anything have survived if even microscopic animals were being killed? Expanding atmospheric shockwaves would have been responsible for at least some significant portion of all of these woes because they would have killed every animal out to a radius where their respective pressures had dropped into the nuisance category.

Unfortunately, dinosaurs that hemorrhaged have not been found. This is basically due to the observation that dinosaurs have mostly been buried in material that enables fossilization and they have become fossilized. Dinosaurs are virtually always found in a fossilized state. Fossilization itself is indicative of unspeakable disaster. The reason for this conclusion is that in order for an animal body to become fossilized, it must be deeply buried as soon as it is killed. If it is only buried in a shallow grave, its remains will be absorbed into its immediate environment and simply disappear. Most of the dinosaurs which have been discovered so far had been buried deep enough to enable fossilization to proceed. It is completely possible that other dinosaurs had also been buried but not deep enough to have been preserved by fossilization. Commonly, the fossilization process is declared to require very long time periods. However this is not known, it is only surmised. If it did, in fact, require long periods of time one wonders why the usual decay processes (which are expected to proceed as soon as any animal is buried) did not operate and simply absorb the animal remains back into the environment before fossilization took place. In this case there would not be any animal remains to examine at all! It might be, on the other hand, that absorption only got as far as the soft parts before fossilization took over! However this doesn't quite solve the riddle of the soft parts (i.e. collagen and certain other parts) recently found in T. Rex remains. (22) Contributing to this dilemma are the fossilized trees of The Petrified Forest of Arizona where the trees have become fossilized so quickly that every small detail of the wood has been preserved. (136) If these trees had been subject to the usual decaying process, fine details of their structure would not have been preserved at all!

The argument for dinosaur remains being ancient is somewhat circular. Since fossilization is a long-term process, dinosaurs must have lived a long time ago. Since dinosaurs lived a long time ago, fossilization must be a long-term process. However these long-standing assumptions have been seriously upset with the recent discoveries of the soft T-Rex tissue which included blood vessels which were still soft and flexible as well as individual cells. (22) Not only has soft tissue from a T-Rex been discovered, an entire animal was recently discovered! A complete hadrosaur was discovered just a few years ago and it appeared as though it had only recently died. A mummified hadrosaur was found in North Dakota with skin and fossilized soft parts. It was fully articulated, apparently in its original form only mummified. (i.e. dried out)It was named "Dakota" and was discovered in 1999.It was exceptionally well preserved which has allowed paleontologists to understand more details about the skin patterns, muscle mass, and body proportions of hadrosaurs. (23) The internal parts were all there. There isn't any burial process known, except freezing, that can preserve a complete animal body for long time periods but this specimen was not frozen. Only the great Ice Age is given credit for preserving animals by freezing but the Ice Age is declared to have been very recent compared to the age of the dinosaurs. This presents a great dilemma which will not be easily resolved. It appears to place the age of the dinosaurs coincident with the age of the wooly mammoth but it is very clear, even to a casual student of animal history that such a scenario would not be readily accepted. Unfortunately, this is a serious possibility in which case the dinosaurs might have died the same day that the wooly mammoths died. If it is readily acknowledged that an asteroid could have killed every last dinosaur on the entire planet, it would seem reasonable to suggest that any wooly mammoths present when an asteroid arrived would meet a similar fate! They were simply caught by a different wave!

Did all types of dinosaurs coexist or not? How is it known that they predated the woolly mammoth? With the discovery of "Dakota" such questions must be entertained. It might very well be that the mammoths happened to be buried in soil which soon thereafter froze (i.e. it is up north where there is permafrost) and the dinosaurs were buried in soil which enabled fossilization to proceed! After all, this is what happened to the trees of Axel Heiberg where some froze and some became fossilized.

2.2.2.3.5   Rhinos, Elephants Buffalo and Horses

Rhinos, elephants, buffalo and horses died with and have been buried with the woolly mammoths. ' … they abounded in mammoth's bones … bones and tusks of elephants … bones of elephants and rhinos in astonishing numbers … the bones of elephants, rhinoceroses, buffaloes and horses …' (24) The mammoths as well as all of these other animals died the same day in the same way. If the expanding shockwave from one or more asteroids killed some of the woolly mammoths, it would seem reasonable to suggest that it killed these other animals as well.

2.2.2.3.6   Hazard Summation

One commentator has described the shockwave as follows; ' … an apocalyptic sight of unimaginable grandeur: a roiling wall of darkness reaching high into the heavens, filling an entire field of view and traveling at thousands of miles per hour. Its approach would be eerily silent since it would be moving far beyond the speed of sound. … (even) people up to a thousand miles away would be knocked off their feet … or clobbered by a blizzard of flying projectiles. ' (50)

2.2.3   Precipitation

Precipitation would be one of the almost-immediate results of an asteroid coming through the atmosphere. Assuming that we are dealing with a large asteroid - one large enough to produce a shockwave - there would be precipitation but the amount of precipitation would depend on the nature and structure of the atmosphere. The atmosphere has not always been constructed the way that it is at the present time as evidenced by the fact that the Earth was once universally warm. Champsosaurs lived in the high Canadian arctic and they were living in a very warm and pleasant climate. (138)

In Part 1 this, as well as other evidence was used to support a vapor canopy hypothesis which, if true, outlines how a layer of water vapor once existed above the current oxygen-nitrogen layer of our present atmosphere. Any such region of water vapor would have been upset and totally destroyed by an incoming asteroid. The destructive element would have been the atmospheric pressure wave. As this wave proceeded outward from ground zero, the atmosphere would have become temporarily compressed. The increase in pressure would have driven the water molecules toward each other and would have formed a spreading layer of water vapor which would track the expanding shockwave outwards for several hundred miles. As the shockwave passed any particular point, the spreading high pressure zone would have been followed immediately by a region of much reduced pressure allowing the water vapor to evaporate back into an invisible state. Unfortunately it wouldn't all evaporate. Some of it would stay in the liquid state and fall to the ground as rain. If the atmospheric load of water vapor had been high, a lot of rain would have fallen. If a region of the atmosphere had been virtually 100% water, a torrential downpour would have resulted. A downpour of this nature would have been similar to the rain that falls during a hurricane. Fortunately, because the shockwave would have been moving quickly, the downpour in any particular location would have been short-lived. However the overall effect would probably have been a downpour over a region which extended for more than one thousand miles from the starting location. A rainfall like this would have been most unpleasant but due to its short duration it would not likely have been life-threatening. The raging wildfires would have been dampened somewhat which would have been a modest positive development amidst so many devastating effects.

2.2.4 Wind

A windstorm would also have resulted from the movement of the asteroid through the atmosphere. In the vicinity of ground zero, this wind would more properly have been called a blast. It would radiate outward with its speed reducing as it went. Within a few diameters of the asteroid it would be a blast of air but the speed would drop off the further it went. While the air in front of the asteroid must move aside to let the asteroid pass, most of this air would have formed the atmospheric shockwave. Following the asteroid there would have been a region of reduced atmospheric pressure. The surrounding air would have moved into this low-pressure region and then followed the asteroid to the ground. At the ground this air would have turned outwards horizontally and it is this movement of air that is being referred to as the wind.

The air movement in this case could properly be compared to the way that air moves near a waterfall. As the water falls downwards, adjacent air is dragged along with it until the water reaches the bottom of the falls. Then it turns and moves outward away from the waterfall horizontally along the ground. This phenomenon can be experienced at any waterfall of any significant height. If one stands near the bottom he/she will always be in a wind which is always heavily loaded with water vapor. Similarly if a person looks down to the base of a large waterfall he/she will notice that a wind is blowing continuously away from the base of the falls. The vapor loaded air moves outwards for a short distance and then turn upwards forming the great vapor plumes that accompany many waterfalls. For our present discussion it is the horizontal movement that is of interest because it would be very similar to the wind from an asteroid. The wind at the base of a waterfalls might be a welcome refreshment on a hot summer day but the horizontal movement of air from a large asteroid would be a disaster. Near ground zero it would be a blast of air which neither trees of buildings could withstand. Further out the speed would drop into the hurricane range and force-feed the spreading wildfires until they were raging like a blast furnace. A little further out the wind could be compared to the firestorms which develop during forest fires. In the forest fire case, the updraft which develops from the burning forest draws air in horizontally around the entire fire area. If the fire is large, the horizontal movement of air towards the fire becomes very intense and might reach hurricane speeds making this factor alone a hazard for any firefighters in the area. The movement of air might become so intense that loose debris becomes swept along and feeds the fire even more. In the asteroid case, the wind would feed the fires that had been started by the blinding flash just a few seconds earlier and the devastation from these fires would be very widespread. Due to this factor alone animal life would be threatened for a radius of hundreds of miles from ground zero. Since only a few animals can survive a forest-fire, only a few animals would survive the wind-fed fires resulting from an incoming asteroid.

2.2.5 Summary

The primary atmospheric activity resulting from the impact of an asteroid would have been devastating for all forms of life – in particular mammals and birds. Most of these creatures would have been killed outright if they happened to have been in the near vicinity (i.e. several hundred miles) of ground zero. This type of devastation would have been comparable to the Tunguska Event which included reports that windows were broken and people were knocked down even when they were several hundred miles from the impact site. (26) Those tragedies were attributed to the atmospheric shockwave which was still detectable even further away in Great Britain. The shockwave in that case would have been quite modest in comparison to the one that a large asteroid would generate and the hazard to life in the asteroid case would have been correspondingly greater.

The blinding flash started the fires which were spread far and wide by the atmospheric shockwave. The wind that followed augmented this particular mode of destruction and fanned the flames for several hundred miles. In the meantime the shockwave instantly killed and blew away every form of animal life that happened to have been living in the area. Being below ground would not have been an advantage either because the shockwave would have followed any tunnel right to its end and killed whatever was hiding there. The net result is that we have utter destruction all around for several hundred miles or even for several thousand miles. Within this area virtually everything would have perished. All of these miseries developed before the asteroid had even touched the ground. Then, when it touched things only got worse.

2.3 Surface Impact Activity

There are, of course, two types of surfaces on the Earth, land and water. It might seem at first that the results of a major impact would be totally different for these two different surfaces but because of the extremely high energy transfer that occurs when an asteroid impacts the surface, there would be a surprising number of similarities. These will be discussed separately with the understanding that the surface of the Earth is only that, the surface. Before a large object will have been at the surface or have gone very far below it, relatively solid material would be encountered anyway. Then when the heat and energy and temperature increase are recognized, solid ground will not seem so solid.

429

2.3.1   Water Impact

The likelihood of an asteroid impacting water is much higher than of impacting land simply because on Earth there is more water surface than land surface. (i.e. approximately 70% water) There is almost a three to one chance that water will be the first relatively solid thing that an asteroid would encounter. Air, even though it is very thin compared to water, will seem to be relatively solid when it is approached at high speed by a small object. For example, if a chunk of rock as large as a car sped into the atmosphere from space, the atmosphere would seem to be relatively solid. Such a rock would heat up and break up and this is a fact very much in our favor because rocks of this size enter the atmosphere regularly. In such cases the atmosphere protects us from them. Similarly the space shuttle, being relatively small, had to enter the atmosphere at a very particular angle or it would have been deflected away from the Earth and the occupants would have perished. However all of the arguments of this nature must be set aside when objects in the very large category are discussed. A stony object the size of a city or even one as small as one mile across will not be affected by the atmosphere and the atmosphere will not provide us with any protection at all. Asteroids this large will reach the surface of the Earth without losing any speed and any discussion of the effects of the atmosphere on such an object must be set aside. The primary atmospheric activity that was discussed previously develops because the asteroid would have an effect on the atmosphere, not the other way around. However, as the solid (relatively) surface of the Earth is encountered the situation changes.

2.3.1.1   Initial Water Splash

For discussion purposes we will consider an asteroid which is several miles in diameter. (For objects this size and even much smaller, it is appropriate to speak of diameter. It has been a popular belief that objects of this size and even much larger, would not have a spherical shape because their gravity would not have enabled this to happen. This idea was based on the theory that the large objects in the solar system have been formed from small chunks of rock which came together and formed large bodies due to the influence of gravity. Promoters of this idea declare that a heavenly body would have to be almost as large as the Moon or its own force of gravity would not be strong enough to cause a spherical shape to result. This idea has been shown to be completely false and must be set aside especially since the recent passing (Nov. 2011) of 2005 YU55 (27). This asteroid was large enough to have caused serious trouble if it had hit us but very small from a self-gravity viewpoint. With a diameter of approximately 1300 feet it would have been much too small to have ever formed itself into a sphere (from rocky chunks) due to its own self-gravity but there it was – perfectly spherical.) Just prior to actual contact, the influence of the asteroid on the surface of the Earth would be felt. Recall from the above discussion that air would be trapped in front of the rushing asteroid and it would have been compressed into a mass with a very high density. It would be this high-pressure air mass that would start to push surface water aside. The extremely high speed that an incoming asteroid is expected to have is relevant to the way that the water would be displaced. Water a little further below the surface would not be able to move out of the way of course and would have an effect which could be compared to a basketball being slammed into a shallow dish of water. With reference to the illustration, 2.3.1.1, Initial Water Contact, it appears that the water further down would be comparable to the solid characteristic of the plate. This means that the water which was being physically forced aside would squirt out almost horizontally and would do so with a speed that was commensurate with the speed of the asteroid, only higher. Nothing would be able to withstand the force of this water. It would cut right through solid objects (like animals and trees) like a knife. They would be destroyed instantly and their remains would be broken into fragments and thrown for miles. The expanding atmospheric shockwave which preceded this water-knife, would have been sufficient to sweep every loose object before it but just in case anything was missed the supersonic speed of the spreading layer of water would finish the job. There isn't any form of life that can withstand this type of abuse.

2.3.1.2 Subsequent Water Splash

The basic problem of getting the water out of the way continues as the asteroid presses further into the surface. The basic thing that changes is the angle at which the displaced water is discharged. As discussed above initial discharge is almost horizontal. As the asteroid plunges deeper and deeper, the angle of discharge becomes more vertical. This raises the question of what becomes of this water. Expectedly most of it will simply splash upwards and then arc over and fall back to the Earth. This would seem to be a reasonable expectation except for the speed factor. The asteroid would be moving at a speed which would be comparable or even greater than the escape speed of the Earth. This means that the water would be ejected from its original resting place at a similar speed.

Escape speed is a term which identifies how fast an object must move in order to leave a planet or a moon or any other object in space and not return to it due to gravitational pull. If, in the upper atmosphere of the Earth, an object is accelerated until it reaches escape speed, that object will leave the vicinity of the Earth and not return due to the gravity of the Earth. Every heavenly body has its own unique escape speed. In particular the escape speed of the Earth is approximately 25,000 miles per hour. This means that if a space capsule in orbit around the Earth is speeded up to 25,000 miles per hour, it will leave the vicinity of the Earth and travel out through space. With this in mind every time that a space module is to be sent to another part of the solar system, it must be accelerated to escape speed. Then it will be able leave the Earth and travel out to another planet or some other destination in the solar system or possibly even beyond the solar system.

2.3.1.1 Initial Water Contact

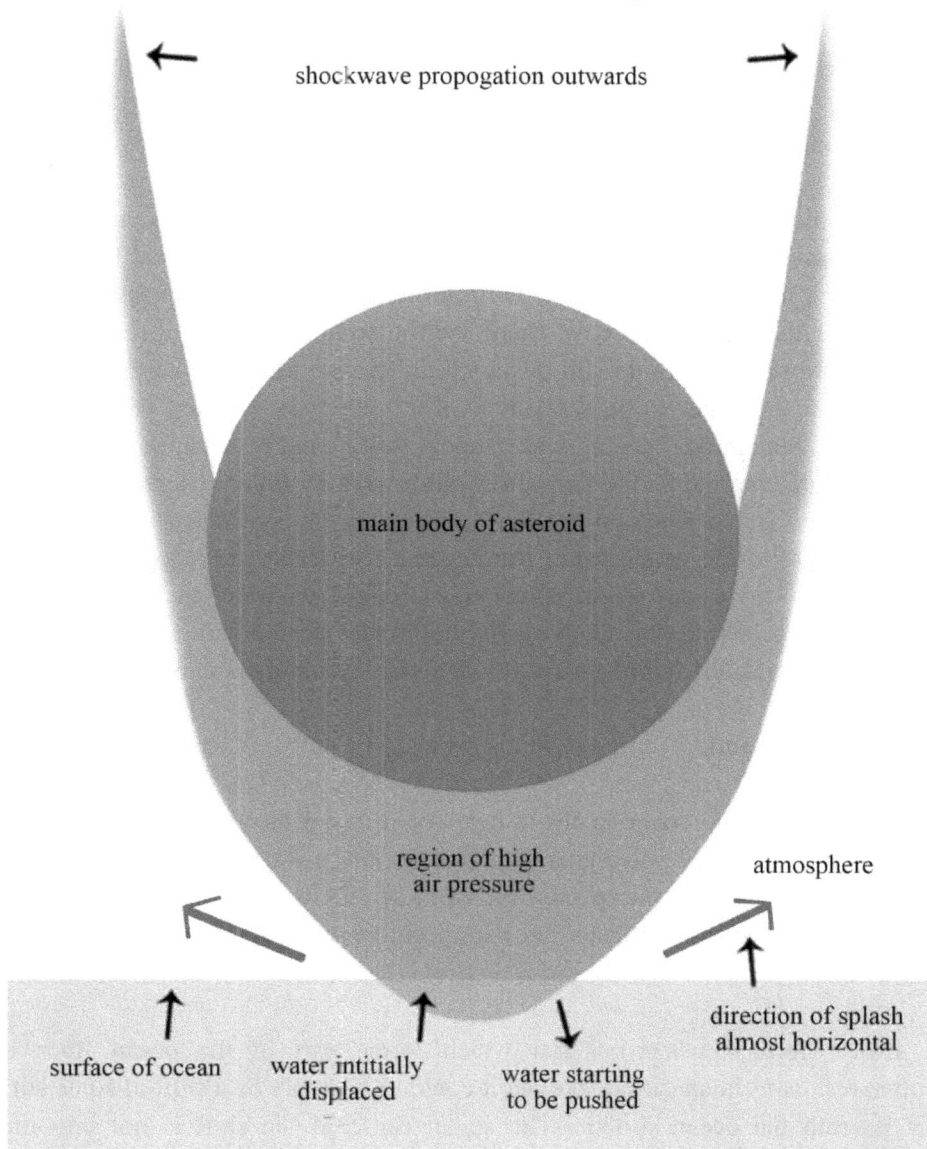

shockwave propogation outwards

main body of asteroid

region of high
air pressure

atmosphere

surface of ocean

water intitially
displaced

water starting
to be pushed

direction of splash
almost horizontal

This raises the question of what would happen to some of the water being ejected from a water impact site by a large asteroid. Could some of it be accelerated to escape speed and leave the Earth? The primary factor on which this depends is the speed of the asteroid. If it had been traveling at a speed that was comparable to the escape speed of the Earth, some of the water would leave the impact site at a similar speed. Therefore some of it could leave the Earth. The rest of it would fall back to Earth in globules which ranged in size from rain drops to huge masses weighing several tons. These large ones would form craters when they landed which makes them comparable to bombs. The area around the primary impact site would therefore be subject to another life-destroying factor and as with the atmospheric shockwave and the fires, this factor would also be hazardous to all forms of life for a radius of several hundred miles.

This hazard would diminish with distance from the impact site but, depending on the size of the asteroid, not drop into the serious-nuisance category until the distance from ground zero had reached several thousand miles. Beyond that the hazard would be reduced to the torrential-downpour category which of course would be survivable if it was not prolonged.

2.3.1.3 Crater Formation

Water impact is the topic but crater formation is the phenomena. A temporary crater would form in the water. The usual idea involved with crater formation is that the impacting object hits the Earth so hard that material is displaced from the surface of the Earth forming a bowl-shaped excavation. Since the Earth is reasonably solid near the surface, the bowl shape will be retained and might remain for thousands of years. With water the bowl shape would still develop but because water is easily moved, gravity would cause any excavation in the water to be short-lived. The temporary opening would fill in again very quickly. However, even though the opening was temporary, it would result in the production of a major world-wide hazard. While it would be the first of several hazards to follow which would also involve the entire world, there would be no place on the Earth that could be considered safe from this particular effect. Further, this hazard to life would result from even a reasonably small asteroid. Any asteroid that could produce a crater which was upwards of a few miles in diameter would generate a worldwide hazard to life from this factor alone! While this temporary crater in the ocean would form quickly, it would also fill in again quickly. Unfortunately the fill-in event would only result in yet another hazard!

2.3.1.4 The Primary Tsunami

The formation of the temporary crater in the water would cause the formation of a world-encircling tsunami. Two characteristics of this wave guarantee that the entire world would be involved. The first of these is the height and the second is the speed. Both of these factors would be dependent on the size and speed of the asteroid where larger and faster are both negative with respect to the survivability of every form of life on the Earth.

Tsunamis are always formed when material which forms part of the ocean floor is displaced. Displacement often results from an earthquake but it could also simply be a shift of some silt or any other material that is beneath the ocean surface. The actual cause of the shift is not important. It is the magnitude and speed of the shift that matters. The recent tsunami that devastated Japan developed when a portion of ocean floor moved. The movement involved an underwater section of rock which extended for several kilometers but the actual displacement was quite small and only involved a few meters. Therefore the resulting tsunami was only a few meters high but because it was a tsunami type of wave and not a wind type of wave, there wasn't any structure available to protect the buildings and other items of the highly structured and civilized society of Japan. The devastation was therefore enormous and numerous lives were lost. (28)

It is the energy of a tsunami that guarantees trouble. Waves on an ocean or lake always form when the wind is blowing. As the wind blows across the surface of the water some of the energy of the wind is transferred to the water. Waves result. The stronger the wind is and the further that it blows the larger the resulting waves will become. This is the basic reason that ocean waves are bigger than waves on a small lake. A wind blowing across the ocean is not restricted and the distance that it blows might be hundreds of miles. Therefore the resulting waves might be tens of feet high. Every so often a so-called killer wave forms and these have been reported as being seventy feet high or even more. Such a wave would be a threat to the largest of ships. Even though ocean waves might become very large, their energy is still contained in a surface layer of water which is only slightly deeper than the actual height of the wave. This is not very comforting to the skipper of a small vessel when he is trying to deal with large ocean waves but it is of great comfort to know that a structure, either natural or man-made, that is about as high as the wave will be quite effective in protecting everything behind it. A little water might crash over the top but any energy that it contains will be quite small. This is not the case at all with a tsunami.

While the energy of a wind-wave is contained in a layer of water slightly greater than the height of the wave, the energy of a tsunami is distributed throughout the entire depth of the body of water where the tsunami develops. If the water was five hundred feet deep, the energy of the tsunami would involve the entire five hundred feet. The surface manifestation of the wave might only be a few feet or even less. A passing tsunami might only raise the ocean surface by eighteen inches in which case it would easily pass under a small ship and not be noticed. With a tsunami, the wave height might be small but the wavelength is large. While a wind wave might be ten feet high and the distance from one peak to another might be thirty feet, a tsunami which was only eighteen inches high might have a wave length of five hundred miles. As Shakespeare would have said, "there's the rub". Or as we would say, there's the problem. When a large wind wave approaches shore a certain type of trouble is expected. Putting up barriers that are more or less the height of the approaching wave will enable most of the trouble that could be caused by such a wave, to be avoided. By contrast there isn't any known barrier that can deal with a tsunami. Japanese society can therefore not be blamed for not having provided proper protection because there really isn't a structure that could have provided the needed protection. In order to understand why this is the case, the energy and water displacement associated with a tsunami must be considered in more detail.

First, reviewing a similar, but much smaller situation would be instructive. Early during WWI a massive explosion involving a fully-loaded munitions ship, occurred in the Halifax harbor. The harbor was destroyed, many people died and buildings near the ship were destroyed. Of even greater interest, for our present discussion, was the observation that the violence of the explosion pushed the water back so forcefully that the floor of the ocean was laid bare. (3) Something similar would have happened if a large meteorite (i.e. small asteroid) had landed there instead.

Suppose that a large asteroid impacts the ocean and forms a crater eighteen miles (29 km) in diameter on the ocean floor. This is actually the reported diameter of the recently discovered Burkle Crater in the South Indian Ocean. Further, for discussion purposes we will assume that all of the water for 50% of this diameter leaves the area as splash water. Continuing with this line of reasoning, the rest of the water, which would be temporarily displaced to form the rest of the crater, would appear as a mountain of water around the impact site. The base of this water mountain would be relatively short, at first, and we will assume that at the beginning it would be spread out for another eighteen miles to form a circular mountain which had a twenty-seven mile outer radius. With these ideas in mind, we can calculate the approximate height that this circular water mountain would reach. This would be the beginning of the great tsunami that would now be propagating out across the ocean. The volume of water above the original ocean level (i.e. forming the tsunami) would be similar to the volume of water that was pushed out to form the temporary crater minus the splash water. With these assumptions and knowing the depth of the ocean, a height calculation can be carried out. (Please refer to the appendix for this calculation which uses the Burkle Crater dimensions.) In this example, the height of this great water mountain would have been about 3460 feet.

While this would be an incredible height for a wave of any kind, the fact that it would be the visible manifestation of the worst kind of wave that is known (i.e. a tsunami), makes clear the nature of this particular hazard. Further, the above assumptions are very probably conservative because of the fact that water is much less dense (or lighter) than the rock of the ocean floor. If the force of the impact could completely dislocate the ocean floor out to a distance of nine miles from ground zero and form a crater which is eighteen miles in diameter, how far out was the water above it really pushed? Since the density of rock can easily be more than four times as dense as water, would the water have been pushed out four times as far as suggested above? Even if it had even been pushed twice as far, the height of the surrounding water mountain could easily have been twice as high as suggested! This would have made the initial formation of the great tsunami wave more than one mile high! Various other commentators have also speculated on the possible initial height of an asteroid-generated tsunami. In particular in the Manson, Iowa case, the crater diameter is given as ' … about 38 km (24 m) in diameter.' (54) 'This impact should have produced a tsunami-like disturbance (steep-fronted waves that travel at velocities greater than 800 kph (about 500 mph). The initial impact may have spawned a huge wave (because Manson, Iowa may have been underwater at the time) … whose initial height may exceed 325 m (1,966 ft).' (55) When the Manson case is compared to other impact sites, it is readily seen that it is quite small. The tsunami-generating capability of the much larger asteroids that have landed would expectedly be much larger. In fact some commentators have suggested 'The impacting explosion was then followed by a lethal cocktail of megatsunamis – thought to have reached thousands of meters high, …' (78)

The next factor of interest is the speed of this tsunami. Tsunamis always move at high speed. Furthermore even though they might develop on a particular ocean shore, they are almost always expected to cross the entire adjacent ocean, even if it is the Pacific Ocean, and reach islands and continents at the far side. Where the Pacific Ocean is concerned, a tsunami is also expected to reach far away shores within a few hours. Actual speeds have been measured. As an example the great Chilean earthquake of 2010, produced a tsunami that crossed all the way to Easter Island in less than six hours. The distance involved was 5310 km (2180 miles) for an apparent speed of 638 km per hour (396 m/hr.). In this development the initial height of the tsunami was recorded at 8 ft, 6 in. (2.6 m) but had dropped to 1.15 ft (0.35 m) by the time it reached there. (29) While this was frightening enough, the Chilean, earthquake-generated tsunami of 1960 produced a giant tsunami that travelled six thousand miles across the Pacific and destroyed downtown Hilo, Hawaii (including five hundred buildings) and killed sixty people. (47)

The nation of Japan (2011) was not as fortunate because the tsunami only formed a comparatively short distance from shore and had not really reduced in height by the time it reached there. Therefore the full force of its destructive power resulted in extensive devastation over many square miles of mainland.

In an asteroid case the great water-mountain/tsunami would have the high speed of a tsunami but also an initial height of thousands of feet. These factors alone indicate that even a distant shoreline would be totally devastated, even if the tsunami's energy was only contained in the apparent height of the wave. However, in a tsunami case, since the entire depth of the ocean is involved, a much different situation develops as soon as the tsunami reaches a shore.

Suppose that a great asteroid-caused tsunami travels across an ocean that is two miles deep. Since the entire depth of water is involved, we understand that there is movement of water throughout this entire depth. All of this movement will continue as the ocean gets shallower near a shoreline. The energy of the tsunami will become concentrated into the shallower water. In the open ocean the forward underwater movement might only have been a few feet. As shorelines are approached this forward movement becomes amplified in proportion to the decrease in depth and might result in the forward movement of water for many tens of feet and then at the shore it might increase to thousands of feet. This type of water movement is called sheet flow. A great sheet of water would be moving forward (i.e. flowing up the beach). In the Japanese case, the sheet of water that flowed past the beach and onto the land was only a few feet deep. In an asteroid case the water that would flow up the beach and onto the land might be several hundred feet deep, extend for a thousand miles along the shore and be moving forward at hundreds of miles per hour. There isn't any structure that can stop this type of movement. Mountains cannot stop it. In the Japanese case the tsunami arrived at an offshore barrier which was several times its own height, climbed over it and continued inshore until a mountainous region was reached. Unfortunately this was several miles from the original shore. The tsunami started to climb the mountain but could only get partway up before the forward momentum could not carry it any higher. Then it stopped. It can therefore be seen that a sheet of water that was several hundreds of feet deep and moving several hundred miles per hour would be able to flow right up the side of a mountain range and quite possibly keep right on going. It might be able to completely cross a continent in which case it would take everything that was the least bit moveable and carry it along or at least relocate it thousands of miles from where it originated.

There isn't any form of animal life or plant life that can deal with a situation like this. Birds could not deal with it because there would be no place to land. Animals on the ground would be swept away the same as the forests would be and anything trying to hide underground would be drowned. There would be no escape. The movement of water alone spells certain death but the entrained debris would make the situation that much worse. Crashing forests being rolled and tumbled along do not present many refuge possibilities. Even crashing animal bodies would be hazardous for any others that seemed to be getting through it. From a Window of Life viewpoint the Window of Life would have closed. Even so more trouble would be following right behind.

2.3.1.5   The Secondary Tsunamis

The secondary tsunami would form and leave ground zero soon after the primary wave had left. The temporary opening in the ocean, all the way to the ocean floor, would indeed be temporary. Water would rush back into the opening and would come in with great force. It would all rush to the center of the opening. But since it would be moving in with considerable speed, it would not stop at the center but would turn upward instead. A pile-up of water would result. It too would be temporary. Since it could readily climb to several thousand feet, it can be appreciated that when it fell down again it would generate another wave. This secondary tsunami would follow the first one but not be nearly as energetic.

The primary tsunami would have been sufficient for total annihilation of life over most of the Earth and the secondary tsunami would have made the situation worse. Unfortunately, the bad news is just beginning. The asteroid would also generate a series of major (i.e. that is, off the Richter scale completely) earthquakes, every one of which would generate tsunamis which would circumnavigate the Earth and could be referred to as tertiary (i.e. third) tsunamis. No matter where the asteroid struck, earthquakes would occur all around the Earth and all of them would generate tsunamis. Then, when these events were well under way, the interior pressure wave would arrive at the antipode location, elevate the ocean floor and generate more massive sheet flow to add to the excitement. While the asteroid would cause the generation of tsunamis from ground zero, the sheet flow which developed at the antipode would ensure that no part of the Earth was spared involvement with massive water flow. All of these waves and water movements would cross each other on their journeys and when they did so the amplitudes at the cross-over points would be additive (as with the more familiar killer waves of the present ocean).Wherever two flowing sheets of water collided, untold mayhem would ensue as the water piled up and surged about. Water, water everywhere and every bit of it trouble!

2.3.1.6   Mountain Climbers

All of these giant waves would travel completely around the Earth repeatedly. It would be of small comfort to life on the Earth that these waves would become smaller every time they circled the Earth, because significant destructive power would still be retained after several passes. It would also be unfortunate that mountains would not be of much assistance in stopping them – at least not until their respective speeds had been considerably reduced. The reason for this is the way that energy is transmitted by this type of wave.

The forward speed of an object determines how high it could climb without using any other power whatsoever. This holds true for a ball tossed into the air, a bullet shot up into the sky or water rushing across the countryside. Furthermore, climbing ability is very nonlinear with speed and is dependent on the speed multiplied by itself. This is called a square law and means in everyday practice that if you double the speed, you get four times the energy and four times the result. For example if a plane was moving at 500 miles per hour and went into a climb without applying any more power, it would climb about 834 feet. If it was going twice as fast at 1000 miles per hour, it would be able to climb about 33,000 feet.

Similarly if a bullet was shot upwards at 1000 miles per hour, it would be able to climb to 33,000 feet (neglecting air friction in all cases). Unfortunately, the forward speed of an asteroid-induced tsunami would be in these high-speed ranges and consequently they would also be able to climb to great heights.

Both wind waves and tsunamis can generate sheet flow. When a wind wave reaches shore, there is forward movement of water which will move up the beach and then recede, due to gravity, back into the lake or ocean. If the beach is shallow and flat, the forward motion might continue for some distance inland. Very definite sheet flow can be observed in such cases. A similar situation holds true for tsunamis except in the tsunami case the energy content of the wave is always very high. As a tsunami approaches shore the same transition occurs as with a wind-wave. When a wind-wave reaches shore, the water is observed to have a definite forward motion. However in the wind-wave case, since the energy is contained in the top few inches (or feet) of water, there will not be enough energy to drive the water very far up the beach. In the tsunami case, since the energy would be contained in the entire depth of the water, at the beach there will be energy available to accelerate the water to high speed and drive it far inland. The water will flow inland as a sheet of water with a depth related to the energy of the wave just before it reaches the shore. Consequently a high- energy tsunami will produce a flowing sheet of considerable depth which will move at high speed. The speed is the factor of primary interest when it comes to climbing mountains or any other obstacles in the way. As the flowing sheet of water goes uphill it will slow down. Then, when a height of land or the ridge of a mountain range is reached, the water will speed up again going down the other side. In this manner its energy will be retained and the sheet of water would continue right over a continent to the next ocean. Table 2.3.1.6 Tsunami Speed-Elevation Relationship shows these relationships. (Please refer to the appendix for the calculation.)

Table 2.3.1.6 Tsunami Speed-Elevation Relationship

Tsunami Speed (mph)	Elevation (feet) (that could be reached by a speeding Tsunami)
50	83
60	120
100	331
500	8342
1000	33,372

In cases where the energy of the moving sheet of water wasn't quite enough to get it over the top, a turbulent accumulation of water would develop as the energy of the moving sheet kept forcing more water in behind the water near the front of the flow. Consequently the water would pile up in the area next to the obstacle. Then, as the kinetic (i.e. moving) energy was converted to the potential energy of the pile-up, the flow would reverse. The water would return back down the slope, recover most of its original energy and speed as it rushed back in the opposite direction. While it is not all clear how any animal life could have survived the first pass of the water, if anything did survive it would now have to contend with the reverse flow. Further, since great tsunamis would be speeding around the world in all directions, one could approach the obstacle from the other side. If it had enough energy, it might flow right over it from the other side and follow the first sheet of water back to the ocean.

How fast would an asteroid-induced tsunami be traveling? A phenomenon of this nature has never been observed but comparison with other events of this nature is instructive. As mentioned above, the earthquake-induced Chilean tsunami crossed the Pacific Ocean at approximately 400 miles per hour. (29) The Moon raises the tides of the world and near the equator and the peaks of these tides are moving at more than 1000 miles per hour. Asteroids move at much higher speeds than the tidal bulge so it would completely reasonable to suggest that they would move at more than 1000 miles per hour. The destructive power available with such water movement is unthinkable. They could move at 2000 miles per hour. With speeds in these higher ranges it is readily seen that they would be excellent mountain climbers. There are simply no mountains or other types of landforms on the Earth that could stop them or even slow them down. From a hazard-to-life view point it is therefore apparent that even a relatively small asteroid (as in the Burkle case), would produce tsunamis that would readily encircle the Earth repeatedly. From a review of the Earth Impact Database, (which is available to everyone online) it is apparent that there have been numerous hits involving objects large enough to produce craters 300 km (187 m) in diameter. But even this great size does not recognize other possible impact sites such as the Congo Basin, Hudson Bay, (39) the Takla Makan Desert of China or the Ishim Meteorite crater in the Tenitz Basin of central Kazakhstan, (38) all of which are much larger still. How high and fast were the waves produced by these monsters? How did anything survive? Other effects of these tsunamis will be discussed in a later chapter.

2.3.2   Ground Impact

Ground impact, whether it was above water or below water, would have several similarities. The major distinction relates to the thickness of the Earth's crust which is usually given as being thicker for land that is above water than it is for land that is below the sea. The water would have some shock absorption function for flying debris but the thinner crust would virtually guarantee that even a medium-sized asteroid would punch right through the crust and expose the hot interior. These and a few other distinctions will be recognized in the following section.

2.3.2.1   The Mud-Rock Splash

As the asteroid presses into the surface, material would be displaced. Two different masses of material cannot be in the same place at the same time. Some of the surface material would be forced aside and form a great mud-rock splash. Material would fly up and arc out over the countryside in a manner which was very similar to the water splash. The high speed of the asteroid ensures that much of the surface material would become virtually fluid whether it started off as soil or gravel or even rock. Some commentators have suggested that rock would actually melt in which case it would splash away just like water. (32) Much of the surface rock would break up and even a portion of the asteroid would break up and join the other rocks flying up from the surface. From a Window of Life viewpoint, being in the surrounding area would mean virtually certain death. At the Chesapeake Bay impact site it has been determined that numerous very large rock fragments have been dislocated, strewn about for miles and have been identified as far away as several crater diameters (i.e. 85 km. or 53 m.) up the coast to the north-north-east. A shower of rocks and mud of such a nature would not have been survivable by any form of animal life!

2.3.2.2   Erratic Boulders

Erratic boulders are boulders that are out of place and are clearly not simply broken parts of the rock where they are found. 'Rocks that differ from the formations on which they lie are called "erratic boulders".' (33) The apparent origin of many erratic boulders has been identified and it appears that they have traveled hundreds of miles in some cases to arrive at their present resting places. 'In North America erratic blocks, broken from the granite of Canada and Labrador, were spread over Maine, New Hampshire, Vermont, Massachusetts, Connecticut, New York, New Jersey, Michigan, Wisconsin and Ohio. … The attentive traveler through the woods wonders at the size of these rocks, brought there and abandoned sometime in the past, frighteningly piled up. (34) Numerous erratic boulders are very large and many are found at much higher elevations than their origins. 'Some erratics are enormous. The block near Conway, New Hampshire, is 90 by 40 by 38 ft and weighs about 10,000 tons, the load of a large cargo ship. … The great flat erratic in Warren County, Ohio, weighs approximately 13,500 tons and covers three quarters of an acre; the Okotoks Erratic, thirty miles south of Calgary, Alberta, consists of two pieces of quartzite "delivered from at least 50 miles to the west," of a calculated weight of over 18,000 tons. Blocks of 250 to 300 feet in circumference, however, are small when compared with a mass of chalkstone near Malmo in southern Sweden, which is "thirty miles long, one thousand feet wide and from one hundred to two hundred feet in thickness and which has been transported an indefinite distance… '. (34) The usual explanation that ice carried these monsters is not credible. The basic problem with that explanation is that most of the ice that formed during the great Ice Age did not move anywhere. Although such a claim is widely held there has never been an explanation of how the ice moved. Further the ice was not strong enough to have been made moveable by anything but to make matters worse no theory of any kind has ever been advanced to explain the source of the enormous force that would have been required for movement. If such a force had been applied, the weakness of the ice would have caused it to structurally fail! So the answer is that in most cases the ice did not move at all!

Erratic boulders that are particularly difficult to explain are those found perched precariously on mountain ledges. ' … they perch on the top of ridges … they balance precariously on the edge of crests.' (34) The photograph 10.4.1 included in Part 1, Adirondack Erratic Boulder, and include here for convenience, was taken in the Adirondack Mountains of upper New York State. In this case, the boulder is overhanging a slope which is too steep for it to remain on if it should move slightly from where it presently sits. This means that if it should be caused to creep forward only another foot or two, it will plummet into Avalanche Lake below. Also, its position so close to the very peak of Mount Colden (i.e. about 100 ft) in an area where tremors occur every few years, is also indicative that it has not been there very long!

2.3.2.2 (1) Adirondack
Erratic Boulder

Another erratic boulder that is much better known and in fact widely recognized as an erratic, is the Okotoks boulder of southern Alberta. Its name is derived, as is the nearby town, from the Blackfoot word for "Big Rock". It is located approximately 45 km. south of Calgary. In this case the rock is thought to have come from the Jasper area which is about 400 km (250 m) to the north-west. Other Erratics are also in the area. 'Noticeably different than the gold-hued sandstone that surrounds it, the Big Rock is the largest remnant of the Foothills Erratic Train … carried down from a Jasper mountain peak.' (133) Other members of this group lie near the towns of Airdrie and Glenwoodville but the Okotoks boulder shown in the picture, 2.3.2.2 (2) Okotoks Erratic Boulder, is the largest at 16,500 tonnes. Erratic boulders require an explanation and the explanation is that an asteroid placed them. In particular, the Ries Crater in the southern plains of Bavaria is surrounded by boulders. The estimated size of the asteroid that formed this crater is a diameter of about a mile and weighing more than a billion tons with the energy of several million megatons of TNT. The force of the explosion is beyond imagination. Huge boulders are found in the surrounding countryside up to distances of more than 35 miles. (39) By contrast, the crater that formed when the great asteroid struck near Sudbury, Canada, is given as 250 km (156 m) across. (35) This means that the material of the Earth's crust was completely dislocated out that far. Then how far did rocks fly from the area? Did they move another 250 km. or even more? If the rocks from the Ries Crater were thrown for two times the asteroid diameter, (or four times the crater radius) it wouldn't be unreasonable to expect that rocks from a much larger object would be thrown a comparatively further distance. If this were the case with the Sudbury impact, we would expect rocks to have been thrown for 4 x 25 or 500 km (310 m) at least. With this possibility in mind we recall the above quote. 'In North America erratic blocks, broken from the granite of Canada and Labrador, were spread over Maine, New Hampshire, Vermont, Massachusetts, Connecticut, New York, New Jersey, Michigan, Wisconsin and Ohio. The attentive traveler through the woods wonders at the size of these rocks, brought there and abandoned sometime in the past, frighteningly piled up. (34)

Even if one should take a hammer which was, for example, three inches in diameter, and drive it into a pile of gravel, some of the gravel will be splashed around for several hammer diameters, possibly 12 inches or more. If a large asteroid hit the Earth at 25,000 miles per hour, material from the crust of the Earth would be widely spread around, expectedly for at least several crater diameters (i.e. for hundreds of miles). Similarly, when the Manicouagan crater in Quebec was formed, rocks ejected from the impact site would have been spread around the countryside for many miles. While both the Manicouagan and the Sudbury craters are in the "very large" category, what are we to make of the chaotic terrain that covers most of Canada, Greenland and the north-eastern portion of the USA. This feature, called the Laurentian Plateau, is roughly centered on Hudson Bay. Across this entire area, which involves more than 3,000,000 square miles, there are no "layers" of the Earth and the crust of the Earth has been broken and split and totally disrupted. (36) If this feature were observed from space without the Earth's ocean being present, it would, in all likelihood, be recognized as an asteroid impact feature. Further, if it was caused by an asteroid, we could expect erratic boulders to be found all over the world because some of them would have gone into sub-orbit and even landed on the opposite side of the world. Indeed, one commentator came to this conclusion by explaining that some of the boulders ejected from an impact site would go into sub-orbit which means that they could come down anyplace on the Earth. Some of the rock ejected from the crater flew right out of the atmosphere rained down destruction on distant continents. Some even escaped into space. (50) Then, when we consider the hazard to life, how could any creature, large or small, have survived when all of these boulders came crashing down to their present resting places? Within an hour, burning rocks would have been pelting down everywhere setting much of the Earth on fire. (51) Since very large asteroid impact sites have been found all over the world and the rocks from even one of them could have spread around the entire world, we would expect erratic boulders to be found all over the world! By association, we would expect mass extinction to have occurred around the entire world from this asteroid-impact factor alone.

2.3.2.2 (2) Okotoks Erratic Boulder

2.3.2.3 The Earthquake

An earthquake is the sudden and repeated movement of the crust or surface of the Earth. Even a movement of a few inches will be quite noticeable but movement of a few feet will be disasterous for tall buildings, bridges and other public works. In earthquake-prone areas, some tall buildings are now designed so that their foundations can move back and forth without actually moving the building itself. In these arrangements the foundation would be able to move several feet back and forth while the building on top would stay in one place. Without such an arrangement, if the foundation moved out from under the upper part of the building, the building would fall over. Of course all buildings can withstand some foundation movement and the upper levels will just sway slightly back and forth. However if there is repeated movement too far in any direction the building will probably just come down. However, even these modern arrangements would not be able to deal with the displacements that would occur with the impact of an asteroid. Since the asteroids of interest in this case would be several miles in diameter, the ground would dislocate a similar distance. There isn't any building that can withstand the displacement of its foundation for 1000 feet and the displacement following an impact would quite readily be even greater than this! The extremely violent shaking would not be a local event but would, for a large asteroid, extend outwards for several thousand miles from ground zero. Within this radius, even well rooted trees would be dislocated and wrenched right out of the ground and tossed around like matches. Other aspects of this phase of the disaster would include the ground progressively breaking open as earthquake waves traveled along the surface. Water would surge out of reservoirs and mix with the rest of the debris that would be surging and flying all around. Other results of the earthquake including mountain formation, chaotic terrain, as well as the generation of a large set of tsunamis, will be discussed in a later chapter. As disasterous as all of these things would be, they would be only the beginning of the convulsion that the Earth would have suffered when it was hit by a large asteroid. Discussion of these other aspects will be included in a later section.

2.3.2.4 Continent-Generated Tsunamis

An asteroid-generated earthquake would cause very violent shaking and it is the magnitude of the shaking that would be intolerable for most forms of animal life. The displacement would be rapid meaning that any object on the surface, whether it was living or non-living, would be horizontally displaced. Rocks and trees and anything on the surface of the ground would be shaken loose and moved. Actually it would be more accurate to say that the rocks and trees would, due to their own inertia, stay almost in the same place while the ground moved out from under them. On a smaller scale it would be similar to a person standing on a rug which is quickly pulled out from under him. Of course, he would move. However, it would be likely that the rug would move further and faster. Similarly, the initial horizontal displacement of the ground following the impact of a large asteroid would be much greater than anything on the surface and because the asteroid might have been several miles in diameter, the displacement of the ground might be in the thousands of feet even a considerable distance from ground zero. This idea has been captured by the following two comments. 'Collision with an asteroid 15 km in diameter would have made the Earth's crust move up and down one hundred meters 1,000 km from the impact site. (53) The earthquake that hit Lisbon, Portugal in 1755 produced a sideways lurch. ' … the city was hit by a sudden sideways lurch …

The convulsive force was so great that the water rushed out of the city's harbor and returned in a wave fifty feet high …' (52)

When we recall the reason that the primary tsunami formed, it is understandable how tsunamis were generated by the earthquake. Any sudden sideways lurch will produce a tsunami if the ground that lurches is in contact with the ocean because that is the way that all tsunamis are generated. An impacting asteroid would cause the ground to lurch sideways in every direction from ground zero. With a large asteroid, the sideways motion would be enormous in the vicinity of ground zero simply because a large asteroid would cause the ground to move a large distance. And it could still be several feet even if the asteroid landed on the far side of the continent. Chesapeake Bay, for example, is on the east coast of the United States but a tsunami would have been generated on the west coast. The following table includes a list of the known major asteroid impact sites together with reasonable expectations of the resulting tsunamis.

Tsunami Expectation Table (xx) indicates the reference

Asteroid,Size	Crater Size	Water/ Land	Dis. to Coast	Ex. Tsunami	Dis. To 2nd Coast	Ex Tsunami
Ches. (56), 2-3 m	53 m	50/50	immediate	huge	3000 m	small
Acra. (55), 3 m	56 m	land	200 m	huge	700 m	medium
Popigai (57) 3 ¼ m	62 m	land	immediate	huge	600 m	medium
Chic., (58), 9.4 m	106 m	50/50	immediate	huge	350 m	huge
Sud., (59),10+ m	134 m	land	550 m	huge	1900 m	medium
Manic., (58), 3 m	62 m	land	300 m	huge	3500 m	small
Vred., (57), 10+ m	188 m	land	250 m	huge	250 m	huge
Czech, (60), 10+m	200 m	land	300 m	huge	300 m	huge

While this table includes the largest of the currently recognized impact sites, there are also a considerable number of craters suggestive of impacting asteroids all the way down from those listed above to those only capable of forming craters like the Arizona Crater (also called Barringer Crater or Meteor Crater). This crater is a little more than three-quarters of a mile in diameter and it might have been formed by an object as small as 50 feet in diameter. (61) (Never-the-less an earthquake would have resulted and a tsunami would have been generated on the Pacific Ocean coast, the Gulf of Mexico coast and the Atlantic Ocean coast. While it might only have been a few feet high, it would have been a threat to every coastal community within its reach. This would have included the west coast of Europe and the east coast of Asia.

As mentioned it is the sudden lurch of a landmass which is in direct contact with ocean water that produces a tsunami. The initial lurch resulting from every one of the impactors in the above table would have resulted in correspondingly high tsunamis. All of them would then have raced away across their respective bodies of water. All of them would have completely encircled the globe and returned to their starting location. Many of them would have circled the world a second and even a third time. Low-lying land forms would have been easily crossed and these waves would have simply carried on. The characteristic of an earthquake includes ongoing shaking or vibration following the initial lurch. Some earthquakes are even quiet for a while and then produce more lurching and shaking. The result of all of this commotion would have been more tsunamis. In every case there would have been a series of waves following each other across the ocean. In every case they would have reached far inland and in some cases, run completely across continents. The aggregate turmoil and devastation for the whole world would have been complete. Entire forests would have been washed away and even the soil that the trees grew in would have been washed away. There isn't any form of animal life that would have been able to survive! In fact mass extinctions of life would be expected, (63) and numerous mass extinctions have been recognized. (64, 65, 66) While the most pessimistic commentator has only allowed for about a dozen mass extinctions, each one of them should be tied to the impact of a large asteroid. If this is really the case however, at least one hundred should be recognized because there have been at least that many major asteroid impacts. (67) Then just to emphasize the point, another large crater is found every few years! So how many were there really?

2.3.2.5   Sedimentary Rock

Sedimentary rocks are 'those which have been deposited as beds … on the floors of rivers, lakes and seas …' (70) Sedimentary rock is found on the continents all over the world and is the type of rock from which most mountains have been formed. In order for this type of rock to have been formed, the material had to have been eroded, transported and deposited and because of the thickness of the various layers of rock, these three processes must have happened on a massive scale (which is clearly not happening at the present time). Asteroid-generated tsunamis which would have produced deep, rapidly-flowing sheet flow with mountain-climbing and continent-crossing power would have been an appropriate mechanism to carry out the erosion, transportation and deposition phases required to form the great sedimentary layers of the Earth and thereby have caused them to come into existence. All three of these factors required very large amounts of the appropriate types of energy and a large asteroid would have been able to supply this energy. Sedimentary rocks are direct evidence that there has been massive continent-crossing sheet flow of water and therefore direct evidence that world-encircling tsunamis have existed. Every large asteroid that has hit the Earth (and there have been many) would have produced three types of tsunami; the primary, the secondary and the continent-generated which means that every large asteroid impact would have resulted in the production of numerous world-encircling tsunamis. Each one of them, dependent on their specific speed at any given location, would have been able to erode loose material, carry a great load of material and deposit all of it at a location where the speed of forward motion decreased enough to allow the sediment to settle out of the flow. A similar type of action is ongoing with many of the great rivers of the world. In these cases deltas are formed and commonly extend for many miles into calmer, deeper water. While the tsunamis from any one particular impact would have been able to circle the world repeatedly, if an asteroid shower had been involved, the process would have continued for months and resulted in regions which would wind up with dozens of layers! Is there any other natural mechanism that could have done this?

Similarly it would have taken the energy of an asteroid to form a mountain range. If anyone is to have a theory concerning how either mountains or sedimentary rock formed, the energy factor must be addressed. In both cases ascribing the required energy to an asteroid is scientifically valid. Unfortunately, conventional theory (i.e. Plate Tectonics) does not address the energy requirement at all.

Some sedimentary rock layers are hundreds of feet thick and when the total assembly in some places is measured, thousands of feet of thickness are observed. How much energy did the placement of this material require? When these massive amounts of energy were in play, how would any form of animal life have survived? When the material was being eroded where would any animal have found refuge, underground? When the material was being transported by the sheet-flow of water – possibly one thousand feet deep, how would any animal have survived? When the material was being deposited, how would any animal have avoided being buried? Unfortunately, we observe that many millions of animals have been buried and have become the fossils that are presently being excavated. We also observe that the placement and arrangement of these fossils is very seldom in neat and orderly layers but more often in closely-packed chaotic arrangements where the bones of any particular animal are broken and often not even connected to other bones but disarticulated instead! Such arrangements indicate that utter chaos was involved in their placement and the chaos of world-encircling tsunamis repeatedly showing up would have been sufficient. 'The dinosaur record is actually rather poor. Intact, articulated, museum-quality skeletons are fairly rare. … In some bone beds there are so many tiny skeletal fragments you'd think the creatures had been dropped from a plane.' (68) Death came violently and suddenly, even to fish. 'Numberless living beings have been the victims of these catastrophes; some, which inhabited the dry land, have been swallowed up by inundation; others which peopled the waters, have been laid dry, the bottom of the sea having been suddenly raised; … a wonderful record of violent death falling at once, not on a few individuals, but on whole tribes … In the red sandstone an abundant aquatic fauna is embedded. The animals are in disturbed positions … The figures are contorted, contracted, curved; the tail in many instances is bent around to the head; the spines stick out … The Pterichthys show its arms extended at the stiffest angle, as if prepared for the enemy.' (68) These creatures sensed that trouble was approaching, tried to prepare for it but then were hit so suddenly that the fright-indicating postures (of their bodies) were preserved by instant burial. Why did all of this happen? The chaotic high-energy event of their placement precluded any semblance of order and is explicable by the crashing of a large asteroid into the Earth!

2.3.2.6 Mountain Formation

As with erratic boulders, mountains are also found all over the world. The usual explanation is that over great periods of time, sections of the crust of the Earth pressed relentlessly against other sections and at the interface, the crust crumbled and broke and piled up into the particular form of chaotic terrain that we call mountains. Explanations are not offered to explain what caused the movement. Indeed it is hard to explain because an appropriate source of energy has not been identified. The popular-science explanation continues with the declaration that while mountains were being formed, other portions of the crust are being forced down into the interior of the Earth, that is, they were being subducted. Unfortunately, the various aspects of the geology of the Earth that are being relied upon as evidence of subduction contradict the idea instead. In particular, submarine trenches are offered as evidence of subduction. But in order to be supportive of the idea, they should be full of sediment. (40) If they are actually sections of ocean

446

floorthat has been migrating across the ocean for thousands of miles, they should be full of sediment, the sediment which had been accumulating on the ocean floor for thousands of years or longer. It is unfortunate for the theory (i.e. the Plate Tectonics Theory) that the trenches are empty. The Chili Trench has almost no sediment, the Middle America Trench and the Tonga trenches are barren of sediment. Also the Kermadec Trench is bare rockand some locations in the Japan Trench have no sediment. (40) In support of these observations we also have the Mariana Trench, in particular its deepest region, the Challenger Deep area which is given as approximately 11,000 meters (6.875 miles) deep. As the deepest location in the ocean, it provides a major challenge for deep sea exploration but because of the numerous difficulties and the high cost involved, only a very few groups attempt to explore it. (62) But why is it there to explore? It couldn't be a subduction zone because if it was, it would not be 11,000 deep. It would be full of sediment instead!

It is even more unfortunate for the theory that the average depth of sediment on the entire deep ocean floor is 2100 ft (0.40 m) and on the entire ocean floor, both shallow and deep, is 2950 ft (0.56 m). (37) Why wouldn't we find at least this much sediment in every one of the so-called subduction trenches, even if the ocean floor had only moved into each trench for one width of itself?

(Two scientific factors indicate that ocean sediment is a Window of Life. The first is that the present average depth of sediment is about 3000 ft and the second is that the rate of sediment entry into the oceans is about 27 billion billion tons per year. (27) At this rate it would only have taken about 30 million years for the present sediment inventory to have accumulated. Therefore, if the present rate were to continue, the ocean would be completely full of sediment in about 100 million years. However, long before that level was reached, the ocean as a source of food would have been reduced to hardly anything. Any form of life which was dependent on the ocean would therefore be hard pressed to survive and the Window of Life for that creature would have narrowed and probably have closed.)

It has also been declared that the entire process has been going on for very long periods of time. So it isn't much better for long-age theories that much of the entire ocean floor is void of sediment because the quantity of sediment-forming material which is continually entering the ocean has been measured and if the present rate of entry has been happening for a very long time, the ocean should be completely overflowing with it by now. (37) So it is unfortunate that none of the declarations associated with Plate Tectonics theory (i.e. the current predominate mountain-formation theory) stand any serious scrutiny whatsoever because they do not accommodate the evidence. So another theory is required. The one being offered herein is that the asteroids caused the mountains to be formed (and along with them the submarine trenches) and that, since all of the happenings associated with the impact of an asteroid occur within a very short time period, the mountains and the trenches formed within a very short time period. But how did it happen?

The aspect of the asteroid impacts that caused the mountains to be formed was the earthquakes (i.e. the ground shockwaves) that spread out in all directions from each impact site. When an asteroid crashes into the crust of the Earth, the crust will be displaced and broken. Then, a portion of the enormous energy associated with the asteroid will be transferred away from the impact site by the surrounding crust as well

447

as the material in the interior of the Earth. The shockwaves that spread out from an impact site use the material of the Earth as an energy transfer medium to distribute energy throughout the entire world. Some of this energy has formed the mountains. Further, it is understood that there would be enough energy in a single asteroid (approximately 8 km (5 m) dia.) to form an entire mountain range more than 200 times. (41) Indeed, it really is obvious that the formation of a mountain range would require energy. In order to move the Adirondack Rock pictured above, up to where it presently rests, would have required energy. It even required energy to climb Mount Colden to get the picture. In particular, in order to form a mountain range, like the Rocky Mountains, for example, would certainly have required a great amount of energy.

Not only was a great quantity of energy necessary, in order to form the mountains before the sedimentary layers became too hard and brittle, the energy had to have been available at a very high power level. That is, it had to have been transferred quickly throughout the entire area where the mountains were being formed! Where could such great quantities of energy with such a high power levels, have come from? Conventional theory offers no explanation whatsoever for this energy requirement! Recognizing that the energy of an asteroid is of the right magnitude and more than sufficient to form a mountain range is therefore a step towards having a theory that is at least internally consistent. But how did the mountains actually form?

One of the disturbing aspects of an earthquake is the way that the energy is propagated. During the Alaska earthquake of 1964, earth-waves were observed traveling along the surface of the ground. 'The land had become the ocean, the streets were rippling like waves … The earth … rolled for five minutes.' (42) As these waves rippled along the ground, like the waves on a lake, the ground continuously broke open and reclosed again. The same thing happened in the Mississippi river valley during the New Madrid earthquakes of 1816. A series of waves about three feet high moved along the ground and the ground opened and closed as they passed. ' … just after midnight … locals found the land rolling in waves up to three feet high and opening in fissures several feet deep.' (43) In the New Madrid case, the local residents feared that they might fall into the temporary openings and be swallowed by the ground.

Transferring energy by means of surface waves is very common and can be observed on any body of water on a breezy day. While transferring energy along the ground in this manner is much less common, we can be thankful that this is the case because life on Earth would not be tenable if it was a common widespread experience. Never-the-less it is quite clear that energy can be propagated by water or land as a wave pattern. In order to transfer energy through the ground, much higher energy levels would be required to achieve the same wave height than would be required for water waves. Mountains formed this way. The difference is that mountains are bigger and it would have required much higher levels of energy to cause a surface wave, the height of a mountain, to have been formed and propagated. But even a relatively small asteroid would have had enough energy to generate land waves of such a magnitude and the folded rolling strata (i.e. the anticlines and synclines or the ups and downs) of the mountains tell us that this was the type of energy that was involved.

Mountain formations consist of anticlines and synclines. The layers of material forming the mountains are curved up and down in the shape of giant waves because that is what they are. The energy has been transmitted but just like the situation when a water wave passes, the energy transmitting material has been left behind. With water waves, the energy travels along and continues to faraway places but the energy transfer material (i.e. the water) remains behind. Even with a tsunami where there is a great amount of energy being transferred, up until the tsunami comes ashore, the energy transfer medium remains behind and in fact never actually moves more than a few feet. With a wind wave the actual water travel is approximately the height of the wave vertically and a portion of this distance horizontally. We are dealing, in both cases with energy transfer, not material transfer.

When mountain ranges are observed from above, they appear as waves would on the ocean. They have all formed into ranges exactly the same way that water waves form on any body of water. One mountain range is followed by another mountain range and the whole assembly looks like a frozen series of great waves. When the great land waves passed through the area, the surface material was in a state that was appropriate to transfer the energy while leaving the material behind but when it had left there wasn't any energy available to return the material back to its original location. So it just stayed put and now we have mountains.

It is also recognized that virtually all of the mountain ranges of the world consist of sedimentary material. This material had to have been soft and unconsolidated when it was deposited and this characteristic made it an excellent energy-transfer medium for earthquake-/shockwaves. So when an asteroid-generated earthquake-/shockwave arrived, its energy was conducted forward by the soft sedimentary material as we might expect and it was conducted forward as a wave. The energy of the shockwave caused giant land waves to form and any remaining energy just petered out into foothills as it moved away. The now-deformed material stayed in place and hardened. In many places the material which had formed the peaks of the waves, broke open and did not reclose.

Mountain ranges occur all over the world. Since they all consist of sedimentary material it is immediately obvious that the mountains had to have formed quickly and almost as soon as the sediments were deposited. Otherwise the sediments would have started to harden. If they had been hard, they could not have been stretched and bent to form the upper regions of the anticlines or the lower regions of the synclines. Neither could they have been compressed and bent to form the lower regions of the anticlines or the upper regions of the synclines. Therefore they had to have been in a soft wax-like workable, bendable-without-breaking state. (45) Unfortunately for conventional theory, soft sediment will not stay soft. It hardens. The amount of time required for it to harden so much that it cannot be bent without breaking is not known. But it has been observed to harden very quickly. There is the case of a person, who lived during a recent time period, being found partially encased in rock. 'In the British Museum there is a skeleton embedded in solid rock which came from the island of Quadaloupe in the West Indies. The rock is a hard limestone and also contains fragments of shells and coral. The skeleton is that of an Indian killed in battle with the British only two centuries ago. Victoria Institute 13:36.' (44) Since this event was historically recent, (i.e. within the last few hundred years) long time periods are obviously not required for sediment hardening. If, in this case, the unfortunate had arrived on the scene a few years or even possibly months later, his body would not have become trapped in the sediment at all but simply stayed on top of it (and we would not have the evidence that sedimentary rock forms quickly). It is also relevant to notice that the petrified forests of Arizona consist of rock which has preserved the exact shape of the wood that was replaced. (136) This means that the wood must have turned into rock before it had a chance to degenerate in the usual way that wood always degenerates. (Other evidence such as the work by Gentry concluding that some rocks have even formed instantaneously, and the finding of a nail in Old Red Sandstone in Scotland (137) further underscores the point that long times for rock formation are not necessary.) Based on this evidence, the mountains, because they had to have formed from soft material (i.e. material that was soft and plastic' (45)) must have formed quickly! (Otherwise mountains would consist of flat parallel strata simply broken and found in haphazard formations instead of waves) This is consistent with the transfer of energy from an asteroid impact. Most of this energy will be dissipated and transferred to wherever it is going, on the same day that the asteroid strikes. If we come back tomorrow, the deed will have been done and hopefully, things will be quieting down.

We must also note that the disruption which was necessary to form the sediments, which are often observed as being thousands of feet thick, could not have simply occurred on one continent at one time and another continent at another time. It had to have happened on every continent as part of a single great event. The great amounts of energy required to place the sediments were also the result of asteroid impact. (as manifested by the tsunamis) Such great disruption would not have been restricted to one continent but had to have been worldwide. It is therefore concluded that the mountains of the world, in particular, all those that consist of sedimentary material, which includes the Alps, the Rocky Mountains, the Andes, and the Himalayas to name a few, formed as a result of the impact of a very large asteroid or possibly even a shower of asteroids and formed within a short time period.

The earthquakes that continue to plague the Earth are often associated with human death on large scales. 'The 1975 earthquake (China) … at least 655,000 dead (Also in China) … in 1556 that killed 830,000 … in Guatemala … 1976, 23,000 … were killed'. (46) While we understand that these particular events occurred in populated areas, an asteroid-generated earthquake would reach populated areas no matter where the asteroid struck. The resulting death toll would consequently number many millions from this factor alone. Even the New Madrid earthquakes were considered a major hazard for human survival. When it is recognized that the magnitude of asteroid-generated mountain waves would be many thousands of times greater than the New Madrid earth-waves, it is understood that the hazard to human and animal survival would have been correspondingly greater. In fact there isn't any conceivable way that anything could have survived disruptions of such magnitude.

2.3.2.7   Continental Shelf Formation

"Continental shelf" is the term applied to submerged areas adjacent to the continents where the depth of the water is less than 600 feet deep. Approximately 15% of the area of the Earth (i.e. 10,000,000 sq. m.) is continental shelf area. (72) Beyond the relatively shallow regions of the shelves, the depth of the ocean increases quite dramatically and the ocean bottom plunges down to the deep abyss which comprises most of the ocean.

The continental shelves have been formed from silt. This is fine-grained material which would have been easily transported by water. River deltas have a similar structure. River deltas have been formed from material which was transported into the sea by the adjacent river. This relationship has never been in doubt. While the material was in transit it only muddied the water and appeared as if it hardly had any substance whatsoever but when it all settled out and had time to dry, it was quite capable of supporting buildings and roads. Consider the Mississippi River delta. At the city of New Orleans the delta material is approximately forty feet deep. (73) In this case it is supporting a city with its streets and buildings. The delta extends far beyond New Orleans and it is inhabited for about another hundred miles further south where numerous people live with their support structures of buildings and roads. This delta continues to grow as more material is brought down and continually settles out of the flow when the speed of the current drops below the level required for transport.

The time for this delta to have developed has been calculated and at only a few thousand years, (74) it is not as long as many commentators would prefer. Further the rate of development is obviously dependent on the flow rate of the river so if the river flow rate had been greater in the past, the delta would have formed quicker. Alternately, even at the present river flow rate, if the delta had been forming for a much longer time period than this it would be half way to South America by now!

Continental shelves formed the same way. They formed when transportable silt was washed from the continents. When the silt-loaded water entered the ocean, it slowed down and the silt settled out. Conversely, when the silt became entrained in the flowing water, the water must have been moving at greater speed. This water washed the silt from the continental areas by going inland and picking up the silt or by washing right across the continent and picking the silt up on the way across. Since continental shelves are found on more than one side of some continents, it is clear that silt-entrained flow occurred in more than one direction. Once this silt had settled out on the ocean bottom it would have been much more difficult to disturb it enough to re-entrain it and subsequent currents would have passed over it without causing further disturbance. This means that the next tsunami could have been going in the opposite direction and passed right over the results of the first one with only minor disturbance. In fact the continental shelf surface location several hundred feet down would have allowed this type of action to recur repeatedly.

The presence of continental shelves around many continental areas testifies to the notion that continent-crossing sheet-flow actually occurred in almost every imaginable direction. This evidence supports the idea that the asteroid impact event was actually a multiple-impact event, that is, it was an asteroid shower. The tsunamis that were generated traveled in numerous different directions, circled the world repeatedly and in the process caused the continental shelves to be formed around most continental areas.

As discussed, in order to form the continental shelves, material had to have been washed off of the continents. If this happened because sheet-flow currents were crossing the continents in all directions, what would have happened to the animal life? Where would the horses have taken shelter? How would the groundhogs have survived if the material forming their burrows was being picked up by passing waterflows and carried away to the ocean? How would a prairie of grazing animals have been able to deal with these great water currents? Is there any way that any animal could have survived content-crossing water flowing in the great sheets that would have been formed by the massive asteroid-generated tsunamis as they came ashore, raced inland and then carried away a massive load of silt to form the continental shelves? Every continent was involved. The current arrangement of continental shelf material indicates that the currents flowed in all directions. If there is to be a theory of survival for this scenario it must recognize and accommodate these continent-crossing monsters.

2.4 Activity from the Interior of the Earth

The asteroid would have rushed through the atmosphere and crashed into the Earth within a very short space of time of very probably less than one second. Numerous hazards would have developed from its passage and Earth contact. As discussed, all of them would have been serious world-wide events, even if the impacting asteroid had been in the relatively small category of one or two miles in diameter. Fires would have been a hazard to survival hundreds of miles from the impact site and erratic boulders would have been spread far and wide and some could easily have landed on the far side of the Earth. However, with full recognition of the world-wide implications, the greatest hazards to survival from the effects of the asteroid rushing through the atmosphere and hitting the ground would have been in the hemisphere where the asteroid came down. This would seem to suggest that the chances of survival would have been better on the far side of the world. Unfortunately this would not have been the case. The troubles on the far side of the world would also have been most hazardous to survival and being on the far side would have not been any better than being on the impact side. The reason for this is the way in which the enormous energy of the impacting asteroid would have been distributed throughout the interior of the Earth. With this reality in mind, the Earth does not seem to be a very big place and it would certainly not have been big enough to provide survival shelter, even if one was on the far side of it. It might even have been worse on the far side – if such a circumstance were possible.

2.4.1 Momentum Transfer

When an asteroid of any size impacts the Earth and becomes absorbed into it, its momentum is transferred to the Earth. Momentum is a concept which includes both material (i.e. mass) and speed. Technically, momentum is the product of mass and speed. (i.e. the mass multiplied by the speed) (77) Obviously, the material of the asteroid would become mixed with and added to the material of the Earth. The speed isn't really added to the speed of the Earth but the effect of the speed is certainly added.

Momentum is discussed whenever two objects collide. If a truck should crash into the side of a house, it is the momentum that the truck is transferring to the house that becomes the cause of the trouble. If the truck was barely moving there would be very little damage to the house. If the truck had been going full speed the damage to the house would be substantial. Similarly, if an asteroid crashed into the Earth while it was barely moving, there would not be any damage and it would be like rolling a large ball along the surface. However, asteroids are never observed to be barely moving. They are always in a hurry and it is their extremely high speed that – like the speeding truck – causes the damage. The incredibly high energy of the crashing asteroid would be recognized in mountain formation, tsunami generation, earthquakes and

452

atmospheric shockwaves. A very large portion of its energy would generate a shockwave which would spread into the interior of the Earth and travel right through to the other side. The Earth would spasm and convulse due to the transfer of so much momentum and the interior shockwave would be the vehicle to carry the energy and momentum to the far side of the Earth. Upon its arrival, a very large area of the Earth's crust would rise up. Numerous catastrophes would result from this development. The interior shockwave would be the means of carrying the energy and momentum to the far side of the Earth and by so doing it becomes the observable aspect of the reality of this energy and momentum as well as the direct physical means of transferring it to a destination so far from the impact site.

Momentum would be transferred to the Earth from the asteroid in basically three steps; initial contact, during shockwave propagation and when the shockwave caused the far side of the Earth to well up in a great mound. There would be several results of momentum transfer including the introduction of a wobble into the rotation of the Earth on its axis. First the interior shockwave will be discussed.

2.4.2 The Interior Shockwave

An asteroid would rush through the atmosphere, crash into the Earth and disappear. The series of life-threatening hazards that would accompany these early phases of the arrival of the asteroid would be overwhelming to most forms of animal life but at least they would be centered around the impact site which would appear to give the far side of the world an advantage. However there would really not be an advantage to being that far away because of the effect that the asteroid would have on the interior of the Earth. The asteroid would certainly crash into the Earth, disappear and within a short time it, or its remains would come to a stop. However the effect of its entry would not come to a stop. A shockwave would develop ahead of the asteroid which would continue travelling into the interior spreading out in all directions. Nothing would stop this shockwave. The internal material of the Earth would slow and eventually stop the material of the asteroid. However in slowing and stopping it the energy of motion of the asteroid is transferred to energy of motion of the material of the Earth. This same material is the means to transmit a large fraction of the energy of the asteroid on into and right through the interior of the Earth all the way to the other side.

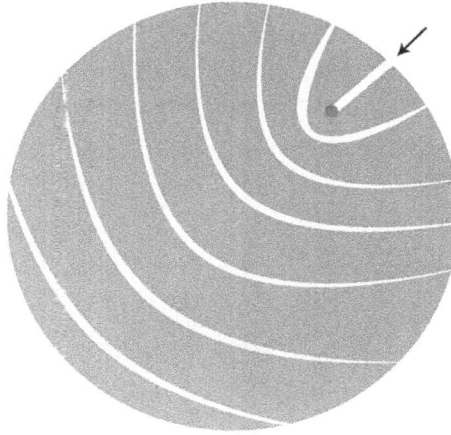

2.4.2 (1) The Interior Shockwave

If an asteroid punched through the crust of the Earth, it would generate
a shockwave which would travel throughout and cause enormous damage
as it resurfaced on the opposite side. In this manner, the overwhelming
energy of the asteroid would be distributed throughout the entire planet.

The diameter of the Earth is slightly less than 8,000 miles. The diameter of the asteroid would not be known but for discussion purposes we can assume that it would be 10 miles. Further, if the asteroid dove into the interior for ten diameters of itself, whatever was left of it would then be about 100 miles deep in the Earth. 100 miles is a significant fraction of the distance through the Earth to the other side. In fact it is more than 1% of the way through. If the asteroid hit the surface at 50,000 miles per hour and if it could have continued at this speed, it would have made it all the way through in less than ten minutes. Further, if the shockwave only traveled at 10% of the asteroid's contact speed, it would arrive at the far side in about an hour. The actual speed of the shockwave would not be known but usually shockwaves move faster in high density material (like the molten rock in the interior of the Earth) than they do in low density material (like air). Even if it only moved at 1% of the asteroid's contact speed, it would still come up under the far side of the Earth in less than one-half of a day. The erratic boulders that were thrown into sub-orbit by the crash would not get there much sooner.

It only requires about 0.5 % of the energy of an asteroid 10 miles in diameter to form a mountain range one km. (3/5 m) high and 480 km (300 m) long (63). Therefore even if several mountain ranges were formed by the surface shockwave and several very large tsunamis were generated, an enormous amount of energy would still be available to do other types of damage. A significant portion of this available energy would be contained in and transmitted by the interior shockwave. Therefore the trouble that would be expected would be sufficient to form several mountain ranges with plenty left over to cause even more trouble. And trouble would result because all of the asteroid's energy would soon appear as some form of chaos and none of it would be left in some type of reserve for a future time.

As a transmitter of energy, the interior material of the Earth would act in some ways like the interior balls in the suspended balls experiment.

2.4.2(2)   Suspended Balls Experiment

The similarity relates to the observation that the interior balls transmit energy but they do not really move very much. This type of transmission is what occurs with an atmospheric shockwave. In this case the material of the atmosphere (i.e. the air) transmits the shockwave but each molecule does not move very far from its original rest position. A water wave is a transmitter of energy but in the water wave case the molecules of water do displace back and forth as well as up and down as the energy of the wave is transmitted. In this example the water wave, whether it is a wind-wave or a tsunami, is more comparable to the effects of the interior shockwave than an atmospheric shockwave would be. The atmospheric shockwave transmits by compression while water waves transmit by displacement. The basic reason that this happens is because air, being a gas, is compressible whereas water is not compressible. If any substance is not compressible it must transmit energy by displacement and this would be the case with the interior shockwave.

2.4.3   The Ground Tsunami

As an asteroid crashes into the crust of the Earth, the material of the crust must move aside or travel along ahead of the asteroid as it forces its way into the Earth. In reality both of these developments would occur. Some material would be forced to the side and some would move ahead, at least for some distance before also moving to the side. None of the material involved is compressible. It cannot compress and bunch up like air can. Therefore motion and displacement would be involved. This motion would have two main consequences one of which would the generation of a ground tsunami. From a survival viewpoint this would be a development with terrible consequences.

If a large asteroid crashed into the ocean, immediately a huge mountain of water would form around the impact site. The magnitude of this mountain would be dependent on the energy of the asteroid. For a large asteroid a tsunami thousands of meters (there are 3 1/3 feet in a meter) high would develop. Could we expect a similar development for a ground tsunami? This would depend on how far material was displaced. If the asteroid was large and energetic enough to punch well into the crust before it stopped, the crust would be laterally displaced and a great earthquake would result but the material from the molten interior would not really move very much. However if the diameter of the asteroid was comparable to or greater than the actual thickness of the crust, such an asteroid would punch right through and continue on into the interior. The displacement of the interior material would be the cause and the reason that a ground tsunami would develop. The further the asteroid or its remains dove into the interior, the greater would be the volume of material in the interior that was temporarily displaced. Several of the already-known asteroids have been ten or more miles in diameter. Since the crust of the Earth, even in continental areas, is hardly twice as thick as this, (88) an asteroid of this magnitude would be expected to punch right through the crust and travel many miles into the interior. All of the material in its pathway would be moved aside. For example, if it was ten miles in diameter and it traveled into the Earth for ten diameters of itself, the volume of material that would be displaced would be approximately 7800 cubic miles. (Please refer to the appendix for this calculation.) Therefore the volume of the ground tsunami (above the zero displacement level until the temporary opening filled in again) would be 7800 cubic miles. The initial pileup of material would be concentrated around the impact site. If we assumed that the initial diameter (at its peak) of the departing tsunami was 100 miles, its circumference would be about 314 miles. If we further assumed that the great temporary mountain of material was 200 miles across at its base, the initial height of it would be greater than 1000 feet. While all of these dimensions are hypothetical, it is clearly seen that an asteroid-generated ground tsunami would be a serious survival problem. Various commentators basically agree with this analysis and offer the following. 'Collision with an asteroid 15 km in diameter would have made the Earth's surface move up and down by hundreds of feet hundreds of miles from the impact site. (79) This up and down motion would be the ground tsunami propagating outward from the impact location. (Repeated up and down is suggestive of a series of tsunamis,)

Over a great enough distance, the crust of the Earth, even though it is basically solid rock, would have some flexibility. If a great displacement pressure wave moved along under the crust and it had a large enough wavelength, the crust would bend a little to accommodate it. While crustal material isn't really flexible it would bend a little as the pressure wave travelled along. In fact because the crust has rigidity it would act like a damper and would be a factor in attenuating vertical movement especially after the pressure wave had traveled outward from ground zero for several hundred miles. However, if a large asteroid was involved, the bending stress introduced by the expanding pressure wave would be too much for the crust to withstand and it would shatter into innumerable jagged blocks. This would occur out to the so-called rim of the crater. Out to this distance the crust of the Earth, whether the entire depth of the crust was involved or not, would be shattered and a crater would form. Relatively small asteroids would only blast away material within the crust and form a crater that would very likely become filled in by other surface material. This is what happened in the Manson, Iowa case. Here a crater has been formed in the Earth's crust and then it has been filled in again by surface material. Consequently there is no surface manifestation of the crater at all. The fill-in agent in the Manson case was probably water because the crater is filled with gravel (71) and the rounded stones in gravel are indicative of water placement. Ice occasionally gets the credit for the fill-in procedure but a mechanism to move the ice of the great Ice Age has never been identified. On the other hand large asteroid impacts are always accompanied by massive water movement through the generation of world-encircling tsunamis so it seems logical to suggest that water was the agent in the Manson case. In any event the crater is filled in as are innumerable other craters around the world.

However, if the asteroid was energetic enough to punch through the crust, a totally different situation would develop. Molten material under the crust would be laterally displaced and a ground tsunami would result. As mentioned, the crust might flex a little as the displacement pressure wave passed. Near the impact site however this would not be the case. The lateral displacement would develop very quickly and the crust would not be able to withstand so much stress. It would shatter and be simply blown out of the area. Any crust that wasn't blown away would be broken and would bob and float on the molten material from underneath which would now be exposed to the surface since the crust had either been blown away or broken and shattered but otherwise remaining close to its original elevation. The molten material from the interior of the Earth would rise until it and the remaining broken pieces of crust re-established a gravity-based equilibrium so that the surface area of the crater followed the generally spherical shape of the Earth's surface. In this manner, the temporary crater opening would fill in again. The new level would be close to the original undisturbed level and would now consist of innumerable chunks of original broken crust floating in the molten material from down below. A chaotic fragmented uneven surface would result (i.e. chaotic terrain).

From a survival viewpoint, being in the vicinity when the ground tsunami was passing would have been like being on a trampoline. The difference is that a trampoline is quite soft and much more forgiving than solid ground. If any animal was caught in this situation it would move up and down with the surface of the ground. However the ground would be moving quickly and on the upward motion the creature would be pressed into the ground as if it weighed much more than it actually did. Then shortly thereafter, when the ground was moving downward, the animal would seem to be lighter and might even leave the surface. This would probably mean certain death because the ground would soon be moving up again and basically smash into the descending animal from underneath. The earthquake would only add to the animals' woes because the ground would be moving back and forth horizontally at the same time. These two components of motion would virtually guarantee death to all animal life in the area. Such a situation can also be compared to being on an elevator that was moving up and down in rapid succession. First the passengers would be smashing into the ceiling and then into the floor. This abuse together with a back and forth sideways motion which would smash them against the walls would result in an unsurvivable situation. Near the impact site within the "crater" area broken crust would be flying away in all directions and hot material from deep in the Earth would be welling up carrying broken bobbing crust with it. There would also be ongoing wave action as the jagged chunks of crust bobbed and twisted before settling down into their new molten support system. Where would any animal locate itself to avoid being tossed around and mixed into this roiling surface? The turbulence would extend for miles in every direction so finding a safe place to stand would not be possible. Certain death would be the fate of every animal within the actual crater zone as well as out to where vertical crustal movement had reduced to just a few inches. Immediately beyond the actual crater there would also have been flying pieces of crater material to deal with so it is safe to suggest that being within several crater diameters from ground zero would have meant certain death for every animal in the area.

2.4.4   The Continent Lifter

As the ground tsunami traveled outward from ground zero, it would basically be the surface manifestation of the interior shockwave which would be propagating deep into the Earth in a three dimensional pattern which would soon involve the entire interior of the world. There isn't any place on the Earth that would have avoided the interior shockwave one way or another but as the tsunami-shockwave converged on the far side of the Earth (i.e. the antipode) it would be coming up over a wide area and an even worse aspect of disaster would result, the continent lifter.

There are several aspects of the interior shockwave that can be compared to the tsunamis discussed in a previous section. First, as mentioned, energy transmission is by displacement rather than compression. Secondly, there would be material motion both back and forth as well as up and down. While the actual displacement of any particular volume of material would not be very great, because the pressure wave would involve so much material it is clear that a very great amount of energy was being transmitted. Thirdly, this energy would be transmitted forward at high speed. In a water-tsunami case the speed of transmission is commonly measured in the five hundred miles per hour range but since this parameter is dependent on the speed at which the tsunami was initiated, it would probably be much faster for an asteroid-generated tsunami. It certainly would not be slower. Fourthly, as the shockwave in both cases runs out of material with which to continue transmitting energy (i.e. the water-tsunamis run out of ocean) the motion changes from back and forth and up and down to just straight forward. All of the energy in both cases is carried by the last portion of material just before the transmission medium either runs out or completely changes. When the water tsunami runs out of ocean and there isn't any more ocean to carry the energy forward, the last volume of water next to the shore carries the energy onward itself because there isn't any more similar material ahead of it to receive and retransmit the energy. The result is that the water next to the beach leaves the ocean and rushes overland in a great forward-moving sheet of water. Similarly, as the ground-tsunami runs out of earth material as it surfaces at the antipode, the last portion of material, having nothing to receive and retransmit its energy simply carries the energy forward itself. This results in the ground at the antipode welling up into a great mound. If a very large asteroid was involved there would be a great amount of energy to deal with and a large area of ground would be affected. In such a case an entire continent would surge upwards. Alternately, a large area of ocean floor would surge upwards and the water on top of it would also rise up and then start racing down the slopes on every side. In this case, sheet flow of a large portion of ocean would result with the initial thickness of this flow being the depth of the ocean at the antipode. In this manner a further form of disaster would result and this one would develop on the far side of the world from where the asteroid actually touched down.

A very large asteroid impact site, Acraman, has been identified in south central Australia. The antipode of the Acraman Crater happens to be the North Atlantic Ocean. There is a major fault line running all the way down the Atlantic Ocean very close to the middle. Further, this fault line is accompanied by numerous side faults called graben (meaning stretch faults (80)). These side faults were formed along with the central fault as the entire ocean floor rose up. As it was forced up from underneath, a great fault formed all the way down the ocean. Material from the interior surged out and piled up on the ocean floor forming the mid-Atlantic Ridge. Since the ocean floor was forced up from below, the material of the ocean floor was stretched and could no longer cover the material underneath so lateral cracks formed – hence the stretch cracks. In order to achieve all of these fault developments, the ocean floor must have risen upwards for a considerable distance. The stretch faults extend for many miles on both sides of the central fault. The diameter of the asteroid which was large enough to cause this damage would have been probably upwards of five miles and could easily have been much more. All of the resulting displacements would have been commensurate with the size of the asteroid and the vertical displacement could, as with the water impact case, have also been measured in miles. Similarly if the pressure wave had come up underneath a continent, the continent would have been displaced upwards for a distance which also might have been measurable in miles or at least in appreciable fractions of a mile. Survival while such things were going on would have been impossible.

2.4.5   Chaotic Terrain

The crust of the Earth has many faults. It also has many major asteroid impact sites. Every major impact would have stressed the crust at the antipode and if the lifting-stretching force was great enough, faults would form as a means of stress relief. Similar developments have occurred on other planets. In particular Mercury has suffered a major asteroid hit and so has Mars. Both of these planets have chaotic terrain at the antipodes of the major impact sites. Venus has also been involved with numerous asteroid hits and the surface is so distorted that much of it could be called chaotic terrain. (89) In all cases including the Earth, the nature of the formations at the antipode would be dependent on the structure of the particular planet.

Mars has been well explored by both orbiting satellites as well as surface exploration machines so there is a great deal of information available regarding its surface features. There are three very large craters on Mars and they all occur on one side. Opposite Hellas, the largest, at approximately 1400 m (2240 km) across (75) there is an elevated region called the Tharsis Bulge. 'The Tharsis Bulge is a large shield or uplift area in the Opposite Hemisphere. (i.e. opposite Hellas) It is approximately 6 ½ miles high in the center relative to the general crust of Mars and it is about 3000 miles in diameter … ' (81) The Bulge includes several very large volcanoes, one of which, Olympus Mons, is the largest in the Solar System at 27 km (88,600 ft) high. (82) There are also several other huge volcanoes located on the Tharsis Bulge. 'Olympus Mons, like Ascraeus Mons, Pavonis Mons, Arsis Mons, Tharsis Tholus, Uranus Patera, and Uranus Tholus, is a volcano on the edge of the Tharsis Bulge. Olympus is on the northwest periphery of Tharsis. Ascraeus is in the north central section of Tharsis, like Pavonis. Arsis Mons is on the Westerly edge.' (80) All of these unusual features are located at the antipode of Mars largest asteroid impact site. There is also much more chaotic terrain on Mars. The Elysian Bulge is at the antipode of the Isidis impact asteroid. (Isidis is smaller than Hellas at 680 m diameter. (1100 km) (75)) While all of these bulges and volcanoes indicate that a great deal of chaos has happened in these areas, there is even more in the form of fissures in the same general area.

'Extending eastward from Tharsis is an immense canyon system, Valles Marimeris, while radiating outward from it are large arrays of tensional faults (graben) ... this 4,000 km (2500 m) long network begins on the east side of the Tharsis Bulge and ends in an immense region of chaotic terrain between Chryse Planitia and Margaritifer Sinus. At the deepest it is some 7 km deep and individual canyons are up to 200 km in width ...' (80)

The volcanic mountains, the bulge, the huge canyons and the rest of the chaotic terrain are located in the antipode region of a very large asteroid impact. It is therefore reasonable to conclude that the entire collection of chaotic features is the result of the monstrous pressure waves that would have traveled all of the way through Mars and come to the surface on the other side of the planet. The great pressure waves uplifted the entire region, caused numerous volcanoes to start belching and actually stretched the surface of the planet forming the canyon system. While the primary hit was around on the other side, being in the antipode region would also have meant certain death. When this type of chaotic terrain was being formed, there is no way that any form of animal life could have survived. This conclusion is in no way a suggestion that there was any animal life on Mars at that time or at any other time. It merely emphasizes that the antipode region of a large asteroid impact would be as dangerous as being right at the impact site itself. If such a terrible development could have occurred on Mars, why couldn't something similar happen on the Earth (if the Earth took a major asteroid hit). The fact is that the Earth also has an immense region of chaotic terrain (i.e. the Laurentian Plateau) and one is therefore inclined to wonder if it too is the result of an asteroid hit – either associated with the hit directly or associated with the developments that would have occurred at the antipode. Survival in either case would not have been any more likely than survival would have been in the Mars case.

Mercury is totally barren of all animal life for several reasons and Mercury also has a region of chaotic terrain. In the Mercury case the chaotic terrain is also at the antipode of a very large impact site. The Caloris Basin is a region on Mercury recognized as an asteroid impact site. It is huge by any comparison but since Mercury is only 3,000 miles in diameter and the Caloris Basin is 960 miles across, it is readily understood that the impact would have been overwhelming to any form of animal life. Of particular interest for the present discussion is the existence of chaotic terrain at the antipode of the Caloris impact site. 'At the exact antipode of the Basin is a large area of hilly, grooved terrain, with few small impact craters that are known as the Chaotic Terrain (also called 'Weird Terrain'). It is thought by some to have been created as seismic waves from the impact converged on the opposite side of the planet.' (83)

In the Mars case, the diameter of the largest crater is given as 1400 miles, whereas the diameter of the planet is only about 4200 miles. In the Mercury case the diameter of the Caloris Basin is about 960 miles compared to the planet diameter of about 3,000 miles. In both of these cases the crater diameters are approximately one-third of their respective host planet's diameters. It is therefore not surprising that so much destruction was produced and it is a wonder that these planets didn't completely fracture under such extreme stress. One can readily appreciate that in both cases the planets would have shuddered and shook and convulsed internally because of so much stress and energy transfer. Survival is unimaginable in both cases. Did this happen on Earth? On the Earth there are several very large craters indicating that the Earth has also suffered major disruption due to asteroid impacts. There is also a roughly circular feature on the

Earth, more or less centered on a large depression and only describable as being chaotic terrain. The Laurentian Plateau is roughly 3500 miles in diameter and if the Mid-Continent Rift (a geological feature buried under gravel, volcanic material, sedimentary rock, lakes and earth) which extends into the Laurentian Plateau on the north side and down into the central states on the south side, (96) is included, the distance across this giant region of chaotic terrain is increased by another 1,000 miles. The Plateau actually extends much further south in other regions as well and is recognized by very uneven rock which pokes through the overburden of gravel and soil in all of Southern Ontario. With all of this in mind, one wonders just how extensive this region of chaotic terrain actually is. The diameter of the Earth is only about 8,000 miles. Therefore it is MORE than one-third of the Earth's diameter across this chaotic region which makes it comparatively larger than either the Mars or Mercury case! Consequently the trauma from any event that was able to produce so much chaotic terrain is about as unthinkable as either of the Mars or Mercury cases. Any event that could totally disrupt the crust of the Earth over such a vast area would have totally disrupted all forms of life and it is hard to imagine how any form of animal life could have survived. Such an extensive area of chaotic terrain is indicative of major disturbance to life on the Earth and is singularly pointing to a mass extinction that would not have spared any form of animal life.

If the Laurentian Plateau is an antipode development we would expect evidence of an asteroid impact on approximately the opposite side of the Earth. This is not really the case although the Burkle Crater is located close to the exact antipodal location and there is a suspicious looking basin on the ocean floor north of Madagascar. The Burkle impactor would probably have been too small to cause the disruption resulting in the Laurentian Plateau. However in the general vicinity of the Burkle Crater there is a monsterous impact site that would be a possible candidate. The Vredefort Crater in South Africa is very large by any comparison. Its diameter is commonly given as 300 km (188 m). (98) However since this impact site includes several concentric mountain formations nobody is quite sure which one to recognize as the actual crater rim. This also begs the size of the asteroid because once an object is large enough to punch through the crust of the Earth, the size of the crater might not be indicative of the size of the asteroid that made it. In any event whatever object made the Vredefort Crater in South Africa was very large and carried with it a tremendous amount of energy which would have become manifest someplace on the Earth. It is possible that the Laurentian Plateau, (more or less centered on Hudson Bay), was the recipient of much of the energy of the Vredefort impactor. In such a case we would expect thousands of square miles of chaotic terrain and that is what we have. Also, something caused the Mid-Continent Rift (or chasm or monster ditch which is attached to the Laurentian Plateau) to form. This great cleavage in the Earth (which includes Lake Superior) is given as 2200 km long, 150 km wide and 32 km deep (1400 m x 94 m x 20 m). How much force would be required to form a ditch 20 times as deep as the Grand Canyon?

If on the other hand the Plateau was an asteroid impact site, chaotic terrain would be expected at the antipode (as there is in the Mars case and the Mercury case) and there is, in fact, a great deal of chaotic terrain all across the floor of the Indian Ocean and it includes numerous fractures, mountains and trenches. (84) There are also several Large Igneous Provinces in the area. In particular the Deccan Traps are located along the west coast of India and are found partly above sea level and partly below. This area could only be described as chaotic terrain as it includes steep-sided mountains, gorges and valleys and is a place where it would be hard to survive even today. As with the other regions of chaotic terrain on the

461

other planets, the evidence of chaos in this case (i.e. the lava that formed the Traps) is spread around over 'an area of nearly 200,000 square miles (500,000 sq. km.) The volume of Basalt (i.e. a type of rock) is estimated to be 12,275 cubic miles (512,000 cubic km.) By comparison, the 1980 eruption of Mount St. Helens produced only 1 cubic km of volcanic material). (99)

The Laurentian Plateau is roughly circular and includes a depression (i.e. Hudson Bay) which, unlike other large salt water areas is not very deep averaging only about 420 feet. When it is recognized that the Bay is 590 miles across (85) it is readily appreciated that it is comparatively very shallow. (i.e. it is about 7500 times as wide as it is deep and it would be very difficult to show such a shallow depth on any diagram drawn to scale) Furthermore, it is rising quite steadily which motion has been attributed to the removal of the great glacier that formed during the Ice Age (86) but the change in elevation could also be due to the great trauma that followed a massive asteroid hit! Again, from a survival viewpoint, as either an impact site or an antipode it is difficult to imagine how anything could have survived for thousands of miles around these areas during the time that such a great amount of chaotic terrain was being formed!

There are two main reasons that it is difficult to match up various features of the Earth and to be absolutely certain how they relate. The first reason is water. The massive tsunamis which circled the Earth as a result of asteroid impact would have modified many features which would help to identify geological relationships. Secondly, there have been numerous impacts. Some of them would have obliterated evidence of others so that a thousand years later getting a clear picture of what happened becomes impossible. Incidentally, both of these factors emphasize the problem of survival!

2.4.6   The Antipode Tsunami

A tsunami would be expected to form at the antipode if the antipode was located in an ocean. Both the primary tsunami and the secondary tsunami would radiate outward from ground zero. The continental tsunamis would start at the coastlines of whatever continent had been hit. Thereafter they would travel outward in all of the directions that the coast of that particular continent could provide. The antipode tsunami would initiate on the far side of the world from ground zero and, as with the continent-generated tsunamis, it would radiate outwards in all of the directions that were available to it. For example, if the antipode was near a shoreline most of the tsunami action would be in directions away from the nearby shore. All of these various high-energy waves would keep traveling until their energy and momentum was spent which in many cases would require several complete circuits of the world. Collectively all of them would travel around the world crisscrossing each other numerous times on every circuit. At the crossover locations their respective amplitudes would add or subtract as expected with wind-waves and occasionally more than two wave fronts would cross resulting in a much greater buildup of water. This also happens with wind-waves resulting in so-called killer waves that can threaten the largest ships. From the tsunami factor alone the entire world would be in chaos – even from the results of a single major asteroid strike. If an asteroid shower was involved possibly hundreds of tsunamis would be traveling around the world frequently crisscrossing each other on their way. Any hope of survival for animal life would completely vanish.

The cause of tsunami formation at the antipode would be due to the interior shockwave. As the interior shockwave came up underneath the crust, the crust would rise up. This would be a major survival hazard on its own because the ground might not be able to withstand the stress. If it couldn't, the crust would fracture and hot magma from the interior would rise up and spread around. As the shockwave came up under an ocean, the ocean floor would rise up. Ocean floor crust is not nearly as strong as continental crust which is understood to average several times as thick. (88) (i.e. 3 miles compared to 18 miles) The logical expectation would therefore be that the ocean floor would rise up more readily and fracture more readily. Both the increase in elevation and the fracturing present serious survival hazards and both of these developments would result in more tsunami activity.

When the ocean floor rises up, the water above it would be resting on a hilltop. It would not stay there. It would start to travel down the slopes on every side resulting in sheet flow involving the entire depth of the ocean at that location. There would not be any delay in the initiation of water movement. As soon as the floor of the ocean started to rise ocean water would start flowing away. The further it rose, the faster it would move but once it started moving there isn't anything that could stop it.

The ocean above the rising mound would start flowing away but pretty soon it would encounter stationary water which had not been accelerated and was still located in its original location. The forces involved are so great that something would have to give. The moving water would flow up on top of the stationary water. An upper layer of water would be moving while a lower layer would still be stationary. (This type of activity is quite common in the ocean today – in particular there is a location in the Pacific Ocean where the surface current is traveling one way, a second current further down is traveling in the opposite direction and a third current even lower is traveling in the same direction as the surface current! (140) The demarcation between the two layers would be quite sharp. Any mountains or underwater mounds of material of any type that protruded into the upper layer would be sheared off down as far as the bottom of the moving water layer. Below that level the water would not be moving and structures at that depth would not be eroded away. The result would be flat-topped under-water mountains which are called gyots. While this would also be considered a serious development, it would not present a hazard to animal life on land.

As the ocean floor was forced up it might not be able to stand the stress. It might fracture. If it fractured by being pushed up, the material that was doing the pushing, the molten material from the interior, would also be forced up and would surge up through the opening. Then it would pile up on the ocean floor on both sides of the opening. However, this material would, like the ocean floor beneath it, also displace ocean water resulting in more water being forced away towards the shores of the ocean. The displacement activity would augment the action of the ocean floor rising. If a large number of cubic miles of ocean were displaced, a similar amount of ocean water would be forced away as a great flowing sheet of water – the antipode tsunami. The speed of the water would be commensurate with the speed of the uprising which would not be slow and casual and would, in turn, be related to the speed of the interior shockwave which we expect would be traveling at several hundred miles per hour at least. We would now have an antipode disaster to deal with and it would only make the entire chaotic scene that much worse. There would be no place to hide. Apparently this type of disaster has occurred repeatedly because there are a large number of underwater mountains associated with faults all around the world. Further and also unfortunately from a survival viewpoint, there is a second type of underwater structure that would also result in massive water movement. However before discussing this second factor the particular characteristic of the underwater mountains must be recognized.

Many of these underwater mountains and in particular those down the middle of the Atlantic Ocean, have a great cleavage right down the middle. This cleavage is a deep sharp-sided trench which occurs all the way down the entire length of the mountain chain. It makes these mountains appear as a great knife cut a trench that appears to be almost as deep as the mountains are high. Material was forced up from below and spilled over onto the uplifted crust on both sides of the fault (i.e. the crack in the ocean floor). Naturally material would pile up on both sides because new material would have been coming up the middle from down below. Expectedly the upsurging material would be coming up with some velocity and probably would have blown up through the ocean water and up into the air as well. From there it spilled over onto the ocean floor on both sides. If any fell straight back it would have been caught in the upthrust and pushed up and out of the way again. The cleavage (or rift or monster ditch) is suggestive of rapid movement of the material from down below. If it had just oozed up it would probably have retained more of a mound shape. Soon the upsurge halted. Some of the material which had gone straight up might have fallen straight back down but by now the upsurge would have petered out and there wasn't enough of it to refill the central opening. So it stayed mostly open. The upsurge might have repeated several times so the same thing might have happened repeatedly until everything quieted down again. The net result was a fissure down the middle of the whole assembly.

The common explanation for these formations is based on the Theory of Plate Tectonics. However that theory is faulty and does not stand up to any scrutiny whatsoever. The main reason for this is that it depends on the notion that material from the interior is continually being pushed up and causes the seafloor to spread apart. In turn the pushing upwards and the spreading require the application of additional pressure which must be continually maintained. The cause of this unusual pressure is never explained. Observation tells us that there isn't any ongoing extraordinary pressure at all. If there was any such pressure underneath the central fault, it would preferentially push up through the fault again and spill material out as before. Further, if there was extra pressure under the ocean floor, the ocean floor would be caused to rise up and keep rising up until the extra pressure could not raise it any further. How would this region of unusually high pressure be maintained only along the fault? Why wouldn't it simply push out in all directions until the pressure was the same all over as would normally be expected? The interior of the Earth is fluid. If pressure developed from any source whatsoever this material would move until that pressure was reduced so far that no further movement could occur. Extra pressure under the ocean floor would raise it wherever that pressure developed. This is not happening so we must conclude that there isn't any extra pressure at all but only the usual pressure due to depth. Of course there is pressure deep in the Earth. In fact there is just enough to raise interior material up to the Earth's surface. If a hole could be drilled down through the crust (i.e. like it almost was on the Kola Peninsula in Russia) material would rise up into the drill hole until it was near the surface. If the interior pressure at that point was a little higher, some material would spill out around the hole and a mound would form. However the opening would not become a blowout hole and material would not explode into the air as happens sometimes with oil wells.

In these cases we have oil under unusually high pressure in the interior because it was supporting an overburden of heavier material and it has become trapped in a pressure tight region which did not let it out until someone drilled a hole into it. So if someone opened a hole into a high pressure accumulation of oil, naturally the oil would rush up through the hole. This type of situation is not happening with respect to the magma underneath the mid-Atlantic fault.

On the other hand the Theory of Plate Tectonics requires exceedingly high magma pressure right in the fault all the way to the surface in order to force the two sides of the fault apart. What formation of ocean floor could possibly contain such incredible pressure? And it would certainly be incredible. The force required to force ocean floor down under continental crust would be exceedingly high. However being exceedingly high would present another problem besides just blowing out. The ocean floor would be the structure that would be required to carry this high pressure all the way to the shore. Compared to the distance involved, the ocean floor is very thin. In fact it is thought to be only about 3 miles thick. It is generally more than 1,000 miles from the mid-Atlantic to any shore so the ocean floor would have to withstand a compressive stress load all the way to the shore. The stress required would be the same that would be required to force the ocean floor under the continents. The slenderness ratio in this example would about 3/1000 or 0.003 which is a value so small that there is no conceivable way the ocean floor could carry any significant compressive stress at all. It would simply buckle and pile up like ice does in the arctic when the wind blows across it. In this case when the stress is a little too great, the ice simply fractures and piles up into pressure ridges. The same thing would happen with the ocean floor and it would be happening all the time. However it is not happening. Therefore there is no unusual pressure at the mid-Atlantic ridge at all.

The other factor associated with the ridge assembly relates to the shape of the ocean floor itself. The ocean floor slopes upwards as the central fault is approached. (88) This is not due to extra pressure in the magma because as discussed such pressure would apply over the entire ocean and not just under the ridge. A more likely explanation is that the ocean floor was forced up when the great interior pressure wave arrived and since material rushed up through the resulting opening repeatedly, some of it likely remained in the fractured opening so when the extraordinary pressure surge was gone, the ocean floor could not quite slump back to its original location. The ocean floor slopes up so there is actually a mound of magma in the region. This assembly of magma mound, raised ocean floor and mound of material on top of it all cause great difficulty for the theory that the crust of the Earth floats on the material underneath. If it was floating, the entire assembly should have slumped back down long ago. But there it rests and doesn't appear to be moving at all. What then is holding it up? Its own structure is holding it up and it isn't floating at all but simply resting on the material below just like a pile of gravel rests on the material below it and nobody questions why is isn't sinking. Why should it sink? Why should the ocean crust sink? It would if it could but since it isn't it must be simply resting where it ended up when all of the commotion settled down. Compared to the diameter of the Earth the thickness of material above the apparent original elevation is practically nothing. The same thing could be said of mountains on land.

The floor of the ocean, as well as many continental areas, is also home to igneous provinces. These are large areas that are covered by material that (like the mountains associated with numerous major faults) has also come up (in a molten state) through a fissure from the interior and spread around. (Hence they are also referred to as fissure eruptions. (100)) Marine geologists have discovered massive basaltic lava flows and defined them as Large Igneous Provinces. (87) The collective volume of all of the igneous provinces on the ocean floor is enormous (i.e. some of them cover several hundred thousand square miles. (100)) and the outpouring of material is understood to have been preceded by enormous uplift of the Earth's surface where the uplift might have reached several thousand feet ((87) or more in oceanic settings. In some places they cover hundreds and hundreds of square miles and every cubic mile has displaced one cubic mile of ocean. Both the uplift preceding the outpouring of material and the outpouring itself would have been rapid. Both of these factors would have resulted in ocean water being quickly uplifted causing it to flow to a different part of the ocean as well as towards shorelines in every direction. On the way these great flowing sheets of water would have collided with other water movements coming from other directions. Whenever these water movements came to a shore, it would be over-run. Whatever the earlier tsunamis had not disturbed would have another chance to be disturbed as these great water flows rushed overland once again. Something forced this material to come up. The interior shockwaves generated by impacting asteroids did this.

In summary, in addition to the generation of chaotic terrain which itself would have been life threatening, the second major result of the upwelling pressure wave would have been a great amount of water displacement in the form of monstrous sheets of water rushing to meet the waves produced at the impact site itself as well as those produced by at continental margins by the sudden movements of the Earth's crust following impact. There would have been water rushing everywhere with no place whatever for any animal to find shelter! For example, numerous dinosaur eggs found in Patagonia (Tableland rising to 3,000 feet above sea level in the southern portion of South America between the Andes and the ocean. (101)) were ' … smothered by a flood . . ' (102) One is inclined to wonder if the eggs were "smothered by a flood" at such elevations, what chance would any other form of animal life, including the dinosaurs themselves have had under such circumstances? Survival of animal life during these massive world-wide flood conditions would not have been possible.

2.4.7   Earth Wobble

The Earth has a wobble. As it rotates, its axis of rotation shifts slightly with the result that the background of stars appears to move north and south slightly as well as in the usual direction and the extended axis of the Earth describes a complex but nearly circular pattern against the sky. The wobble of the Earth is sometimes called the Chandler wobble in recognition of the American astronomer who discovered it. A wobble was predicted theoretically years before by several scientists including Isaac Newton and Leonhard Euler and was expected due to several features of the Earth including the flattening of the Earth at the poles (caused by its rotation). Several other physical characteristics of the Earth were also thought to contribute to it.

The Earth's wobble can be changed. While the Earth is very large, it will respond to every force that is exerted on it even when such a force may seem small compared to the Earth's size. There was a magnitude 9 earthquake off the coast of Japan in the spring of 2011. A change in the wobble was expected theoretically and in this case it might even be measureable. ' ... the wobble is large enough that scientists might be able to measure it, by looking for small changes in the Earth's tilt. ' (90) Other earthquakes including the one in Chili in 2010, would have had similar results. In this case, ' ... it may have shortened the length of the day by 1.26 microseconds and moved the Earth's axis of rotation by 3 inches or 2.7 milliarcseconds.' (91)

While such earthquakes are serious business, they pall in comparison to the energy involved with an incoming asteroid. For example, the asteroid that caused the Arizona Crater in Arizona is recognized as having the same order of magnitude of energy as an earlier earthquake in Chili as well as one in Alaska. (92) The Arizona crater might have been caused by an object as small as 50 feet in diameter. (93) What then might be expected from an object 5 miles in diameter and having an energy level which was several million times as high? The Earth would measurably recoil from such an impact and the wobble of the Earth would certainly change. Since the Earth has been struck by numerous objects this large and even larger, a measurable compound wobble should be evident. While it must be recognized that numerous factors would contribute to a wobble, the evidence of the asteroids, because of their very high energy levels absolutely must be included.

While any wobble change is of academic interest, would there be any threat to life directly? The wobble is understood to produce a very small tide called the Pole tide. (94) While this tide is very small as is the wobble itself, threats to animal safety are not apparent. However the Earth and all of its interacting systems are in equilibrium so that upsetting one of them always upsets the others to some degree. If the wobble or the great ancestor of the present wobble was caused by an asteroid impact, it would have shown up suddenly. Generally things that happen suddenly are not welcome. A sudden shift in the rotational axis of the Earth means that other factors must readjust. In this case the center of the rotational bulge would shift. This bulge is quite significant in size as the surface of the Earth is about 26 miles further from the center of the Earth at the equator than it is at the poles. The water in the ocean would be caused to shift a little and the material in the interior of the Earth would react with all of these events. The crust of the Earth would be slightly stressed until stability was reestablished. Would any of this affect survival? It would depend on the magnitude. We can only observe the present situation and from this conclude that there probably would not be any direct threat to animal survival. However, how large was the wobble when it first appeared? If the Congo Basin, the Taklimakan Desert or the Laurentian Plateau represent asteroid impacts as does the 720 km Ishim Meteorite crater in the Teniz Basin in Kazakhstan (95) there would have been implications for survival. The entire crust of the Earth would have readjusted. The ocean level would have shifted and the entire crust of the Earth would have been stressed. Would any fissures have opened in the crust? Would the changing ocean level have produced any tsunamis? These things are a matter of degree and a point could easily have been reached where animal life, particularly near the sea, could have been threatened.

2.4.8 Volcanism

It is absolutely certain that volcanic activity will follow an asteroid impact. While there are several different types of volcano, including shield, fault-centered and igneous provinces, all of them would be involved. Shield volcanoes are also called cone volcanoes because the material that oozes out of the top runs down all of the sides and forms a cone-shaped shield. The fault-centered type forms along a fault line and material from the interior of the Earth piles up on both sides of the fault. With this type a rift or cleavage is found right over the fault line and separates the ejected material into two parallel mounds which follow the fault line for most or all of its extent. Igneous provinces, on the other hand, are formed when material from the interior oozes up through a fissure (i.e. a crack or fault in the Earth's crust.) and spreads around sometimes actually forming mountains as with the Deccan Traps of India which rise about 4,000 feet in places. (99) In Iceland both the fissure type and the shield type are found.

Cone volcanoes come in every conceivable size from very small ones to 'Mauna Loa in Hawaii, which towers 30,000 feet from the floor of the Pacific Ocean to its summit.' (103) Mauna Loa is the tallest mountain on Earth and it could be compared to the tallest one on Mars, Olympus Mons at 88,500 feet. (104) Giants such as these are comprised of a tremendous volume of material, all of which came from the interiors of their respective planets.

The great underwater mountain ridges that generally follow most of the major faults of the Earth are not lacking for volume either. Many of them are several thousand feet high and carry on for thousands of miles along the floor of the sea.

All types of volcanoes involve hot material which is commonly accompanied by water and other gaseous substances from the interior. The water could present a significant nuisance as it precipitates out in the form of heavy rain but otherwise it would not be a hazard to one's health. However, in addition to the lava which is sometimes driven into the air by exploding water pockets in the volcano but which more commonly just runs down the side of the volcano, there is also a type of material which turns into rock after it rains down as heavy dust. The type of rock which forms from this second type of material is called tuft. When it first starts to fall it is bothersome factor but in a short space of time it can accumulate into a layer many feet thick. This is what happened at Pompeii in southern Italy in the first century of the Common Era when Mount Vesuvius erupted and buried the entire town so quickly that hardly anybody escaped. Within a time frame that could have been measured in minutes, the town was buried. It happened so fast that people could not even launch boats and get away. Everything was blanketed and buried and there was no escape.

An excellent example of shield volcanoes forming as a result of asteroid impact is found on Mars. In this example numerous volcanoes are found at the antipode area of the Hellas Basis, the largest impact basin at 1400 miles, in the Solar system. A very large asteroid landed there and would have sent shockwaves right through Mars to the other side. There we find numerous volcanoes including Olympus Mons, the largest volcano yet discovered. It rises more than 88,000 feet above the surrounding plain and is spread out over an area as large as Iowa. It has excellent company with three other monster volcanoes, Pavonis Mons (23,000 feet high), Ascraeus Mons (36,000 feet high), and Arsia Mons (30,000 feet high). The Tharsis Bulge where all of these monsters is located is also home to several other volcanoes including Tharsis Tholus, Uranus Patera, and Uranus Tholus. (97) If such activity happened on Mars due to an asteroid impact, one would expect something similar to have happened on Earth.

The Earth has a very large number of volcanoes and a significant fraction of them are considered active. This doesn't mean that that the active ones erupt all the time but simply that they do erupt from time to time with quiet periods in between that might last for several years. With the massive pressure waves from impacting asteroids traveling back and forth through the interior of the Earth, virtually all of these volcanoes would have been active. When it is recognized that there are some ten thousand visible volcanoes on Earth (109) one can understand what a complete disaster it would be if all of them or even a significant fraction of them were active simultaneously. Even now there 'are 516 active volcanoes waiting to loose their violence. An eruption begins somewhere every 15 days.' (111) Fortunately, only a very small number are actually active at the same time. Also, this number does not include the fissure type of volcano that formed the Large Igneous Provinces. While this type does not form a cone, the thousands of cubic miles of hot lava (i.e. the volume of the Deccan Traps may have exceeded two million cubic kilometers (112) on their own.) that they spewed out would have been accompanied by other survival problems including billions of tons of sulfuric acid. (112)

The trauma associated with the placement of all of these materials would have been too much for any type of life. The material was always hot, moving quickly and accompanied by other substances including nocuous fumes which would have restricted breathing or even made it impossible. All forms of animal life are quite fragile compared to developments such as this and if the actual placement of material didn't cause trouble, the breath-robbing pollution of the atmosphere would have. However, where volcanoes are involved, these aspects of the situation would only have been the beginning of the hazard to survival for the animals of the Earth. Exploding volcanoes, dangerous fumes, falling rock-dust, and flowing magma are all dangerous activities and they would be a serious threat to all forms of animal life but these activities would be very closely associated with the location of the volcano itself and not really have world-wide consequences. However, there are two other aspects of volcanic eruption that would have world-wide consequences, no matter where any particular volcano was located. Further, these additional factors would continue long after the volcanoes had quieted down. In fact certain atmospheric results from Krakatoa, a volcanic event that happened more than 100 years ago, continue to the present time. (110)

2.5   Secondary Atmospheric Activity

2.5.1   Volcanic Clouds

Volcanoes release dust into the air. In some cases they release a very large quantity of dust and scientists try to estimate the quantity of dust and give their estimates in cubic miles. One cubic mile of dust is a very great deal of dust! The following list gives the estimated volumes of dust for a few of the significant volcanic eruptions of past years.

Table 2.5.1   Dust Cloud Volume

Number	Year	Volcano	Quantity of Dust (Cubic miles)	Reference
1.	1883	Krakatoa	one	(106)
2.	1815	Tambora	several	(107)
3.	1912	Katmai	several	(108)
4.	1980	St. Helens	one cu km	(139)
5.	ancient	Toba	many	(113)

The dust that is emitted from a volcano is not restricted to the vicinity of the volcano. Unlike the ash, water magma and exploding lave bombs that fly out of some volcanoes and endanger life in the immediate area, the dust from a volcano rises into the atmosphere and completely leaves the vicinity. In this manner it becomes a hazard to animal life in far-away places and can even threaten life around the entire world. This is thought to have happened long ago when Toba, a volcano in northern Sumatra, erupted. 'No one knows how big it was other than it was a whopper. ... the Toba blast was followed by at least six years of "volcanic winter" ... The event may have carried humans right to the brink of extinction, reducing the human population to no more than a few thousand individuals.' (113) The reason that volcanic dust could do this is because it would have blocked sunlight from reaching the Earth and plants would not have been able to grow properly. Also the lack of sunlight means that the weather would have been cold. Mid-latitudes would have had snow every month of the year. Bodies of water that would normally have been ice-free for several months during the summer would have remained frozen. All of these conditions are thought to have been brought about by a single massive volcanic eruption. While Toba is understood to have been an ancient event and hence more open to various interpretations concerning the results, there is a much more recent and datable event which was witnessed by many people leaving very little uncertainty regarding the consequences of a major eruption. Tambora is a volcano in Indonesia. Its eruption in 1815 was preceded by two other eruptions; Soufriere on Saint Vincent in 1812 and Mayon in the Philippines in 1814. 'There was frost every month. The crops were cut off. ... The inhabitants saw famine approaching... The cold was not local but worldwide. ... volcanic dust persisted for several years.' (117) Another example which emphasizes the cooling factor which accompanies a major volcanic eruption is Krakatoa which exploded in 1883. As a result of the Krakatoa eruption the solar radiation reaching the Earth's surface reduced to less than 88% in 1884 and 1885. (118)

Volcanic eruption and asteroid impact are directly connected. This is clearly seen in the Mars case where a field of volcanoes is located directly opposite the major asteroid impact sites. The connection is understood to have been the shockwaves that were produced by the asteroid on impact and which quickly traveled right through the planet to the far side. 'So immense were the pressure waves from the Hellas impact ... (that they) were hitting the opposite side of Mars ... in about 28 or 29 minutes ... and caused the uplift in the Opposite Hemisphere.' (115) A similar connection between asteroid impact and volcanic activity has been made for the Earth. Further, the volcanoes of the Siberian Traps are thought to have erupted at about this time and they would have spread lava over several million square mils. (116) The "time" in this case is referring to the impact of the asteroid, Chicxulub, on the tip of the Yucatan Peninsula in Mexico. While the Siberian traps are not at the exact antipode of Chicxulub, they are well around on the far side of the Earth where the interior pressure waves would have been coming up underneath the crust. There is no doubt that the entire country of Russia would have been elevated by those pressure waves so if the crust of the Earth split open and lava oozed out as it would have when the Siberian traps were forming it would not be a surprise to anyone. The massive pressure waves from the impact of a large asteroid would have stressed the Earth's crust into failure mode and an outpouring of material would have followed.

The Earth is home to several very large asteroid impact sites including, Chicxulub, The Check Republic, Vredforte, Popigai, Acraman, Sudbury, Manicouagan, and Ishim. In addition there are several dozen more that are considered "large". If the pressure waves from Chicxulub could cause the crust of the Earth to rupture, it would be reasonable to expect serious turmoil from the smaller objects as well. Everything happens for a reason. Something caused the material forming the Igneous Provinces to come up onto the surface of the Earth. As with the cone volcanoes of Mars, it is reasonable to expect similar activity on the Earth to result from the many impacts that the Earth has suffered. If we are dealing with a large asteroid, numerous volcanoes of every type would be expected to erupt. If we are dealing with an asteroid shower, all ten thousand (109) of them might have erupted simultaneously!

2.5.2 The Impact Cloud

When a large asteroid hits the Earth, a very large cloud develops. One commentator describes it as follows (paraphrased). As the fireball reached the top of the atmosphere it turned an impenetrable black and spread out over the entire world. Of the trillion tons of material blasted from the crater, perhaps a billion tons of fine dust remained suspended in the atmosphere, darkening the sky all around the globe. The Earth remained dark for several months. When the dust finally settled out there was enough of it to form a layer of it over the entire world. (114) Another commentator offers a similarly pessimistic scenario. ' … the impact would be damaging … but more important would be the 100 trillion tons of dust lofted into the atmosphere … (which) would have produced a pall over the whole globe … lasting several months. During this period of global darkness, neither light nor heat was available to sustain life. With photosynthesis at a standstill and the Earth plunged into deep cold, it is no wonder there was a great dying.' (50) Situations such as this are quite comparable to the eruption of Toba, the giant volcano in northern Sumatra. It has been declared that the Toba blast was followed by six years of "volcanic winter" and goodness knows how many poor growing seasons after that and reduced the global population to a few thousand individuals. (142) Unfortunately the commentator, in this case, did not suggest how anyone could survive "six years of volcanic winter".

As discussed above, when Krakatoa exploded in 1883 a great deal of dust was blown into the atmosphere. Water vapor was also involved in the explosion and a great quantity of it would also have been lofted very high above the Earth. It is consequently a little more than curious that the noctilucent clouds were first noticed shortly after Krakatoa exploded. (127) These night-glowing clouds are very high and they remain at about the 50 mile-high level. It has been confirmed that they consist of specks of dust surrounded by water atoms. (121) Of course there are various explanations for these clouds including the water from small comets. (121) However, the timing of their first appearance coincides with Krakatoa seriously suggesting that the two phenomena are connected. The main point for the present discussion is that they are still there. It has been more than 100 years since the Krakatoa event and one therefore wonders how long atmospheric contamination would remain after a major asteroid impact. In any event, the months immediately following impact would have been a period which would have been hard to survive. If people cannot grow food for even one season, they will be in serious trouble. Any longer than one season would mean certain starvation for a large percentage of humanity which percentage is referring only to those that managed to survive the events which immediately followed the impact. Without any plants to eat, animal life would not fare any better than human life and there is no place on the Earth that would have been immune from these tragic consequences.

2.5.3 Moisture Clouds

As the entire Earth became darkened by a great assembly of volcanic clouds together with numerous impact clouds, a second type of cloud, a moisture cloud, developed and added to the darkness which in most cases wasn't really possible but would have guaranteed that any light that did manage to penetrate through the dust layer would have been blocked by this second layer. The vast dust cloud would have been generally high in the atmosphere. Moisture clouds would have been low in the atmosphere.

The moisture clouds alone would have been sufficient to block most of the sunlight that normally would reach the surface of the Earth. Moisture clouds are not strange to the Earth but are the dominant type of cloud which develops regularly in the atmosphere. As long as they remain relatively thin, sunlight can indirectly get through by repeatedly reflecting back and forth through the cloud until some of it emerges on the lower side and the cloud appears white. White fluffy clouds do not frighten anyone. However if they become thick they usually indicate a thunderstorm and during a storm they commonly become thick enough to appear black and cause the surface of the Earth to darken significantly. Everyone has experienced the darkness that accompanies a storm. While it is the storm that is the immediate threat when this happens, it is the thickness of the cloud that causes the darkness. Some thunderclouds increase in thickness until they are 20,000 or 30,000 feet high. In recent years storm clouds have been observed to be more than 40,000 feet high. Very little light penetrates a cloud that has become that thick causing the cloud to appear black. Directly underneath it will be as dark as late evening after the Sun has set. There was a case in recent years where a vast storm cloud developed near the Rocky Mountains. It became very thick and widespread and remained in place for several hours. Underneath it had become as dark as night and all the street lights had turned on. (122) It is expected that the moisture clouds that developed following the impact of an asteroid would have been similarly thick and would have, all on their own, blocked sunlight from reaching the Earth. A more dismal scene is hard to imagine. But what caused the moisture clouds to form?

When water evaporates from the surface of the Earth, it becomes part of the atmosphere. It might be completely obvious that water evaporates from the surface of lakes and rivers and ponds but in actuality water evaporates from almost everything on the Earth. It even evaporates from trees! The warmer the water becomes the more it evaporates. When a pot of water is placed on a stove it will only evaporate at a very slow rate and might take several days for the pot to dry up. However as soon as the burner is turned on, it will evaporate quickly and the pot might dry up in a few minutes. It is the heat and the increase in temperature that quickens the evaporation process. The same process will develop with any body of water. As a shallow pond heats up in the summer it will evaporate and if the water is not replenished, the pond will dry up. In this case the heat from the Sun heats the pond water as well as the ground underneath so the water warms up and evaporates away. Alternately, if the pond could have been placed on a hot stove, it might have been caused to dry up in a few minutes. The same thing would happen to a lake. If a lake could be placed on a hot stove, it would heat up and evaporate and would simply dry up like the pond. What if the ocean was placed on a hot stove? Would it also dry up? It might not dry up but the evaporation rate would increase dramatically! Following the impact of the asteroids there was world-wide volcanic activity. Much of this activity was beneath the ocean surface and it had the effect of placing the ocean on a hot stove. Wherever igneous provinces, cone volcanoes or underwater ridges associated with faults were forming, hot molten material from the interior of the Earth came into direct contact with the ocean water. At all of these locations the ocean boiled. This caused vast moisture clouds to form and they became so thick that most incoming sunlight would have been blocked except that the dust clouds had already blocked it. With this development there were two layers of cloud and together they virtually guaranteed that the entire surface of the world would be dark. In this case dark also means cold. So the Earth became cold and dark. Could a scenario any worse be imagined?

When an asteroid hits the Earth, great pressure waves propagate back and forth through the interior. As they come up underneath the crust, it fractures and volcanic activity follows. Molten material comes up and spreads around on top of the original material. Igneous provinces form and cone volcanoes develop. Flowing lava resulting from the pressure waves would have occurred all around the world. If an asteroid shower was involved, there would have been lava flowing everywhere. Recognized Igneous provinces are all over the Earth. Cone volcanoes are also all over the Earth and as with igneous provinces, many of them are underwater. These developments would have heated the ocean and in many locations it would have boiled. Following asteroid impact the ocean would have been caused to boil at hundreds of locations so the steam clouds (i.e. moisture clouds) that developed were very thick and widespread and it would even have been hard to breath whenever they crossed though an area. The atmosphere had become overloaded with moisture. More was coming continually. There would have been enough heat available, as noted by the volume of the igneous provinces alone, to cause the evaporation of hundreds of feet of ocean water. (Please refer to the appendix for this calculation.) Where then did all of this moisture go?

2.6 The Great Ice Age

2.6.1 The Ice Age Cloud

The two volcanic factors of world-wide significance are; the dust cloud and the thermal energy. Further, it is these two factors working together with the dust cloud produced directly by the impact that caused the great Ice Age. In particular, it is the volume and type of dust that is of interest. It would be more accurate to refer to the volume of the solid material in the cloud than to the volume of the cloud itself because it is this parameter that will have the most far-reaching consequences around the entire world. The world-encircling aggregate cloud that resulted from the impact of the asteroids and from the numerous volcanoes that were activated by the earthquakes and the interior pressure waves separated the heat of the Sun from the surface of the Earth. This was the first necessary factor to cause and enable the great Ice Age to commence. The cloud caused the Earth to cool. Even moderate cooling simply is not tolerable but will result in the Window of Life closing. Life would not have been able to survive this double-edged factor (i.e. darkness and cooling) because the one factor means that plants cannot grow and the other means that plants will freeze. The hazard is also double-sided and ensures that even before the great Ice Age got started widespread animal death was ensured.

2.6.2 A Too-Warm Ocean

These two factors; the chilling and the heating, are both necessary and together they provide sufficiency for an ice age to commence. The first factor caused the Earth to cool. The second factor caused the oceans to heat and in many areas to boil (as is happening now south of Hawaii in the Pacific Ocean where a new island-volcano, Loihi, is building (105)). Unusually warm water not only caused vast quantities of moisture to be released into the air but made life for many sea creatures unbearable. In particular those creatures that have a controlled body temperature would not have been able to tolerate any significant increase in water temperature. They would simply perish. Many sea creatures have body temperatures which closely match the water temperature. In these cases their body temperatures would have risen with the water temperature but there would have been a limit that they could have tolerated. Whenever and wherever this limit was reached they too would have perished. An ocean that is too warm would have been a significant hazard to most life-forms in the ocean.

2.6.3 Permafrost

A frozen region will continue to deepen as long as the temperature above ground remains below freezing. Extreme cold is not required. If extreme cold was present it would hasten the process of heat transfer but there isn't any particular rate that is necessary for permafrost to develop. An example of heat transfer gradually forming a great depth of ice can be understood from the way that ice arenas work. As the mechanical equipment removes heat from the ice slab, heat flows out of the ground into the equipment and is transferred outside and dissipated into the atmosphere. As this happens, the ground below the slab of ice will freeze and the frozen region will continue to deepen. Normally insulation is installed to prevent freezing. The insulation will reduce the rate of heat transfer from the ground into the mechanical equipment so the ground will not freeze. However, if insulation is not installed, ice will form continually and the frozen region will get deeper and deeper. There was a recent case in Mexico where an arena in Monterrey was built without insulation under the ice. After a few seasons a large bulge formed in the ice and a consultant (i.e. the author) from Canada was called in to investigate. As part of the investigation a hole was dug into the ground near the center of the ice pad area and it was discovered that the ice had formed down to about the six foot level. It would have kept right on forming except that the investigation had been called. Since it is difficult to skate over an ice mound, the situation was recognized as requiring correction. In this case extreme cold was not involved as the operating refrigerant was only a few degrees below freezing. From this observation and reasonable common sense it can be concluded that the only necessary condition to form permafrost is that the surface temperature be kept continually several degrees below freezing. As long as this condition is met, within a few years a considerable depth of permafrost will appear and even if the heat removal process is then terminated it will require time for the ground to thaw out again.

Quite commonly when the Ice Age is discussed the requirement for rapid heat transfer comes up with respect to the preserved stomach contents of woolly mammoths. Several of these animals have been found frozen solid with recognizable food in their stomachs. As is well documented they have been found with undigested food in both their mouths and their stomachs. How cold would it have to have been for the seeds and plants to be preserved against the attack of normal stomach acid? The key concern here is "normal". Unfortunately things were not the least bit normal. The woolly mammoths were killed instantly by the atmospheric shockwaves from incoming asteroids. These shockwaves would have been comparable to being hit by a giant steel paddle which slammed into their large bodies and knocked them off their feet and sent them flying for hundreds of feet, in many cases completely disarticulating their bodies. (128) They were hit so suddenly and so hard that all of their body processes were interrupted. Blood circulation stopped instantly. They hemorrhaged internally and the blood gathered into large pools. (124) Their mouths were full of food. Death was instantaneous. Would normal stomach secretions therefore have continued? No they would not. In fact the compression and distortion of their bodies from the blast would have squeezed all of their body fluids into strange places. All of this is suggestive that carrying out an experiment with "normal" stomach conditions would not really be valid. All body processes depend on the intake of oxygen and the maintenance of blood-flow. Both all of these processes would have been terminated at the time of their instant death. Since many of these massive animal bodies were also quickly isolated from oxygen by either burial or drowning, what aspect of normal process could be identified and used to support the conclusion that the temperature plummeted to lower than minus 150 F. (- 101 C) (123)

Such an extreme temperature is not credible so an alternate explanation is warranted. First we recall that the ocean was warm and being heated even warmer. Great tsunamis generated by the asteroids were washing across the continents repeatedly keeping them relatively warm. The volcanic activity on land would have further delayed the freezing process. However the heat from the Sun had been cut off so as soon as these temperature moderating factors subsided, the air temperature over the land fell below freezing and snow started to fall. Since the ocean was still being warmed by volcanic activity, massive moisture clouds continued to transfer water from the ocean to the land and the great conveyor belt of snow buried everything. Under these conditions the ground would start to freeze and since the Sun did not shine again for months and months the ground simply froze and kept on freezing. In fact, in some places it has remained frozen ever since and would continue to remain frozen except that now the world is warming up and permafrost that has persisted for years is melting. Further, some permafrost that still isn't quite warm enough to melt has warmed up which observation is cited as evidence that the entire world is on a warming trend!

Permafrost is therefore a hazard which would have developed with the onset of the Ice Age.

2.6.4   Snow and Ice

As polar regions are usually cold there isn't any trouble understanding that snow could fall there and if the conditions remained cold, it could build up into substantial depths. Mid-latitudes do get snow and many areas are commonly covered by a layer of it most of the winter. If summer didn't come it is understandable that the snow could linger on into spring and quite possibly right through the summer. After all, this almost happened when the three large volcanoes erupted in the early 1800's. Snow and ice in June is not the least bit welcome and can lead to near starvation for both animals and humans. This abysmal development occurred even though the Sun was not completely blocked and its intensity was only reduced by a few percentage points. It is therefore readily seen that if it had been blocked out by a few more percentage points – even down to fifty percent of its usual intensity – snow and ice would have persisted right through the summer. This, of course, would not be acceptable because crops could not be grown and the food supply would simply run out. Trees and shrubs would not be able to develop their normal leaf arrangements. Unfortunately some plants cannot tolerate such a development. If their normal leaf arrangements do not develop all the way through a complete growing season, they might not be able to produce leaves the following season. In any event if there isn't reasonably normal leaf development for a whole summer season, the animals that depended on such growth would starve. But with the snow piling up, very few animals could have gotten to a food supply anyway. They could not have waded through even ten feet of snow so it isn't conceivable how they would have managed several hundred feet of it. Humans would not have fared any better.

Would conditions have been any better near the equator? The simple fact is that snow and ice persist at the equator at the present time. However, currently, this only happens at higher elevations. For example Mount Kilimanjaro is located barely three degrees from the equator but it remains snow covered all year. However what if the entire area was dark and the Sun did not shine at all? It is the heat from the Sun that makes the equatorial regions hot. If this heat was not available, there is no reason to believe that the tropics would be any hotter than the poles. While this is the reasoning side of the argument, is there an observational side? Near the equator there are several countries where signs of the great Ice Age persist right to the present time. 'In … equatorial Brazil … an ice cover … on the very equator … scratched rocks … fluted valleys and the smooth surface of tillite … region … covered by ice several thousand feet thick … likewise in British Guiana … equatorial Africa and … Madagascar … and India.' (132) Such evidence supports the reasoning. While great thicknesses of snow might not have persisted in the tropics quite as long as they did further north, even a few months with one hundred feet of snow would have been intolerable for all types of animal life.

2.6.5   No Ice Age Mammals

Asteroid showers and ice ages are very rare events. Animal life would not have been able to survive either one of these events and this might be the reason that the survival topic is never discussed. No one can survive being buried in 1,000 feet of snow. However, in spite of such realities it has been declared that numerous large "ice age mammals" existed and thrived right through the great Ice Age! (130) Apparently they were able to paw threw the snow and eat the nutrient rich grass and thereby survive quite well. On the other hand, it has been declared that the snow accumulated until it formed ice and the ice accumulated until it was thousands of feet thick. Is there any possibility that any animal could paw through 10,000 feet of snow? Therefore the idea of "ice age mammals" surviving right through the great Ice Age must be set aside! In fact, a theory has never been advanced to explain how such animals survived through such a difficult time. It is simply declared that they did!

2.7 Assessing the Hazards to Life

2.7.1 Primary Atmospheric Hazards

2.7.1.1 The Blinding Flash

For every incoming asteroid, greater than a significant fraction of a kilometer in diameter expectedly, there would be a blinding flash. A pulse of heat and light would radiate outward from the impact site just as the asteroid approached surface contact. While this flash would be devastating to any person looking in that direction, realistically this hazard would only be dangerous in the immediate vicinity of ground zero out to possibly 50 or 100 miles (with exceptions of course). The larger the asteroid the greater the area of damage but since there were relatively few really large asteroids, the overall damage to life on the entire planet would be quite minimal. Nevertheless being far from ground zero would be preferred.

2.7.1.2 The Atmospheric Shockwave

While the atmospheric shockwaves would also have radiated outward from every ground zero, they would have had a much more devastating effect. Atmospheric shockwaves would have radiated outward from every ground zero with the effect of an expanding steel plate. The entire herd of Woolly Mammoths, which herd actually included numerous additional types of animals including rhinoceros, elephant, buffaloes and horses, (134) were killed so fast that they could not even swallow their last bite. From this evidence it can be reasoned that human beings would not have fared any better. Since millions of animals perished, the devastation was obviously widespread. Death was instantaneous and would have been for any humans had any been in these areas at that time.

Trees as well as animals were blown away and piled up. If the entire forest is being blown right past you along with uncountable animal carcasses, where would anyone find shelter and how would that be accomplished since there was no warning and no time to relocate yourself anyway. Death would have been as certain for human beings as it was for the animals.

Dinosaurs were, especially the larger varieties, much stronger and much better able to withstand rough handing than any human being could have. However the dinosaurs perished as one result of an incoming asteroid. Some of them would have died from the atmospheric shockwave just as the Wooly Mammoths did. Also it must be recognized that the dinosaurs all died – right to the last one – but that most of them

477

were also disarticulated. They were smashed to pieces! It is well recognized that it is quite unusual to find an intact dinosaur skeleton. (128) How much force would it have taken to hit a dinosaur so hard that it came apart? In fact when it comes to dinosaurs it is apparently quite difficult to find one that is intact. Such disarticulation confirms that these creatures did not simply lie down and die. They were killed so violently that their bodies came apart. Some great force hit the entire herd, which would probably have numbered into the millions, so hard that individual dinosaur bodies came completely apart. If that is what happened to the dinosaurs, what would have happened to any other animal in the vicinity or even to a human being? Of course it isn't popular to think that humans were alive at the same time as the dinosaurs but a reasonable amount of evidence suggests otherwise. (135) Neither was widespread death restricted to the creatures already mentioned. Many other fine specimens from the past died similarly. As an example, the baluchitherium appeared to be as robust as any animal we could mention and any one of them could have pulled more plows than a team which included a dozen horses. However they perished as well. Where would any human have taken shelter if these great creatures were being killed nearby? Obviously they wouldn't have fared any better and would have died at the same time. (With these great calamities in mind, it isn't very helpful to insist that mammals lived at a very different time than the dinosaurs or that human beings lived at some other time altogether because death and destruction was going on at such a viscous pace that it doesn't really avoid the argument to insist on great temporal separations. The present argument is explicitly suggesting that death and destruction was so widespread and rampant that arguing for several different time frames is of no consequence. Everything that was there died! How was anything to be alive for another round at a later time?)

2.7.1.3   Wind

A great blast of wind would follow every asteroid. It would radiate outward from ground zero with great speed and everything in its path would have been swept away. Of course the atmospheric shockwave would have preceded the blast of wind so there probably wasn't very much left to relocate. Therefore the argument that the wind would have swept almost everything away is somewhat moot. It is of little comfort though that the atmosphere packed a one-two punch as if the first one wasn't quite sufficient. First the atmospheric shockwave rushed past and very soon thereafter the blast of wind swept by. Perhaps it only carried a wall of dust and dirt but such would have been a hazard of its own and not of any comfort to any possible would-be survivor.

2.7.2   Surface Impact Hazards

2.7.2.1   Water Splash

If an asteroid should land in the ocean there would obviously be a water splash. The hazard related to this splash depends on the amount of water involved which in turn is dependent on the size of the asteroid and the depth of the water. If, for example, an asteroid landed where the water was one mile deep and the asteroid was one mile in diameter, it can readily be seen that the water splash would involve several cubic miles of water. If this amount of water came back down as rain there would be a great downpour for several minutes and then it would be over. Unfortunately only a small portion of the total amount that was dislodged would come down as rain. Some of it would return to the Earth as great globules of water. If these great masses came down on the water there would be a secondary splash. More water would splashup but unless there was someone in the immediate vicinity there would not be any significant threat to human life. However if the globules came down on land there would be a threat to both human and animal life because the globules would not have time to break up before landing and would in fact land with the effect of exploding bombs. They would present a threat to all forms of life in the entire area. These globules would hit the ground so hard they would create small craters. Material from these craters would splash away just as if a small meteorite had landed. Both the incoming water globule and the ejecta from the crater it would form would be a threat to animal life. No animal can stand being slammed into the ground by a mass of anything that weighed several tons so being at the landing sites where the globules came down would mean certain death. As hazardous as this situation would be near ground zero, the risk to animal life would only extend outward for a few miles and thereafter peter out.

2.7.2.2   CraterFormation

Whether an asteroid landed on water or solid ground a crater would form. A water landing would result in the formation of a temporary crater in the water. Some of the water that was displaced to form this crater would be the splash water just discussed. The remainder would form the primary tsunami which will be discussed in a following section. The crater of greater immediate interest however would be the one that would be formed on either the bottom of the ocean or on the land. The material that would be dislodged to form this crater would rain down on the surrounding area and every bit of it would be a threat to every form of animal life. Great amounts of mud would be involved. All sizes of broken rock would fly outwards and shower down on the entire countryside for miles around. Being in such an area would mean certain death. There simply isn't any type of animal that could have withstood being hit by a piece of rock that was traveling at several hundred miles per hour. Crater material would come down all around the crater site out to a distance of several crater diameters. For example, if an asteroid two miles in diameter formed a crater that was twenty miles in diameter, expectedly crater material would rain down out to a distance of two or three hundred miles. All of this material would pose an immediate threat to every form of animal life and if any was in the area it isn't likely that it would survive.

2.7.2.3   Tsunamis

Of all the hazards to life that would develop following a major asteroid hit the tsunamis could readily be referred to as the worst because they would involve the entire world. There would be three types of tsunamis resulting from an impact and all of them would be of a magnitude that would ensure that they completely encircled the entire world. The reason for this is because of their size. The primary tsunami spreading out from ground zero would be expected to start off several thousand feet high. Also it would be moving at least several hundred miles per hour. Where would anything, birds, mammals or human beings find refuge? Waves of this magnitude would pass right over continents and their respective velocities would ensure that they kept going right across the next ocean. The continents would be washed clean of vegetation and all types of soil. Since the vegetation would be entrained in the flow, potential survivors would have to deal with rushing turbulent water that was full of debris. Just being able to stay afloat would not be any guaranty of success. The tsunamis which developed from continent shaking would probably not be as high as the primary tsunami but there would be an extensive sequence of them which would continue as long as the continents kept shaking. The shaking could last for hours, days or even weeks because the entire thickness of the crust of the Earth would be involved guaranteeing that aftershocks would repeatedly develop long after the initiating event, (i.e. the impact) was over. Further, since these waves would be moving at high speed, they would have the energy to climb mountains so having a mountain range as a potential barrier would be of little comfort. This means that at any one location a train of waves would show up from one direction and other trains from other directions. Further, since an asteroid shower was involved and not just a single hit, every location on Earth would be close to an impact location. There simply would not be any way to survive a situation of this nature. All forms of animal life would perish!

2.7.2.4   The Mud/Rock Splash

As the asteroid drives into the ground a mud/rock splash would develop. While it can be considered a relatively local event, mud, rocks and broken pieces of the asteroid would be propelled away from the impact site and could travel outwards for several crater diameters. It will be recalled that the crater rim defines an area where the surface of the Earth has been totally disrupted and for a large asteroid this disruption would extend right through the crust to the fluid interior of the Earth. Because of the influence of the fluid magma the area within the crater rim would thereafter assume a shape that followed the general curvature of the Earth. With this much disruption within the crater rim it seems reasonable to expect that material from this area would also be found outside of the rim area. Expecting to find material that had splashed out of the crater area for several crater diameters would be in keeping with the incredible trauma of the impact. It follows that being within the area defined by several crater diameters would be extremely hazardous because of the great shower mixture of solid and semi-solid material raining down. Since its total weight would exceed many thousands of tons, there isn't any form of animal life that could withstand such abuse and any animals caught in the area would certainly not survive.

2.7.2.5   Erratic Boulders

Erratic boulders are part of the mud/rock splash. Since they have been identified hundreds of miles from their apparent sources, it is obvious that it would have been extremely hazardous to have been in the area at all when they were arching up from the impact location and then raining down on the entire countryside for hundreds of miles. What type of animal could have survived their descent and impact?

2.7.2.6   The Earthquake

Even a modest earthquake is hazardous wherever buildings, bridges, overpasses or any other types of structures are involved that would be dangerous if they became dislocated. A major asteroid impact would dislocate the very bedrock of the Earth. Therefore whether a person was located near a building or not would not make any difference whatsoever. The entire countryside would be moving horizontally as well as vertically and for several hundred miles of the impact site this movement would involve hundreds or even thousands of feet. Entire continents would be involved. The impact in South Africa, for example, would have shaken the entire continent even to the Mediterranean Sea and solid footing would not be found. Further, the speed of motion would guaranty that nothing could stand - not even trees which would be shaken right out of the ground. If trees and boulders were being tossed around what chance would any form of animal life have had?

2.7.2.7 Sedimentary Rock

The formation of sedimentary rock is directly connected to the tsunamis discussed above. It was the tsunamis that removed and then deposited the material that would later become sedimentary rock. Material was deposited, swept away, redeposited and swept away again before being left in place by the last major sheet-flow. This chaos continued as long as the currents had enough energy to remove and transport the sediments. Many waves and currents were involved. Untold quantities of material were swept away, deposited and swept away again. After the energy of the currents was spent, the sediments remained in place. In the meantime it is obvious that with all of this commotion going on there would not have been any safe place for man or beast. If the very earth under one's feet is being taken away it is obvious that survival would not have been possible. Since this was happening on every continent on the Earth, it is also obvious that no particular region would have been favored for survival. Therefore it can be concluded that survival while sedimentary material was being deposited would not have been possible.

2.7.2.8 Mountain Formation

Contrary to the current popular science viewpoint mountain formation happened quickly and it was not a docile affair. As discussed previously all of the mountains of the world formed within a one-year time period. This conclusion has been reached because they had to have formed in between the time that the sedimentary material was deposited and the time that it hardened. The strata in mountains are all bent into anticlines and synclines. This indicates that some of the material stretched as the entire assembly was bent and some compressed as the assembly was bent. The only way that this could have happened is if the entire assembly was soft and both stretchable and compressible. Otherwise bending would not have been possible. However that is the way that mountains presently exist. They are bent into great waves with crests and valleys. When viewed from above the wave formation is much more visible and the waves always run parallel to continental margins indicating that the source of the earthquake energy was the ocean floor. Energy was required to form the mountains and this energy was supplied by the asteroids. Energy transmission through the Earth resulting from an earthquake is commonly in the form of waves and mountains are simply very large waves – the waves that developed in the material as it was transmitting the asteroid-generated earthquakes. Large waves formed because a great amount of energy was available. It follows immediately that being in the region would have been extremely hazardous and survival would not have been possible. The ground was not only shaking, it was being distorted into giant waves. Some of these waves would have broken open at the top and some would have become submerged as they dipped beneath the average elevation. Animal life would not have been able to withstand that much trauma. All animal life that was in any region where mountains were being formed would have been extinguished.

2.7.2.9   Continental Shelf Formation

Continental shelves consist of soft sediment that has been washed off of the adjacent continent. A great sheet of water must have washed it off and carried it into the ocean. This is, of course, exactly the same way that deltas are formed at the mouths of large rivers. The water flows down the river channel carrying with it a load of sediment. As soon as the depth of the water increases at the mouth of the river, the current slows down and the sediment settles out. As the continental shelves were being formed great sheets of water were flowing off of the continents and then slowing down when the greater depth of the ocean was encountered. If the water flow had been extensive enough to carry such a massive load of material, it implies that it was at least several feet deep and might even have been several hundred feet deep. Animal life would not have been able to deal with this much water flow and consequently would have perished. Sometimes animals become caught in the flow of a river and cannot deal with the current, become tired and succumb to the water. When water was flowing off of a continent in the form of a massive sheet, animals would not have been able to deal with it and they would have died. Similarly any person caught in the situation would not have faired any better. If a river is flowing slow enough and it is narrow enough an animal might be able to swim to shore but when continental shelves were being formed there wasn't any shore to swim to and any hope of survival would have vanished. It can be concluded that being on any continent from which the soil was being washed away to the ocean to form the continental shelves would not have been a place where survival would have been possible. It can further be concluded that such massive flow would not have been confined to any one continent but would have been happening to all of the continents at the same time. (i.e. it is impossible to have such massive flow confined to a single continent.) Consequently there would not have been any continent that would have been safe for animal life. When the continental shelves were being formed survival for animal life around the entire world would have been very difficult and probably not possible.

2.7.3   Hazards due to Activity Coming from the Interior

2.7.3.1   The Ground Tsunami

If a large asteroid (i.e. one which had a diameter comparable to the thickness of the Earth's crust) hit the Earth, it would punch right through the crust and dive deep into the interior. The molten magma under the crust would be violently forced aside. This type of situation would be comparable to asteroid impact on water where the water is forced aside to allow the asteroid to pass. The magma in the interior would respond in a similar way. This means that a great circular mound of magma would start propagating outward under the crust in all directions. It would be a magma wave. In response to the passing of this wave the crust would rise up and then settle back down when this great magma pressure wave had passed. The surface of the Earth would move up and down and continue moving up and down for some time. A small vertical displacement would not be a hazard for animal survival but in a large asteroid case, the vertical movement near ground zero could easily be several thousand feet. Furthermore the movement would be sudden. At the same time that vertical movement was occurring horizontal movement would also be happening. The magnitude of the movement and the vertical and horizontal combination of movement would present an unsurvivable hazard for all forms of animal life. Creatures would not be able to stand in any one place because the ground would be repeatedly and quickly moving out from under them. In a circumstance like this, animals could be compared to dice in a dealer's hand. As the dice would be shaken so would any animals in the area and of course human beings would not fare any better. For several thousand miles from ground zero death, by being tossed around on the surface of the Earth, would be a certainty.

2.7.3.2   Continent Lifter

When the massive pressure wave had traveled right through the interior of the Earth, it would well up under the crust on the far side. The crust would rise up. If ocean-bottom crust was involved, the ocean would rise up. The water in the elevated area would commence to flow down from the elevated region and would flow outward in all directions. This water movement would involve thousands of square miles and present a hazard to everything in its pathway. Sheet flow of varying depths and involving entire oceans could be involved and this would result in certain underwater mounds being sheared off. Unfortunately this would not attenuate these flow patterns significantly and they would continue to all of the shorelines in their pathways. Shorelines would not be able to slow these flows significantly either so great sheets of rapidly moving water would wash right over many continental areas and continue on around the world. Since many of these water flows would be globe-trotters, it can be appreciated that all of this activity would be hazardous to animal life and would present a situation where there simply would not be any place on the surface of the Earth that could be called safe. Widespread death to animal life would be a certainty.

2.7.3.3   Volcanism

As the earthquakes rattled the crust of the Earth, the interior pressure waves repeatedly swept back and forth through the interior. As a result the crust of the Earth was repeatedly stressed into failure mode and it faulted (or cracked) in many places. Either directly on these faults or very close by, material from the interior came to the surface and volcanoes were born. Unfortunately from a survival viewpoint, volcanoes are not good news. Volcanoes produce a variety of material including; dust, water, grit, obnoxious gases and material that will harden into a type of rock called tuft. The rock-forming material usually falls out nearby and can pile up until it is several feet deep as was the case at Pompeii. Also the water will expectedly fall out within a few miles of the eruption site and is often accompanied by thunder and lightning like a regular thunderstorm. The grit consists of sharp-sided particles varying in size from clearly visible to microscopic. If there is enough of it, vegetation can be buried but even if only a light coating falls to the surface of the Earth, vegetation becomes difficult to eat. It would be like trying to get a mouthful of food but getting one-half of it as sand instead. The obnoxious gases will generally dissipate into the atmosphere but can, as in the Mount St. Helens case, erupt from the ground in such a concentrated form that they topple trees as they roll down the side of the mountain. However once they diffuse into the atmosphere, they become world travelers as would much of the fine-grained dust. Both the dust and the gases often rise to great heights and stay in the atmosphere for months or even years. It has not been uncommon for the dust from a single eruption to spread throughout the atmosphere to such an extent that it completely encircles the Earth and remains aloft for years.

There are several ways for a volcanic eruption to result in death for animals in the area. They can be buried outright as the heavier materials fall to the Earth. Their source of food can be buried but even if it is not buried just having a coating of grit or dust on it would render it unfit as animal food. Most animals require plants to eat and if they do not get enough of them they will simply die of starvation. Hopefully this would only be the case within a few miles of the actual volcano so being further away would have improved the chances of survival. The problem would have been compounded however if numerous volcanoes erupted at the same time or if some of them were exceedingly large. Indeed this would have been the case when the asteroid shower occurred. In fact since the earthquakes and interior pressure waves faulted the crust of the Earth in numerous places all around the world, it follows that numerous volcanoes would have erupted within the same very short time span. It is being suggested here that all 10,000+ volcanoes erupted at this time and some of them kept right on erupting for months. The problems involving grit, dust and other survival hazards would consequently have been multiplied raising the spectacle that there would not have been very many places on the Earth that would have been free enough from these volcanic effects to have provided a safe place for animals to find food. Therefore starvation from the volcanic effect would have been worldwide.

2.7.3.4 Igneous Provinces

The material that formed the igneous provinces has come to the surface of the Earth from beneath the crust. Its arrival on the surface is a type of volcano described as 'massive basaltic lava flows'. (141) It oozed, boiled up and spread around leaving hard rock surfaces where there had previously been a variety of surfaces including much softer sedimentary rock and even sand, gravel, soil and other materials normally expected on the surface. Unfortunately igneous provinces are not rare. They actually cover a large percentage of the Earth's surface – possibly more than 5% of it. Some areas in the Pacific Ocean are larger than many European countries. Of course survival would not have been possible wherever they were developing because the material would have been hot – as hot as all material is when it comes out of a volcano and where there was water contact the water would have boiled.

2.7.4 Secondary Atmospheric Hazards

2.7.4.1 The Impact Cloud

One result of an asteroid impact is a cloud. A great amount of dust would result directly from the impact of the asteroid with the surface of the Earth. It would be comparable to the cloud produced by an atomic bomb where the cloud structure includes a circular "stem" with a large "mushroom" top – hence "mushroom cloud". In the atomic bomb case the cloud will form and rise and the stem will keep feeding the top until it spreads out over a large area. Something similar would happen if an asteroid hit the Earth. Various commentators suggest that the cloud would rise very high into the atmosphere and then spread out. In the asteroid case, the spreading-out activity would continue for days and soon cover the entire world. The world would become dark! Further, the present work suggests that there was an asteroid shower and not just a single asteroid. Consequently there would have been an aggregate cloud consisting of the dust produced by every single asteroid. Consequently there would be no question that the world would have become dark. It would, in fact, have become as dark as night and stay that way for months. This darkness factor would mean certain death for all animal life that was dependent on plants for food. The plants would all wither and die. There would be nothing for any animal to eat. Since the cloud would be worldwide, death by starvation would also have been worldwide.

2.7.4.2   The Volcanic Cloud

As a result of a single asteroid impact, volcanoes would have erupted all over the world. As with the asteroids, the clouds produced by the volcanoes would also have formed an aggregate cloud which would also have covered the entire world. In actual fact the cloud-forming material from the volcanoes (i.e. the dust) would have become mixed with the dust stirred up by the asteroids resulting in a thick continuous cloud completely enveloping the entire Earth. Animal survival would not have been possible. Any small opening in the dust from one of the two sources would have been filled in with the dust from the other source. The entire Earth would have become dark and would have remained dark for months. Animal survival would not have been possible.

2.7.5.3   The Moisture Cloud

The volcanoes not only produced a great quantity of dust, they also heated the water in the ocean. At the same time that dust was being released into the atmosphere, molten material from the interior of the Earthwas oozing out and flowing up onto the floor of the ocean. In many places the water boiled. All over the world the over-heated ocean evaporated tons and tons of water into the air. In fact, before it cooled down many months later more than 2000 ft of ocean had evaporated. The air became loaded to the saturation point with moisture. All of these great banks of moisture drifted across the ocean and mixed with other moisture clouds until a virtually continuous layer of moisture cloud enveloped the Earth. Expectedly this moisture cloud would have been lower in the atmosphere than the dust cloud but the effect was the same. The entire world was darkened. The two layers of cloud guaranteed that sunlight would not get through and that the entire world would be dark and also guaranteed that the surface of the Earth would chill due to the blockage of sunlight. By reinforcing the hazard of the dust cloud the moisture cloud sealed the fate of animal life so that any possibility of survival was hopeless.

2.7.5   Ice Age Hazards

2.7.5.1   Introduction

The stage had now been set. The Earth had been darkened resulting in a worldwide chill. The oceans are being heated resulting in prodigious evaporation. When the great banks of moisture cloud drifted over the chilled land, precipitation resulted. At first this precipitation was mostly rain but as the temperature drifted downwards, it snowed instead. Meanwhile the ocean kept producing vast quantities of moisture.

As the temperature over the land drifted down past the freezing point, the rain changed to snow and the great Ice Age was underway.

2.7.5.2   Rain

The rain fell in great torrents. It was the type of rain that is expected from a hurricane. It is not the least bit uncommon for a hurricane to produce several inches of rain every hour for several hours. As the great Ice Age got underway hurricane-like rain fell but instead of falling for a few hours, it fell for weeks. A rainfall of this magnitude would not have been any more survivable than the rain from a hurricane would be today if it kept on raining with similar intensity for weeks. As land and trees and other forms of natural protection were being washed away, where would any animal have found shelter? There simply would not have been any. As hiding places in the ground were washed away and trees were washed away, the number of safe places was reduced. Perhaps shelter could have been found in a cave for a while but pretty soon the cave would fill with water and wouldn't be a suitable shelter either. Birds do not need a cave to hide in but they certainly need a place to land. As the great Ice Age got underway, there would not have been any place for a bird to find shelter from the rain either. Consequently they would have all died. Animals on the ground would not have had it any better and they would have been washed away with everything else. The rainfall that ensued as the Ice Age started would not have been survivable.

2.7.5.3   Snow

From a survival viewpoint, rain is more survivable than snow. At least rain will flow to the lowest location. Snow will not flow anywhere and it will just keep piling up where it lands. How is anything to survive when the snow piles up until it is ten feet deep. An animal cannot move through that much snowand it certainly cannot dig through that much to find food. As the snow continued to deepen it is readily understood that the chances of survival continually reduced until they simply vanished. What food can be found when the snow is a hundred feet deep. During the great Ice Age the snow was thousands of feet deep. In fact it kept right on snowing until the heating effect of all of the underwater volcanoes and igneous provinces abated and all of these heat sources had cooled down. Since the cooling would have taken many months or even years, it can readily be seen that the survival question is unquestionably answered. Because of the snow which fell during the beginning of the great Ice Age, survival of animal life was not possible.

2.7.5.4   Temperature Drop

The temperature drop which caused snow to fall instead of rain would not have been survivable even if it had not snowed because nothing can survive when all sources of drinking water are frozen solid. With the Ice Age the period of cold was prolonged and even if it was only a few degrees below freezing, it would not have been survivable. Every type of animal requires drinking water. If they do not get any they die of thirst. It is therefore submitted that the temperature drop during the great Ice Age was singularly unsurvivable and simply augmented the survival hazard presented by the rain and snow. Animal life could not have survived the temperature drop that occurred as the Ice Age got underway.

2.7.5.5   Plant Death

As already mentioned, the plants would have died from the rain, the snow or the temperature drop or some combination of all three. Of course without the plants animals cannot survive and would also have died.

2.7.5.6   Melting Snow, River Formation, Submarine Canyons, Erosion and Coulee Development

These topics are tightly interconnected and in fact represent various aspects of the situation as the great Ice Age came to an end. The terminal events of the Ice Age were as hazardous as the earlier events and because they were manifest over the entire world, the associated hazards were also manifest over the entire world.

Snow accumulating for hundreds and thousands of feet is certainly a life-threatening hazard. Melting snow is not much better. When the snow melted, it melted very quickly and flooding resulted. One commentator has declared that the melting snow caused the Hudson River in the northern USA to swell its banks until it ran more than two hundred feet higher than it does at present. (86) The current would have been commensurate with the increase in water level with the result that being in the vicinity would have been very hazardous for all types of animal life. While it can be asserted that the hazard would have been restricted to the immediate area, the hazard level within the flood area would have been overwhelming. Further, as the Hudson River was in flood stage, other rivers across the country would also have been in flood stage. The great Mississippi River would have been experiencing massive flow as well and would have flooded the entire central region of the continent. A similar situation would have occurred with the Saint Lawrence River. Flood stage on this river would also have involved the region on both sides of the river for hundreds of miles. Europe has the Danube River and the Rhine River and other regions of the world also have very large rivers As the Ice Age abated they would all have been in excessive flood stage, so being in any of these areas would have been extremely hazardous.

This was the time that all of the rivers of the world were being formed. It was a recent time as indicated by the size of the deltas of all major rivers. When flow rates and sediment loads are used to calculate delta formation it is clear that these deltas have accumulated within a few thousand years making their respective rivers not very old. In fact they are only a few thousand years old which time in coincident with the time when the great Ice Age abated. As it ended numerous troubles became manifest and only intensified during the ensuing weeks.

Submarine canyons have formed as a result of river extension into the sea. While the formation of submarine canyons would not have presented a direct threat to life on land, since they are indicative of massive cold-water flow, their presence reinforces the threat that the melting snow presented to life on the land. The canyons tell us that the rivers were in excessive flood stage and that their cold melt-water was slightly colder than the ocean which had not yet cooled to its maximum-density temperature. In some cases, they also tell us that rivers once flowed where there aren't any at the present time. No doubt these abandoned rivers would also have been experiencing heavy flow and so would have presented the same hazard as the others. When the submarine canyons were being formed the hazard to life due to floodwater would have involved the entire world.

Excessive river flow always involves erosion and the erosion associated with the end of the Ice Age included not only the usually-expected developments along rivers but also included the formation of all of the great canyons and coulees of the world. Coulees are the side canyons and ditches associated with most of the river systems of North America. They are very clearly visible in the North American west where the lack of forest cover enables them to be clearly viewed. Coulees are identical in appearance to much larger canyons – in particular the Grand Canyon of Arizona. Canyons and coulees were formed when the snow of the Ice Age melted. They formed at the same time that the rivers of the world were formed. It is very evident from observing these formations that the water flow must have been enormous because the missing material is not to be found. It has completely left the area and neither is it found downstream. It is simply gone. A massive water flow was required to do this. A massive water flow was exactly what would have been expected when the snow ceased to accumulate and started to melt. A beginning trickle turned into a torrent and the flows became so huge that the entire volume of all of the great canyons and coulees of the entire continent of North America was completely washed away never to be found again. Such massive flow has not been experienced since that time which is of course a good thing from a survival perspective because when the material which was removed to form the canyons and coulees was being taken away it would have been extremely hazardous for all forms of animal life. Any animal that got caught in any of these flow patterns would not have survived but would simply have been washed away to the ocean with everything else that was being removed.

2.7.5.7 Volcanism

As mentioned above in section 1.4.8, a major asteroid impact would have resulted in a great deal of volcanic activity. The earthquakes, with their resulting shockwaves, would have caused the eruption of numerous volcanoes. An asteroid shower would probably have resulted in all 10,000+ volcanoes erupting within a short period of time. Unfortunately, as witnessed by the layers of ash which are distributed throughout the remaining glacial ice it is clear that volcanism continued well into the Ice Age. (131) In fact volcanism has continued right to the present time and has never otherwise been very well explained. Even if only 100 volcanoes erupted within a short time span, a significant hazard to life around the entire world would be the result. Besides the dust, which would probably drift away to become someone else's problem, volcanoes spew out a host of undesirable material including noxious gases. Being in the vicinity of an erupting volcano is never good for one's health and the volcanism that continued into the Ice Age would have only compounded the already overwhelming hazard to life that the Ice Age presented.

2.8 Summary of Hazards

Type	Characteristic	Extent
Blinding Flash	Extremely bright light	Localized within a few miles

Type	Characteristic	Extent
		around every impact site (GZ)
Atmospheric Shockwave	Rapidly expanding hard wall	Dangerous within several hundred miles of every GZ
Precipitation	Repeated waves of hurricane-like rain	Spreading outward for several hundred miles of GZ
Wind	Multiple thousands of miles/hour	Dangerous for a thousand miles from every GZ
Initial Water splash	Large globules of falling water	Dangerous within several hundred miles of GZ
Subsequent Water Splash;	Smaller globules of falling water	Dangerous within a hundred miles of GZ
Crater Formation	Flying mud and rocks	Several crater diameters
Primary Tsunami	Mountainous expanding circular wave moving hundreds of miles per hour	Circling the entire world repeatedly
Secondary Tsunami	Mountainous expanding circular wave moving hundreds of miles per hour	Circling the entire world repeatedly
Mountain Climbers	Water moving in sheet-flow hundreds of feet deep	Moving fast enough to climb right over mountains thousands of feet high
Mud-Rock Splash	Masses of mud and rock falling from the sky	Localized within several hundred miles of every GZ
Erratic Boulders	Boulders weighing from thousands of tons to a few pounds falling from sky	Hundreds of miles from every GZ
Earthquake	Back and forth and up and down ground movement	World-wide from every impact
Continent-generated Tsunamis	High-speed ocean-crossing waves	Circling entire world repeatedly
Sedimentary Rock	Thick layers of material deposited by moving water	Every continent of the world

Type	Characteristic	Extent
Mountain Formation	Soft sedimentary layers distorted into wave-like formations by shockwaves	Every continent of the world
Continental Shelf Formation	Great quantities of sediment washed off of continents	Every continent of the world
Momentum Transfer	Small realignment of Earth's axis	Entire world affected
Interior Shockwave	Crust of the Earth lifted upwards	Thousands of square miles centered on every antipode – above or below water
Ground Tsunami	Ground moves up and down in wave pattern spreading from GZ	Thousands of miles from every GZ
Continent Lifter	Entire continent is elevated by hundreds of feet	Thousands of square miles centered on every antipode
Chaotic Terrain	Crags, escarpments, valleys, bluffs cracks, blocks, rivers& lakes	Within crater rims, also at antipodes
Antipode Tsunami	Sheet-flow wave spreading from every antipode	Repeated encircling of the world
Earth Wobble	One-time sideways lurching of Earth's surface	Mostly noticeable at north and south poles
Volcanism	Ash, gases and lava	Area around volcano
Volcanic Clouds	Sky-darkening layer high in atmosphere	Encircling entire world
Impact Cloud	Sky-darkening layer high in atmosphere	Encircling entire world
Moisture Clouds	Dense, moving, precipitating fog blanket hundreds of feet high	Thousands of square miles of both ocean and land
Rain	Hurricane-like, averaging several inches every hour	World-wide, lasting several weeks
Snow	Heavy, blizzard conditions	World-wide lasting for months/years
Temperature Drop	Below freezing temperatures year round	Within hours near GZ, within weeks worldwide
Plant Death	Follows temperature drop	World-wide
Melting Snow	Water running everywhere	Entire world
Erosion, River Formation and Coulee Development	Soft rocks/soil washed away, river valleys developed	Entire world

Type	Characteristic	Extent
Submarine Canyons	Deep valleys cut through under-water silt	Extending from every major river into the deep ocean

3. Multiple Impacts – Single Event

3.1 Sedimentary Rock

Sedimentary rock is a dominant factor in the geology of the Earth. Sedimentary rock is widespread and found in abundance on every continent. In fact, sedimentary rock is the type of rock that is the most common in mountain formation and of course, mountains are found all over the world. There are basically two types of sedimentary rock; the type formed by wind and the type formed by water. The first type is, while being widespread, is not really very abundant in extent or volume. The type formed by water is very widespread being found on every continent in abundance. The individual layers are commonly quite thick and might measure from several meters thick to hundreds of meters thick. Even where each layer is thick, there are often several layers involved making for total thicknesses of several thousand feet. It appears as though each layer represents a particular event of moving water. This immediately implies that the water flow that was involved in the placement would have been of exceptional magnitude. It would have been deep, fast and widespread. Then, when one particular flow stalled or disappeared, another one entered the scene carrying another load of material which was subsequently deposited. This process was repeated until the entire formation was in place. Perhaps there was a time interval between water flows but if such a time interval had been of significant extent one would expect to find evidence of the upper surfaces of each layer being weathered. This type of feature is not found leading to the conclusion that subsequent layers were placed before any weathering could take place. While the placement of even one layer could properly be considered a major event, the placement of numerous layers makes the entire episode a catastrophe of unimaginable proportions.

Such a conclusion is reinforced by the types of layers found in the Grand Canyon of Arizona, USA. The features of the Grand Canyon have been studied quite thoroughly by numerous competent geologists and some of them have offered the following; '… bed-forms deposited by high-velocity currents …flows of high-density turbulent flow … rapid deposition in continuous currents …massive flooding event.' (5)

The energy that would have been involved when a large asteroid struck the Earth would have been exceedingly great and properly recognized as being more than enough to produce an Earth-encircling sequence of water waves. The second and subsequent times that such water flows appeared in any given place would expectedly result in the placement of subsequent layers of material. Would these subsequent material-placing waves also have been sufficiently energetic enough to repeatedly carry and place enough material to form layers thousands of feet thick on every continent around the world? Probably not! It is therefore concluded that the event taking place involved more than one asteroid. It was an asteroid shower and the asteroids rained down on the entire world over a period of several weeks. Assuming that the major hits were reasonably spread around the world, (which they are) every continent would have been adjacent to a place where giant waves were being produced and would therefore have been over-run by the water-

491

waves shortly after they had formed. They would therefore have still been so energy-loaded that they could entrain several thousand feet of sediment-rock forming material. An asteroid shower was required to do this.

3.2 Mountain Formation and Features

Sedimentary rock and mountains are geologically directly connected. Mountains consist of sedimentary rock, albeit bent, broken and twisted but non-the-less sedimentary rock. This indicates that sediment-entrained water was involved, then another wave-fronted sheet of sediment-entrained water flowed over the same area and left another layer. After this had happened repeatedly the water flows petered out and the material just stayed in place to harden. Apparently, the hardening process does not require very much time because the material involved has certain cement-like characteristics. Soon it was firm enough to walk on and soon after that it was too solid to be readily penetrated by hand-held objects. Within a few months it would have required a hammer to make any impression whatsoever. At this stage it would not have been bendable. As more time passed the rock just got harder and the time-threshold had passed when any type of bending could occur.

In order to bend anything, from a piece of paper to a steel beam, some portion of the material must be able to stretch and some must be able to compress. That's what bending involves. If one bends a round rod, for example, the material on the outer side of the bend will stretch. The material on the inner side of the bent section will compress. If, either the stretching phenomenon or the compression phenomenon, do not occur the material in these locations will fail and split open.

When mountains are examined, one finds that great frozen rock waves are in place. Clearly therefore, the material on the outer surface of these bends has stretched while the material on the inner surface has compressed. Otherwise bending would not have occurred but only fracturing. It is also true that the folds are often accompanied by fractures. This indicates that the bending action was too severe for the particular material at that particular time. Fracturing therefore resulted. Fracturing indicates that the bending stress was too much either because the material had hardened a little too much or the magnitude of the bending activity was simply too great. However, since bending has occurred and, in fact, is widespread, it is clear that some of the rock-forming material has been stretched and some has compressed. Also in many regions of the mountains, bent regions are accompanied by fractured regions. In both of these circumstances (i.e. bending and fracturing) a very great amount of energy was involved. However, since bending is widespread and bending associated with fracturing is widespread it must be concluded that the bending-energy-carrying shockwaves passed through the area before the sediments had hardened too much to allow any bending at all. It is also clear that the bending occurred nearly simultaneously around the entire Earth because such vast amounts of material simply cannot be placed on a continent by continent basis (at vastly different times) without something similar happening on other continents as well. Neither would the unusually energetic shockwaves have been limited to any one continent but would have been manifest all the way around the entire world. Similarly, the massive sediment-carrying water-flows would not have been isolated to any particular continent. The entire world would have been involved.

It is therefore concluded that the sediment-placement process must have occurred on a world-wide basis as a near-simultaneous event, Further, since the bending phenomenon must happen before hardening sets in and since this too is a world-wide observation one must conclude again that mountain formation must have occurred on a worldwide basis as a near-simultaneous event. Both the sediment-placement process and the sediment-bending process, being so time-dependent, indicate that the massive energy involved was also manifest around the entire world which points to an asteroid shower as the source of all of the upheaval.

3.3   Ice Age Improbability

It has proven most difficult to explain how an ice age could happen. One commentator has suggested that there have been at least sixty attempts to explain it with capable people holding widely different convictions. (1) Furthermore it is unfortunate that no explanation stands up to very much scrutiny so that readers are left without any real consensus in which to place their trust. Several scientists have recognized the need for huge amounts of heat to be applied to the oceans but there is a complete absence of suggestion as to the source of such heat. At least such recognition is a cause for some encouragement to the average person. In spite of so much disagreement it is probably fair to conclude that an ice age would be a rare event. How would several hundred feet of ocean find its way out of the ocean and onto the land and be caused to accumulate as ice far from any shoreline? Of course we understand that water is leaving the ocean all the time and sometimes it finds its way over the land and falls as snow. What would cause it to keep on snowing right through summer? Wouldn't the Sun shine and simply melt it? So much ice could not have built up in a single winter so it is clear that more than one winter was involved. Why wouldn't it simply melt again in the spring as it does now? Some agent had to keep it from melting so the temperature must have been held below freezing. However, if it was a little too cold, the source of the moisture (i.e. the ocean) would have crusted over and no water would have escaped to the land. Therefore, the minimum requirement is that the ocean be ice-free but that the land be cold enough for snow to accumulate prodigiously. These basic criteria appear to be contradictory.

Would the landing of a single asteroid have been able to produce both the world-wide ocean heating requirement as well as the world-wide land-chilling requirement? Undoubtedly a large asteroid would have produced a world-encircling cloud which would have chilled the land. It would also have caused the eruption of numerous volcanoes, some of which would have been underwater and these volcanoes would have heated the ocean. However it is clear that the ice age involved the removal of several hundred feet of ocean water (2) which somehow made its way onto the great ice fields. Further there is still more than two hundred feet of ocean tied up in existing glaciers around the world. (3) These two factors indicate that possibly five or six hundred feet of ocean was caused to evaporate and find its way into fields of ice. Unfortunately to this point the story is only half told. In order to get that much ocean water into the ice fields a quantity of water several times that amount would have had to evaporate because it is most unlikely that every drop of evaporant found its way into ice. It is much more likely that some of it simply fell right back into the ocean and never made it to land at all. It is also probable that some of the evaporate fell as rain instead of snow and would therefore not contribute to the massive buildup of ice. In recognition of these probabilities it is likely that the total amount of ocean that actually evaporated to have the great Ice Age was several times the above-recognized five or six hundred feet. It more likely would have been several thousand feet. Further the evaporation of several thousand feet of ocean would require a very large amount of thermal energy to be released. Would a single asteroid impact have been capable of causing the release of so much thermal energy? Clearly this is not very likely. Therefore it is concluded that an asteroid shower that involved the entire Earth was what happened.

3.4   Dust Cloud Demise

It is absolutely certain that the great Ice Age involved a sun-blocking dust cloud. Very simply, if there had not been a cloud it would not have been cool enough to enable the ice to accumulate. Furthermore even though accumulation was rapid, since a great amount of ice was involved, the accumulation process must have been prolonged over at least several months and even longer in those areas where great thicknesses of ice were enabled to build up. It is also absolutely certain that the dust cloud would have been a world-encircling cloud because the atmosphere is always moving and within a few weeks air that is over one continent will be over another continent. If this dust cloud had permitted the Sun to shine through, accumulated snow would have melted – particularly during the summer. While this is probably what happened pretty soon anyway, it would not have been the near-term scenario over regions far from the equator. The ice that built up in the polar and sub-polar regions was very thick. This means that the cooling effect of the cloud was extended for much longer than it was further south and altogether likely involved several years instead of just a few months. From a survival viewpoint this would have been very bad news for all types of animal life. In fact survival would not have been possible if the sun-blocking cloud lasted for years. Even if snow did not accumulate during such a condition, nothing would have been able to grow anyway. Plants need sunlight. If sunlight is withheld for years, plants will not grow, Instead they will die. When the plants are dead, what will the animals feed on? It can therefore be concluded that survival remote from the equator, probably from about half-way to the poles and extending into the high latitude regions, life beneath the never-ending cloud, would not have been possible at all. It would have been completely dark all of the time! It was a world-encircling dust cloud that was responsible for all of these miseries.

Survival near the equator would not have been possible either as long as the cloud persisted in that area. Perpetual dark, even for several months, is not survivable. A minimum of several months is assumed because the evidence suggests that there was snow accumulation in equatorial areas as well as far from the equator. (e.g. There is still snow on Kilimanjaro which is only 3 degrees from the equator. (7) The ice-age accumulation on Kilimanjaro must have been considerable because it has been melting rapidly for years and there is still a great amount left.) While it did not match the massive accumulations of the higher latitudes, there would have been more than just a few snowstorms. It simply isn't possible to imagine a world where the upper latitudes are cloud-bound for years where snow continues to accumulate while the lower latitudes were cloud-free and snow-free. Such extensive cloud cover would not have been restricted to any particular part of the world. Even the explosion of Mount Tambora in 1815 was an event that produced cloud cover for the entire world. (5) The impact of even a single asteroid, together with the numerous volcanoes that would have been caused to erupt certainly would have produced an Earth-enclosing cloud. The vital question is to be asked is; how long would the composite dust cloud from a single asteroid impact have been able to persist continuously enough to enable the unspeakable quantities of snow of the great Ice Age to accumulate? Without a persistent world-encircling cloud, accumulations like this would not have happened. After all, as already proclaimed above, ice ages are extremely rare events which require extremely rare conditions. Would the cloud from a single asteroid hit have persisted long enough to enable the huge amounts of snow to accumulate to have an ice age? (10,000 ft of ice is still piled up in some places which represents 100,000 ft of snow. Even if this snow fell at the rate of 10 feet every day it would have taken thirty-seven years to get this thickness.)The conclusion is that it would not have been possible and that more than one hit would have been required. The simplest explanation is that the event was an asteroid shower which could have lasted for several weeks and which was able to produce a world-encircling sun-blocking cloud that persisted for at least several months in the tropical regions and probably for several years in the upper latitude regions. Further it is clear that if the cloud had not met its demise - at least in some places within a few months – there would have been no hope of survival at all. But since the cloud must have persisted for years in the upper latitudes, an asteroid shower is the most logical explanation.   What happened was an extremely rare event which agrees philosophically with the notion that ice ages are also extremely rare events. The final conclusion is that the nature of the demise of the dust cloud is indicative of an asteroid shower and not just a single hit. A single hit would, no doubt, have been devastating but would probably not have produced a world-encircling cloud that lasted for years.

3.5   Greenhouse Gases

Greenhouse gases are the portions of the atmosphere that trap the heat from the Sun and prevent it from radiating back into space. These gases are absolutely necessary for our survival and without them the Earth would be much colder. The energy from the Sun on its own is not sufficient to keep the surface temperature of the Earth in the comfort zone and without the greenhouse gases the average surface temperature would be about -18C. (Please refer to Climate and the Oceans by Wallis. P. 165) Without the greenhouse gases the Earth would be frozen solid.

It would be perilous to alter the quantity of either one of the two main gases, CO_2 or water vapor. The amount of water vapor is directly dependent on temperature so if the temperature of the surface of the Earth were lowered for any reason the amount of water vapor in the air would be reduced. Therefore the temperature would drop even further and keep dropping until any further change had no effect on the amount of water vapor. (Hence the -18C estimate)

The opposite would also be true. This reality underlies the current concern over increasing CO_2. In this case the situation is further exasperated because an increase in temperature due to an increase in CO_2 will enable the air to hold more water vapor thereby causing even further heating.

If a large asteroid hit the Earth a globe-encircling dust cloud would be expected. Volcanoes would also be activated with their effluent adding to the cloud cover. The Earth would chill. Underwater volcanoes would cause the ocean to heat and so the question is raised; would the heating of the ocean be enough to counteract the chilling of the land so that the overall effect of the greenhouse gases would not be compromised? The answer is; not very likely. The ocean is already very cold – only a few degrees above freezing except for a few feet near the surface. It would take a lot of volcanic activity to raise the temperature of the ocean enough to counteract the chilling of the land. A single asteroid hit would not generate enough volcanic activity to warm the ocean significantly to ensure that the greenhouse gas heating effect would not be lost. Therefore a single asteroid hit would not likely be survivable because the Earth would simply freeze solid and without the greenhouse gases to trap the Sun's heat when the clouds parted again, there isn't any way that the Earth would be able to heat up again. This means that the great event that produced the world-wide flooding and the great Ice Age was an asteroid shower.

3.6   Other Heavenly Bodies

3.6.1   The Moon

The Moon, being so close to the Earth, has been readily observable in considerable detail on the side that faces the Earth and since various space vehicles have been sent there, it has also been observed on the far side. There are some very large impact sites on the Moon but they are not randomly distributed as would reasonably be expected as a result of events that were widely separated in time.

This assumes that the maria on the Moon are evidence of asteroid impact. The reasoning for this assumption relates to two main features of the maria. Firstly, mass concentrations are located beneath the surface of most maria suggesting that large objects hit the surface, broke through into a molten or semi-molten region, became embedded, remained in place and are now measureable as gravity-increasing mass concentrations. Secondly, immediately subsequent to impact, molten material flowed out across the surface forming the "seas" (i.e. the maria) and after that happened, the impacts ceased. It is safe to assume that most impact activity ceased because there are virtually no craters in the maria. Any other objects that might have landed there had to have landed before the lava solidified because once it became solid any impacting object would have left a mark. All of this is suggestive that the entire event took place before the Moon could rotate to cause these large impact sites to be more evenly spread out. An asteroid shower would explain this evidence.

3.6.2   Mars

Mars provides evidence of a similar event taking place. As with the Moon, Mars is also heavily cratered. In fact some of the largest craters in the entire solar system are found on Mars, including Hellas, the largest, with a crater more than one thousand miles in diameter. Curiously two other very large craters are also found on Mars but the main point of interest for the present discussion is that all three of these monster craters are located on one side. Is this coincidental? Further the side which includes the very large craters also includes more than three thousand other craters more than twenty miles in diameter. (6) This would, independently, not be conclusive of anything except that the opposite side has only about ten percent as many. It cannot be coincidental that one side has three monsters and almost a thousand lesser impact sites while the opposite side has only a small fraction. The logical explanation is that an asteroid shower occurred and was over before Mars could rotate to cause the impact sites to be spread out more evenly geographically as well as temporarily.

Evidence like this, involving three heavenly bodies, supports the suggestion that an asteroid shower occurred in the solar system. Since Mars rotates at approximately the same rate as the Earth one would wonder why a similar concentration wouldn't be found on Earth. It might be however, that the object that broke up and caused these asteroid showers in the first place, broke up in the vicinity of Mars. In such a case there would have been a much greater concentration of objects near Mars than there would have been when the shower had traveled further into the solar system and encountered both the Earth and the Moon. By then it would have spread out allowing impacts on the Earth to be more randomly spread out. The Moon on the other hand does not rotate nearly as fast as the Earth and so the objects which hit it would have appeared more concentrated in location for a similar time period.

3.7 One Mass Extinction

3.7.1 Earth Impact Database

The Earth Impact Database is a website (www.unb.ca/passe/ImpactDatabase/ that is maintained by the University of New Brunswick, Planetary and Space Science Centre, in Maritime Canada. Therein are listed at least 175 impacts and more are added as they are discovered and confirmed. As the list is examined it seems that Australia has a disproportionate number of impact sites in comparison to its size. This visibility is probably due to the lake of vegetation and it makes one wonder how many would show up in other countries if they had a similar lack of trees. Also Sudbury is only listed once whereas there is the Sudbury Basin, which is about 40 miles in diameter and centered close to the city of Sudbury, as well as the major impact (listed and centered further to the north-west) that was much larger.

The list of impacts has been discussed in Part 1, chapter 12.4.1 but it is worth noting here that there are 75 greater than 10 km. (6 miles) in diameter including 35 with diameters greater than 25 km. (16 miles). All of the impacts which produced these craters would have had life-threatening consequences around the entire world. Even many of those smaller than 10 km., each having the destructive power of hundreds of large bombs, would have produced world-wide trouble and destroyed life over very wide areas.

3.7.2 Surviving Cohort

The surviving cohort would have been the total group of animals that made it through the disaster. (Cohort simply means some particular group.) Some commentators suggest that there would be a 5% survival following a major asteroid impact. (As mentioned previously, this type of statement is never explained or justified.) If we allow that this is a reasonable assertion, it must, at the same time, be recognized that the 5% would not necessarily be referring to 5% of each and every animal group or species all the way around the world. For example, it would seem logical to suggest that near the impact site survival would have been much less than 5% and probably closer to zero. This would simply happen because there would be much more trauma near ground zero than a few thousand miles away. This means that whatever animals happened to have been on the impact continent would probably all have been annihilated. The great Vredefort Crater in South Africa is given as 300 kilometers (185 miles) in diameter. When this crater was formed the entire continent would have convulsed repeatedly. Those animals that weren't blown away by the shockwave would have been shaken and tumbled until their relatively fragile animal bodies were broken. There simply would not have been any animal survival on Africa after the giant asteroid hit South Africa. While only two major survival hazards are mentioned here, from the earlier discussions it is clear that there were more than two to deal with and any animals that escaped the first two would have still had to deal with several more. With this limitation in mind the 5% that survived must have been elsewhere and therefore more relatively concentrated.

The other relevant problem relates to the social characteristic of most of the animals around the world. There is no species of animal that is spread evenly around the world. Even after all of these thousands of years every animal group is only found in certain areas and not in all areas. This is most obvious from continent to continent but it is also quite glaring within every continental boundary. For example, if one is to observe all of the animals of North America it become quite obvious that most animals occur in groups in relatively restricted areas. There are obviously numerous reasons for this but included in the list would be the type of habitant and the specific nature of the climate. All of this improves the probability that some species would be wiped out completely if a large asteroid struck their particular continent. The suggestion therefore that 5% survived requires an explanation. Which species had a 5% survival and which ones had less?

At the same time when it is recognized that there have been numerous impacts by large objects from space, one is forced to deal with the surviving cohort problem. This problem has occasionally been explained away by stating that most of the large impacts occurred before there was any significant life on the Earth. Unfortunately, this type of explanation does not really help very much. Even if this was the case, it still means that the entire cohort that presently exists (or that existed up until recently as we recognize the present epidemic of both disappearing and seriously-endangered species) descended from the group that was left after the last wipeout. Similarly, the total group of animals large or small that existed immediately prior to the last impact had to have evolved from the cohort that survived the impact before that. And so on. Referring back to the Earth Impact Database it can be clearly seen that there have been numerous impacts and that many of them would have wiped out most of the animals on the Earth. Further, this does not even count the impacts that occurred and are not listed in part because they have not yet been discovered. (e.g. The Earth has a 70% water surface so we would expect that at least as many impacts occurred on the ocean floor as did on the land.) With all of this in mind, preferring a scenario which includes major impacts widely separated in time makes the dilemma (of trying to explain how inter-impact evolution occurred) worse.

As discussed in Part 1, declaring that there have been numerous mass extinctions occurring widely separated in time leads to bizarre conclusions. It is clear that in the case of repeated impacts separated in time, the surviving cohort would not have been the same every time. This factor becomes more evident when it is recognized that most of the survivors of every major impact would tended to have been the smaller species like mice and rats. Large animals would have had a much smaller chance of survival. This means that most of the large animals of the Earth had to have descended from a group of small animals. Further, this would have been the case between every two impacts. While this is the logical conclusion (if major impacts were widely separated in time) it is certainly not very attractive. Therefore it is concluded that such explanations do not really recognize the problem and are therefore not credible.

The remaining possibility is that there was one overwhelming traumatic event and then it was over. How anything got through the first few months after either a single major impact or a shower of impacts has never been explained by anybody (apart from the accounts given in the Epic of Gilgamesh and the Book of Genesis) and certainly will not be attempted here either. However if they did somehow manage to get through the first few months, or possibly years if necessary, perhaps there might have been a chance to carry on from there. While it is relatively easy to destroy any one animal there really is a great resilience to life and given a chance it will thrive. To suggest however that there have been numerous nearly complete annihilations of life is not the least bit credible.

3.8 Survival

The hazards associated with an asteroid impact have been listed above in section 1.8. Each and every one of these hazards would have had an impact on the survival of animal life on the Earth.

Type of Hazard	Survival Scenario
Blinding Flash	Survivable except within a few miles of GZ (ground zero)
Atmospheric Shockwave	Completely devastating to all forms of animal life within at least hundreds and possibly thousands of miles of GZ.
Precipitation	Survivable if shelter could be found and the rain did not completely wash away both plants and topsoil where plants grow.
Wind	Unsurvivable within hundreds and probably thousands of miles of GZ due to the force of the wind as well as the debris it carried.
Initial Water Splash	Unsurvivable within several hundred miles of GZ due to falling globules of water.
Subsequent Water Splash	Survivable but dangerous within several hundred miles of GZ
Crater Formation	Unsurvivable within crater rim as well as outwards from the rim for at least one or two crater diameters.
Primary Tsunami	Unsurvivable world-wide due to Earth-encircling waves of water washing right over continents repeatedly.
Secondary Tsunami	Unsurvivable world-wide due to Earth-encircling waves of water washing right over continents repeatedly.
Mountain Climbers	Unsurvivable world-wide due to the speed of the water and the magnitude of the flow.
Mud-Rock Splash	Unsurvivable within several crater diameters of GZ due to falling mud and rock.
Erratic Boulders	Unsurvivable within several crater diameters of GZ due to rocks falling

Type of Hazard	Survival Scenario
	from the sky.
Earthquake	Probably unsurvivable on the entire continent where the impact occurred. Also, dangerous world-wide
Continent-Generated Tsunamis	Unsurvivable world-wide due to a sequence of numerous waves washing right over continents repeatedly
Sedimentary Rock	Unsurvivable wherever the material for sedimentary rock was has being deposited which includes every continent of the world.
Mountain Formation	Mountains were formed by massive shockwaves. Survival would not have been possible either near the mountains or within several hundred miles.
Continental Shelf Formation	Unsurvivable due to massive sheet water flow entering the oceans from the continents.
Momentum Transfer	Survivable
Interior Shockwave	Possibly survivable if antipode was continent-centered. Unsurvivable if antipode was ocean-centered due to water cascading towards shore.
Ground Tsunami	Unsurvivable if ground displacement was too great or ground broke open too much and too often.
Continent Lifter	Same hazard as interior shockwave
Chaotic Terrain	Unsurvivable due to unpredictable fracturing, lifting, sinking, horizontal displacement and distortion of bedrock.
Antipode Tsunami	Unsurvivable due to repeated world-encircling waves.
Earth Wobble	Survivable
Volcanism	Unsurvivable for a few miles around wherever volcanoes occurred as well as for several miles downwind.
Volcanic Clouds	Unsurvivable if numerous volcanoes were involved due to the world-wide darkness which would have caused plant death.
Impact Cloud	Unsurvivable due to sunlight-blocking clouds which would have resulted in plant death.
Moisture Clouds	Unsurvivable due to blocking of sunlight and carrying of moisture from ocean to land causing snow to precipitate in abundance.
Rain	Survivable only if shelter could be found and rain did not wash away the crop-producing soil too much.
Snow	Unsurvivable due to accumulations reaching heights of thousands of feet in height and covering thousands of square miles.
Temperature Drop	Unsurvivable due to freezing of plants and water.
Plant Death	Unsurvivable due to extinguishment of animal food sources.
Melting Snow	Unsurvivable due to extensive flooding including excessive river flows.
Erosion, River Formation and Coulee Formation	Unsurvivable due to water movement and material entrained in water.
Submarine Canyons	Unsurvivable wherever water flowed from the continents into the ocean to form the canyons.

Several of these hazards would not have been survivable on a world-wide basis for even a single asteroid hit. Several more of them would not have been survivable within a few thousand miles of GZ and several more within a few miles of GZ. If an asteroid shower occurred instead, and the impacts were spread around the world, most, and not just a few, of the hazards listed above would not have been survivable at all and some of them would have so repeatedly endangered the entire world that the survival question becomes completely moot.

4.   Hypothesizing About Survival

4.1   Introduction

It is curious that so many commentators have offered suggestions concerning the percentage of animal life that might have survived a major asteroid impact, but none have offered any discussion explaining how survival was achieved. Instead it is usually simply declared that when a mass extinction occurred, a small percentage of animal life made it through. It is completely unthinkable that everything died. After all, life is here now isn't it? Rather than actually dealing with the matter, it is simply set aside. At the same time it is admitted that there have been several mass extinctions on the Earth. (one commentator suggests 12. (2)) Generally the suggestion that there have been several, matches with the evidence that there have been impacts by several very large objects. The situation only becomes worse with such an admission because now an explanation is needed in every case.

The compounding problem comes into focus when it is realized that the surviving cohort in every case would have been dominated by small animals. Since it is understood that small animals, like rats, would have had a better chance of survival than large animals like horses and camels, how did the present cohort of large animals develop from rats, mice, raccoons and gophers?

Further, almost all the animals around the world are found in relatively restricted areas. We never find all of the animals spread evenly around the entire world. For some reason animals always cluster and it doesn't seem to relate necessarily to climate. For example, in Canada the Magpie is found across the southern prairies where open field and grassland predominate. However, it is also found in Ontario in the forested region around the Lake of the Woods. Why wouldn't it also be found in other parts of Ontario? While it is not a very large bird, being in between a robin and a crow in size, it is unusually aggressive not hesitating to attach a domestic cat whenever opportunity allows. Similarly the Pelican is seldom seen in Ontario which is directly north of Florida where there are plenty of them. At the same time it is found in Alberta which is not at all like Ontario. This clumping characteristic is quite common throughout nature and ensures that the cohort that survived would have been different every time a mass extinction occurred but it would still have been dominated by small animals.

The survival of animal life on the Earth has always been a precarious situation for several reasons including the necessity that there must always be an appropriate environment to enable survival. Each type of creature has particular needs and if these needs are not met, survival will not be possible. A tragic example of the extinction of an apparently secure and very well populated species is the Passenger Pigeon. Numerous comments have been offered to describe the incredible size of Passenger Pigeon flocks such as; 'Roaring past like a tornado, countless numbers, more than three billion creatures, … ceaselessly for two and a half days, etc.' (1) These birds numbered many millions up until the late eighteen hundreds. They were very easy targets for hunters who shot them in large numbers shipping three boxcars every day of a forty-day hunt. (3) In this case the environment included the hazard of hunters. Anytime that hunters show up intent on extermination of some particular creature, it becomes very difficult for such a creature to survive. In addition the forest habitant was also rapidly being destroyed thereby reducing possible nesting sites. One would still expect a few to survive somehow but coupled with these first two factors is the fact that they had a particular characteristic which involved being in large flocks all the time. Apparently, it was against their nature to exist as individual pairs. This made them easy targets for complete extinction and that is what happened by 1914. Survival of any species requires a particular set of conditions which is easy to take for granted until these conditions are upset. Fates similar to the passenger pigeons' continue to happen but at least at the present time there is more awareness and positive efforts applied to try to remedy deficiencies in particular environments so that survival will be ongoing.

However, in spite of good intensions there is only so much that can be done. For example, it is difficult to change the mixture of gases in the atmosphere or the level of water in the ocean. It would seem to be impossible to change the massive amounts of material that these two entities involve but this seems to be happening anyway. After all it only takes a small amount of pollutant to contaminate a vast amount of either water or air so the problem of pollution is being increasingly recognized as a very serious problem which involves our very survival. The problem of ocean water level involves both melting glaciers and a warming environment. The warmth melts the glaciers thereby filling the ocean. Warmth also causes ocean water to expand presenting a double-headed problem.

Even a cursory review of the world-wide hazards to life discussed at length in chapter 1 and summary listed in paragraph 2.7 makes it clear that even a single large asteroid hit would have been virtually impossible to survive. Great tsunamis really would have encircled the globe and swept away all reference points repeatedly. With such reality in mind it doesn't seem logical to simply declare that some significant percentage of world-wide animal life survived. Perhaps this is the reason that a scientific explanation has never been offered. What particular hiding place could have been used? Animals not only require shelter, they also require food every day. Being washed by massive currents all the way across a continent would have violated both of these necessities. In fact, there is no telling just how far they might have been carried. The waves kept coming repeatedly until even the soil was washed away. Fish would also have been in trouble because of the violent disturbances in the water movement in the ocean. In addition the heat from the igneous provinces and cone volcanoes would have been intolerable for many of them. A great many fish would have been buried (and in fact they were (4)) and many more would have been overheated and cooked alive. Expectedly, the possibility of survival would have been better in the ocean than it would have been on the land but with the violence of the water movement, it wouldn't have been either a safe or pleasant place to have been at all! Some rationalization for survival underwater can be entertained but extending any such discussion to the land is not credible. The situation only became worse after the waves had passed because then the temperature dropped. Soon thereafter the snow came. It came in great blizzards and it kept snowing almost continuously until it had piled up for thousands of feet. In recognition of this reality, the over-heated ocean doesn't seem quite so bad. With the snow accumulating to such an extreme it is clear that simply declaring that a certain percentage survived has no foundation in rational scientific thinking. Nothing can survive in a thousand feet of snow!

Neither is it good enough to simply say that animal life is here isn't it. Therefore the animals must have survived. Such declarations might be allowed at the popular science level but this doesn't make them really valid scientifically. Science demands a plausible scenario and such a discussion has never been brought forth. In recognition of this dilemma, the present discussion acknowledges that there isn't any explanation in science and that one isn't very likely to appear. How then is one able to proceed? While science does not offer any explanations it does provide us with an idea that is useful.

4.2 The Tunneling Concept

What is a tunnel? A tunnel in the most common sense is an opening through the ground. Animals make tunnels. These are openings in the ground leading down to a larger opening where the animal lives. In this case the opening will probably only be a few feet long but some animals make openings that extend much farther. There can also be natural tunnels which were carved out by some non-living agent of nature – water being the most common. Sometimes water works its way through the ground and forms an opening that might continue for a considerable distance. A cave could be referred to as a tunnel except that the concept most often applies to an opening that is smaller than some enlargement at its end. For example, if a small passageway through the ground led to a large room farther in, the passageway would be referred to as a tunnel and the larger opening as a room.

Tunnels are also man-made. In this case some obstacle would have been blocking the way. It is quite common in mountainous areas for tunnels to be opened through the ground to enable a highway to connect two different parts of the country. While constructing a tunnel through a mountain range is a serious undertaking, it has been done on numerous occasions to get past mountains or even to get to the other side of a waterway. Of course bridges are commonly built to cross rivers but in some instances a tunnel might be more appropriate or economically feasible or simply be desired for some other reason. There are also tunnels for trains. Trains have the limitation of not being able to climb steep hills. A train realistically can only climb a grade that is quite shallow whereas a car can climb a much steeper grade. Where first-class highways are required however, the requirement for shallow grades becomes more demanding so in mountainous regions resort is often made to a tunnel.

Tunnels are also commonly used to enable water to move from one location to another. In the Niagara region of Ontario, Canada, there is a tunnel to conduct water from the Niagara River to a large reservoir so that it can be used later to turn the generators at the power station and generate electricity. In this case water is conducted through the tunnel to a storage area and held for some time before being released for electricity production. Gravity causes the water to flow through such tunnels and then gravity comes into play again later when the water falls downhill and rotates the generators. Tunnels such as this are very useful and are an integral part of the power system of numerous countries.

All of the tunnels mentioned so far are real. They physically exist and can be observed, measured and modified. They can also be opened or closed or even filled in. They can be enlarged or lengthened as well. However there is a type of tunnel used in the study of physics which is not real at all. Never-the-less it is quite useful even though no one has ever observed one. This type of tunnel exists only as a mental construct to help explain certain observations that are otherwise difficult to explain. Tunnels of this nature enable discussion of a situation but they really do not provide an explanation.

In most of the cases where this abstract type of tunnel has been included in the discussion, certain observations have preceded its use and these observations do not obey the relationships that are expected from a historical understanding of the situation. Referring back to the car situation, if a car travels along a highway and comes to the shear vertical face of a mountain we would expect the car to stop. However if it is later observed to be on the other side of the mountain, it clearly had to get there somehow so even if nothing more was known of the situation it might be explained as having been accomplishable by using a tunnel. This tunnel might not have been observed. Perhaps the only observation was the car first on one side of the mountain and then later on the other side. Even with this much information it could be concluded that the car did not climb the mountain and descend down the other side. If it had climbed, it would have been observed to have climbed. But climbing was not observed. Further, we understand that the car could not theoretically have accomplished the climb because of the steepness of the hill. Therefore climbing is ruled out. It could not have flown over the mountain either because flying would have been observed as well. The car however, is on the other side and an explanation is required.

In the more abstract case, a physical mountain is not involved but an energy mountain is involved. There are cases in the study of physics where a particle is observed to be on one side of an energy barrier and then to be on the other side of the same barrier. How could this be? Were the measurements incorrect? The particle is observed on the other side of the energy barrier but this is not easy to explain at all. Physics cannot argue against itself. There has to be consistency in all relationships. The situation has to make sense. Rather than stall the entire conversation, it is simply declared that the particle tunneled through the barrier and has appeared on the other side. This is not the same as magic. Where magic is introduced, any bizarre situation can be dreamed up without regard to physics at all or any other branch of serious study. Whenever the tunneling concept has been introduced into a discussion in physics, there is no hint or intention to be resorting to magic. In many cases in the study of the observable world, situations might be unexplainable by logic but magic would never be invoked as an explanation.

4,3   The Explanation Void

Surviving a major asteroid impact or a shower of impacts followed by an ice age has proven to be difficult to explain. (In fact explaining the ice age has attracted numerous explanations none of which withstand very much serious scrutiny.) Otherwise various explanations for this combination of events would have been offered by now. On the other hand we do have a relative abundance of animal life on the Earth so we clearly have a dilemma on our hands. Therefore the tunneling concept is introduced. If, somehow, animal life could have survived for the first period of time after the major upheaval began, then perhaps it could have had a chance to carry on from that point. Survival for the immediate period following impact cannot really be properly explained but after some time had passed and things had settled down it is conceivable that life could have carried on from that point. As time went on after that first period, supposedly circumstances would have improved. All that was necessary was to get to that critical point in time after which survival can be rationalized. So we simply say that somehow it happened but without a realistic explanation, the situation becomes similar to the tunnel idea. Somehow life got through a completely impossible-to-explain situation to a point where it is conceivable that it could carry on. It tunneled through. However even after such a fill-in vacuous explanation is put forth the ongoing state of the environment must be hypothesized as having been able to support life.

The Ice Age must have lasted for more than one year – at least where it built up into such monstrous thicknesses. Further the Ice Age would not have spared any particular part of the world. As long as the world-encircling darkness-generating cloud persisted, it would have kept the ground below freezing all over the world, so that all of the moisture that precipitated would have been snow. This only adds to the explanation dilemma. Therefore, in order to have any hope of ongoing survival the cloud cover had to dissipate in a timely manner. In some part of the world the Sun had to shine through, warm the Earth and enable life to regenerate. It would not have been necessary for this to have happened over widespread areas but it certainly had to have happened someplace. Some feature of the Earth enabled the clouds to break up and dissipate much sooner than they did over the great snowfields of the upper latitudes. With such a development, life would have been enabled to carry on and gradually become re-established. Some location on the Earth must have been far enough from the moisture-producing ocean to have avoided the great buildups of snow. There must have been an area that was just beyond the reach of the snow clouds, far enough away to have been sheltered from the relentless arrival of snow. The place that was required would not have been extremely cold or extremely hot; a place where life, if given a chance, would have had a chance. Where was this place? How did it happen?

4.4 Location Criteria

During the Ice Age, in order to have significant accumulations of snow, the distance from any area in question to the open ocean, must be less than the critical distance beyond which the air would have dumped its moisture load and not have been able to produce significant amounts of snow. Beyond this distance the moisture load would have been spent with the consequences that any remaining clouds would have thinned out and would not have been able to produce significant precipitation of any kind. The air mass would conceivably have drifted on but without its moisture load, it simply could not produce any more snow. Conversely, this implies that there must have been a substantial, virtually continuous, moisture-filled cloud prior to this hypothetical line. Scattered clouds, even if each of them carried some moisture, would not have been able to block out the sunlight enough and continuity of cloud cover would have been necessary to prevent the Sun from simply melting the snow even as it does at the present time. (Further, as discussed in Part 1, the necessity of a continuous cloud immediately implies that the moisture-supplying bodies of water were warm enough to evaporate moisture at or above the critical rate to enable the snow to actually accumulate on a year-round basis.) If the snow only came during the winter all we would have had would have been snowy winters and snowy winters are not an ice age.

Many counties experience snowy winters every year but they are not having an ice age. It was therefore absolutely necessary for the moisture-supplying ocean to have been within the critical snow-producing distance of any region which would experience an accumulation of snow all through the year to the extent that an ice age could actually happen.

After the great Ice Age reached maximum, the snow stopped piling up continually and snowstorms became intermittent before petering out completely. However this would not have been the case simultaneously over the entire world. It is reasonably expected that the snow would have tapered off much sooner in tropical and semi-tropical regions than it would have at higher latitudes simply because the temperature at lower latitudes would be expected to have recovered much sooner than at more extreme latitudes. The further one goes from the equator the more indirect sunlight becomes, Up north, for example, the Sun never shines straight down at any time of the year and the weather is expectedly colder the further north one goes. Ice from the great Ice Age is still found in numerous places in the northern regions, in particular in Greenland, Iceland and some of the Mountains of Scandinavia. Also, a great deal of Ice Age ice is found in Antarctica in the extreme south. It is therefore reasoned that the buildup of ice and snow would have continued for much longer at higher latitudes than it would have closer to the equator.

We must recall that the climate during the great Ice Age would not necessarily have been extremely cold. Snow falls quite readily as soon as the freezing point is reached. In fact more snow will fall when it is just below freezing than it will when it is much colder. This is simply because cold air cannot hold as much moisture as warm air. The expected scenario is therefore that areas at high latitudes would have still been gripped by the Ice Age when regions closer to the equator were returning to a much more normal level of precipitation and temperature. Any survivors of the great Ice Age would have had a much better chance at low latitudes than they would had at high latitudes because the Ice Age peaked and abated sooner there.

Even though the snow storms would have abated sooner at low latitudes, it would still be reasonable to expect that there would have still been patches of snow and ice here and there all summer just as there is now in many places around the world. Greenland is a good example of this type of situation. Within a few miles of the coast of Greenland a massive Ice Age glacier still exists. However it does not deter the people who live along the coastal strips of bare ground from carrying on with farming operations and other activities that one would expect in any civilized country. As long as the land right here can produce a crop every summer – even if it is only a crop of grass – people can find a way to sustain themselves and carry on with a reasonably normal life, even virtually within site of a massive glacier. Mountains provide another example where Ice Age ice is still found on mountain peaks while life carries on quite normally in the valleys and foothills. There is even an example of this sort of thing in Africa which is a hot country by anyone's standards. Mount Kilimanjaro is within three degrees of the equator but it remains snow-covered all year.

We must recall that when the Ice Age had peaked and started to wane, the snow and ice would have started to melt. Soon there would have been massive torrents of water flowing to the sea and being involved in one of these situations would have been dangerous. High melting rates would have become increasingly common as more time went by. Soon rivers would have swollen past their banks and there would have been an endless supply of ponds and lakes which would have formed and then drained away within weeks, months or even after a few years. Being in areas where any of these things were happening would have been very unwise. For example, the Hudson River in the northern USA is understood to have flowed as much as three hundred feet higher at that time than it does now. (6)

Being in any region where massive melt-water was being produced would have been a difficult place for animals to have survived. This implies that being in regions which had experienced the great accumulations of snow during the Ice Age would have been regions where survival immediately after the Ice Age would also have been difficult. After all of life's processes had been so violently upset, getting re-established would have been much easier elsewhere.

The criteria, therefore, for the ongoing possibility of survival would have been a region in the lower latitudes and a place where there had not been any great accumulation of snow. It would have been a place where the Ice Age was short-lived. It must have been a place that was far from the moisture producing ocean and a place that was beyond the moisture shadow of all surrounding regions. It would also have been a place where any smaller bodies of water would not have been able to generate the excessive moisture being produced by the oceans, either because they were too small or possibly because they had cooled down too much and had become crusted over with ice and were not evaporating any significant moisture. The land might have warmed up, but some ice remained on the pond.

Wind direction would have been another important factor. If the wind simply did not blow from an area of relatively nearby ocean, it would have been the same as being too far away for the moisture to get there. Finally a mild climate during the worst of the Ice Age would have been necessary. This is in some ways similar to stating that lower latitudes were a better possibility than high latitudes because the climate at lower latitudes would never have been nearly as cold as we would expect further north. Extreme persistent cold weather was not a universal scenario during the Ice Age. At lower latitudes the climate would have been mild and damp and just cold enough to cause snow to fall instead of rain. There is a possible location on the Earth where all of these criteria could have been met.

4.5 The Shadow of Africa

Africa casts a shadow. This is not an optical shadow like the one cast by a tree when the Sun is shining but it is a shadow never-the-less. Africa is a large continent and because of this the moisture shadow from the Atlantic Ocean would not have quite covered it. This means that the middle-eastern countries would have been out of reach of this shadow. To the north and slightly to the east of Africa there are several countries including; Israel, Syria, Jordon, Turkey, Iraq, Iran, Saudi Arabia and Kuwait as well as several others which are quite small. Egypt, although part of Africa proper, could also be included in this list. All of these countries lie beyond the moisture shadow of Africa. It is so far across Africa from the South Atlantic Ocean that any moisture that was being carried across would have been dissipated long before the Middle East region had been reached. A similar comment could be made for all of the other regions which isolate the Middle East from ocean water.

If the moisture carrying clouds had to approach the Middle East from either the south, the south-west or the west it is clear that they would have to have travelled further from the ocean than virtually any other place on the whole world. From all of these directions the Middle East lies beyond any moisture shadow that was generated by the moisture-filled clouds that were leaving the ocean. The Middle East lies beyond the shadow of Africa. It is also a considerable distance across Europe to the Middle East so it is also beyond the shadow of Europe. Europe isolates the Middle East from the North Atlantic Ocean just as Africa does from the South Atlantic Ocean. In order to get to the Middle East from the Arctic Ocean there is also a considerable distance to cover but the Arctic Ocean would not have been involved anyway because it would have been frozen over.

Further if the smaller bodies of water in the region including the Black sea, the Red Sea and the Mediterranean Sea had been too cool or partially crusted over with ice the only remaining candidate for moisture supply would have been the Indian Ocean. Wind direction could have played a part in this case. If the wind from the Indian Ocean generally drifted eastward, (as it does now) the Indian Ocean would not have been able to supply the necessary snow-making moisture to the Middle East thereby helping to make it a place where the Ice Age was short-lived and making it a place where animal survival would have had a better chance than almost any other place on the surface of the Earth.

Snow and ice would have accumulated to a much lesser extent in the Middle East and the restricted shadow of Africa was part of the reason. There is no other location on the Earth where a land mass is so extensive that even though it might cast a great shadow, there will be a region beyond that is out of reach. The Middle East represents such a region with respect to Africa. The great Ice Age was short-lived in the Middle East and lasted much longer elsewhere and part of the reason was the extent of the landmass of Africa. Otherwise the Middle East region would have had a more extreme climate and survival of life on the Earth after the Ice Age would have had a much lower probability of happening.

5. Review, Summary and Conclusion

The "Window of Life" involves an extensive set of conditions, all of which must be in place and operating within their respective life-enabling ranges in order for all forms of life to thrive. Several of these windows were listed and discussed in Part 1 and only a few are repeated here. Temperature is a well understood window and is itself dependent on a further list of other parameters being within their proper operating ranges as well. For example, the orbit of the Earth around the Sun must be stable and predictable and stay within a narrow range of distance from the Sun. It has even been conjectured that the Earth must stay within 5% of its present median distance from the Sun or it will simply be too hot or too cold for life on Earth to thrive. While the orbit of the Earth would be difficult to change, the amount of change that would spell disaster is really quite small. This makes us aware that it might not require large changes in other temperature-controlling parameters before the inhabitants of the Earth would be in serious trouble. One of these other factors is the mix of gases in the atmosphere. Carbon dioxide, for example, constitutes a very small portion of the atmosphere but changing the amount of carbon dioxide would result in significant changes in temperature. Could life exist if the amount carbon dioxide in the air doubled? If it doubled, it would still only constitute a very small percentage of the total atmosphere but the anticipated heating effect would be very noticeable and would have very serious consequences.

Another factor affecting temperature is the amount of ice cover on ocean water. Ice reflects much more sunlight than open water so if the ice-cover was reduced, less sunlight would be reflected and more of the Sun's heat would be absorbed into the water increasing its temperature. This factor is far from academic because the pack ice in the far north is quickly disappearing. As it continues to melt, the water will absorb more heat which will cause it to warm up and melt more ice. This type of activity is called a viscous cycle (or more formally a positive-feedback loop). With this type of arrangement, the results make the cause increase which makes the result increase. In this example, a warmer ocean makes a warmer ocean.

Another "Window of Life" is the location of ocean water. There is a great amount of water in the ocean and we require that everygallon of it remain right there in the ocean. We cannot tolerate having the ocean sweep over the land. If it did, the "Window of Life" would close. Neither can the Earth tolerate continuous cloud cover because if this happened, the heat from the Sun would be prevented from reaching the surface of the Earth which would cool down as it has done in the recent past following major volcanic eruptions. How would life exist if everything was frozen?

Further, how would life exist if everything was washed away, which is just what would happen if a large asteroid hit the Earth. Activities of this nature are totally unacceptable and would result in widespread death around the entire world. What if it started to snow and kept right on snowing until it was a thousand feet deep? How could anything survive a catastrophe of this nature? How much volcanic activity could life on Earth tolerate? One of the products of volcanic eruptions is usually obnoxious gases. Thousands of tons of obnoxious gases are commonly released into the atmosphere when a volcano erupts. Also the ground becomes covered with grit or even worse becomes buried in ash. The dust from some volcanoes has gone completely around the world from the Far East and caused cold weather during the growing season resulting in famine in both North America and Europe. Close to a volcano hot molten rock is often

509

present. While the molten rock does not usually cover more than a few square miles, it is totally devastating wherever it goes. The eruption of numerous volcanoes simultaneously around the entire world would be a recipe for widespread annihilation of every form of life.

Unfortunately, whenever the life-threatening effects of an asteroid impact are discussed, all of these calamities are included and more besides. How much punishment could life on the Earth tolerate before it would all be wiped out? If a large asteroid impacted the Earth, all of these woes would come to pass. Disappointingly, the evidence indicates that an asteroid shower occurred instead of just a single hit which would only ensure that the chaos introduced by the first hit would have been compounded and multiplied and spread around the entire world. Then just to make things worse, if this was possible, an ice age would follow. (While any one of many of the results of impact would singularly have been devastating, having to deal with all of them would have been too much.) As the Ice Age developed it brought another list of life-eliminating factors. Therefore it is readily seen that it is very hard to explain how anything could have survived. This is most likely the reason that theories of survival have never been offered. Where would one begin?

In recognition of these difficulties, the approach that has been taken herein is to just accept that life somehow got through the immediate devastation to a time when things had settled down enough to have been tolerable. After this first unexplainable period of time, what was needed was an oasis-of-survival area. While the required area did not have to be very extensive, it had to have been large enough (but certainly not continent sized) to enable life to start getting re-established. While it is impossible to rationalize how life could have survived the first period of time after the asteroids stopped falling, suggesting that this impossible-to-survive period stretched out over dozens of years is totally unrealistic. At the same time it must be recognized that over large regions of the Earth, the Ice Age lasted much longer than just a few months or years. Therefore it can only be concluded that it did not last very long in at least one area that otherwise offered a good possibility for survival. How else can we proceed? Life had to somehow get through the immediate devastation so that there could be any chance of rationalizing the situation from that time onwards.

The oasis-of-survival area certainly did not happen in the north because it is completely obvious that the Ice Age continued for many years in the north and a period of years is just too long to try to offer any discussion of survival. Would it have been near the equator? This is not very likely either because one of the main factors involved in producing the Ice Age was an overheated ocean. If the ocean had not been supplied with an enormous amount of heat in a timely manner, the water would not have evaporated above the critical rate to produce the continuous cloud cover which was necessary to have an ice age in the first place. The water near the equator would have been the last to cool down (terminating prodigious evaporation) and almost certainly would not have cooled down within the time required for life to have a chance of regenerating. It would have still been evaporating at a prodigious rate and producing abundant amounts of snow-producing cloud past the time when an area was absolutely needed for life to start regenerating. While the "survival" area was enjoying a cessation of catastrophic events, (in particular where the clouds had parted and the Sun was shining again) near the equator the Ice Age would have still been in progress. All of the evaporated water had to go someplace so basically this would have caused the

Ice Age near the equator to continue past the "any chance of survival time". The place required had to be remote from the moisture-producing ocean which basically excludes areas near the equator.

We must recall that the prodigious quantities of snow that fell had to have been supplied by virtually continuous clouds which not only would have supplied the snow but would have caused the darkness that chilled the Earth causing snow to fall instead of rain. But even if it had warmed up somewhat, continually falling torrential rain would not have been much easier to deal with than constant blizzards. While rain does not pile up, how is one to plant crops or harvest fruit? Torrential rain that continues for weeks is hardly more survivable than continually falling snow. Survival under such conditions cannot be explained either and this would have ensured that areas near the equator would not have qualified as "oasis of survival" areas.

The conclusion of all of this reasoning is that;

a.   The impact of a major asteroid would not have been survivable.
b.   The impact of a shower of asteroids would have been even less survivable.
c.   The subsequent Ice Age was not survivable either.
d.   The asteroids must have stopped falling within a few months at the most.
e.   An "oasis of survival" area was required necessitating that in at least one region of the Earth the Ice Age must have been brief.
f.   Such a survival area could not have been in the north because the Ice Age continued too long in the north for survival to have been possible.
g.   Also it had to be remote from the moisture-producing ocean which excludes areas near the equator.
h.   Thirdly, the survival area had to be beyond the precipitation-producing shadows cast by lands adjacent to oceans.
i.   The Middle East area lies beyond all of the moisture-producing shadows cast by all surrounding lands with ocean coastlines and fulfills the above criteria providing a possible "oasis of survival" area after the impossible-to-explain period had passed.
j.   There isn't any other place on the Earth which can meet these criteria.

Trying to explain how the "Window of Life" closed and how life subsequently regenerated, has proven to be most difficult. However it did survive so some discussion of the problem seemed warranted. Therefore, herein, a tenuous thread of survival scenario has been offered with the full recognition that it basically only offers a window-of-possibility even more restricted than the "Window of Life" was in the first place. Surviving the asteroids and the Ice Age is difficult to explain on a good day but even more difficult in reality.

The End

APPENDIX

Thought experiment number one. Bullet impacting a solid lead ball.

Ball: Diameter 22 inches

Material lead

Volume V = 4/3 x pi x R x R x R = 4/3 x 3.14 x 11 x 11 x 11
5574 cu. in.

Density of lead The density of water is 1 gm./cu. cm. and there are 16.386 cu. cm./cu. in.
The specific gravity of lead is 11.34 (Funk & Wagnalls Vol. 16 p 17) which means that lead is 11.34 times as heavy as water.
Therefore the density of lead is 16.386 x 11.34 = 185.8 gm./cu. in.

Weight (mass) M2 = 5574 cu. in. x 185.8 gr./cu. in.   =   1,035,649 gr. or 1,035 kg.

Bullet Diameter 0.22 inches

Material Lead

Volume V = 4/3 x pi x 0.11 x 0.11 x 0.11
= 0.00557 cu. in.

Density of lead. = 185.8 gr./ cu. in

Weight (mass) M1 = 0.00557 cu. in x 185.8 gr./cu. in. = 1.0356 gr.
= 0.0010356 kg.

Bullet speed = 1600 ft./sec. = 1090 miles per hour = 484 meters per sec.

High speed bullet = 16,000 ft./sec. = 10,900 miles per hour = 4848 meters per sec.

From basic physics it is understood that: M2 x G x H = 1/2 x M1 x V x V

Where: M1 = mass of bullet (kg.)

M2 = mass of ball   (kg.)

G   = acceleration of gravity = 9.8 m/sec.(2)

H   = vertical displacement of ball   (meters)

V   = speed of bullet (meters per second)

Therefore: H = 1/2 x 0.0010356 kg. x 4848 meters per sec. x 4848 meters per sec.
Divided by 1,035 kg. x 9.8 m/sec.(2)

H = 1.199 meters or 47.23 inches

In this thought experiment, the high speed bullet, even though it is only 1% of the diameter of the ball, would be able to push the ball sideways far enough to raise it 47.23 inches. (This would be easier to visualize if the bullet was fired straight up into the bottom of the ball in which case the ball would be lifted up 47.23 inches) By comparison, if the bullet were only traveling at bullet speed, the ball would only be raised about 0.47 inches.

Thought experiment number two: Bullet impacting liquid crust-covered ball

The volumes, speeds and energies for this case are the same as for case number one. The difference is that the ball may not be pushed sideways (or lifted). The bullet would punch through the crust and its energy would be distributed throughout the ball by an interior pressure wave. It is expected that the entire ball would become involved in this pressure wave and that the entire crust of the ball would be transiently distorted. It is also possible that the bullet and/or its pressure wave would have enough energy to blast right out through the other side.

Thought experiment number three   BB impacting solid ball

Ball   Same as case number one

BB   Diameter   0.11 inches
            Material   lead
            Volume   V = 4/3 x pi x 0.055 x 0.055 x 0.055
                              = 0.000695 cu. in.
            Density   185.8 gr./ cu. in.
            Weight (mass)   0,000695 cu. in. x 185.8 gr./cu. in. = 0.1293 gr.
                                               =   0.0001293 kg.
            Velocity or speed   4848 meters per sec.

Using the above formula to relate the speed of the BB to the displacement of the ball we have:

H = 1/2 x 0.0001293 kg. x 4848 meters per sec x 4848 meters per sec./(1,035 kg. x 9.8 m/sec(2))

        H = 0.1498 m or 5.89 in.

This means that a BB, which was travelling at 10,000 miles per hour, would be able to push a solid lead 22" dia. ball sideways far enough for it to be lifted 5.89 inches. When compared to the ball, the BB is hardly visible but if it were to impact the ball at high speed, the ball would be noticeably displaced. If the BB were traveling at normal bullet speed, the displacement of the ball would only be about 50 thousands of an inch.

Thought experiment number four   BB impacting a liquid crust-covered lead ball.

The physics of this thought experiment would be the same as for thought experiment number three but in this case the BB would punch through the crust of the ball and its energy would be distributed throughout the interior of the ball by a pressure wave. The crust would undergo transient distortion.

Temperature increase for water vapor

$$T = \text{heat input (BTU)} / (\text{weight of water vapor x specific heat})$$

The weight of the canopy water vapor is dependent on the example chosen. For a 25% canopy the equivalent thickness of water would be 8.46 feet. A one square foot cross section from bottom to top would weigh   w = 8.46 cu. ft. x 62.5 lbs/cu. ft. = 529 lbs.

A similar square foot portion from a 50% canopy would weigh about 1058 lbs and from a 100% canopy would weigh about 2117 lbs.

The energy available from the Sun (for a whole day) at the equator on Mar 21 (where the Sun passes directly overhead) is approx. 1924 BTU/sq. ft. Of this approximately 53% occurs in the near infrared or heat region of the spectrum. (ASHRAE Handbook Fundamentals pages 27.31, 27.33, 27.35 & 27.2) The specific heat of water is one BTU./lb./degree F.

Therefore the temperature increase for the 25% canopy would be about;

$$\text{Temp. increase} = (1924 \text{ BTU x } 0.53) / (529 \text{ lbs. x } 1.0 \text{ BTU/lb./degree f})$$
$$= 1.93 \text{ degrees F.}$$

Similarly, the increase for the 50% canopy would be 0.96 degrees F. and for the 100% canopy it would be 0.48 degrees F.

Several additional examples are listed below. As the distance from direct overhead increases, the available solar energy drops off.

Sun Pathway	Heat BTU/ sq. ft.	Temp increase for various canopies (degrees F)			
		12.5%	25%	50%	100%
Direct Overhead	1020	3.86	1.93	0.96	0.48
16 degrees from overhead	810	3.06	1.53	0.76	0.38
32 degrees from overhead	520	1.96	0.90	0.49	0.25

These data have been plotted on the diagram, 'Vapor Canopy Temperature Increase in One Day' 5.10 (1)

515

Atmospheric Carbon Load

Atmospheric pressure is 14.7 lbs. per sq. in. (College Physics p 168). There are 14.7 lbs. of atmosphere sitting on every square inch of the Earth's surface. The carbon dioxide portion of the atmosphere is 387 parts per million. (Climate Wars p 14)

Therefore $CO_2 = 14.7 \times 387 / 1,000,000 = 0.00569$ lbs. per sq. in.

The carbon portion is: $12 / (12 + (16 \times 2)) \times 0.00569 = 0.00155$ lbs. per sq. in.

The surface of the Earth is: $196.95 \times 10(6)$ sq. miles.x $4.01 \times 10(9)$ sq. in. / sq. mile (Handbook of Chemistry and Physics 52nd Edition p F 163, DOSCO Tables of Conversion Factors p 55)

Therefore the surface of the Earth = $789.77 \times 10(15)$ sq. in.

Therefore total weight of carbon = 0.00155 lbs. per sq. in. x $789.77 \times 10(15)$ sq. in.

$= 1.22 \times 10(15)$ lbs.   (lbs. x $4.5359 \times 10(-4)$ = metric tons DOSCO Tables p 44)

or $5.55 \times 10(11)$ tons(metric)

Crosschecking:
Atmospheric carbon load (from CRSQ Vol. 15 p 219) = $0.68 \times 10(18)$ grams or
$0.68 \times 10(18) \times 10(-6) = 6.8 \times 10(11)$ tons (metric)

Atmospheric carbon load (from CRSQ Vol. 29 p 170) = 0.12 gram/cm(sq) or
0.12 gram/cm(sq) x 6.45 cm(2) / in (2) x $790 \times 10(15$ in(2) $= 611 \times 10(15)$ gm
Or $611 \times 10(15) \times 10(-6) = 6.11 \times 10(11)$ tons (metric)

Coal reserves around the world $15 \times 10(18)$ grams = $15 \times 10(12)$ tons (metric CRSQ Vol. 20, p212)

There is therefore $150 / 5.55 = 27$ times as much carbon in the coal reserves as there is in the entire atmosphere. In order to provide enough carbon for the coal reserves from an ancient atmosphere, there would need to have been 27×387 parts per million or 10,459 parts per million CO_2 in the ancient atmosphere. As a percent this is 1.04 % which would have been barely tolerable for breathing. It is therefore feasible – but not very likely - that an ancient atmosphere could have supplied the carbon to grow the plants for the present coal reserves simply by being depleted. However, the present oceans include most of the present CO_2 inventory of the world and approximately 90 % of all the CO_2 in the world is presently dissolved in the oceans. If this had been absorbed from the atmosphere as well, the original atmosphericload of CO_2 would have been about $10,459 + 3870 = 14,329$ parts per million or about 1.4%. A value this high would have been very noticeable and would have made breathing difficult.

The present biosphere is estimated to include $2.33 \times 10(11)$ tons (metric) of carbon.(CRSQ Vol. 15 p 219) If the ancient coal-forming biosphere had all been alive at the same time, it would have had: $27 \times 5.55/2.33 = 64.3$ times as much carbon in it as the present biosphere.

Water Vapor Layer

The diagrams 5.8 (1) and 5.8 (2) 'Solar Energy Distribution' show 38% of the energy from the Sun as visible and 53% as infrared. Another 9% of the Sun's energy is in the ultraviolet region. (ASHRAE 1981 Fundamentals p 27.2)

More than one attempt has been made to mathematically model the atmosphere with a vapor layer. All such models are only as valid as the assumptions and algorithms employed. Some have predicted that the water layer would become too hot and get hotter with increasing elevation. However, in a whole day, the Sun could only increase the temperature of a 50% canopy by slightly less than 1 degree F. As the vapor rose and drifted poleward in the upper elevations, a similar temperature decrease would result. Radiation heat loss from the vapor to space is a reasonable expectation because space appears dark and cold and would be a good receptacle for this energy. This type of heat loss operates in the atmosphere at the present time without any difficulty. When the Sun goes down and the sky is clear, even desert areas can drop to below freezing. Also it should be noted that the total area involved in heat loss is greater than the area involved in heat gain. The entire sphere of water vapor would be losing heat all of the time. Only the area, which was directly under the Sun, would have more gain than loss. Then, at night, heat loss would occur (without heat gain) in this area as well. A further complication involving heat gain and loss relates to the tilt of the Earth's axis of rotation. While the diagram shows the case when the Sun is shining straight down on the equator, most of the time it would be shining more on either the northern or southern hemisphere. During these times more heat would be received by the sunnier region and less by the other region. The area of the world where loss would exceed gain would therefore be maintained while global circulation patterns in the water vapor layer would shift to accommodate the different locations where the heat was received.

The diagram also assumes that two-thirds of the incoming visible light would be scattered by the water vapor. Scattering would certainly be dependent on how much water was in the vapor region. Scattering of light by water is very non-linear. A layer of water twice as thick will not scatter twice as much light. Otherwise the level of light would soon reduce to zero.

1 cm of water scatters 2-4% of incoming visible light which value assumes atmospheric pressure. However, with any amount of material in the vapor layer, the pressure would decrease exponentially with elevation so as higher elevations are reached, less scatter would result because the molecules would be further apart at the lower pressures. Averages probably do not apply so some reduced scatter rate will be

considered for discussion purposes. Assuming the scatter function is an exponential, with 1% scatter for 1 cm. the amount of scatter for other amounts may be determined. At the rate of 1% for the first cm we have 100% at 100 cm. or 1 m. Since we are working with an exponential, this identifies the 63% level. Therefore the equivalent of 1 m of water vapor would scatter about 63% of incoming light. Two meters would scatter 86.5%. Two thirds of the incoming light would be scattered by a thickness of approximately 108 cm or 3.54 ft. This would be equivalent to about an 11% canopy and two meters would be slightly less than a 25% canopy. The scattering effect would be more pronounced at the higher frequencies. It is expected therefore that the ultraviolet end would be scattered much more than the infrared end of the visible spectrum and that the ultraviolet portion above the visible spectrum would be practically all scattered. Some visible light would certainly get through. Light will pass through liquid water as well. Even at 200 feet down in a lake there will be some light. If light is either scattered or absorbed with an exponential function as it passes through water, the attenuation factor may be determined. At 1000 feet down it has been reported that 0.2% of incident light was still present. From this the attenuation rate may be calculated and may be determined to be 6.2 time constant equivalents. With this terminology, 6.2 is related to 1000 feet. Therefore one time constant is equivalent to 1/6.2 x 1000 ft. = 161.5 feet. Following this line of reasoning, 63% of incoming light would be scattered by 161.5 feet of pure water.

Since water vapor would be concentrated in the lower level of the canopy, it might be expected that whatever heating effect was taking place would be more concentrated in these regions. As mentioned previously, modeling of the atmosphere is a most formidable task and can probably never be done precisely. In light of this reality the following assumptions will be made to enable a discussion to proceed.

1. Ultraviolet light would be completely scattered by a vapor canopy equivalent of 12.5%.
2. The infrared portion of the solar spectrum will be completely absorbed by a vapor canopy equivalent of 12.5% or more.
3. The visible light portion will be either
    i.   partially scattered according to 63% scatter with a one meter equivalent of water vapor canopy with an exponential function of scatter.
    ii. partially absorbed according to 63% absorption with 161.5 ft. of water equivalent.

With option i., light levels at the surface of the Earth will be reduced. This would be an advantage because cloudless sunshine is too bright for naked eyes. The temperature of the canopy would not be affected in this case. However if light was absorbed, the temperature of the canopy would be affected. Surface light levels would still be reduced, but not as dramatically as they would be with the vapor function. Since approximately the same amount of energy is in the visible portion of the Sun's energy, as in the infrared portion, the greatest possible effect is for the increase in temperature of the canopy to be about twice as much as shown in the diagram 'Vapor Canopy Temperature Increase'. The temperature increase after a full day of exposure would still be quite modest at approximately two times 1.93 degrees or 3.86 degrees f for a 25% canopy. With either option, is may be said for certain that light levels at the surface of the Earth would be reduced if a vapor canopy was in place. Suggesting reductions in the range of 50% to two-thirds would be realistic. It is also certain that the canopy would be warmed by the infrared portion as well as possibly by part of the visible portion. Therefore, the extent of warming is uncertain but would be between the respective levels shown on the chart and twice these levels.

The diagram, 5.4 'Atmospheric Pressure Profile' indicates how atmospheric pressure decreases with increase in altitude. However to prepare this diagram it must be assumed that specific weight of the atmospheric material is the same all the way up. It would be if the atmosphere was all air, which is the case with line no. 1. For the other lines however, above the air portion of the atmosphere lies the water vapor region and it will have a specific weight, which is slightly more than one-half of the air portion. Therefore the pressure in the upper portion would drop off at a reduced rate.

The diagram, 5.12 'Atmospheric Temperature Profile' has been included to illustrate how the temperature of the atmosphere might vary with height. Since the water vapor layer may absorb one-half of the incoming solar energy (i.e. the infrared portion) as well as some portion of the visible portion, and since most of the material in the water vapor layer is concentrated near the bottom, it is expected that the lower regions of the vapor canopy would heat up the most and become hotter than the air lower down. This would certainly create a 'no pass zone'. Actually the temperature at the base of the vapor layer would be a 'no pass zone' even if it was slightly cooler than the atmosphere lower down because if any air should attempt to rise, it will cool by expansion. As long as some region higher up was still a little warmer, any such rising air would not be able to rise and the 'no pass zone' barrier would still be operating. Vertical stability would thereby be ensured. However vertical stability within the vapor region would not be ensured. Since so much of the vapor layer is near the bottom, some decrease in temperature with increasing altitude is expected but it would not require very much to encourage circulation patterns to develop. The horizontal temperature expected and shown on the diagram 5.10 (1) 'Vapor Canopy Temperature Increase' would create a situation where vapor on either side of the area right under the Sun, would have a slightly higher pressure than the area right under the Sun. Vapor in both of these areas would be pushing towards the central area. A circulation pattern would be expected. Vapor from the lower central region would rise. Replacement vapor would move in from the sides. The rising vapor would soon be inclined to move horizontally away from the central region because more vapor would be welling up under it. It would probably start to move horizontally before it had risen very far. However, in rising upward it would cool by expansion. This may create a second 'no pass zone' within the vapor layer where vapor even higher up was slightly warmer. Consequently this higher vapor may not become involved in the circulation pattern. As horizontal movement continued, an area would be encountered where the upper regions were cooler because all of these upper regions are adjacent to the chilling effect of space. Further heat loss of the moving vapor would therefore be expected. Lower down, near the bottom of the vapor layer, the movement would be the other way towards the directly heated zone. As the upper moving layer approached the polar regions, it would start to sink in behind the lower layer going the other way. As it sank, its temperature would increase due to the increase in pressure closer to the Earth. Further, the upper layer would always be at a lower density than the lower layer. If it were twice the density, it would have to move at twice the speed of the lower layer in order to supply the replacement material. If the lower layer is moving a certain speed with twice as much material as the upper layer, the upper layer must still transfer the same amount of mass in the same amount of time so it would have to move faster (or a greater height of material would be involved). A second speed differential would also be involved due to the curvature of the Earth. The fact that it is a sphere and not a flat or cylindrical shape, means that movement in both the upper region and the lower region would be faster near the poles than near the equator. Near the equator, there would be a greater mass of material rising up because the equator has a circumference of 26,000 miles. As this mass moves northward, the length of the latitudes is decreasing. Therefore the material would expectedly be moving faster. This curvature effect would also influence short circuiting and some vapor would not travel the entire maximum route. While all this was going on, high up in the vapor layer there may not be any motion at all. The nonlinear movement of vapor would further increase the effectiveness of heat transfer from the tropics to the poles.

Further, we recall that any movement in the atmosphere must recognize the Coriolis Effect which of course, acts on all material that moves on the Earth whether it is in the atmosphere or in the ocean or even a truck driving down the freeway. This force ensures that if the specific material is not restricted by friction or some mechanical obstacle any movement will not be restricted to the north-south direction only but will include an east-west component as well. Simply put, the Earth is rotating and the rotation will always mean that the Coriolis effect is at work. Mathematical modelers of ocean currents, for example, always recognize the rotation of the Earth. At the same time, the underlying assumption included with the idea of a vapor canopy is that because the Earth was warm from equator to pole there really wasn't significant global circulation in the lower atmosphere. Temperature difference is the major factor causing movement in the first place. Just when things couldn't get any more complicated we have the atmospheric bending/refraction/reflection factor that would be integral with the vapor canopy option. If light was bent to the polar regions, heat would also be bent. If the equator at ground level had never become overheated in the first place how much heat would it require for it to just stay pleasant? If heat wasn't being so easily lost in the polar regions, how much heat would be required to keep it pleasant? If light had been bent around and illuminated Axel Heiberg Island, how much heat was also bent around and heated it?

The bottom line remains that Axel Heiberg was warm and that an oxygen/nitrogen atmosphere would not have been able to have done this. Therefore any difficulty or inability to explain it is not a fault of nature but only a limitation in our own understanding. The vapor canopy hypothesis appears to offer a possible explanation.

Cooling by Expansion

When a parcel of air rises it cools. It cools simply because it expands as it rises. The following table shows how air cools as elevation increases. (ASHRAE Handbook 1981 Fundamentals p 5.1)

Altitude		Temperature		Altitude		Temperature	
Meters	Feet	Celsius	Fahrenheit	Meters	Feet	Celsius	Fahrenheit
0	0	18.2	64.8	5000	16404	-17.5	0.5
500	1640	15.0	59.0	6000	19685	-24.0	-11.2
1000	3281	8.5	47.3	7000	22966	-30.5	-22.9

Altitude		Temperature		Altitude		Temperature	
Meters	Feet	Celsius	Fahrenheit	Meters	Feet	Celsius	Fahrenheit
2000	6562	2.0	35.8	8000	26247	-37.0	-34.6
3000	9843	-4.5	23.9	9000	29528	-43.5	-46.3
4000	13123	-11.0	12.2	10000	32808	-50.0	-58.0

Evidence Recognition

Many items of evidence are available to support a theory of the Earth. In particular, the following have been recognized by the particular theory of the Earth being discussed herein.

1. Mountain Formation
   a. Many of the continental mountains of the world are parallel to ocean coastlines.
   b. Layers of rock in mountainous areas are found folded and bent like they would be if they were formed by earthquake pressure waves!
   c. The surface of the Earth (i.e. at the basement rock level) is extensively fractured, buckled and broken.
   d. Only a portion of the ocean coastline of the world has an adjacent submarine trench.

2. Mid-Atlantic Fault Formation
   a. The Atlantic Ocean floor appears to have failed in buckling parallel to the central mountain ridge.
   b. The Atlantic Ocean has a fault or crack in the ocean floor, which is approximately one-half way between the continents on either side.
   c. Radiating laterally from the central fault are numerous other faults, which extend most of the way to both shores.
   d. Piled up adjacent to the central crack are mountainous accumulations of rock.
   e. Measurements indicate that the central fault and the ocean floor are not increasing in width.
   f. Certain underwater mountains are flat-topped - particularly in the Pacific Ocean - and appear to have had their tops sheared off.

3. World-Wide Fault Formation
   a. The crust of the Earth has a world-wide pattern of inter-connected faults.
   b. The vast majority of volcanic activity of the Earth occurs near the fault lines.
   c. The major faults do not necessarily separate land from ocean floor. As in the case of the North Atlantic, the ocean floor is continuous with the North American continent.

4. Himalayan Mountain Formation
   a. The Himalayan Mountain formation occurs partially concentric to the Taklimakan Basin and is separated from it by the Plateau of Tibet and the Kunlun Mountains.

5. Asteroid Crustal Punch-through
   a. The Earth includes numerous impact features, which indicate that large objects have collided with the Earth.
   b. The elevation of the central areas of these impact features is not significantly different from the surrounding areas.

6. Continental Shelf Formation
   a. Immediately adjacent to many ocean shorelines there are underwater accumulations of silt, which form underwater plains apparently from material which was washed off of the continents.
   b. Continental shelves have gorges cut into them.

7. Warm World-Wide Climate
   a. Coal is found at Spitzbergen Island in the far north.
   b. Logs from large trees have been found at Axel Heiberg Island in Northern Canada.
   c. The remains of numerous creatures - both ancient and modern - have been found in the frozen ground in the far north of Canada.
   d. The remains of numerous creatures have also been found in the frozen muck of Alaska.
   e. The remains of the Woolly Mammoth have been found frozen in both the ground and in the ice of the far north.
   f. The bones of innumerable creatures have been found in the sand of the islands north of Russia.
   g. Ocean temperature dropped at end of 'chalk' period. (CSRQ Vol. 9 page 218)

8. Greater Historical Atmospheric Pressure
   a. Fossil dragonflies have been found, which had wingspans greater than 2 feet.
   b. The Teradactl had wings and appeared to be perfectly suited for flying if atmospheric pressure had been greater.
   c. Oxygen analysis of small submarine shelled fossils indicates that oxygen pressure was greater in the past. (A Short History of Nearly Everything page 338)
   d. Many creatures of the past were very large compared to their present counterparts which is explained by the Hypothalamus Disregulation Theory if it is valid.

9. Glacier Formation
   a. Several planes from WWII have been found buried under 250 feet of blue ice in Northern Greenland. (Thelostsquadron.com)
   b. Conjectured glacial flow-lines radiate outward from central nodal areas in Northern Canada.
   c. Ice Age required the evaporation of at least 2000 feet of ocean water.
   d. Glacial till or 'terminal moraines' are completely free of vegetation. For example, the extensive gravel deposits of the so-called terminal moraine 'The Oak Ridges Moraine' are mined and the gravel is used to make roads, concrete and asphalt, which would not be possible if the material was contaminated with logs, leaves and grass. While an occasional piece of wood is found in 'glacial till' (carbon dated at 3300 years! (CRSQ Vol. 9 page 210)) it is otherwise completely free of any vegetable matter.

10. Earth Interior Pressure increase
   a. The central fault system together with the faults which radiate outward from it indicate that the Atlantic Ocean floor has been elevated.
   b. Rift valleys are areas, which appear to have dropped.
   c. Escarpments are areas, which appear to have risen.
   d. Certain mountain-type formations appear without reference to any fault line and look as if they were formed by material oozing up from underneath.
   e. The central Atlantic mountain range is apparently not offset by thicker crust but appears as if material from the interior was thrust up through a central opening.
   f. There are numerous volcanoes around the world, which are formed from material which has been thrust up from below the surface of the Earth.
   g. Most volcanoes of the world are at major faults in the crust of the Earth.
   h. In particular, Iceland is located on the central Atlantic fault, and appears to be completely formed from volcanic activity.
   i. Further, the Hawaiian Islands of the Pacific Ocean occur on a 'hot spot' system and are the largest mountains and volcanic structures in the world.

11. Coal Formation
   a. Coal deposits are found all around the world.
   b. Coal appears to have formed from plants.
   c. Coal is often found in layers separated from other layers by rocky material.
   d. Fossils of trees are occasionally found vertically right up through several layers of rock and coal.

12. Igneous Provinces - Igneous provinces cover large areas of the world – in particular they are distributed throughout the ocean and in total cover hundreds of thousands of square miles. They also occur on land – in particular the Deccan Plateau of India at 200,000 thousand square miles and the Parana Plateau of Brazil and Paraguay at 300,000 square miles. (The Violent Face of Nature p 195)

13. Sudden Death of Mammals - The Wooly Mammoth died instantly while it was still eating its lunch. (Earth in Upheaval page 18)

14. Submarine Canyons always occur as the extension of rivers and continue right to the bottom of the ocean – even to 15,000 feet. (The Genesis Flood, page 125)

15. The Earth has a wobble which shows up as the North Pole describes a circle with a radius of approx. 12 meters 'completing a single loop every 433 days.' (Maclean's/August 14, 2000)

Energy Comparisons

Event	Energy Level (ergs)	Reference
Mountain Range Formation	10(29)	CRSQ Vol. 21 p 84

Event	Energy Level (ergs)	Reference
Asteroid 10 mile dia. (16km)	2.1 x 10(31)	by calculation
Asteroid 20 mile dia. (32km)	1.75 x 10(32)	CRSQ Vol. 21 p 84
Asteroid 40 mile dia. (64 km)	1.4 x 10(33)	CRSQ Vol. 21 p 84
Barringer Crater, Arizona	10(24)	CRSQ Vol. 21 p 84
Earthquake (Richter 8.5)	10(24)	The Violent Face of Nature p 232

The asteroids energies given assume the asteroid was traveling at 24.5 km /sec when it hit the Earth. This is about 55,000 miles per hour. Several asteroids that have recently come close to the Earth were traveling twice this speed. (refer to chapter 4)

Hammer and Punch

Suppose there was a hammer 40 miles dia. and 135 miles long and that it consisted of material similar in density to an asteroid. Suppose it was driven into a punch 8 miles in diameter, which was pressing against the surface of the Earth. Suppose that it was driven into this punch at 1000 miles per hour. Recalling that the crust of the Earth is only up to 20 miles thick, it would be expected that such a hammer would be able to drive the punch right through the crust and into the interior. The energy for this hammer is given by;

Hammer Energy	=	½ x mass x velocity(2)
=		½ x volume x SG x velocity(2)
=		½ x (pi x r(2) x l) x SG x velocity(2)
=		½ x 3.14 x 20 x 20 x 135 x SG x 1000 x 1000
=		8.48 x 10(10) x SG

Where SG = the specific gravity of the hammer

Using the same energy units, which are arbitrary and used herein for comparison purposes only, this energy level may be compared to an asteroid 8 miles in diameter, moving at 25,000 miles per hour.

Asteroid Energy	=	½ x mass x velocity(2)
=		½ x volume x SG x velocity(2)
=		½ x (4/3 x pi x r(3)) x SG x velocity(2)
=		½ x (4/3 x 3.14 x 4 x 4 x 4) x SG x 25,000 x 25,000
=		½ x 268 x SG x 625 x 10(6)
=		83,750 x SG x 10(6)
=		8.4 x 10(10) x SG

Where SG = the specific gravity of the asteroid = the specific gravity of the hammer.

The units of energy do not matter in this case because we are doing a simple comparison. The speed of an asteroid would not, of course, be known. It has been suggested in the literature, that asteroid speeds are expected upwards of 25,000 miles per hour. (see article by Terence Dickinson in Canadian Geographic May/June 1995)

If the hypothetical hammer could drive the hypothetical punch right through the crust, the asteroid postulated in this example (because it would have as much energy as the hammer) would also be able to drive right through the crust of the Earth.

Light Curvature Around the Earth

If light curved around the Earth from the Arctic Circle to Axel Heiberg Island, it would have to curve for 15 degrees of latitude. If a beam of sunlight were shining parallel to the Earth at the Arctic Circle, it would be;

D   =   Earth radius/cos(15) – Earth radius
=   4000/0.965 - 4000
=   145 miles above the Earth at Axel Heiberg

This is much too far above the atmosphere for sunlight to be reflected from anything in the atmosphere. At one-third of the way from the Arctic Circle to Axel Heiberg, only 5 degrees of latitude are involved. In this case the parallel light beam would only be 16 miles above the Earth. Some reflection might be possible if there was anything at this elevation that could reflect light.

In the case of London, England, a light beam parallel to the ground at the Arctic Circle would be 494 miles above the ground. It is not known where the observer was that wrote the letter to the London Times but if they were on the same parallel of latitude as London, the light from the Arctic Circle would have had to curve around the Earth for 17 degrees of latitude. A combination of reflection and refraction may have worked.

Noctilucent clouds are 50 miles above the Earth. The Sun would be visible at this altitude even if it was 9 degrees of latitude over the horizon (as shown by the following calculation).

The cosine of the degrees of latitude =   Earth radius / (height + Earth radius)
=   3963 / (50 + 3963)
=   3963 / 4013
=   0.9875
The number of degrees of latitude is therefore = 9.

In the Axel Heiberg case, a layer of ice crystals high in the atmosphere above the Arctic Circle would probably have been able to refract or bend sunlight and cause it to reach Axel Heiberg at noon on Dec 21, the shortest day of the year. Similarly, a layer of fog at a similar elevation may have achieved the same result. Skylight would extend all of these results.

Human Gigantism

Even a preliminary investigation into human gigantism produces unreliable evidence. In particular, a photograph showing two human skeletons in the foreground and two people in the background is misleading. (www.davidicke.com) It may simply be that the picture was taken without any particular motive but the arrangement is suggestive that the human skeletons are very large. In reality they may just be closer to the camera. In any event this type of presentation is not helpful in the pursuit of truth. There are other photographs showing an apparently large human skeleton with one or more people alongside. These are undoubtedly fraud. The human would, in this case, need to be 35-40 feet high and there is no independent information against which such a claim can be checked. There are, on the other hand, numerous reports of the remains of very large humans having been found which do have credibility. Such claims place the size of the humans represented by their remains in the nine to twelve foot high category. Credibility relates in part to the large number of claims as well as to the fact that claims to gigantism have been advanced in historical time as well as quite recently. In particular, the Bible speaks of Goliath 'whose height was six cubits and a span' (i.e. about 9 ft. 6 in. high, I Sam. 17 vs. 4). Also Og, king of Bashan had a bed 13 ft. long and 6 ft. wide (Det. 3,11).

Volume of Ocean Water

Vol. of water at 0 degrees Celsius = 1.00 cm.(3)/gm
Vol. of water at 25 degrees Celsius = 1.0025 cm(3)/gm   (Physics, Resnick and Halliday p 462

The mean depth of the ocean is 3800 m = 12,540 ft.   (En Br Vol. 16, p 682)

If all ocean water warmed from 0 degrees to 25 degrees without changing the boundaries of the oceans, the change in level would be:

L   =   3800 m x 0.0025
=   9.5 m
=   31.35 ft.

With the expected increase in surface area, the level increase would be less – possibly 20 to 25 feet.

Evaporation of Ocean Water

Heat to evaporate water = 540 cal./gm.
Heat to change temperature of water one degree Celsius = 1.0 cal./gm.

Therefore 15 degrees Celsius temperature change requires 15.0 cal./gm.

It requires 540/15 = 36 times as much energy to evaporate one gram of water as it does to change its temperature 15 degrees. Therefore if we had 36 times as much warm water available as we wished to evaporate, by cooling the water 15 degrees, enough energy would be available to carry out the evaporation procedure.

The amount of ocean water evaporated during the Ice Age may have been upwards of 2000 feet. In order for the warmth of the water to evaporate this much water;

D = 2000 x 36 = 72,000 feet of ocean would be required

However, only 12,540 feet are available. Therefore a warm ocean could not have supplied enough heat to evaporate the water required to form the great ice fields of the Ice Age and more heat would have been needed.

Earth-Moon Distance

The current rate of separation of the Moon from the Earth is given as approximately 3.8 cm/yr. (The Rough Guide to the Universe p 49) Also the distance from the Earth to the Moon is 384,400 km. If the rate of separation has always been the same, the time back to zero separation would be;

Distance = 384,400 km = 384,400 km x 100,000 cm/km = 3.84 x 10(10) cm
Therefore time back to zero separation is 3.84 x 10(10) cm / 3.8 cm/yr. = 10 billion years.

The time reduces when a more realistic calculation is made because the rate of separation of the Moon from the Earth would certainly not be constant but would continually reduce. This means that as we go back in time, the rate of separation would increasingly be greater the farther back we go. The separation function would be closer to an exponential than it would be to a constant.

Suppose that the Moon is already 63% of its way to its final distance. If an exponential relationship is actually involved, the final separation distance will be:

Final distance = 384,400 km / 0.63 = 609,000 km.

Since the rate of change at the 63% distance is 37% of the initial rate of change (i.e. it is the complement – refer to the 10.1.2 Time Constant Curve) the initial rate of change would have been;

Initial rate of change = 3.8 cm/yr / 0.37 = 10.27 cm/yr. and time back to zero separation would be;

Time = 3.84 x 10(10 cm/10.27 cm/yr = 3.76 billion years

When factors like tidal friction are factored in, this time might be reduced to 1.78 billion years. (The Moon – Its Creation, Form and Significance, p 39)

It has been estimated that the final distance to the Moon will involve an orbital period of about 46 days and the Moon will remain stationary over one particular location on the Earth. 'Assuming you're on the side of the Earth where you can see it, the Moon will appear to hang immobile in the sky and the length of the "day" will be over 1100 hours (46 days) long.' (The Rough Guide to the Universe page 49)

Residency Periods for Chemicals in Ocean Solution (years) Reference CRSQ Vol. 11 p 42

Aluminum	100
Titanium	160
Chromium	350
Manganese	1400
Iron	140
Nickel	18,000
Copper	50,000
Silver	2,100,000
Mercury	42,000
Sodium	260,000,000

Igneous Provinces Heat Release

The volume of material that was ejected from the interior of the Earth when the Deccan Traps were formed is;

Volume = 12,275 cubic miles or 512,000 cubic kilometers   ref. Wikipedia, Deccan Traps

The temperature of the molten rock is not known but we do know that clay glows red around 700 F and becomes bright cherry red at about 1200 F. Molten rock from the interior would have been at least this hot.

The specific heat of rock can conservatively be taken as 10% that of water. This means that it takes ten times as much heat to heat a volume of water as it does a similar volume of rock.

It requires 970 BTU/lb to boil water and another 200 BTU/lb to raise its temperature from freezing to boiling. This makes a total of 1170 BTU/lb.

From this simple calculation it is clear that the heat available in the cooling rock could have boiled about 10% as much ocean water. Therefore 1,227.5 cubic miles of ocean could have been evaporated by the Deccan traps alone. However, the Deccan traps represent only a small fraction of the total volume of the

Igneous provinces. For example if they represent 1%, then 122,270 cubic miles of ocean could have been evaporated by the Igneous Provinces alone.

The area of the ocean is given as 139,000 square miles. Therefore the depth of ocean that could have been evaporated is about 90 feet.

Blinding Flash Calculation

Diameter of the Earth   7926.6 miles
Radius of the Earth   3963.3 miles

Suppose that the blinding flash started to form when the asteroid was 10 miles up.

With reference to the diagram, what is the distance (d in the diagram) to the horizon from a point that is 10 miles up in the air? The angle a in the diagram may be determined from the trigonometric relationships of a triangle.

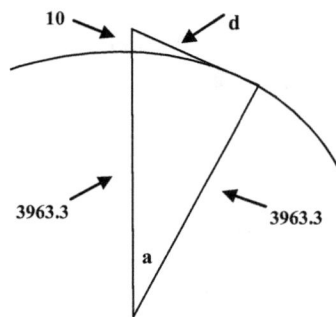

Cosine 1	=	$3963.3/(3963.3 + 10)$
	=	3963.3/3973.3
	=	0.9975
Therefore a	=	4 degrees, 3 min.
Therefore tan a =		0.0708
Therefore d	=	0.0708 x 3963.3 miles
	=	280.6 miles

At an elevation of ten miles the distance to the horizon would be 280.6 miles. The blinding flash would be devastating even beyond this distance but the shadow of the earth would increasingly restrict the damage to material above the ground. Roughly speaking, the area of devastation would extend for approximately 300 miles from ground zero. The area of devastation may also be determined. A circle with a radius of 300 miles has an area of:

$$Area \quad = \quad 3.14 \text{ x } 300 \text{ x } 300 \text{ square miles}$$
$$= \quad 282,743 \text{ square miles}$$

The minimum effect at the outer fringes of this area would be brilliant illumination, enough to cause instant blindness. The general effect over most of the area would be instant combustion and instant melting away of exposed flesh or even hair-covered flesh.

Water Mountain Formation

The water that would be displaced when the temporary crater in the ocean was formed may be divided into splash water and mountain-forming water. Using the Burkle Crater dimensions and a few assumptions a calculation may be carried out. The first assumption will be that all of the water out to about a four and one-half mile radius would become splash water and leave the area completely. Secondly, we assume that the crater would form out to the nine mile radius. At the Burkle Crater location the depth of the ocean is 12,500 feet (2.36 m). Finally we assume that the initial base of the water mountain would be 18 miles across and that the cross-section of the mountain would be triangular. All of these assumptions are reasonable and will enable us to form a rough idea concerning how high the circular water mountain might have become. This water mountain would be the tsunami as it formed and started to leave the area. All of the water between the four and one-half mile radius and the nine mile radius would contribute to water mountain formation. Therefore the water between these two radii for a depth of 2.36 miles would equal the volume of the water mountain which we have assumed was eighteen miles across at the base. Therefore h, the height of the mountain may be determined.

A   =   total crater volume
B   =   splash water volume
C   =   total volume out to the nine plus eighteen mile radius minus the crater volume
D   =   total volume out to the nine plus eighteen mile radius
H   =   mountain height

$$C = A - B \text{ and } C = (D - A)/2 \qquad \text{where } D = 3.14 \times 27 \times 27 \times H$$

Therefore   H $= (2 (A - B) + A) / (3.14 \times 27 \times 27)$ miles
$= ((2 \times 2.36 \times 3.14 ((9 \times 9) - (4.5 \times 4.5)) + (9 \times 9))) / (3.14 \times 27 \times 27)$ miles
$= (2 \times 2.36 \times (81 - 20.25) + 81 \times 2.36) / 729$ miles
$= 0.655$ miles or 3460 feet

Mountain Climbers

Kinetic energy   =   potential energy

½ x m x v x v   =   m x g x h

m is the mass, which occurs on both sides and can be ignored
v is the velocity in feet per second
g is the acceleration of gravity
h is the height in feet

Note: other systems of units could readily be used but the British system has been chosen. Using the above relationship, Table, 2.3.1.6 Tsunami Speed Height relationship, has been generated.

Asteroid Speed

The speed of passing asteroids is commonly given in kilometers per second. The speed can also be expressed in miles per hour.

e.g.   10 km/sec.   =   10 x (5/8 miles/km x 3600 sec./hr.)
                      =   10 x 2250 miles/hour
                      =   22,500 miles/hour

Therefore 100 km/sec. =   225,000 miles/hour

Ground Material Displacement

Suppose an asteroid was ten miles in diameter and travels into the interior of the Earth for ten diameters of itself. The volume of potential displacement would be:

$$V = \pi R^2 \times 1$$
When R = 5 miles
    1 = 100 miles

Therefore V = π x 5 x 5 x 100
            = 7354 cubic miles

Initial ground tsunami cross section 200 miles
Initial diameter 100 miles

The volume would be roughly

$$V = H \times \frac{314 \times 200}{2}$$

Therefore the height H would be

$$H = \frac{V \times 2}{314 \times 200}$$

    = (7354 x 2)/(314 x 200) = 0.23 miles or 1237 feet

References – Source

	Reference	Abbreviation
1.	A Short History of Nearly Everything, By Bill Bryson, Anchor Canada, a division of Random House of Canada Limited	Short
2.	American Petroleum Geologist Bulletin 56 No 2 1972,	Am Pet1
3.	An Ice Age Caused by the Genesis Flood by Michael J. Oard, Institute for Creation Research, P.O. Box 2667, El Cajon, California 92021	Ice Age
4.	Apocalypse When? Cosmic Catastrophe and the Fate of the Universe by Frank Close William Morrow and Company Inc., New York	Comets
5.	ASHRAE Handbook Fundamentals, American Society of Heating Refrigeration and Air-Conditioning Engineers Inc., 1791 Tullie Circle, NE Atlanta GA 30329	Fun 1981
6.	By Design, By Jonathan Sarfati PhD, Creation Book Publishers, Atlanta Georgia	By Design
7.	Canada from Space, By Brian Banks, Camden House Publishing, Suite 100, 25 Shepherd Ave. West, North York, ON M2N 6S7	Canada
8.	Cassell's Atlas of Evolution, By Andromeda, Weidenfield & Nicolsen, London UK	A of E
9.	Chemistry by James V. Quagliano Prentice – Hall Inc., Englewood Cliffs, New Jersey, USA	Chem
10.	Climate Wars, by Gwynne Dyer, Random House Canada	Climate Wars
11.	College Physics by Weber, White and Manning, McGraw-Hill Book Company, New York NY	College
12.	Comets and Asteroids and Future Cosmological Catastrophes compiled by Glen W. Chapman, www.2s2.com/chapmanresearch	Cosmic
13.	Creation Matters, Creation Research Society, P.O. Box 8263, St. Joseph MO 64508-8263 USA	CM
14.	Creation Research Society Quarterly, 6801 N. Hwy 89, Chino Valley AZ 86323	CRSQ
15.	Design and Origins in Astronomy, By George Mulfinger, Jr., Creation Research Society Books	Design
16.	Earth Impact Database, www.unb.ca/passc/Impact Database	EID
17.	Earth in Upheaval by Immanuel Velikovsky, Dell Publishing Co., Inc., 1 Dag Hammarskjold Plaza, New York NY 10017	E in U

	Reference	Abbreviation
18.	Encyclopedia Britannica 1958, Published by William Benton, Chicago London Toronto	En Br
19.	Engineering Mechanics, Dynamics, by Meriam & Kraige, John Wiley & Sons Inc, New York, London	Eng Mech
20.	Field Notes from a Catastrophe, By Elizabeth Kolbert 2007, Bloomsbury, USA	Notes
21.	1st International Conference on Creationism Vol II	Conf 1
22.	Funk & Wagnalls New Encyclopedia, Edited by Robert S. Phillips, Funk and Wagnalls	F & W
23.	Grand Canyon, The Story Behind the Scenery, By Merrill D. Beal, KC Publications Inc., P.O. Box 14883, Las Vegas, Nevada 89114	Grand
24.	Handbook of Chemistry and Physics, 52nd Edition, The Chemical Rubber Publishing Company, 18901 Cranwood Parkway, Cleveland Ohio 44128	Handbook
25.	Historical Geology, By Carl Owen Dunbar, John Wiley & Sons Inc., New York London	Geology
26.	How It Works, The Magazine That feeds Minds	HIW
27.	In the Minds of Men by Ian T. Taylor, TFE Publishing, Toronto	In the Minds
28.	Kronos Press, PO Box 313, Wynnewood PA 19096	Kronos
29.	MacLeans, 11th Floor, One Mount Pleasant Road, Toronto, ON M4Y 2Y5, Vol. 122, Number 24, June 29, 2009	Mac1
30.	Macleans, Aug. 14, 2000	Mac3
31.	Macleans, Aug. 24, 2009	Mac2
32.	Modern University Physics, By Richards, Sears, Wehr & Zemansky, Addison-Wesley Publishing Company Inc., Reading Mass., USA	Modern
33.	National Geographic Society, 17th and M Streets, NW Washington DC 20036	Nat Geo
34.	Nature Alberta, By James Cavanagh, Lone Pine Publishing, Edmonton, Alberta	Nature
35.	Pensee, Student Academic Freedom Forum, P.O. Box 414, Portland Oregon 97207	Pensee
36.	Peoples of the Sea, by Immanuel Velikovsky, Doubleday & Company Inc., Garden City, New York	Peoples
37.	Petrified Forest, The Story Behind the Scenery, By Sidney R. Ash and David D. May, Petrified Forest Museum Association, Petrified Forest	Pet For

	Reference	Abbreviation
	national Park, Holsbrook Arizona 86025	
38.	Physiology and Biophysics by Ruch and Patton, Nineteenth edition, W. B. Saunders Company, Philadelphia and Company	P&B
39.	Postcards from Mars, by Jim Bell, Penquin Group (USA), 375 Hudson Street, New York NY 10014	Postcards
40.	Principles of Microbiology Eighth Edition, By Alice Lorraine Smith, The C. V. Mosby Company, 11830 Westline Industrial Drive, Saint Lewis Missouri 63141	Principles
41.	Scientific American Inc., 415 Madison Ave., New York NY	Sci Am
42.	Silent Snow by Marla Cone, Grove Press, 841 Broadway, New York NY 10003	Silent
43.	National Geographic Atlas of the World, seventh Edition	Nat Geo At
44.	The Beothucks or Red Indians, By James P. Howley, Prospero Canadian Collection	Beo
45.	The Big Splash, by Dr, Louis A. Frank, Avon Books,New York	Splash
46.	The Concise Oxford Dictionary, edited by H.W. Fowler and F.G. Fowler, Oxford	Oxford at the Clarendon Press
47.	The Concise Oxford Dictionary, Oxford University Press, Walton Street, Oxford, 0X2 6DP	Oxford
48.	The End of the World, by John Leslie, Routledge	The End
49.	The Genesis Flood, by Whitcomb & Morris, The Presbyterian and Reformed Publishing Company, Philadelphia, Pennsylvania, 29 West 35th Street, New York, NY 10001	The Flood
50.	The Greatest Show on Earth, By Richard Dawkins, Free Press,New York	The Greatest
51.	The Living Cosmos by Chris Impey, Random House New York	Living
52.	The Moon Its Creation Form and Significance, By John C. Whitcomb/Donald B. DeYoung, BMH Books, Winona Lake, Indiana 46590	The Moon
53.	The New World of the Oceans, Men and Oceanography, by Daniel Behrman, Little Brown & Company	New
54.	The Oceans, By Sylvia Earle & Ellen Prager, McGraw-Hill Book Company, New York NY	Oceans

	Reference	Abbreviation
55.	The Rough Guide to the Universe by John Scalzi, Rough Guide Ltd., 80 Strand, London, WCR2 ORL	Rough
56.	The Scientific American Book of the Cosmos, Daniel H. Levy Editor, St. Martin's Press, New York, New York	Cosmos
57.	The Sea Around Us by Rachel Carson, Oxford University Press Inc. 2003, 198 Madison Ave., New York NY 10016	The Sea
58.	The Sun and Stars, By J C Brandt. 1966, McGraw Hill Book Company, New York NY	The Sun
59.	The Trouble with Physics by Lee Smolin, Houghton Mifflin Company, 215 Park Avenue South, New York, New York 10003	Trouble
60.	The Violent Face of Nature, by Kendrick Frazier, William Morrow and Company Inc., New York 1979	Violent
61.	Time Upside Down, By Erich A. Von Fange, 460 Pine Brae Drive, Ann Arbor MI 48105	Time
62.	Weather, A Visual Guide by Buckley, Hopkins and Whitaker, Firefly Books Ltd. 2008	Weather
63.	Worlds in Collision by Immanuel Velikovsky, Doubleday & Company, Garden City, New York	W in C
64.	In The Hills, published by MonoLog Communications Inc. R. R. 1, Orangeville, ON	The Hills
65.	Discover, The Magazine of Science, published by Time Inc., 3435 Wilshire Boulevard, Los Angeles, California	Discover
66.	American Society of Heating, Refrigeration & Air Conditioning Engineers	Ashrae
67.	Climate and the Oceans, by Geoffey K. Wallis, Princeton University Press	Climate

References By Chapter – Part 1	Topic
Chapter 1 – Introduction	
1. By Design p150	life too complex to explain
2. By Design p 152	DNA very great amount of information
3. By Design p 153	DNA has meaning due to physical indeterminacy
4. By Design p 186	life probability
5. CRSQ Vol 46 p 219	billions of parts properly assembled needed for life
Chapter 2 - Whoppers or Fairy Tales for Adults	
1. Oxford p 120	definition of theory
2. Chem p 31	definition of theory
3. En Br Vol 10 p 376	soft ice admits stones
4. CRSQ Vol 43 p232	polystrate fossils
5. CRSQ Vol 38 p181	mutations neutral or harmful
6. Nat Geo News Mar. 24,2005	soft dino tissue from Montana
7. C M Vol 7 O/N 2002 p 1 & Vol 8 M/J 2003 p 8	immune system self attacks
8. Rough p 10	stellar cloud collapsed
9. Trouble p 181	string theory is only conjecture
10. CRSQ Vol 20 p 215	marine shells, erratic boulders found in coal
11. CRSQ Vol 13 p 27	many ice age theories
12. CRSQ Vol 45 p 71	Proto-Avis 75m years before archaeopteryx
13. CRSQ Vol 39 p 153	CPT faulty
14. CRSQ Vol 16 p185	Archaeopteryx
15. Rough p 31	Sun 99% of matter
16. not used	
17. CRSQ Vol. 18 p 46	Geologic Column
18. E in U p 57	bones in caves
19. CRSQ Vol 24 p 46, Vol 23 p133	wrong order fossil formations
20. CRSQ Vol. 18 p 111	Hoatzin, archaeopteryx
21. C M Jan/Feb 2006 p 4	archaeoraptor fraud
22. C M May/June 2007 p 11	collagen fiber not bird feathers
23. C M M/J 2013 p 8	truth discussion
24. CRSQ Vol. 11 p 39	ocean water
25. P & B p 881	plasma constituents
26. C M May/June 2008 p 1	self-organization
27. Trouble p 179	no experiments to confirm string theory
28. A of E p 280	evolution
29. C M May/June 2000 p4	Haeckel fraud

References By Chapter – Part 1	Topic
30. CRSQ Vol 37 p 111	Earnst Haeckel fraud
31. CRSQ Vol 28 p 31	tropical forests world-wide before flood
32. CRSQ Vol 37 p145	trench sediment undeformed
33. CRSQ Vol 20 p215	Erratics in coal
34. CRSQ Vol 14 p101	radiohalos in coal
35. CRSQ Vol 14 p103	radiohalos
36. CRSQ Vol 7 p 56	C14
37. CRSQ Vol 12 p 24	50 billion tons of coal mined
38. Notes p 201	CO2 380 ppm
39. CRSQ Vol 16, cover	drumlins in Alberta
40. CRSQ Vol 7 p13	hotter cloud needed to collapse inner cloud
41. CRSQ Vol 13 p 29	ice slab more than 7 m long too long
42. Cosmos p 189	Solar System from gas cloud
43. CRSQ Vol 30 p 74	Solar output increase in 4.5 B yrs 25%
44. CRSQ Vol 32 p76	Habitable Zone 5% closer & 1% further
45. Nat Geo June 1973 p6	Gondwanaland
46. Nat Geo June 1973 p 13	1-2 inches per year
47. www. Science News.org Vol 123, Jan 8,1983 p20,21 Atlantic Ocean not widening	
48. CRSQ Vol 38 p 93	Subduction questionable
49. CRSQ Vol 26 p 134	ocean floor 5-6 km thick, continents 18-20 m thick
50. CRSQ Vol 25 p 29	no continuity of fossils
51. Nat Geo March 2003 p 18	complete fossil dinosaurs rare, usually broken
52. Nat Geo April 2003 p 12	Megatherium, Indrictherium
53. C M July/Aug 2001 p 2	Piltdown, Java, Nebraska, Peking man
54. C M July/Aug. 2001 p1	evolution fraud in Time magazine
55. Nat Geo Nov 1985 p 585	'Lucy' assembled from several hundred bone pieces
56. Nat Geo Nov 1985 p 608	Java Man
57. Nat Geo Nov 1985 p 609	Peking Man disappeared during WWII
58. CRSQ Vol 8 p 106	Hyrax 4 toes front, 3 toes rear
59. Wikipedia, Baluchitherium	Baluchitherium
60. Nat Geo At p 5	planetary rotation times
61. Nat Geo At p 19	Lunar Maria, mountains, temperature
62. Nat Geo At p 5	Ice Age
63. Nat Geo Atp 5	continents adrift
64. Nat Geo At p 6	'growing fingernail'
65. C M May/June 2007 p 6	Pangea fragmented, CPT
66. E in U p 202	sea shells in coal
67. E in U p 204	12' peat = 1' coal, coal layers 50' thick, seams divided
68. W in C p 11	La Place Theory

References By Chapter – Part 1	Topic
69. W in C p 8	Planets more angular momentum than Sun
70. W in C p 5	no similarity among planetary orbits
71. W in C p 7	no regularity in Solar System
72. E in U p 29	Old Red Sandstone
73. E in U p 26	rhinos, hippos, giant swine, & thousands of bones
74. E in U p 19	Wood Hills of New Siberia 300' high
75. E in U p 70	La Brea tar pits
76. E in U p 71	human skeleton in La Brea tar
77. E in U p 72	giant swine 6' high
78. CRSQ Vol 39 p150	ocean floor stopped moving
79. CRSQ Vol 39 p 152	CPT ocean floor vanished
80. CRSQ Vol 39 p 151	CPT ocean floor subducted during flood
81. Oceans p 157	Pacific Ring of Fire
82. CRSQ Vol 37 p144	no sediment in trenches
83. CRSQ Vol 8 p 107	horse evolution not supported from evidence
84. En. Br. Vol 12 p 26B	hyrax
85. In the Minds p 222	Java Man fraud
86. In the Minds p 236	Chinese Man fragments mixed with animal fragments
87. In the Minds p 234	Peking Man
88. Geology 1960 p 447	animal bones with human bones
89. In the Minds p 238	Peking Man
90. In the Minds p 229	Piltdown fraud
91. CRSQ v 28, p30	1" of coal, 1000 years
92. CRSQ Vol 41 p 319	gene transfer
93. E in U p 56	fossil skeletons incomplete and bones greatly broken
94. CRSQ Vol 29 p 13	seafloor 'spreading' conjecture
95. New p 209	seafloor spreading
96. International Geology Review Vol10 p775 continental drift	
97. Oil and Gas Journal Vol 84 p 115 ocean floor magnetism	
98. In the Minds p 88	dead fish dissolve
99. CRSQ Vol 26 p 133magnetic patterns in lava	
100. Design p 100	Triton retrograde orbit, Neptune radiating heat
101. CRSQ Vol 18 p 86	outward gas force prevents solar system from forming
102. CRSQ Vol 37 p 79	Cambrian Explosion
103. Ostrom, John 1985, p 174	Archaeopteryx
104. CRSQ Vol 32 p 19	Archaeopteryx, flew
105. CRSQ Vol 32 p 20	five specimens of Archaeopteryx found, all birds
106. C M Sept/ Oct 2005 p 6	Archaeopteryx wing feathers indicate flight ability
107. The Toronto Star Mon Feb 21, 2000 A15 dinosaur with 'feathers'	

References By Chapter – Part 1	Topic
108. C M July/Aug 2001 p 2	Piltdown Man, Java Man
109. C M Nov/Dec 2007 p 11	'Dakota' hadrosaur found in N. Dakota
110. CRSQ Vol 37 p 143	Solar behavior
111. C M May/June 2007 p 10	Mangrove plants in coal
112. The Sun p 111	must account for stars somehow
113. CRSQ Vol 7 p 9	Protostars
114. CRSQ Vol 7 p 11	gas cloud pressure
115. Design p 16	Jean's Length
116. CRSQ Vol 34 p 187	bacteria in 'old' rock
117. CRSQ Vol 20 p 212	coal 50x carbon in atmosphere
118. CRSQ Vol 38 p 97	Hotspot definition
119. Files.usgwarchives.net	Pacific 'ring of fire'
120. www.soest.hawaii.edu	Hawii Hotspot
121. C M July/August 2002 p 1	Hawii hotspot moving
122. En Br Vol 10 p 379	sea level 300' lower at ice age maximum
123. CRSQ Vol 38 p 43	gas cloud collapse
124. CRSQ Vol 13 p 28	ice too soft for hilly land
125.   Howorth, Sir Henry H. 1905 Op. Cit p xxxvii   glacier up & down hills
126. rst.gsfc.nasa.gov	Manson Crater
127. Wikipedia.org Manson Crater	Manson Crater
128. CRSQ Vol 32 p 76	Earth barely able to support life
129. Nat Geo May 2009 p 53	mammoth - elephant 400,000 DNA differences
130. C M Nov/Dec 2001 p 1	irreducible complexity
131. C M Jan/Feb 2009 p 4	chirality instability
132. E in U p 139	magnetism in rocks
133.   www.NASA.GOV/WORLDBOOK/STAR   most stars are double
134. CRSQ Vol 18 p 16	binary stars unequal ages
135. Cosmos p 136	white dwarf, mass of Sun size of Earth
136. Cosmos p 126	most stars binary
137. CRSQ Vol 21 p 174	magnetostriction
138. Pet For p 8	Petrified Forest
139. Grand p 42	fossil material must be buried quickly
140. CRSQ Vol 29, p17	sulphide problem
141. Mac2 p 41	dinosaur from chicken
142. E in U p 118, 119	no wandering continents
143. Am Pet1 p 264	no taped magnetic record
144. CRSQ Vol 9 p 49	lightning causes magnetism
145. CRSQ Vol 9 p 50	mag netostriction
146. E in U p 208	no buffalo fossils

References By Chapter – Part 1	Topic
147. Can Geo May/June 95 p 32	Tunguska Event, asteroids move 11 – 100 km/sec
148. C M Sept/Oct 2008 p 3	bacteria complexity
149. Can Geo Mar/April 1999 p 38	large animal bones from Yukon
150. E in U p 13	Alaska muck full of bones
151. Nat Geo April 03 p 31	Ice Age mammals
152. CRSQ Vol 9 p 213	no ice age in northern Siberia
153. Ice Age p 125	Ice age mammals
154. Can Geo Mar/April 1999 p 44	Yukon bones
155. CRSQ Vol 45 p 241	Lucy's structure
156. CRSQ Vol 45 p 243	Chimpanzee DNA
157. www.lindsay-archive.org/creation	trees fossilized quickly
158. Living p 281	scientists speculate when data is rare
159. Living p 293	RNA reduces
160. The Greatest p 416	life very improbable
161. By Design p 13	irreducable complexity
162. By Design p 17	info in cell = 120 vol of En Br
163. By Design p 18	mutations
164. The Greatest p 377	energy from sunlight
165. The Greatest p 410	genetic code
166. The Greatest p 36	bulldog birthing
167. The Greatest p 134	guppies
168. Nature p 15	evolution from fish
169. By Design p 147	Origin of Life prize
170. By Design p 148	natural selection
171. By Design p 149	Origin of Life prize
172. By Design p 2314 giga-years needed	
173. Pensee Fall 72 p 20	coal
174. In The Minds p 433	Uranium decay
175. CRSQ Vol 14 p 105	radiohaloes
176. En. Br. Vol 10, p 376	ice easily deflected
177. By Design p 160	DNA
178. CRSQ Vol 47 p 22	horse fossils
179. Trouble p 181	string theory
180. HIW Issue 26 p 062	asteroid shapes
181. www.space.com	asteroid 2005 YU 55
182. IEEE Spectrum 8.12 p 46	brain
183. CM July/Aug. 2014 p 9	soft dinosaur skin
Chapter 3 - Some Windows of Life	
1. Rough p 49	day 1100 hrs long

References By Chapter – Part 1		Topic
2.	CRSQ Vol 8 p 24	magnetic field
3.	Wikipedia QuebecLandslide	Quebec Landslide
4.	News.bbc.co.uk	astronauts died while still on the ground
5.	P & B p 784	O2 in blood
6.	CRSQ Vol 17 p108	O2 in blood
7.	Peoples p 49	Egyptian battles
8.	Wikipedia, Extinction	Extinction
9.	Rough p 82	Venus surface 900F
10.	Rough p 74	Mercury
11.	Rough p 145	Neptune, Uranus
12.	Design p 78	Moon – Earth relation
13.	Cosmos p 319	pH of water on distant planets
14.	Chem p 468	acids and bases
15.	CRSQ Vol 38 p 187	mutations
16.	CRSQ Vol 38 p 181	mutations damaging
17.	CRSQ Vol 14 p 56	protein improbability
18.	Hurontarian June 5, 2008 B1	honey bees
19.	En Br Vol 7 p 99	Dead Sea
20.	En Br Vol 10 p 374	Glacial Epic
21.	WSHVAC Design Manual p C23	ground water temperature
22.	Rough p 67	Earth's core
23.	CRSQ Vol 32 p 76	life conditions difficult to meet
24.	CM Jan/Feb 2005 p 1	Binary stars 60 to 90%
25.	The Toronto Star Apr 16 1999 A3	Ist extra-solar system
26.	Cosmos p 191	Jupiter necessary for Earth
27.	Rough p 68	Earth very dense 5 ½ x water
28.	Design p 78	Moon touching Earth , 2 billion years
29.	Climate p 14	Greenhouse Effect needed
30.	CRSQ Vol 30 p 196	CO2 5000 ppm upper limit
31.	Ice Age p 204	ocean temp 4C
32.	Fun 1981 p 27.2	Solar Energy
33.	CRSQ Vol 26 p 137	Noctilucent Clouds
34.	Rough p 74	Mercury Day
35.	Rough p 49	Earth Day 1100 hrs long
36.	The Moon p 141	Moon orbit offset by 5 degrees
37.	CRSQ Vol 33 p 85	Moon needed for axial tilt
38.	The Moon p 138	pull of Moon on Earth
39.	Design p 78	Moon touching Earth , 1 billion yrs
40.	En Br Vol 2 p 640	atmospheric pressure

References By Chapter – Part 1	Topic
41. Postcards p 52-57	photos of Mars
42. Climate p 206	Coriolis 'force'
43. Short p 249	Moon stabilizes axial tilt
44. Time p 25	Ice Age
45. By Design p 238	Sun located at co-rotation radius
46. HIW issue 22 p 007	Jupiter moon Io
47. HIW issue 21 p 050	1 in 3 bites of food is from bees
48. http://tycho/usno.navy.mil/leapsec	leap-sec
49. The End p 81	Howard-Kooman-Mickels1979 into Sun

Chapter 4 - Chaos and Bad News

1. Can Geo May/June 95	more than 1 billion asteroids
2. The End p 84	asteroid missed Earth by 6 hours
3. Wikipedia 1991 BA	1991 BA
4. Can Geo Sept/Oct 1999 p 71	asteroids to hit Earth
5. E in U p 51	Spitzbergen coal
6. Can Geo Nov/Dec 1999 p 43	Axel Heiberg
7. CRSQ Vol 20 p 212	coal carbon 50x biosphere
8. Can Geo may/June 1995 p 32	Tunguska 1908
9. Violent p 89	lightning voltage
10. CRSQ V 21 p 84	Tunguska 1908
11. Weather p 101	ice crystals
12. Short p 202	Comet Shoemaker-Levy into Jupiter
13. CRSQ Vol 37 p 187	asteroid shower 3 ½ months after 1178 impact
14. http://museum.gov.ns.ca	Halifax explosion
15. Cosmic p 5	10 km asteroid, kill one hemisphere
16. Violent p 225	Richter scale 10x for each integer
17. Violent p 219	earthquakes
18. Violent p 206	Krakatoa darkness
19. En Br Vol 13 p 498	Krakatoa waves
20. CRSQ Vol 21 p 84	Tunguska 1908
21. The Moon p 105	Moon hot and active
22. Wikipedia 1991 BA	1991 BA
23. The Cosmic Serpent by Clube and Napier (Cosmic p5) windstorm from impact	
24. Wikipedia Tunguska Event	Tunguska Event
25. Short p 194	asteroids near earth
26. Can Geo May/June 95 p 32	Tunguska 1908
27. Short p 203	blinding flash 60,000K
28. Can Geo May/June p 38	Comet Shoemaker-Levy9 into Jupiter
29. Eng Mech p 391	escape speed

References By Chapter – Part 1	Topic
30. Toronto Star Oct 26, 2003	Hermes asteroid
31. Cosmic p 5	tidal waves & vapourized water from impact
32. Cosmic p 4	future damage from asteroids
33. Short p 203	blinding flash
34. EID/images/manicouagan.htm	Manicouagan Crater
35. Nat Geo Map of the World Physical	Takla Makan Basin
36. E in U p 24, 25	Paris Limestone
37. E in U p 55	Georgia marine deposits
38. E in U p 56	English limestone full of bones
39. E in U p 58	Gibraltar rock
40. E in U p 61	water destruction
41. En Br World Atlas Plate 21	land elevations
42. CRSQ Vol 25 p 121	river beds on top of earlier sediments
43. E in U p 159	Florida fossils
44. In the Minds p 459	seals on Antarctica
45. CRSQ Vol 44 p 68	Burckle Crater
46. EID p 5	Vredeforte Crater
47. CRSQ Vol 41 p 206	Chesapeake Bay impact
48. CRSQ Vol 14 p 6	mammoth blood
49. E in U p 19	red blood cells
50. En Br Vol # p 672	countless million of buffalo
51. Cosmic p 2	Popigai Crater energy
52. EID p 3	Kamensk Crater
53. CRSQ Vol 21 p 83	Tenitz Basin
54. CRSQ Vol 26 p 144	Czech crater
55. Short p 212	50' wave Portugal earthquake
56. The Sea p 160	earthquake & 50 ft wave
57. The Sea p 163	tsunami speed 470 m/hr
58. Violent p 241	Chile earthquake & tsunami
59. CRSQ Vol 16 p 154	drumlins
60. Mac1 p 42	asteroids
61. Oceans p 107	tsunami speed
62. E in U p 191	glacial lakes , fossils
63. E in U p 176	flood evidence
64. Ein U p 30	fish died in fear
65. Mac3 p 43	Earth wobble
66. Weather p 246	Earth wobble
67. http://rst.gsfc.nasa.gov/SECT 18	major asteroid more than energy than all bombs
68. CRSQ Vol 21 p 82	asteroid craters

References By Chapter – Part 1		**Topic**
69.	CRSQ Vol 38 p 13	Cypress hills
70.	http://www.thestar.com/NASApp	Tunguska Event
71.	http://rst.gsfc.nasa.gov	Sudbury Crater
72.	Comets p 40	asteroids near Earth
73.	Comets p 51	comet into Sun, Howard-Kooman-Mickels1979
74.	Comets p 37	Howard-Kooman-Mickels 1979
75.	Handbook p F-154	asteroid data
76.	Pensee Fall 1972 p 43	Asteroid Toro
77.	Wikipedia Asteroid Toro	Asteroid Toro
78.	www.space.com	Asteroid 2010 XC15 passed Earth ½ way to the Moon

Chapter 5 - The Vapor Canopy Hypothesis

1.	CRSQ Vol 15 p 31	high O2 positive effect on healing
2.	Wikipedia Pterosaur	Pterosaur
3.	Toronto Star Sept. 16, 1980 A16	Giant Teratorn
4.	Can Geo Nov/Dec 1999 p 43	Axel Heiberg
5.	http://www.avalanchepress.com/svalbard.php	coal on Spitzbergen
6.	Wikipedia Baluchitherium	Baluchitherium
7.	En Br Vol 7, p 391	Dinosaurs
8.	An Ice Age p 28	palm trees in Alaska
9.	Short p 347	very large animals
10.	CRSQ Vol 19, p 40	Hypothalamus Disregulation Theory
11.	CRSQ Vol 17, p 108	atmospheric pressure
12.	CRSQ Vol 29, p 190	Axel Heiberg
13.	Rough p 169	Tunguska 1908, letter writer
14.	Can Geo. Nov/Dec 1999 p 45	Axel Heiberg flora and fauna
15.	College p 489	looming
16.	CRSQ Vol 15 p 3	Mars
17.	Short p 338	O2 35%
18.	Maine.gov July 10, 2008	hurricane rain
19.	CRSQ Vol 26 p 137	noctilucent clouds
20.	CRSQ Vol 17 p 65	Vapor Canopy
21.	CRSQ Vol 29 p189	Axel Heiberg
22.	Ice Age p27	Vapor Canopy = uniform climate
23.	Ice Age p 71-73	warm ocean
24.	s8int.com Out of place Artifacts	Out of place artifacts
25.	CRSQ Vol 28 p 126	canopy collapse = atmosphere expansion
26.	CRSQ Vol 28 p 60	higher O2 = better healing
27.	CRSQ Vol 28 p 31	coal from biomass
28.	CRSQ Vol 32 p 47	dinosaurs up north

References By Chapter – Part 1		Topic
29.	National Post Dec 18,1998	Axel Heiberg
30.	Wikipedia Pterosaur	Pterosaur
31.	Nat Geo News June 25, 2007	penquins in Peru
32.	www.genesispark.com	giants found in tombs in Ireland
33.	CRSQ Vol 7, p 35	coal 50,000 years old
34.	CRSQ Vol 14, p 101	radiohalos
35.	Wikipedia Pterosaur	Pterosaur
36.	Nat Geo April 2003, p6	evolution
37.	Modern p 934	nuclear reaction
38.	E in U p 51	Spitzbergen flora
39.	E in U p 19	wood hills, animals
40.	E in U p 52	warm Earth in the past
41.	E in U p 72	rhinos, horses, giant swine
42.	Splash p 103	comets into atmosphere
43.	CRSQ Vol 26 p 137	Noctilucent clouds
44.	CRSQ Vol 26 p 15	Noctilucent clouds
45.	W in C p 89	sky fell & killed humans
46.	Violent p 189	Tambora, 35 cu m of rock
47.	Sci Am. April 1977 p 60	14" of rain from thunderstorm
48.	E in U p 94	Sahara pasture
49.	Ein U p 97	Arabian desert bloomed
50.	E in U p 98	deserts once populated
51.	Weather p 32	Hadley Cells
52.	Wiki, Hadley Cells	Hadley Cells
53.	CRSQ Vol 15 p 32	canopy, shield for nitrogen
54.	E in U p 16	complete edible mammoth found in Siberia
55.	The Sea p 82	Submarine Canyons
56.	S & T p 82	higher CO2 with warm Earth
57.	CRSQ Vol 29 p 170	carbon dating
58.	CRSQ Vol 17 p 65	visible water heaven
59.	CRSQ Vol 17 p 67	Canopus
60.	In the Minds p 319	pterosaur, pteranodon
61.	CRSQ Vol 24 p 133	expanse divided waters
62.	CRSQ Vol 26 p 137	Noctilucent Clouds
63.	En Br Vol 15 p 519	insects
64.	CRSQ Vol 15 p 153	Vapor Canopy
65.	ASHRAE 1981 p 27.2	solar radiation
66.	F & W Vol 3 p 42	Hellas
67.	The Flood p 375	Vapor Canopy

References By Chapter – Part 1	Topic
68. Silent p 128	Faroe Islands
69. By Design p 76	Pteradactyl
70. By Design p 78	30% increased lift
71. HIW issue 21 p 051	Spiracles
72. CM Jan/Feb 2914 p 9	Tropical trees on Antarctica
73. The Greatest p 268	giant marsupials

Chapter 6 - More Bad News

1.	E in U p 21	Erratic Boulders
2.	CRSQ Vol 13 p 28	Erratic Boulders
3.	CRSQ Vol 20 p 215	Erratic Boulders in coal
4.	CSRQ V20 p 216	trees found in coal
5.	E in U p 22	Erratic Boulders
6.	Ice Age p 196	Erratic Boulders
7.	E in U p 23	Erratic Boulders on ocean islands
8.	E in U p 21	Erratic Boulders
9.	Sci. Am. Jan. 1973 p 49	Mars craters
10.	Cosmos p 231	Mars rocks to Earth
11.	Kronos XI Vol 1 p 65	Valles Marineris, 4,000 km long
12.	Rough p 77	Calosis Basin, weird terrain opposite
13.	Cosmos p 198	Neptune
14.	E in U p 22	Erratic boulders
15.	Nat Geo May/June 1995 p 40	Meteorite into house
16.	Cosmos p 144	one million asteroids > 1 km dia
17.	Wikipedia Hellas Planitia	Hellas Basin
18.	Can Geo Sept/Oct 1999 p 73	Asteroid 1994XM1
19.	Can Geo May/ June 1995 p 32	Tunguska Event
20.	CRSQ Vol 31 p 153	Arizona Crater
21.	Wikipedia Hellas Planitia	Hellas planitia
22.	CRSQ Vol 26 p 134	crust thickness, mountain formation
23.	Wikipedia Sudbury Basin	Sudbury Basin
24.	Can Geo Sept/Oct 99 p 78	Manicouagan Crater
25.	Rough p 105	Antarctica Meteorite ALH 84001
26.	CRSQ Vol 31 p 153	Arizona Crater
27.	Can Geo Sept/Oct 1999 p 78	Manicouagan Crater
28.	Can Geo Sept/Oct 1999 p 76	14 mass extinctions
29.	Design p 81	Lunar Mascons
30.	CRSQ Vol 21 p 84	10 km asteroid destroy life in one hemisphere
31.	CRSQ Vol 26 p 134	asteroid impact results
32.	Can Geo May/June 95 p 38	Mass Extinctions, NEO threaten Earth

References By Chapter – Part 1	Topic
33. www.cat.com/cda/files/1060444/7/	Sudbury mine > 1 mile deep
34. Short p 215	Kola Borehole
35. Wikipedia Olympus Mons	Olympus Mons
36. Rough p 68	Earth material very dense
37. Rough p 168	Ceres, asteroid composition
38. Kronos X Vol 3 p 25	Mars craters 91% on one side
39. Sci Am Jan. 1973 p 55, Kronos Vol X, 3 p 31 asteroid list	
40. Kronos Vol XI, 1 p 61	Hellas impact
41. Violent p226	Earthquakes every year
42. Rough p 67	Earth solid core
43. Can Geo May/June 1995 p 32	Tunguska Event, unnoticed asteroids
44. E in U p175	North Sea inhabited recently
45. CRSQ Vol 37 p 191	ocean includes continental crust
46. Kronos Vol X,3 p 32	Mars craters, asteroid list
47. Kronos XI 1 p 61	Hellas pressure waves
48. E in U p 61	sudden destruction
49. E in U p 104	waves against continents
50. Oceans p 107	ocean waves move quickly
51. CRSQ Vol 38, p 8	gyots
52. The End p 82	Chicxulub, Tambora eruption
53. Can Geo May/June 1995 p 37	Chicxulub 70% extinction
54. Cosmic p 3	impact cloud covered Earth
55. Nat Geo Atp 3	mapmaking
56. Violent p 239	New Madrid earthquakes
57. CRSQ Vol 21 p 84	mountain range formation
58. Violent p 225	Richter scale
59. CRSQ Vol 37 p 185	Giodano Bruno Crater, 1178 lunar impact
60. C M Nov/Dec 2007 p 7	Igneous Provinces
61. E in U p 138	marooned whales, Erratic boulders widespread
62. Oil and Gas Journal Vol 84 p 115	magnetism varies both vertically & horizontally
63. E in U p 139	magnetism in rock 100 x Earth's magnetic field
64. CRSQ Vol 21, p 174	magnetic field
65. E in U p 141	magnetism in clay pots
66. CRSQ Vol 26 p 134	ocean floor fractures, ocean floor thickness
67. Violent p 195	Igneous Provinces, Deccan Plateau
68. Short p 192	more than 1 billion asteroids
69. Cosmos p 144	1 m asteroids > 1 km dia
70. CRSQ Vol 37 p 142	Submarine Trenches void of material
71. Cosmic p 3	San Francisco earthquake 1906

References By Chapter – Part 1	Topic
72. CRSQ Vol 18 p 113	mountains from sedimentary rock
73. En Br Vol 10 p 166	folded sedimentary rock in mountains
74. Violent p 203	Alaska earthquake, Mt Katmai
75. EID	impact sites > I km dia
76. Cosmic p 2	Earth recoiled from impact
77. HIW issue 22 p 016	lunar wobble, libration
78. Ice Age p 31	depressing world
79. The Oceans p 198	Chinese tidal bore

Chapter 7 - Post-Impact developments

1.	Violent p 189	Tambora, 35 cu mi of rock, other major earthquakes
2.	Violent p 188	famine in New York, 1816, frost every month
3.	Violent p 204	Katmai, days of darkness
4.	Violent p 207	Krakatoa, solar energy drop to 88%
5.	Violent p 206	pressure wave around Earth 3 x
6.	E in U p 143	Krakatoa
7.	Violent p 198	Mt Peelee
8.	W in C p 129	Santorini, Egypt dark, shadow of death
9.	The End p 82	Tambora
10.	Short p 222	Mount St Helens, smoke to 60,000 ft
11.	Cosmic p 3	havoc on distant continents
12.	Can Geo Nov/Dec 1999 p 43	Axel Heiberg
13.	Can. Geo. May/June 1995 p 37	Chicxulub, 70% extinction
14.	Ice Age p 42	clouds affect surface temperature
15.	CRSQ Vol 41 p 212	Chesapeake Bay impact
16.	The End p 83	large impacts result in surface deflections
17.	Can Geo. Sept/Oct 1999 p 76	Tunguska Event, Sudbury Crater
18.	Cosmic p 5	tektite fields from impacts
19.	Can Geo May/ June 1995 p 38	mass extinctions, NEO threaten life
20.	CRSQ Vol 11 p 213	ocean cooled after the Chalk period
21.	Ice Age p 45	more forest with higher CO2
22.	Ice Age p 69	Ice Age volcanoes
23.	Ice Age p 37	nuclear winter
24.	Wikipedia Submarine Canyons	glaciers used 125 m of ocean water
25.	National Post Dec 18, 1998	ice breakup starves bears
26.	Ice Age p 71	warm ocean
27.	CRSQ Vol 9 p 213	far north never glaciated
28.	CRSQ Vol 11 p 215	continental drift ?
29.	Coleman, A. P. Ice Ages Recent and Ancient, Macmillan, N York (p 216) ice age theories	
30.	Robin, Gordon de Quetteville, 1966 Origin of the Ice Ages, more than 60 theories	

References By Chapter – Part 1	Topic
31. Comets p 5	mankind very improbable
32. Cosmic p 5	great dying from cold & darkness
33. Violent p 201	volcanic cloud produces darkness ?
34. Short p 273	3% fresh water
35. En Br Vol 10 p 379	ocean level during Ice Age 300 ft lower
36. In The Minds p 97	Antarctic coal
37. Discover July/August 2013, p 55	Lake Vostok
38. F & W Vol 5 p 294	CO2 dissolves in water
39. En Br Vol 4 p 838	CO2 into water
40. CRSQ Vol 21, p 69	surface heat transfer
41. CM Sept/Oct p 3	ice getting deeper in Greenland
42. CRSQ Vol 11, p 71	high ocean temp = more snowfall
43. E in U p 158	massive floods at end of ice age
44. CRSQ Vol 21 p 73	warm ocean = more snowfall
45. Beo p 150	3 ft of snow in one night
46. College p 207	relative humidity
47. Violent p 188	famine in N. America & Europe
48. Ice Age p 31	ocean level
49. CRSQ Vol 9, p 213	no ice on northern Greenland
50. Weather p 250	warm ocean = increased rainfall
51. Can Geo Nov/Dec 1999 p 45	Earth warm 15 to 20C
52. Ice Age p 204	ocean temp 4C
53. The Lost Squadron.com	Greenland planes found under 300' of ice
54. C M Sept/Oct 2005 p 2	ice cores 110,000 yrs
55. Violent p 171	Buffalo snowstorm 3'
56. CRSQ Vol 21 p 70	snow albedo 0.7
57. Ice Age p 83	climate wet in winter during Ice Age
58. Sci Am April 1977 p 60	Colorado thunderstorm
59. Climate p 16	Earth losing heat
60. Principles p 273	UV light kills bacteria
61. National Post Dec 18 1998 p A2	Axel Heiberg champsosaur
62. En Br Vol 10 p 375	glacial flow patterns
63. C M Nov/Dec 2007 p 7	Igneous Provinces map
64. En Br World Atlas plate 21	land elevations
65. The Sea p 177	Atlantic has a deep south flowing current
66. Ice Age p 176	Lake Mead depressed land 20 cm
67. En Br Vol 10 p 374	28% land covered at glacier max
68. Ice Age p 99	ice depth 700 m average

References By Chapter – Part 1		Topic
69.	The Sea p 177	southward ocean current under Gulf Stream
70.	E in U p 131	no ice would form if Earth cooled
71.	E in U p 158	Miss. river delta 5,000 yrs old
72.	Ice Age p 16	Astronomical Theory of Ice Age
73.	Ice Age p 116	rapid ice age
74.	The Flood p 126	explaining submarine canyons is a problem
75.	The Flood p 125	many submarine canyons, no explanation
76.	Can Geo Sept/Oct 1999 p 75	decade long winter from 1.5 km asteroid
77.	Wikipedia, Submarine Canyons	Submarine Canyons
78.	En Br Vol 14 p 83	UV causes skin cancer
79.	CRSQ Vol 41 p 212	Chesapeake Bay, 100x nuclear arsenal
80.	Short p 273	90% of ice on Antarctica, 200' ocean rise
81.	Violent p 207	Santorini, 1400 BC, massive tsunamis
82.	Violent p 172	Buffalo snowstorm, drifts 25' high 1977
83.	E in U p 50	no glaciations in far north
84.	E in U p 128	heat needed for ice age
85.	E in U p 153	Lake Agassiz < 1000 yrs
86.	Nat Geo Atlas of the World 7th ed p 5	Magnetic Field, Ice Age
87.	Ice Age p 98	more ice with volcanic activity
88.	Wikipedia Ice Age	Ice Age
89.	CRSQ Vol 38 p 88	canyons from continental runoff
90.	www.Britannica.com	density induced underflow
91.	Nat Geo Apr 2013 p 51	fresh mammoth tusk, 150 lbs, $60,000

Chapter 8 - Implications for the Windows of Life

1.	Beo p 150	3' snow in one night
2.	CRSQ Vol 21 p 84	impact wind speed 2500 km/hr 2,000 km from impact
3.	Mac1 p 43	asteroids near Earth
4.	Weather p 246	Earth wobble
5.	CRSQ Vol 18 p 40	magnetic field gone in 3180 AD
6.	F & W Vol 5 p 250	radiation causes cancer
7.	Short p 247	1% further, 5% closer
8.	Cosmos p 144	remaining asteroids = 0.0005 mass of the Earth
9.	Kronos X 3 p 25	91% of Mars craters on one side
10.	Rough p 96	Mars orbit
11.	Rough p 74	Mercury day = 88 Earth-days, night same
12.	En Br Vol 16 p 1001	ozone production
13.	En Br Vol 14 p 83	UV causes cancer. Less UV on cloudy days
14.	En Br Vol 2 p 640	atmospheric data
15.	F & W Vol 22 p 69	effects of radiation

References By Chapter – Part 1	Topic
16. CRSQ Vol 39 p 4	photochemical reactions
17. CRSQ Vol 39 p 3	magnetic field
18. CRSQ Vol 21 p 85	magnetic currents disturbed
19. Wikipedia 2010 Chili Earthquake	Chili Earthquake
20. EIW Issue 24 p 007	Vesta
Chapter 9 - Evidence of Asteroid Impact	
1. Kronos X 3 p 33	Mars craters compared to asteroids remaining
2. Can Geo May/June 95 p 35	Jupiter's effect on asteroids
3. Kronos X 3 p 26	Jupiter's effect on asteroids
4. Short p 194	many asteroids near Earth
5. Can Geo May/June 95 p 38	2,000 near-Earth objects
6. CRSQ Vol 41 p 206	Chesapeake Bay impact
7. www.igsb.uiowa.edu/browse/manson99	Manson Crater
8. Can Geo may/June 95 p 36	Canadian Craters
9. Canada p 83	Manicouagan Crater
10. Can Geo May/June 95 p 37	mass extinction from object 5 km or more, Chicxulub
11. CRSQ Vol 21 p 82	Caloris Basin
12. CRSQ Vol 21 p 83	Tenitz Basin
13. Wikipedia – Acraman Crater	Acraman Crater
14. EID p 5	Woodleigh Crater
15. EID p 5	Yarrabba Crater
16. EID p 4	Tookoonooka Crater
17. Wikipedia Popigai Crater	Popigai Crater
18. EID p 4	Puchezh-Katunki Crater
19. EID p 3	Kara Crater
20. CRSQ Vol 26 p 144	Czechoslovakia Crater
21. Wikipedia Mass Concentrations	Mass Concentrations
22. CRSQ Vol 27 p 123	artifacts & skull under lava beds
23. CRSQ Vol 11 p 56	Lewis Overthrust
24. CRSQ Vol 27 p 125	mortar & pestle under lava
25. The Moon p 86	Lunar mountains
26. Wikipedia Yellowstone Park	Yellowstone Park
27. The Face p 195	Deccan Plateau
28. Oceans p 84	Coriolis 'force'
29. CRSQ v 21 p 82	Vredeforte Crater
30. Wiki, Mimas	Mimas
31. CM N/D 07 p 7	Large Igneous Provinces
Chapter 10 - Evidence of Chaos and Catastrophe	
1. CRSQ Vol 7 p 62	C14 for coal, natural gas and petroleum, 50,000 yrs

References By Chapter – Part 1	Topic
2. CRSQ Vol 29 p 173	C14, SPR & SDR
3. not used	
4. Violent p 209	Mexican volcano
5. Wikipedia Yellowstone Park	Yellowstone Park ?
6. CRSQ Vol 25 p 122	Mississippi River delta
7. In the Minds p 83	Niagara Falls
8. Sci Am Jan 1973 p 54	Nix Olympia
9. CRSQ Vol 42 p 1	Rim Gravel
10. E in U p 23	sea over land
11. En Br Vol 23 p 113	Vesuvius
12. E in U p 29	Old Red Sandstone gravel
13. E in U p 213	bones in rocks
14. CRSQ Vol 23 p 133	Wrong order fossils
15. E in U p 56, Time p 18	English limestone full of bones
16. E in U p 16	ground froze quickly
17. E in U p 13	Alaska, Yukon muck
18. Can Geo Mar/April 99 p 37	Yukon bones
19. CRSQ Vol 43 p 232	Polystrate fossils
20. CRSQ Vol 41 p 93	dinosaur 'nests'
21. Ice Age p 109	current ocean temp. 4C
22. E in U p 13	Alaska muck
23. Kronos XI Vol1 p 59	200,000 craters on Moon
24. Design p 81	Mascons
25. E in U p 245, 246	summary of catastrophe
26. W in C p 28	Niagara gorge, 5,000 yrs
27. The Moon p 86	200,000 craters on Moon
28. Kronos XI 1 p 59	Tharsus Bulge
29. CRSQ Vol 38 p 13	Cypress Hills
30. E in U p 246	Summary of chaos
31. CRSQ Vol 14 p 5	Woolly Mammoth fresh
32. E in U p 215	grass in mouth, clotted blood = sudden death
33. E in U p 18	elephant, hippos bones
34. In the Minds p 318	C14 production and decay
35. CRSQ Vol 29 p 171	Radiocarbon dating
36. CSRQ v 14 p 5	not used
37. Wikipedia Canadian Shield	Canadian Shield
38. CRSQ Vol 47 p 33	trilobites
39. Discover July/Aug 2013 p 55	Antarctica Flood
40. http:www.universetoday.com	lunar bulge

References By Chapter – Part 1	Topic

Chapter 11 - Predictions for the Future

1.	Sci Am April 1977 p 98	Sun-like stars mostly doubles
2.	Cosmos p 126	inter-stellar cloud momentum, single stars rare
3.	C M Jan/Feb 2005 p 1	binaries 60 – 90 %
4.	Nat Geo At p 20	Solar System
5.	The Toronto Star Fri. April 16, 1999 A3	Ist extra-solar system found
6.	CRSQ Vol 44 p 76,77	star-planet definition murky
7.	CRSQ Vol 14 p 54	probability of protein formation low
8.	Climate Wars p 94	CO2 exits warm ocean
9.	Climate Wars p 90	sea level rise
10.	Climate Wars p 95	methane in Siberian bog
11.	Climate Wars p 60	Hadley cell expansion
12.	Can Geo Oct 2008 p 42	Hudson Bay warming
13.	Climate Wars p 4	starvation coming
14.	Climate Wars p 2	CO2 & methane from bogs too much
15.	Climate Wars p 50	seed sterility with temp rise
16.	Climate Wars p 47	trouble accompanies higher temp
17.	National Post Dec 9, 1998 A6	Polar Bears losing weight
18.	Weather p 275	Antarctic ice sheets
19.	www.nasa.gov/worldbook/star	75% of stars are binaries
20.	Rough p 189	Alpha Centura A & B
21.	CRSQ Vol 32 p 76	Mars 'canals', Earth habitable zone
22.	The Toronto Star Jan 17, 1999 p F8	extra-solar planets
23.	CRSQ Vol 37 p 185	1178 impact on Moon
24.	Can Geo Sept/Oct 1999 p 78	Manicouagan Crater
25.	Nat Geo Atp 6	most volcanoes near faults
26.	Can Geo May/June 1995 p 35	Jupiter modifies asteroid orbits
27.	Can Geo May/June 1995 p 38	mass extinctions from asteroid impacts
28.	Rough p 83	Venus
29.	Living p 281	scientists speculate when data short
30.	Living p 282	galactic habitable zone
31.	Nat Geo Dec 2009 p92	planet detection
32.	Nat Geo Dec 2009 p 91	370 exo-planets as of 2010
33.	Nat Geo Dec 2009 p 93	Gliese 581 d, red dwarf
34.	CM Mar/April 2011 p 10	2nd Sun needed for 2nd Earth
35.	HIW Issue 37, p 60	Jet Streams
36.	http://tycho/usno.navy.mil/leapsec	leap-sec

Chapter 12 – Conclusions

1.	The End p 83	mass extinctions, Tunguska Event

References By Chapter – Part 1		**Topic**
2.	Can Geo Sept/Oct 1999 p 76	Earth wobble, Tunguska, 14 mass extinctions
3.	Climate p 65	force discussion
4.	Short p 273	Antarctica melt = ocean rise 200'
5.	Climate p 46	Coriolis 'force' is not a real force
6.	Design p 81	Mascons on Moon
7.	CRSQ Vol 21 p 82	asteroid impacts
8.	not used	
9.	Nat Geo Oct 2003 p 91	Chicxulub impact and darkness
10.	Wikipedia Manson Crater	Manson Crater
11.	www.igsb.uiowa.edu	Manson Crater
12.	CRSQ Vol 44 p 68	Burkle Crater
13.	C M Nov/Dec 2007 p 7	Igneous Provinces
14.	Kronos XI Vol 1 p 59	Tharsus Bulge
15.	Wikipedia Olympus Mons	Olympus Mons
16.	Kronos XI Vol 1 p 63	Valles Marineris
17.	Wikipedia Mass Concentrations	Mass Concentrations
18.	Can Geo May/June 1995 p 37	Chicxulub Impact
19.	Cosmos p 185	1 km asteroid would kill most life, Tunguska Event
20.	The End p 83	Tunguska Event, crust deflection from impacts
21.	not used	
22.	Can Geo Set/Oct 1999 p 74	Chicxulub, asteroid discoveries
23.	Violent p 206	Krakatoa
24.	CRSQ Vol 34 p 187	bacteria in 'old' rock
25.	The End p 83	Tunguska Event, crust deflection
26.	Cosmic p 3	large animals < chance of survival
27.	Climate p 50	Coriolis 'Force'
28.	Kronos X 3 p 36	Mars craters
29.	EID	70 impacts with craters > 1 km
30.	CRSQ Vol 21 p 84	10 km comet destroy life in one hemisphere
31.	CM Nov/Dec 2010 p 3	Late Heavy Bombardment
Chapter 13 – A Theory of the Earth Based on Asteroid Impact		
1.	CRSQ Vol 47 p 266	log mats during Flood
2.	W in C p 330	360 day year
3.	CRSQ Vol 50, p 40	360 day year

References by Chapter   The Window of Life Part 2

References by Chapter – Part 2	Topic
Chapter 1, Introduction	
1. The End p 83	Chicxulub impact, Tambora ash
2. Cosmic p 5	a great dying
3. The End p 83	Tunguska Event
4. Nat Geo Oct 03 p 91	asteroid impact results
5. Can Geo May/June 1995 p 37	90% extinction from Chixzulub
6. CRSQ Vol 11 p 216	60 ice age theories
7. Short p 273	Antarctica = 200 ft of ocean
8. Wikipedia/Ice age	400 ft of ocean in ice age glaciers
9. Comets p 5	Mankind appearance very unlikely
10. Comets p 53	mass extinction 95%
11. Comets p 58	comet speed 25 m/sec
12. Comets p 15	impact energy
13. HIW Issue 38 p 60	space junk danger
Chapter 2, The Hazards of Asteroid Impact	
1. Comets p 51	comet into the Sun
2. Comets p 15	meteor speed 50 m/sec
3. http://museum.gov.ns.caHalifax explosion	
4. Violent p 296	Krakatoa
5. Can Geo May/June 1995 p 32	asteroid speed 50 miles per sec.
6. Violent p 206	Krakatoa rumble
7. Wikipedia Halifax explosion	Halifax explosion opened a hole in the ocean
8. Wikipedia, TunguskaEvent	Tunguska Event
9. CRSQ Vol 21 p 84	30 mile radius of destruction from Tunguska
10. Wikipedia, Tunguska eve	human knocked down
11. CRSQ Vol 14 p 5	death of Woolly mammoth
12. CRSQ Vol 14 p 6,	sudden freezing of mammoth
13. E in U p 16,	Arctic islands bones
14. E in U p 19	piles of trees 300 ft high
15. EID,	Tenitz Basin
16. Cosmos p 41	Chicxulub
17. Ein U p 19	rhinos, buffalo, elephants, bison
18. Cosmos p 54	galaxy interaction
19. Cosmic p 2	Woolly Mammoth with modern animals
20. Nat Geo Nov 85, p 592	Lucy disarticulated
21. E in U p 14	broken bones
22. CM Mar/Apr 2005 p 10	soft dinosaur tissue

References by Chapter – Part 2		Topic
23.	CM Nov/Dec 2007	Hadrosaur discovery
24.	E in U p 16	Siberian islands composed of bones
25.	Ein U p 13	Alaska muck full of animal remains
26.	Wikipedia Tunguska event	Tunguska Event
27.	Wikipedia asteroid	asteroid
28.	www.cbc.ca/news/world/story/2011/03/11/japan-quake-tsunami	
29.	Wikipedia 2010 Chili earthquake	Chili Earthquake
30.	CRSQ Vol 44 p 68	Burkle Crater
31.	CRSQ Vol 41 p 213	Chesapeake Bay impact
32.	Comets p 2	age of universe
33.	E in U p 21	Erratic Boulders
34.	E in U p 22	Erratic Boulders
35.	EID	Sudbury Crater
36.	Wikipedia Canadian Shield	Canadian Shield
37.	www.icr.orgocean sediment	
38.	CRSQ Vol 21 p 83	Ishim Crater
39.	CRSQ Vol 24 p 126	Ries Crater
40.	CRSQ Vol 37 p 144	Submarine Trenches
41.	CRSQ Vol 21 p 84	Energy to form a mountain range
42.	Violent p 233	Alaska earthquake
43.	Short p 213	New Madrid earthquakes
44.	Time p 16	Man in limestone
45.	CRSQ Vol 38 p 121	Material soft and plastic
46.	Violent p 225	Richter Scale, massive death tolls
47.	Short p 212	50' wave from Lisbon earthquake
48.	Short p 203	blinding flash
49.	Short p 201	Chicxulub dimensions
50.	Cosmic p 3	'flaming'rocks set world afire, great dying
51.	Short p 204	planet dark within one hour
52.	Short p 213	New Madrid earthquakes 1811=1812
53.	The End p 83	Earthquake felt far away
54.	Wikipedia MansonCrater	Manson Crater
55.	http://rst.gsfc.nasa.gov	wave from Manson impact
56.	CRSQ Vol 41 p 208	Chesapeake Bay impact
57.	EID	Popigai Crater
58.	Can Geo Sept/Oct 1999 p 74	Chicxulub Crater
59.	http:rst.gsfc.nasa.gov	Sudbury Crater
60.	CRSQ Vol 26 p 144	Czechoslovakia Crater
61.	CRSQ Vol 31 p 154	Arizona Crater

References by Chapter – Part 2	Topic
62. MacLeans Apr 2012 p 50	Mariana Trench costly to explore
63. CRSQ Vol 21 p 85	impacts disturb magnetic currents
64. Can Geo May/June 95 p 38	asteroids, mass extinctions
65. Short p 343	mass ex. 65, 210, 245, 365, 440. 75 – 90% ex
66. Can Geo Sept/Oct 1999 p 76	numerous mass extinctions
67. EID	more than 100 mass extinctions
68. E in U p 30	fish died in agony
69. Nat Geo Mar 2003 p 18	sedimentary rocks
70. En Br Vol 20 p 272	sedimentary rocks
71. Wikipedia MansonCrater	Manson Crater
72. En Br Vol 6 p 338	continental shelf
73. CRSQ Vol 25, p 122	delta depth at New Orleans
74. CRSQ Vol 25, p 122	delta formed in 4500 years
75. Adler Planetarium	Tharsus Bulge
76. Kronos XI 1 p 65	Mars chaotic terrain, trench 4,000 km long 200 km across
77. College p 103	momentum
78. HIW Issue 32 p 015	tsunami
79. The End p 83	Tnuguska Event
80. Kronos XI 1 p 63	tension faults (graben) out from main trench
81. Kronos XI 1 p 59	Tharsus bulge, 6 ½ miles high, 3000 m dia
82. Wikipedia Olympus Mons	Olympus Mons
83. Wikipedia Caloris Basin	Caloris Basin
84. Nat Geo maps	ocean floor fractured
85. En Br Vol 11 p 859	Hudson Bay
86. Ice Age p 176	Hudson Bay rising
87. CM Nov/Dec 2007 p 7	Igneous provinces
88. CRSQ Vol 26 p 134	crustal thickness
89. HIW Issue 32 p 067	Venus Landscapes
90. Nat Geo News 2011/03/110316-japan-earthquake Japan Earthquake	
91. Wikipedia Chile Earthquake	Chili Earthquake
92. CRSQ Vol 21 p 84	earthquake, asteroid energy
93. CRSQ Vol 31 p 155	Arizona Crater
94. Wikipedia Chandler wobble	Chandler Wobble
95. CRSQ Vol 21 p 83	Tenitz Basin
96. Journal of Creation August 2001	Mid-Continent Rift
97. Kronos XI 1 p 66	Martian volcanoes
98. EID	Vredeforte Crater
99. http://volcano state.edu	Deccan Traps
100. Violent p 195	Igneous Provinces very large

References by Chapter – Part 2	Topic
101. En Br Vol 17 p 366	Patagonia
102. Nat Geo March 2003 p 18	dinosaur nests flooded, rare articulated skeletons
103. Violent p 191	Earth crust very thin
104. Wikipedia mountains on Mars	mountains on Mars
105. Wiki Loihi	Loihi
106. Violent p 207	Krakatoa
107. Violent p 189	Tambora
108. Violent p 204	Katmai
109. Short p 224	10,000 volcanoes
110. CRSQ Vol 26 p 137	Noctilucent Clouds
111. Violent p 14	thunderstorm 500,000 tons of water, earthquake every 15 days
112. The End p 82	volume of Deccan Traps
113. Short p 217	litho means stone
114. Cosmic p 315	dust layer over the world
115. Kronos XI 1 p 61	Hellas impact devastated Mars
116. The End p 83	Siberian Traps, 75 96% extinction, Tunguska Event
117. Violent p 188	1816 cold, starvation
118. Violent p 207	Krakatoa reduced solar radiation to 88%
119. Cosmic p 5	Ries Crater, great dying from impacts
120. Short p 227	Trieste descent into Mariana Trench,
121. CRSQ Vol 26 p 137	Noctilucent Clouds
122. Sci Am April 1977 p 60	Storm clouds causing darkness
123. In the Minds p 99	extreme cold preserved dinosaurs
124. E in U p 215	mammoth sudden death
125. CRSQ Vol 21 p 69	Astronomical theory 2F only
126. Ice Age p 13	60 Ice Age theories
127. CRSQ Vol 26 p 137	Noctilucent Clouds
128. E in U p 57	disarticulated Woolly Mammoths
129. Wikipedia Ice Age	400' of ocean in glaciers
130. Nat Geo April 03 p 31	mega-fauna vanished all at once
131. Ice Age p 69	Volcanic ash mixed with snow
132. E in U p 48	Ice Age down south
133. Westworld June 2011 p 70	Okotoks Erratic boulder
134. E in U p 18	mammoth sudden death
135. In the Minds p 448	dinos alive with humans
136. Pet For p 16, 17	trees fossilized quickly
137. Time p 25 & 30	nail in sandstone
138. National Post Dec 18, 1998	Champsosaurs up north
139. Wikipedia Deccan Traps	Deccan Traps

References by Chapter – Part 2	Topic
140. The Sea p 177	deep Atlantic current
141. C M Nov/Dec 2007 p 7	Igneous Provinces
142 Violent p	Toba volcano

Chapter 3,   Multiple Impacts – Single Event

1.	Robin, Gordon de Quetteville, 1966 Origin of the Ice Ages, Science p 53	
2.	Wikipedia, Submarine canyons	300 ft of ocean used during ice age
3.	Short p 273	200 ft in present glaciers
4.	En Br Vol 13, Kilimanjaro	snow on Kilimanjaro
5.	Violent p 188	Tambora
6.	Kronos x 3 p 36	Mars crater count > 3000
7.	En Br Vol 13, p 377	Kilimanjaro

Chapter 4,   Hypothesizing About Survival

1.	Hills, Vol. 19, Number 2, P 60	Carrier Pigeons
2.	Can Geo Sept/Oct 1999 p 76	Jupiter shifts asteroids, Tunguska, 14 mass extinctions
3.	Hills, Vol. 19, Number 2, p 60	Carrier Pigeons
4.	E in U,p 28	Old Red Sandstone full of bones
5.	CRSQ Vol. 48, P 288	Grand Canyon layers
6.	E in U, p 158 serious flooding as ice melted	

Illustrations – Part 1

Illustrations   Part 2

Photos

Index

www.ingramcontent.com/pod-product-compliance
Lightning Source LLC
Chambersburg PA
CBHW061126210326
41518CB00034B/2498